Welding Print Reading

Sixth Edition
Write-In Text

By
John R. Walker
W. Richard Polanin

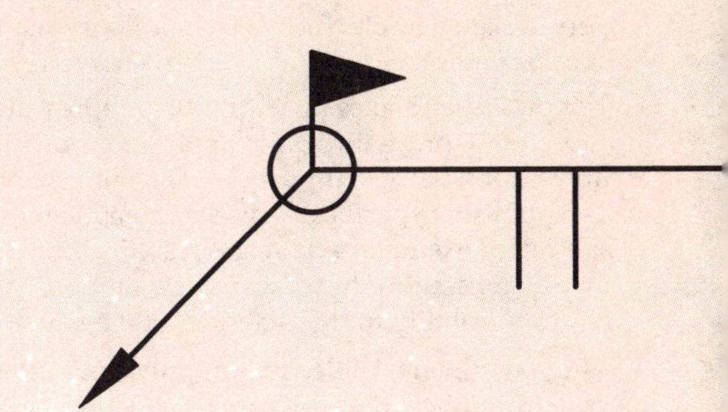

Publisher
The Goodheart-Willcox Company, Inc.
Tinley Park, Illinois
www.g-w.com

Manufactured in the United States of America.

Library of Congress Catalog Card Number 2012002990

ISBN 978-1-60525-911-6

3 4 5 6 7 8 9 – 13 – 17 16 15 14

Cover Image: Richard Thornton/Shutterstock

Library of Congress Cataloging-in-Publication Data

Walker, John R.
Welding print reading : write-in text edition / by John R. Walker, W. Richard Polanin. -- 6th ed.
p. cm.
Includes index.
ISBN 978-1-60525-911-6
1. Blueprints. 2. Welding--Drawings. I. Polanin, W. Richard. II. Title.

T379.W28 2013
671.5--dc23

2012002990

Introduction

Just about every manufactured product uses welding, either directly or indirectly. Over the years, a system of symbols and notations has been developed to convey exact weld specifications. Welding symbols and notations allow a large amount of data about a weld to be condensed into a small amount of space on a print. They simplify communications between the designer/engineer and the welder and also between other workers associated with the production of a weldment. Symbols and notations help assure that welds meet design requirements.

A welder, or anyone else (technician, engineer, drafter, etc.) working with welding prints, must know how to use the welding symbols and notations. This text is designed to help you grasp this information as quickly and easily as possible.

Text Design

Welding Print Reading provides instruction on interpreting and using the type of engineering drawings and prints found in the welding trade. It is a write-in text, or text-workbook, that starts out with the basics and progresses to more specialized coverage of specific welding symbols and notations. The information in this text follows to the most recent standards set up by the American Welding Society (AWS) and the American National Standards Institute (ANSI). However, this text is based on actual prints that are used in industry. While most industry practices conform to the national standards, any variation in a particular print has been retained for realistic experience. This will prepare you to work with prints being used in industry today.

This text is intended for students in high schools, vocational/technical schools, community colleges, for apprentices, and for workers on the job. It may also be used as a self-study course for those unable to attend print reading classes.

Unit Format

Each unit is designed to deliver complete coverage of specific welding print reading topics. Example prints, illustrations, symbols, and notations are used throughout each unit to reinforce these topics. At the end of each unit, there are problems that deal with the topics just covered in the unit. These problems are used to review the key concepts learned in the unit.

Unit 25, *Print Reading Activities,* consists of additional prints and related questions. These activities are designed to give you the opportunity for added practice of your welding print reading skills. It is suggested that these activities be performed after the completion of the first 24 units, but they may be used anytime as a review.

Notice to User

The procedures and practices described in this text are effective methods of performing given tasks. However, note that this information is general and applies to most situations. You must make sure all weld requirements are being fulfilled. It is also important for the resulting weldments to meet design specification. You must also make sure you are following all safety rules.

This text contains the most complete and accurate information able to be obtained from various authoritative sources at the time of publication. Goodheart-Willcox Co., Inc. cannot assume responsibility for changes, errors, or omissions.

About the Authors

Each author has many years of experience in the teaching, welding, and print reading fields. They are confident that you will find this text a tremendous tool for learning how to read and interpret welding prints.

John R. Walker

John R. Walker is the author of thirteen textbooks and has written many magazine articles. Mr. Walker did his undergraduate studies at Millersville University and has a Master of Science degree in Industrial Education from the University of Maryland. He taught industrial arts and vocational education for thirty-two years and was Supervisor of Industrial Education for five years. He also worked as a machinist for the U.S. Air Force and as a draftsman at the U.S. Army Aberdeen Proving Grounds.

W. Richard Polanin

W. Richard Polanin is a professor at Illinois Central College, as well as the coordinator of the Manufacturing Engineering Technology and Welding Technology programs. Dr. Polanin has a bachelor's and master's degree from Illinois State University and a doctorate degree from the University of Illinois. In addition to his twenty-five years of teaching, he is an active consultant in welding and manufacturing. He is an AWS Certified Welding Inspector, AWS Certified Welding Educator, and a SME Certified Manufacturing Engineer. He has published numerous technical papers and has made many technical presentations in the areas of welding, manufacturing, robotics, and manufacturing education.

Contents at a Glance

Contents

Section 1 Introduction

Unit 1
Prints—The Language of Industry

After completing Unit 1, you will be able to:

- ❍ Describe the various processes for making original drawings.
- ❍ Define and describe the parts of a print.
- ❍ Explain the various methods used to reproduce a drawing.
- ❍ Describe the advantages of developing drawings with a CAD system.
- ❍ Explain traditional printmaking processes.
- ❍ Explain the importance of welding information found on a print.
- ❍ List precautions in the care of prints.

Key Words

blow-backs
bluelines
blueprint
blueprint process
CAD system
computer-aided drafting (CAD)
computer-aided manufacturing (CAM)
diazo process
drawings
electrostatic process
engineering copiers
imaging systems
ink jet printers
laser printers
microfilm process
plotters
print
reproductions
symbols
tracing
white print
xerography

Drawings are used by the welding industry to communicate ideas in graphic or picture form. Drawings are often the only means of showing the construction of a complex product and structures, Figure 1-1. Drawings are called the "language of industry." It is a precise language that utilizes symbols. ***Symbols*** (lines and figures) have specific meanings to accurately describe the shape, size, material, finish, and fabrication of an object. The symbols are standardized and used as a universal language around the world. This standardization makes it possible to interpret and understand drawings made in other countries.

Original drawings are seldom found on a job site. They would soon become worn or soiled, making them difficult to read. Also, workers at different locations producing the various parts or subassemblies of a product or structure each need a set of drawings. This means that several sets of identical drawings are needed at the same time. Drawing a set of plans for each person who needs them is impractical because of the costs. Creating new originals to replace drawings that are damaged or ruined is also costly. ***Reproductions,*** or prints, of the original drawings are used in these situations.

Figure 1-1.
Drawings are used to communicate design of complex products and structures. A—Thrill rides at amusement parks are fabricated using welding processes. B—Complex steel structures such as this building are welded rather than riveted.

Making Original Drawings

Before a print can be made, a high-quality original drawing must first be created. Original drawings must have dense and uniform lines for good reproductions. Drawings are most typically created using the tools of a ***computer-aided drafting (CAD)*** system. A ***CAD system*** is a combination of hardware and software that allows drawings to be drawn, viewed, and output (printed or plotted) electronically. CAD software programs provide drawing, editing, and dimensioning commands and other functions for use in developing original drawings.

Manual drawing techniques can also be used. However, the process is slower, less accurate, and the quality of the drawing is dependent upon the drafter's technique.

CAD Systems

The growth in the computer field has had a direct and significant impact on the drafting field. The CAD system allows the designer to modify and manipulate the original drawing quickly. Instead of paper and pencil, the changes are made on a computer screen and saved on the storage media of the computer. The saved drawings may be transmitted electronically or retrieved and made into paper prints, Figure 1-2.

CAD systems can be used to produce simple part drawings, but the modern CAD system is used to produce a complex three-dimensional model of a part or assembly. The drafter is able to design the part and then view it from all sides. Assemblies can be viewed to see if all of the parts fit together properly, or the CAD system can be used to test the assembly for strength and proper operation. In some design phases, drawings of CAD models are used in the production of prototypes for testing and evaluation purposes.

CAD drawing data is also used by machine tools in automated manufacturing applications. In ***computer-aided manufacturing (CAM)***, machining systems use information from drawings to produce parts and assemblies. In this process, data from a CAD drawing is calculated by the system in order to determine the movement and operation of tools.

Printers and Plotters

As computers become less expensive, so do the output devices attached to them. Personal computer packages are commonly purchased with a printer. In addition, the price of printers and plotters continues to drop as computers advance. Printers and plotters produce drawings that have dense lines and other features. Sometimes, printing or plotting multiple copies is the easiest and least expensive way to reproduce drawings. However, in situations where dozens of prints must be made, using one of the printmaking processes is best.

Ink jet printers are inexpensive devices that produce good quality output. Most are small enough to fit on a desktop, Figure 1-3. Many ink jet printers can print colors.

Laser printers work on the electrostatic process. These printers typically produce a very high quality image. Many laser printers are capable of printing large size drawings while maintaining a high-quality image, Figure 1-4. Color laser printers are common.

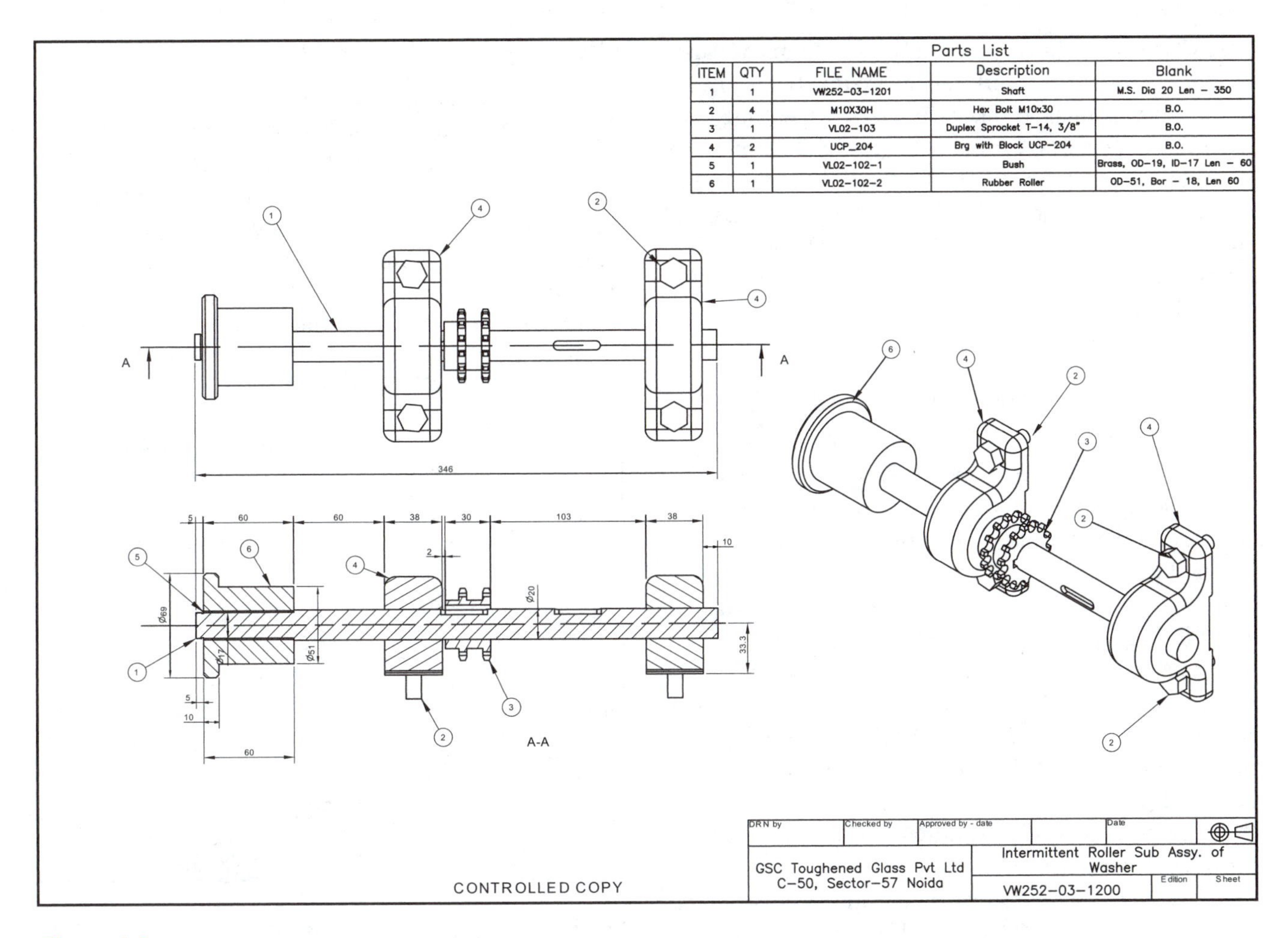

Figure 1-2.
Changes are easily made on screen and then stored in a CAD system.

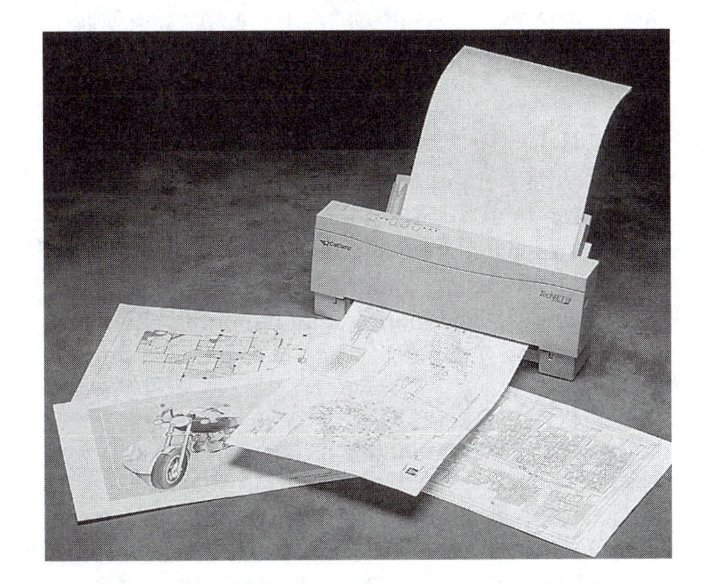

Figure 1-3.
Ink jet printers are often small enough to fit on a desk. These printers can produce good quality prints from a CAD system. (Calcomp)

Figure 1-4.
Laser printers can produce high quality prints from a CAD system. (Calcomp)

Plotters are most commonly used to make paper drawings from CAD work. There are two types. *Ink jet plotters* are similar to ink jet printers, except they are used for large-format prints. *Pen plotters* use pens to reproduce object lines. These devices duplicate the motion that a human hand might make while drawing. In other words, they draw lines instead of printing "dots" on the page. Pen plotters produce high-quality work, but they are fairly slow. Both ink jet and pen plotters are capable of color output.

Making Tracings for Reproduction

Tracings are commonly made for manual drafting when prints are to be made. The first step in making a tracing is to draw it on a translucent (somewhat see-through) material. Many types of this material are available in sheet or roll form.

Pencil Tracings

Sharp, clear reproductions are made from pencil tracings with modern reproduction equipment and materials. Many drafters make original drawings on the tracing material that will be used to produce the copies. For more complex jobs, the layout is usually "traced" from the original drawing. This is where the term ***tracing*** comes from.

Tracings for reproduction are made in the same way as conventional drawings. The drafter completes the drawing, adds notes, and dimensions the drawing. Lines are kept uniformly sharp and dense so they will reproduce well.

Every effort must be made to keep the tracing clean. As with prints, handle tracings carefully if you come into contact with them.

What Is a Print?

A ***print*** is a reproduction, or duplicate, of an original drawing. Basically, a print consists of the following elements:

- Lines—show part surfaces and points of machining
- Dimensions—give the size of the part
- Notes—provide information not given by lines and dimensions
- Specifications—special notes for standards, type materials, or specific processes to be used
- Views—normally front, side, and top of part

When several sets of prints are needed, the original drawings are reproduced or duplicated. The reproduction method must be quick and cost-effective. It must produce accurate prints and not destroy the original drawings.

Why Are Prints Used?

It would be very expensive and impractical to draw a set of plans for each worker who needed them. It is also impractical to replace original drawings when they wear out or become damaged during normal shop use. Instead, reproduction techniques are used to make accurate copies of the original drawings quickly and inexpensively. These copies are usually ***white prints*** or ***bluelines*** (dark lines on a light background). In the shop, they are known as prints, drawings, or ***blueprints***, Figure 1-5.

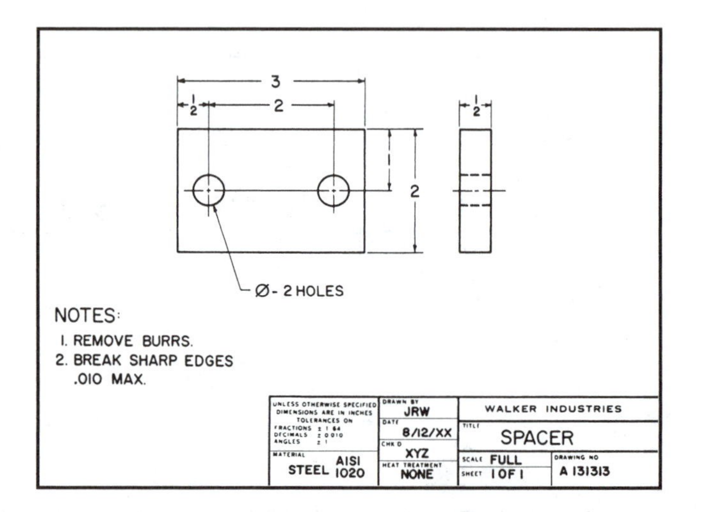

Figure 1-5.
Prints are most commonly a white print, or dark lines on a light background. The term "blueprint" is commonly used when referring to all types of prints regardless of color of lines or background. A true blueprint is a print with white lines on a blue background.

Reproduction Techniques

After an original drawing is completed, it may then be reproduced. Prints of CAD drawings are generated electronically by computer. Drawings can also be reproduced by using xerography or making microfilm from drawings. The more traditional printmaking processes employ a chemical reaction to produce prints.

Xerography (Electrostatic) Process

Xerography (pronounced ze-rog-ra-fee) is a printmaking process that uses an electrostatic charge to duplicate an original, Figure 1-6. Xerography process is commonly called ***electrostatic process***. This process is based on the scientific principle that like electrical charges repel and unlike charges attract, Figure 1-7.

The electrostatic process exactly duplicates the original. The copy can be enlarged or reduced in size from the original if needed. Unlike traditional processes, the original drawing does not have to be on translucent material.

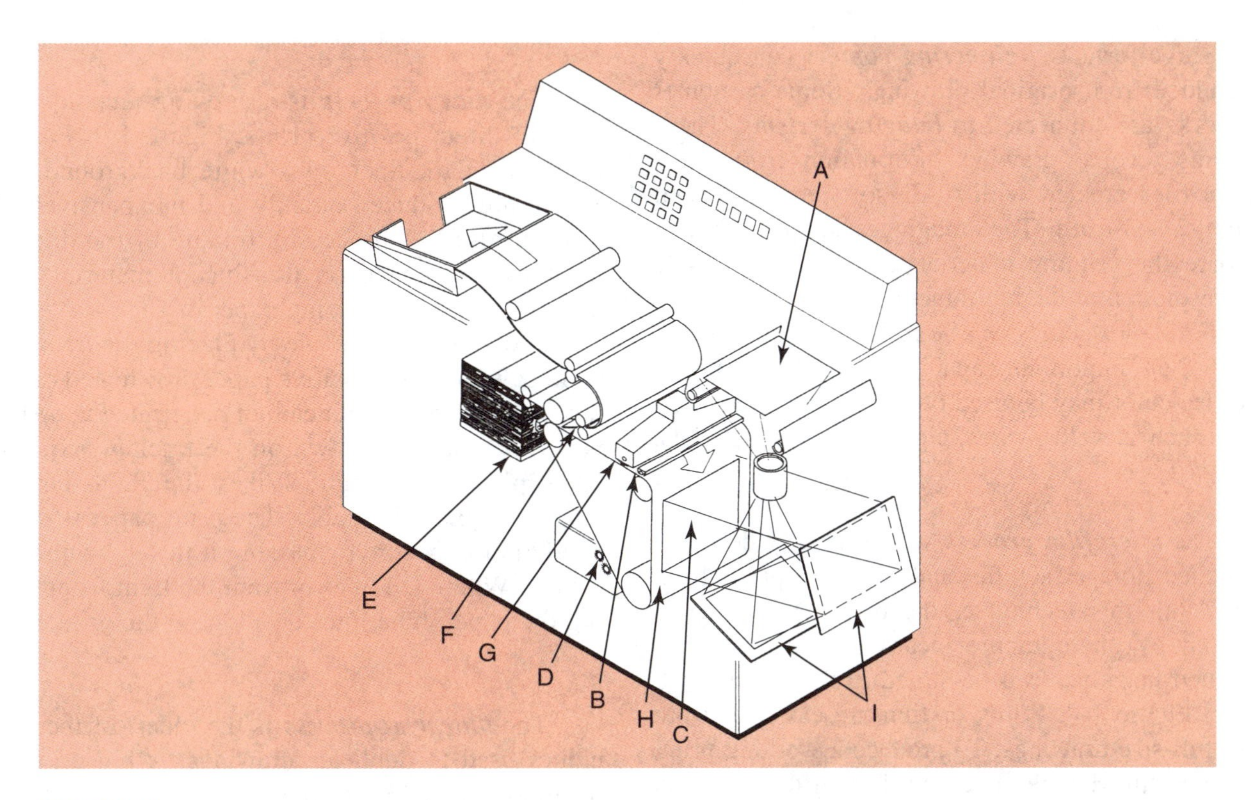

Figure 1-6.
The various parts of an electrostatic (xerographic) printer are shown. A—Original drawing. B—A positive charge is placed on the photoconductor. C—The positive charge is removed from the non-image areas on the photoconductor. D—Negatively-charged toner ("ink") is placed on the photoconductor. E—The photoconductor presses against the copy paper. F—Rollers fuse the toner to the paper. G—Brushes and a vacuum remove remaining toner. H—Photoconductor belt. I—Mirrors and lens.

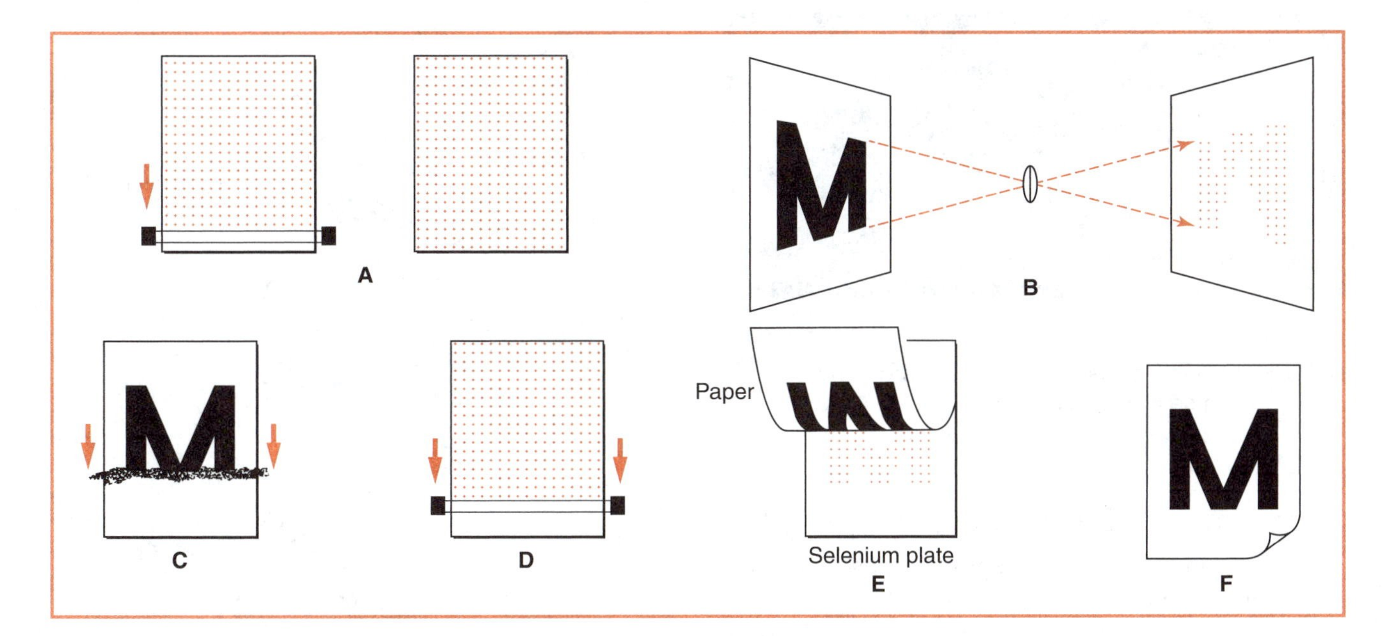

Figure 1-7.
This is the sequence for electrostatic printing. A—The conductor is positively charged. B—The positive charge is removed from the conductor except where the image will be. C—Negatively-charged toner is passed over the conductor. D—The paper is positively charged. E—The paper is pressed against the conductor and the image is transferred. F—The toner is fused to the paper for the final product.

Large-format ***engineering copiers*** can quickly reproduce from original drawings. Some computer networks are connected to ***imaging systems***. These systems receive drawing information from electronic files created with CAD software or special graphics software. The imaging system is able to prepare the printing materials and control the size and resolution of the reproduction.

Full-color copies can be made on color copiers. This is an important capability in CAD drafting, because color may be used on final prints to identify items such as welds, wires, piping, or other materials.

The Microfilm Process

The ***microfilm process*** was originally designed to reduce storage facilities and to protect prints from loss. With this technique, the original drawing is reduced by photographic means to negative form. Finished negatives can be stored in roll form or on cards, Figure 1-8. Cards or film are easier to store than full-size drawings. To produce a working print, the microfilm image is retrieved from the files and printed on photographic paper. These prints are called ***blow-backs.*** The print is often discarded or destroyed when it is no longer needed. Microfilm can also be viewed on a microfilm reader to check details without making a blow-back.

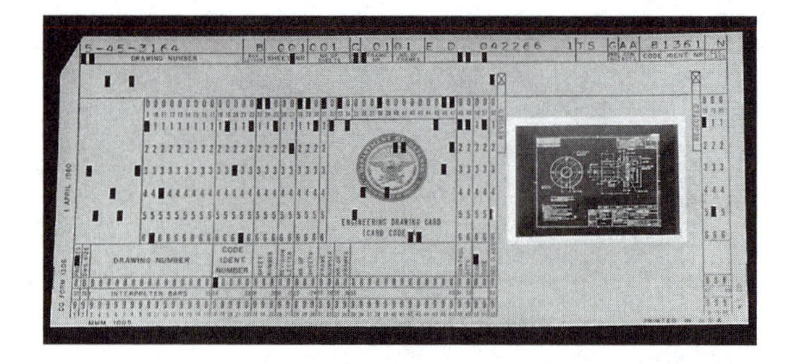

Figure 1-8.
Microfilmed drawings can be stored on cards. The drawings can be enlarged and printed later.

Traditional Printmaking Processes

There are two traditional printmaking processes, called the diazo process and the blueprint process. As CAD drafting becomes more prevalent, more prints are generated electronically. This has caused a decline in the use of these traditional processes. However, you should be familiar with these processes.

Diazo Process

The ***diazo process*** is a copying technique for making direct positive prints, Figure 1-9. Positive prints are dark lines on a white background. The copies are produced quickly and inexpensively. In order to use this process, a tracing of the drawing must first be made on translucent material. This material is usually tracing paper or film.

A diazo print is made by placing the tracing in contact with light-sensitive paper or film and exposing them to light. Light cannot penetrate the opaque pencil or ink lines drawn on the tracing. Exposure takes place when light strikes the light-sensitive coating of the print paper. The print paper is developed after exposure by passing it through ammonia vapors. Where the lines prevent the light from striking the paper, blue lines develop on the print.

Blueprints

The ***blueprint process*** is the oldest of the techniques used to duplicate drawings. A blueprint has white lines on a blue background.

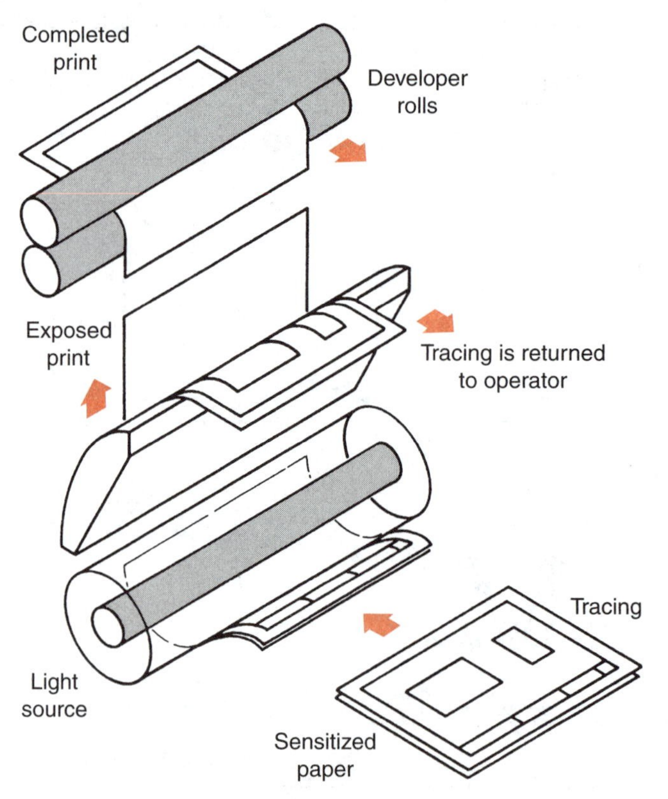

Figure 1-9.
For diazo process prints, the original drawing is placed with light-sensitive paper into the machine. The two sheets are exposed to light, and ammonia vapors develop the exposed image.

This process is similar to the diazo process. First, a tracing is made. However, the print must be developed, "fixed" in a chemical solution, washed, and then dried. The blueprint process is not used much anymore. Many times, the term "blueprint" is used to refer to any print.

Importance of Prints to Welders

Prints have special importance to welders. They show where and how the various components to be welded fit together. Prints provide the welder with all information needed to make the weld(s). Information may include the following:

- Welding process to be used
- Weld size
- Weld type
- Kind/Type of filler metal
- How the weld is to be finished
- Other pertinent data needed to make welds meet design specifications

In addition to having welding skills, welders must know how to read and interpret prints. Otherwise, there is little chance the welds will be made to specifications, Figure 1-10.

The inability to read and interpret prints could result in higher cost parts because more welding was done than necessary. Unsafe or dangerous welds are possible since incorrect print interpretation could result in weak or rejected welds that do not meet design specifications.

Care of Prints

If given proper care, a print will have a long, useful life before it must be discarded. The following is a list of precautions used when handling prints:

- Keep prints clean. Avoid setting food or drinks on prints. Dirty prints are difficult to read and can cause errors when welding.
- Avoid tearing prints. Fold and unfold prints carefully when you use them.
- Avoid laying tools and other objects on the prints.
- Do *not* make revisions (changes) on a print unless you have written authority to do so.
- Some prints are concerned with work that is secret or classified in nature. Handle and store these prints according to company or agency security policies. Penalties for ignoring security precautions can prove costly to you.

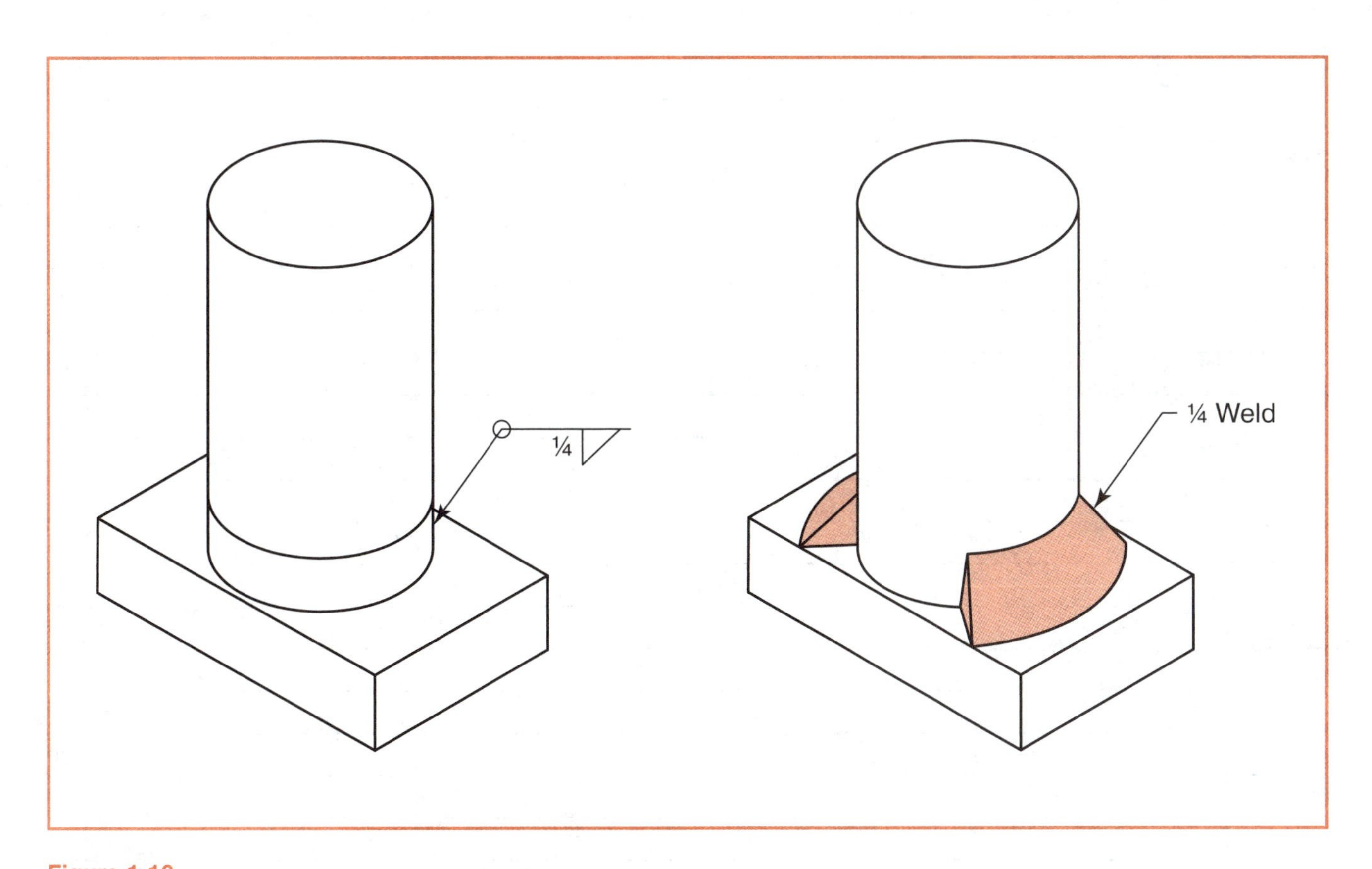

Figure 1-10.
Without a print describing the welding process, weld size and type, kind of filler metal, method of finish, and other pertinent data, there is no way a welder can be sure of making a weld that will meet design specifications.

Notes

Name ______________________ Date __________ Class __________

Review Questions

1. How are drawings used by industry? ________

2. Drawings are often called the ____ of industry.

3. List the reasons why it is impractical to use the original drawings in the shop. ________

4. Define a CAD system. ________

5. Name two advantages of using CAD for drawings. ________

6. Name three devices used to make prints from drawings created with computers. ________

7. What is a print? ________

8. What is used in the shop in place of the original drawings? ________

9. Define a white print. ________

10. What specific information do prints provide the welder? ________

11. Microfilm prints retrieved from stored files and printed on photographic paper are called ____.

12. Summarize the two traditional methods of making prints. ________

13. What might occur if a welder is unable to read prints? ________

14. What is proper care for prints? ________

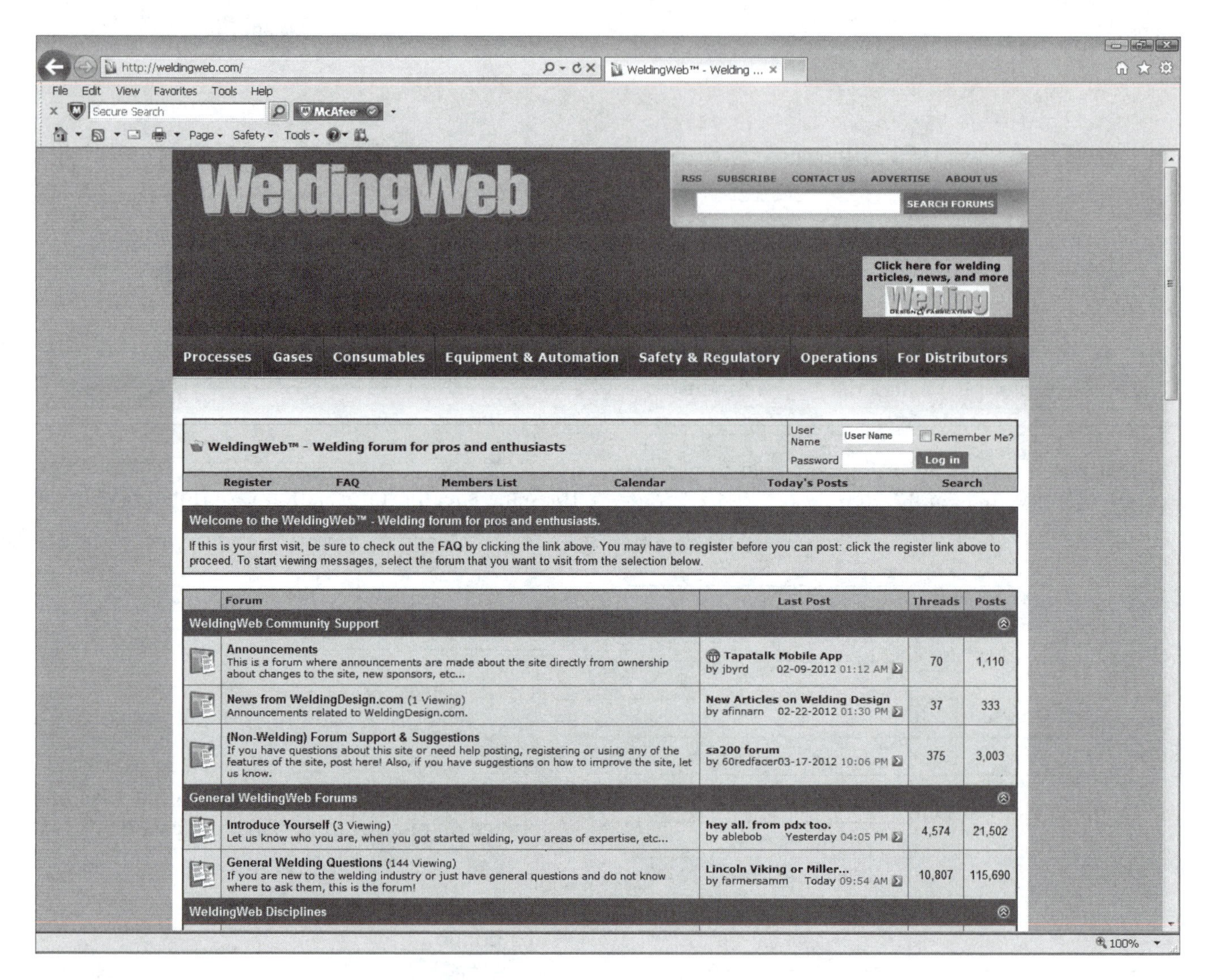

Welding Web (www.weldingweb.com) offers numerous discussion forums, articles, and videos for both students and professionals. The site is especially useful because it provides tips and tricks for making proper measurements.

Unit 2 Review of Measurement

After completing Unit 2, you will be able to:

- Identify the two measurement systems used on prints.
- Read a fractional inch, decimal inch, and metric graduated rule.
- Convert linear measurements between units.
- Convert between decimal fractions and common fractions.
- Interpret measurement information on drawings.
- Use a fractional inch, decimal inch, and metric rule to make linear measurements.

Key Words

bracket method	inch rule
common factor	linear measure
decimal fraction	lowest terms
decimal rule	metric rule
dividend	millimeter (mm)
dual dimensions	note
graduation	position method
hole diameter	shaft diameter

Precise measurement is extremely important in welding. Mistakes made in measuring are very costly in both time and material. A welder must know how to make measurements accurately and quickly. He or she must also be familiar with the proper way to use the measuring tools of the welding trade, Figure 2-1.

U.S. Customary and Metric Measuring Systems

A welder must be able to read measuring tools using the U.S. customary (English) system and the SI metric system. For the U.S. customary system, the unit of measure is the inch. For the SI metric system, the unit of measure is the ***millimeter (mm)***, which is equal to 1/1000 of a meter. For both systems, it is common to have measurements that are less than an inch (expressed as fractional inch and decimal inch) or millimeter.

Reading a Fractional Inch Graduated Rule

Linear measure refers to measuring the straight-line distance between two points. Secure a rule or tape measure and examine the edges used for measuring. Each mark or division on the edge of the rule is called a ***graduation.***

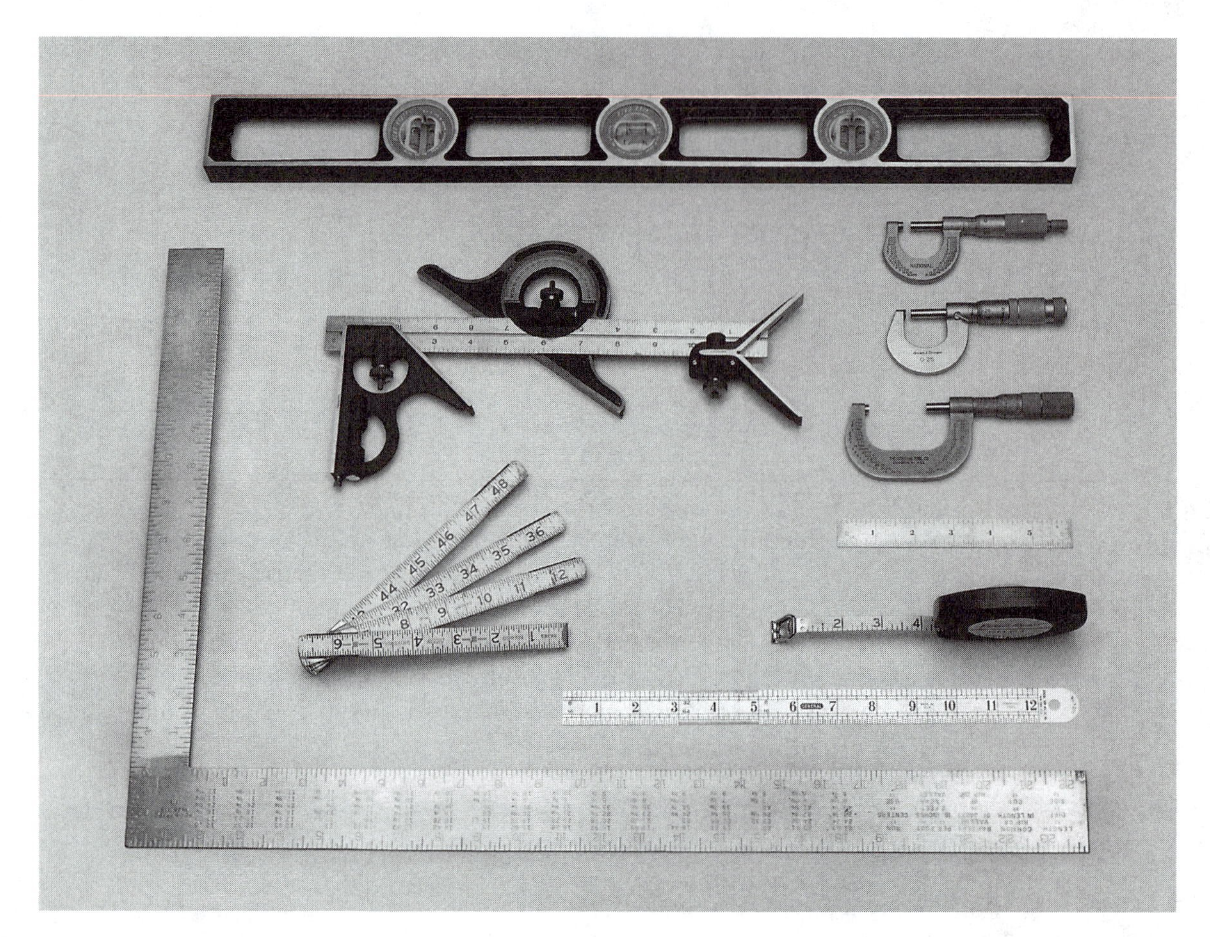

Figure 2-1.
Welders need to know how to use basic measuring tools.

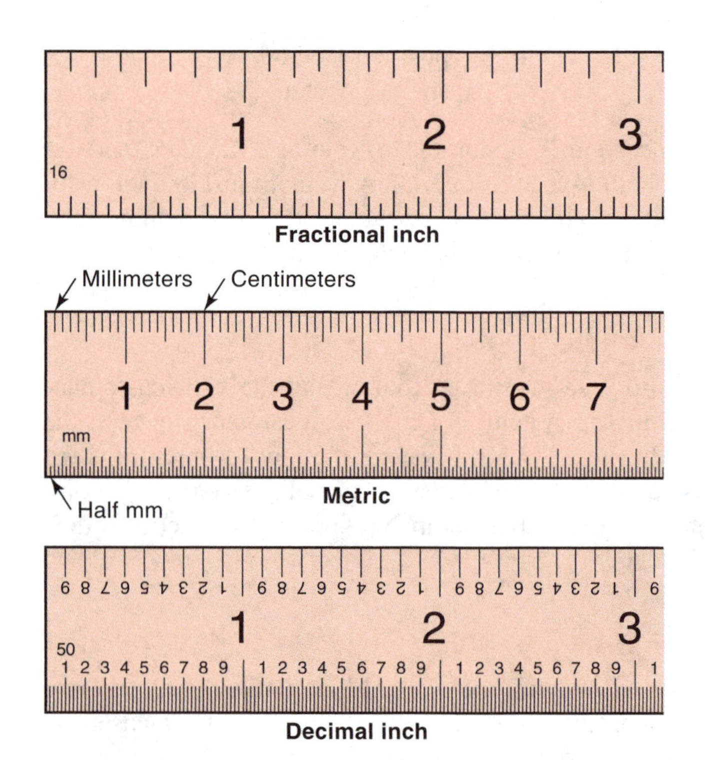

Figure 2-2.
Study these rules. Compare the divisions of each carefully.

The ***inch rule*** can be graduated in divisions as small as 1/64″ (one sixty-fourth of an inch). For general welding work, you should be able to read a rule to 1/16″ (one-sixteenth of an inch). Precise weldments require more accuracy. Figure 2-2 shows inch, metric, and decimal inch rules.

Look carefully at the rule edge used to measure 1/16″. Many rules have 16 or 1/16 stamped or engraved on this edge. If your rule does *not,* count the graduations. For a 1/16 graduated rule, there will be 16 of them per inch. This is shown in Figure 2-3.

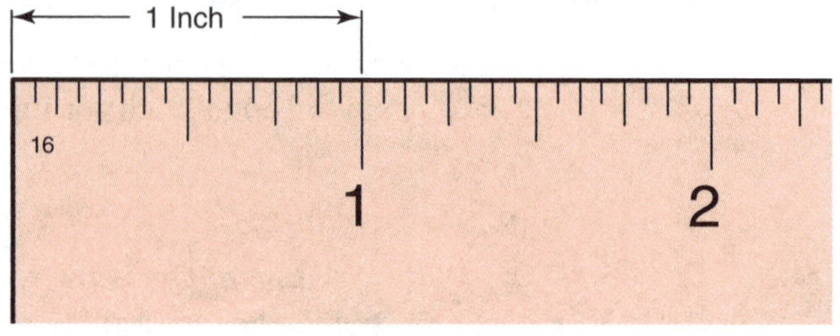

Figure 2-3.
There are 16 equal divisions in each inch of this rule. Each division or graduation is equal to 1/16″.

Note that the 1″ graduation is the longest. The 1/2″ graduation is the next longest, and so on down to the 1/16″ graduation, which is the shortest. Until you are familiar with the rule, visualize or imagine that each 1/16″ graduation is numbered as shown in Figure 2-4. You may find it easier to count the graduations or divisions when you first start to use the rule. After some practice, however, this should not be necessary.

Remember fractional measurements are always reduced to lowest terms. ***Lowest terms*** means there is no common factor for the numerator and denominator. A ***common factor*** is a number that will divide evenly into both the numerator and the denominator of the fraction. For example, a measurement of 8/16 is not in lowest terms. Divide the numerator and denominator by 8, which is a number that will divide evenly into both. Then the fraction is reduced to 1/2. Another example is 2/16 (divide by two), which is read as 1/8. See Chapter 3 for more information on fractions.

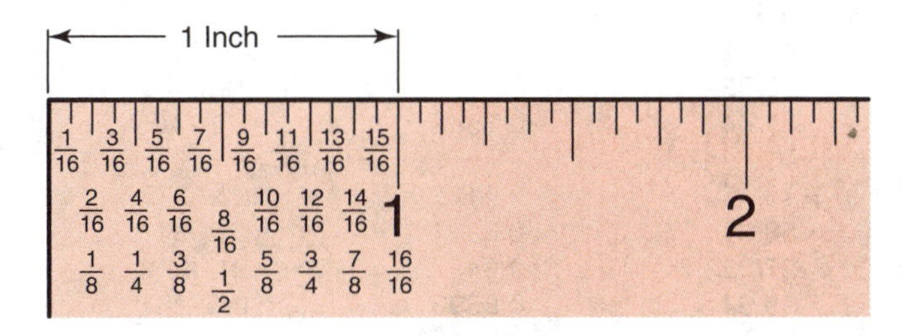

Figure 2-4.
Until you are familiar with the rule, memorize the value of each division. Study this scale closely.

Reading a Decimal Inch Graduated Rule

The more common ***decimal rule*** has 10 divisions per inch (one division equals 0.1 inch) on one edge. The opposite edge has either 50 divisions (one division equals 0.02 inch) or 100 divisions (one division equals 0.01 inch). Figure 2-5 illustrates a decimal inch rule.

General welding layout seldom requires measurements as close as 1/50″ (0.02″) or 1/100″ (0.01″). Only special welding jobs need this much accuracy.

Reading a Metric Graduated Rule

A ***metric rule*** is graduated in millimeters (mm). In addition to using fractional and decimal inch measurements, the welder must also be familiar with metric measure.

You may find the metric unit rule easier to read than a fractional inch rule. There are no fractions to complicate your calculations. Numbers in the metric system are added, subtracted, multiplied, and divided as with decimal inch numbers.

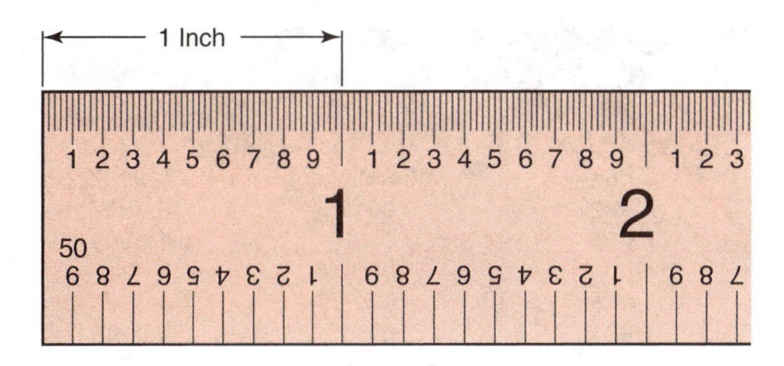

Figure 2-5.
The more common decimal rule has 10 divisions per inch with each division equaling 0.1″ on one edge. On the second edge, 50 divisions, each equaling 0.02″, are shown.

A metric rule or metric tape is graduated in millimeters (mm) on one edge and with one-half millimeter (0.5 mm) graduations on the second edge. Every ten millimeters (ten millimeters equal one centimeter) is numbered. This is shown in Figure 2-2.

Converting Linear Measurements

A welder may need to change figures back and forth between fractional inches, decimal inches, and millimeters. Common conversion formulas follow:

- Converting feet to inches—Multiply the number of feet by 12
  For example, change 5′ to inches by multiplying $5 \times 12 = 60$.
  5′ = 60″
- Converting inches to feet—Divide the number of inches by 12
  For example, change 18″ to feet by dividing $18 \div 12 = 1'6''$.
  18″ = 1′6″
- Converting inches to millimeters—Multiply the number of inches by 25.4
  For example, change 7″ to millimeters by multiplying $7 \times 25.4 = 177.8$.
  7″ = 177.8 mm
- Converting millimeters to inches—Divide the number of millimeters by 25.4
  For example, change 88 mm to inches by dividing $88 \div 25.4 = 3.46$.
  88 mm = 3.46″

Figure 2-6 shows some common linear conversions. Also see the Conversion Charts in the reference section.

The United States is in the process of adopting the International System of Units (the SI metric system). The measurement units of the International System (abbreviated SI) are shown in Figure 2-7. Many companies, especially those selling products outside the United States, have converted to the metric system.

Conversion Table: Fractional Inches into Decimals and Millimeters					
Inch	Decimal Inch	Millimeter	Inch	Decimal Inch	Millimeter
1/64	0.0156	0.3967	33/64	0.5162	13.0968
1/32	0.0312	0.7937	17/32	0.5312	13.4937
3/64	0.0468	1.1906	35/64	0.5468	13.8906
1/16	0.0625	1.5875	9/16	0.5625	14.2875
5/64	0.0781	1.9843	37/64	0.5781	14.6843
3/32	0.0937	2.3812	19/32	0.5937	15.0812
7/64	0.1093	2.7781	39/64	0.6093	15.4781
1/8	0.125	3.175	5/8	0.625	15.875
9/64	0.1406	3.5718	41/64	0.6406	16.2718
5/32	0.1562	3.9687	21/32	0.6562	16.6687
11/64	0.1718	4.3656	43/64	0.6718	17.0656
3/16	0.1875	4.7625	11/16	0.6875	17.4625
13/64	0.2031	5.1593	45/64	0.7031	17.8593
7/32	0.2187	5.5562	23/32	0.7187	18.2562
15/64	0.2343	5.9531	47/64	0.7343	18.6531
1/4	0.25	6.35	3/4	0.75	19.05
17/64	0.2656	6.7468	49/64	0.7656	19.4468
9/32	0.2812	7.1437	25/32	0.7812	19.8437
19/64	0.2968	7.5406	51/64	0.7968	20.2406
5/16	0.3125	7.9375	13/16	0.8125	20.6375
21/64	0.3281	8.3343	53/64	0.8281	21.0343
11/32	0.3437	8.7312	27/32	0.8437	21.4312
23/64	0.3593	9.1281	55/64	0.8593	21.8281
3/8	0.375	9.525	7/8	0.875	22.225
25/64	0.3906	9.9218	57/64	0.8906	22.6218
13/32	0.4062	10.3187	29/32	0.9062	23.0187
27/64	0.4218	10.7156	59/64	0.9218	23.4156
7/16	0.4375	11.1125	15/16	0.9375	23.8125
29/64	0.4531	11.5093	61/64	0.9531	24.2093
15/32	0.4687	11.9062	31/32	0.9687	24.6062
31/64	0.4843	12.3031	63/64	0.9843	25.0031
1/2	0.50	12.7	1	1.0000	25.4

Figure 2-6.
It is helpful for a welder to know how to convert linear measurements between units.

Converting Decimal Fractions to Common Fractions

A ***decimal fraction*** expresses a number *less* than a whole number. Numbers to the left of the decimal point are whole numbers. Numbers to the right are fractional parts. Positions to the right of the decimal point have different place values.

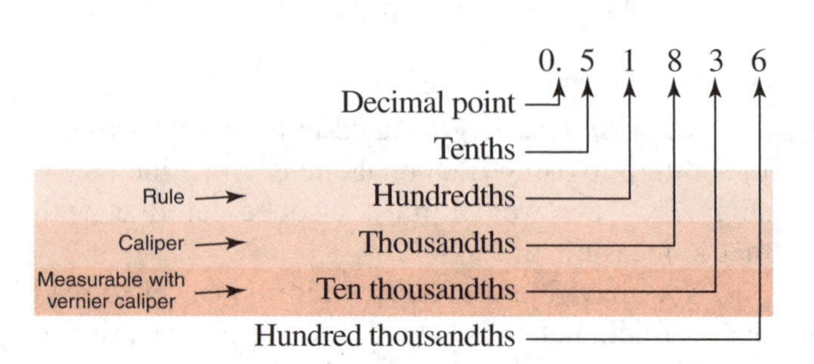

Metric Prefixes, Exponents, and Symbols					
Decimal Form	**Exponent or Power**	**Prefix**	**Pronunciation**	**Symbol**	**Meaning**
1 000 000 000 000 000 000	$= 10^{18}$	exa	ex'a	E	quintillion
1 000 000 000 000 000	$= 10^{15}$	peta	pet'a	P	quadrillion
1 000 000 000 000	$= 10^{12}$	tera	tĕr'á	T	trillion
1 000 000 000	$= 10^{9}$	giga	jĭ'gá	G	billion
1 000 000	$= 10^{6}$	mega	mĕg'á	M	million
1 000	$= 10^{3}$	kilo	kĭl'ō	k	thousand
100	$= 10^{2}$	hecto	hĕk'to	h	hundred
10	$= 10^{1}$	deka	dĕk'a	da	ten
1					base unit
0.1	$= 10^{-1}$	deci	dĕs'ĭ	d	tenth
0.01	$= 10^{-2}$	centi	sĕn'tĭ	c	hundredth
0.001	$= 10^{-3}$	milli	mĭl'ĭ	m	thousandth
0.000 001	$= 10^{-6}$	micro	mi'krō	μ	millionth
0.000 000 001	$= 10^{-9}$	nano	năn'ō	n	billionth
0.000 000 000 001	$= 10^{-12}$	pico	péc'ō	p	trillionth
0.000 000 000 000 001	$= 10^{-15}$	femto	fĕm'tō	f	quadrillionth
0.000 000 000 000 000 001	$= 10^{-18}$	atto	ăt'tō	a	quintillionth

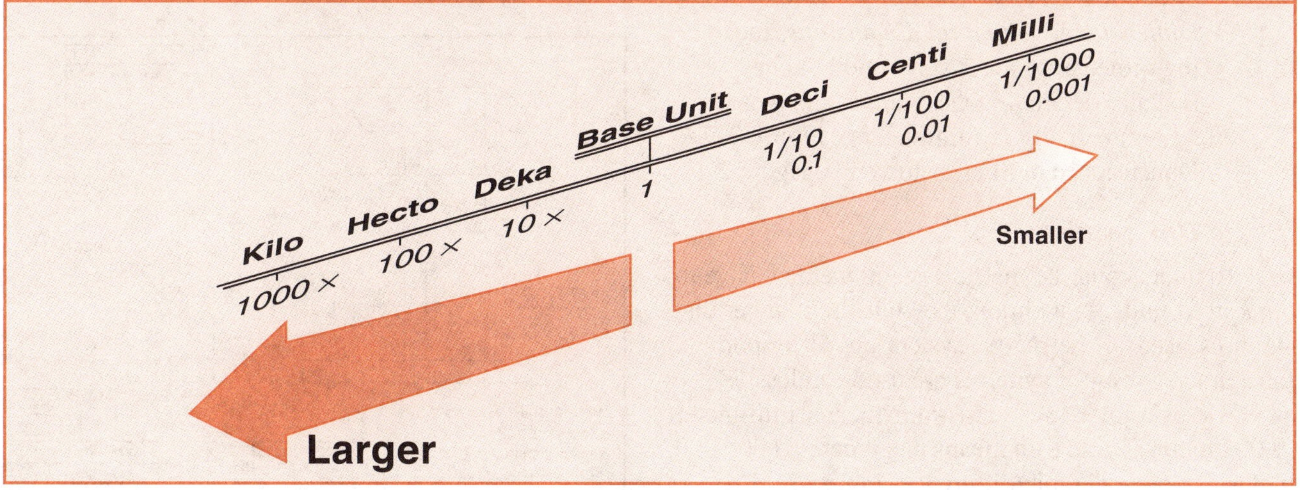

Kilo
1000 ×
Hecto
100 ×
Deka
10 ×
Base Unit
1
Deci
1/10
0.1
Centi
1/100
0.01
Milli
1/1000
0.001
Smaller
Larger

Figure 2-7.
The United States is in the process of adopting the International System of Units (also called metric or SI system). Study the basic units of this system.

Use the last place value as the denominator and the decimal fraction numbers as the numerator when changing a decimal fraction to a common fraction.

For example, 0.8 = 8/10

$$0.34 = \frac{34}{100}$$

$$0.759 = \frac{759}{1000}$$

Denominators are always multiples of 10. Once the change has been made, reduce the fraction to lowest terms.

$$0.0220 = \frac{220}{10000} = \frac{11}{500}$$

Converting Common Fractions to Decimal Fractions

Change a common fraction to a decimal fraction by dividing the numerator by the denominator.

$$\frac{7}{8} = 8\overline{)7}$$

It is important for the decimal to be properly located in the answer. Place a decimal to the right of the dividend (7). The ***dividend*** is the number being divided. Place a second decimal directly above this one in the answer.

$$\frac{7}{8} = 8\overline{)7.}$$

Then, add a zero to the right of the decimal in the dividend and begin to divide. Continue adding zeros to the dividend as needed until the problem is solved.

$$\begin{array}{r} .875 \\ 8\overline{)7.000} \\ 6\,4 \\ \hline 60 \\ 56 \\ \hline 40 \\ 40 \\ \hline 0 \end{array}$$

Interpreting Dimensions on Drawings

Many large firms that use prints are presently converting to the metric system. During the transition period that will take many years, welders and other workers will use some drawings that are:

- Dimensioned with inch units (fractional and/or decimal), Figure 2-8.
- *Dimensioned with* ***dual dimensions,*** meaning dimensions are given in both inches (usually decimal inches) *and* in the metric system (usually in millimeters), Figure 2-9.
- Dimensioned in SI or metric units, Figure 2-10.

Metric Unit Symbols

To understand the metric system, welders have to *think* in SI units. In addition, they must remember that symbols used in the metric system are all important. Inaccuracies result if symbols are used carelessly.

For example, lowercase mm means millimeter (0.001 meter) while Mm means megameter (1 000 000 meters), a considerable difference, Figure 2-11.

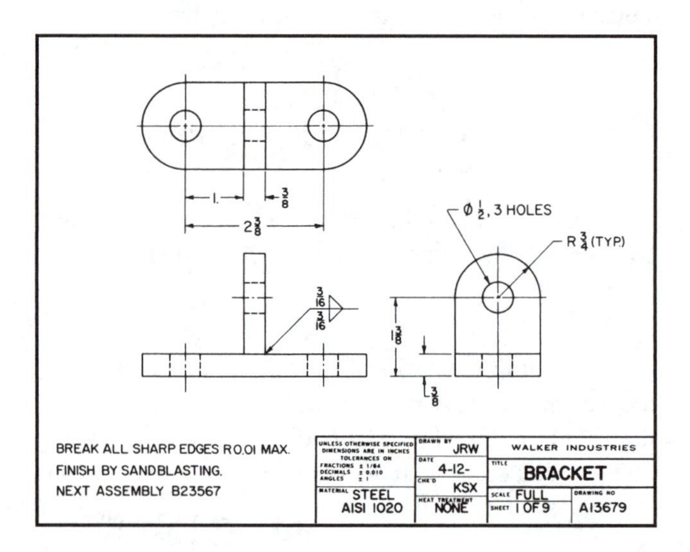

Figure 2-8.
This drawing is dimensioned with inch units.

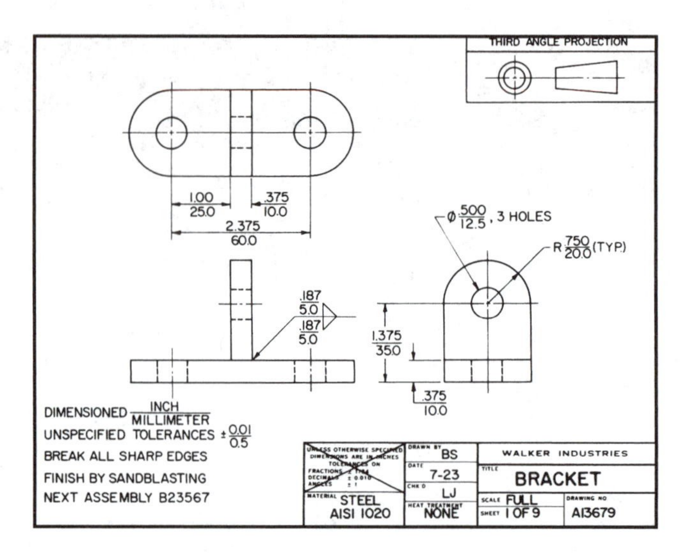

Figure 2-9.
This is a dual dimensioned drawing. Inches are on top and millimeters are on bottom.

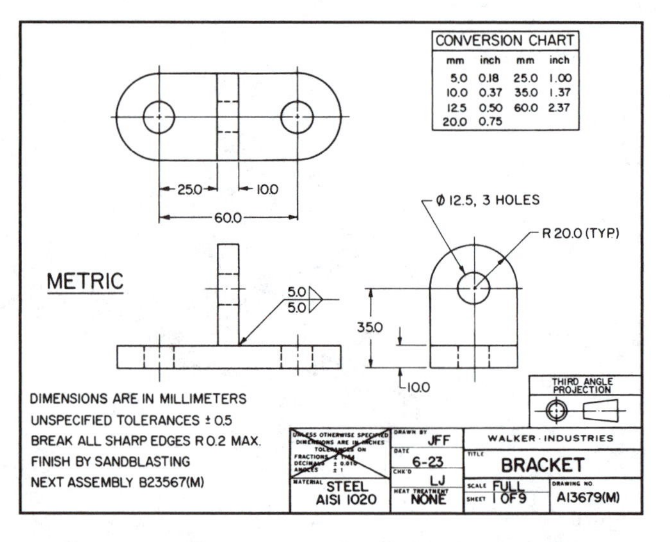

Figure 2-10.
Note how this drawing is dimensioned in metric units. The conversion chart is rounded.

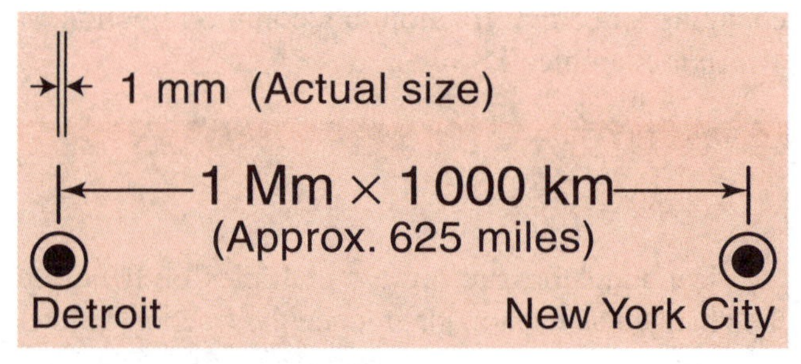

Figure 2-11.
Symbols in the metric system are important. Mistakes result if they are used improperly. Note the difference made by upper- and lowercase letters.

The millimeter (mm) is the basic unit for linear dimensions on metric drawings used by manufacturing industries. In building construction, the meter (m) is normally used.

Reading Inch and Metric Measurements on Drawings

Two methods have been designated to display dual dimensions on a drawing, Figure 2-12. Drawings using the ***position method*** (primarily the United States) display the inch dimension above or to the left of the millimeter dimension. The ***bracket method*** encloses the metric dimension in square brackets if the drawing is to be used in the United States.

A ***note*** is placed on the drawing to show how the inch and millimeter dimension can be identified, Figure 2-13. ***Hole diameter*** or ***shaft diameter*** is indicated on SI unit drawings with the number and symbol ∅.

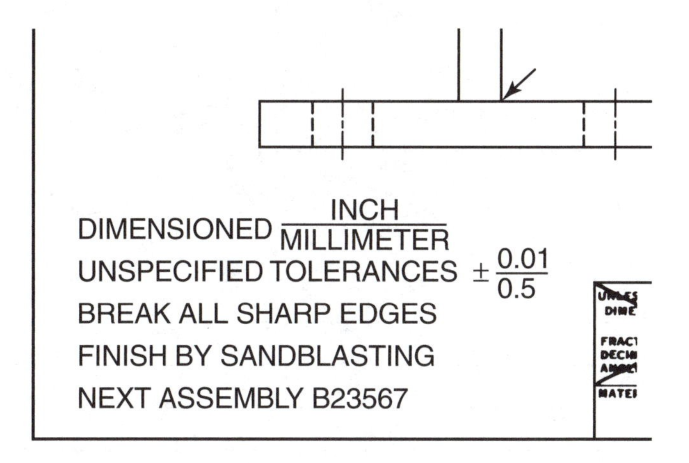

Figure 2-13.
A note is placed on a drawing to show how inch and millimeter dimensions can be identified.

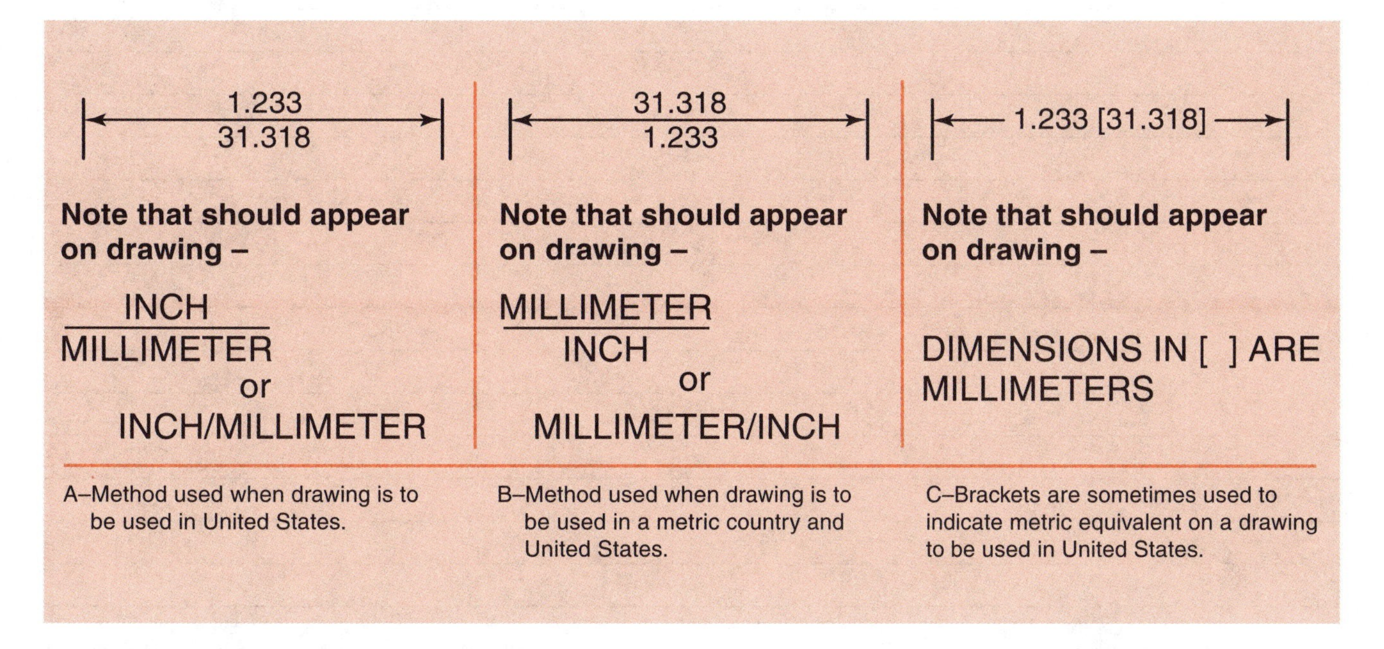

Figure 2-12.
Study how inch and metric dimensions are indicated on a dual dimensioned drawing.

Notes

Name ______________________________ Date ______________ Class ______________

Review Questions

Part I

1. For general work, welders should be able to read an inch rule to graduations of ____.

   ____

2. When reading a decimal inch graduated rule with 10 divisions per inch, each division is equal to ____.

   ____

3. Metric rules are usually graduated in ____.

   ____

4. Convert the measurement 3′4″ into inches. ____

   ____

5. The value 0.5 is an example of a decimal ____.

   ____

Part II

Use Figure 2-6 to make the following conversions.

1. 3/8″ into decimal inches ____
2. 1/4″ into millimeters ____
3. .875″ into fractional inches ____
4. 5/8″ into millimeters ____
5. .125″ into fractional inches ____

Print Reading Activities

Part I

Place the correct reading for each inch measurement in the blank space provided. Reduce fractions to their lowest terms. For example, $^{10}/_{16} = ^{5}/_{8}$.

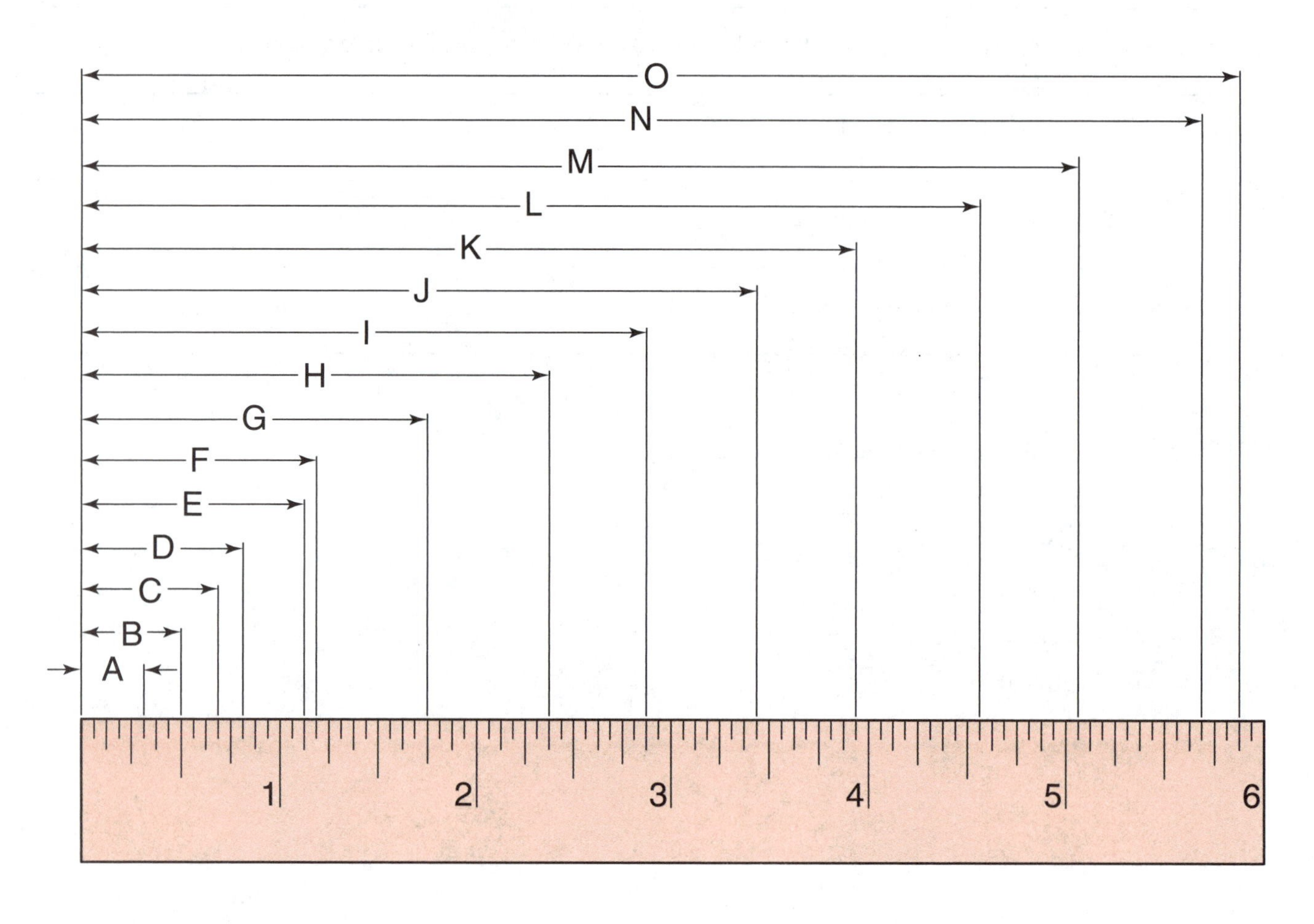

A ____________ D ____________ G ____________ J ____________ M ____________

B ____________ E ____________ H ____________ K ____________ N ____________

C ____________ F ____________ I ____________ L ____________ O ____________

Name ______________________________

Part II

Place the correct decimal for each measurement in the blank space provided. Be careful to place the decimal point in the proper location.

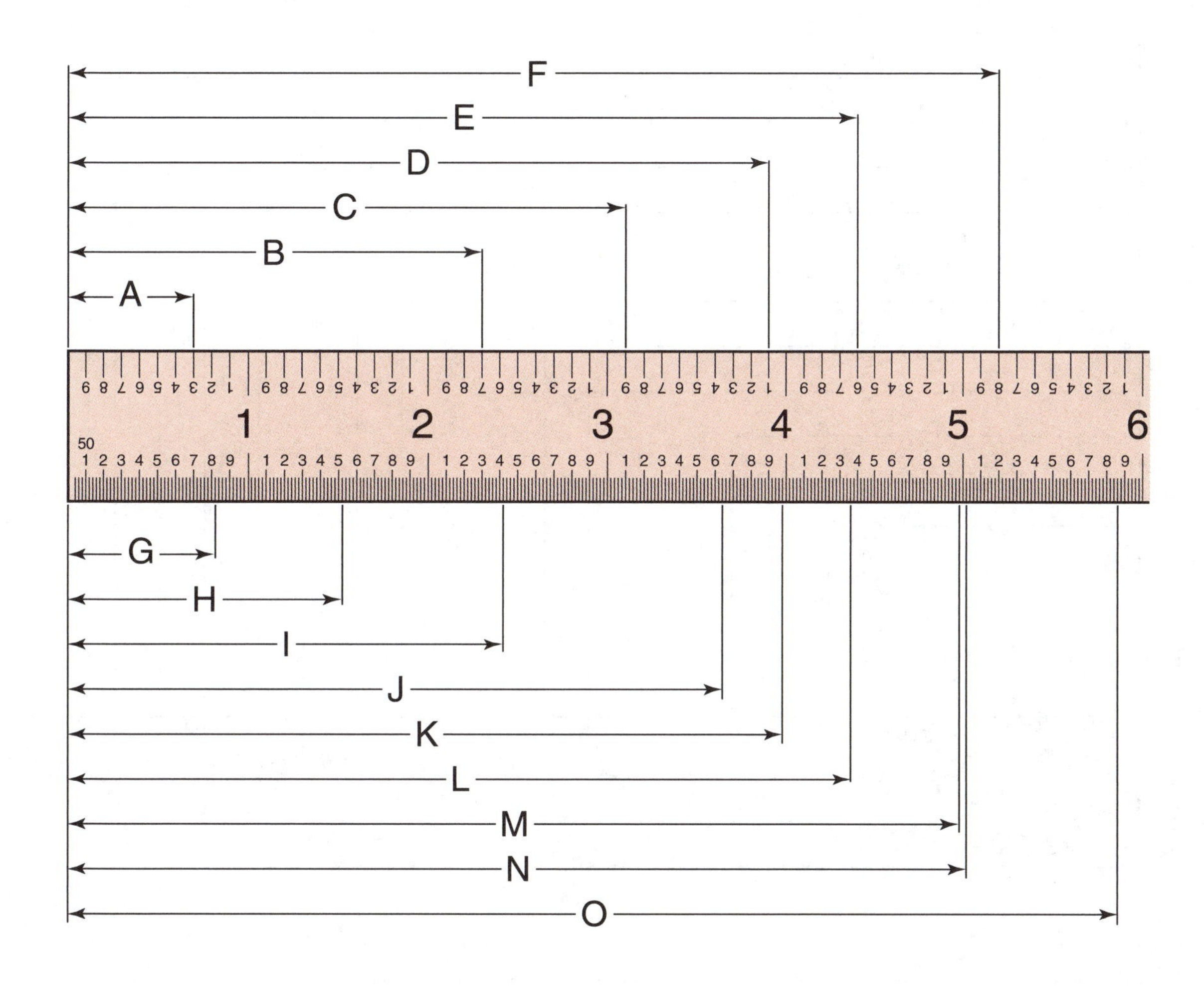

A 0.7″ D ________ G 0.82″ J ________ M ________

B ________ E ________ H ________ K ________ N ________

C ________ F ________ I ________ L ________ O ________

Part III

Place the correct metric reading for each measurement in the blank provided.

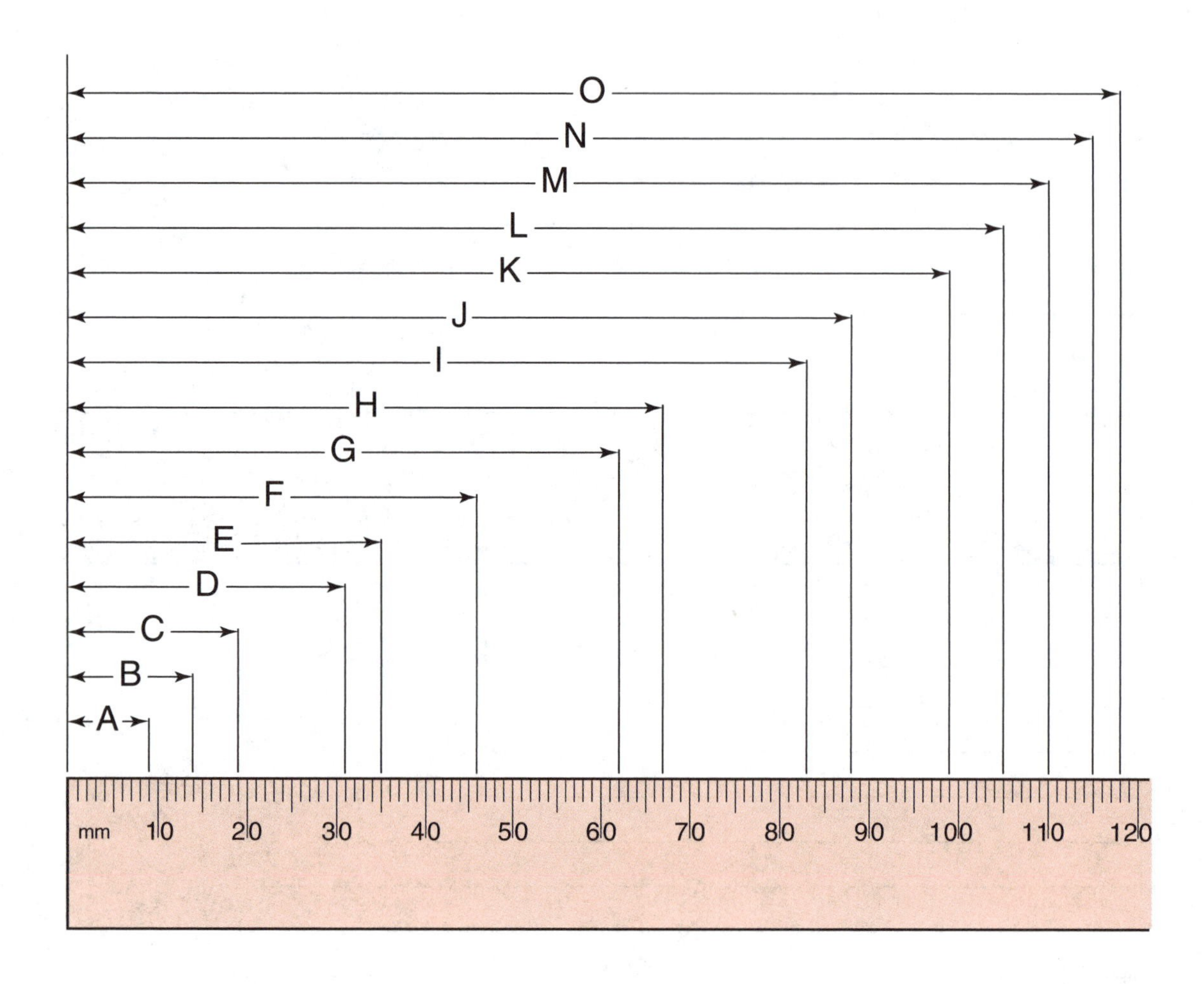

A ____________ D ____________ G ____________ J ____________ M ____________

B ____________ E ____________ H ____________ K ____________ N ____________

C ____________ F ____________ I ____________ L ____________ O ____________

Name ____________________________________

Part IV

Use a rule with 1⁄16″ graduations and measure the length of each line. Place your answer in the space to the left of each line. Reduce fractions to their lowest common denominator. For example, 6⁄8 = ¾.

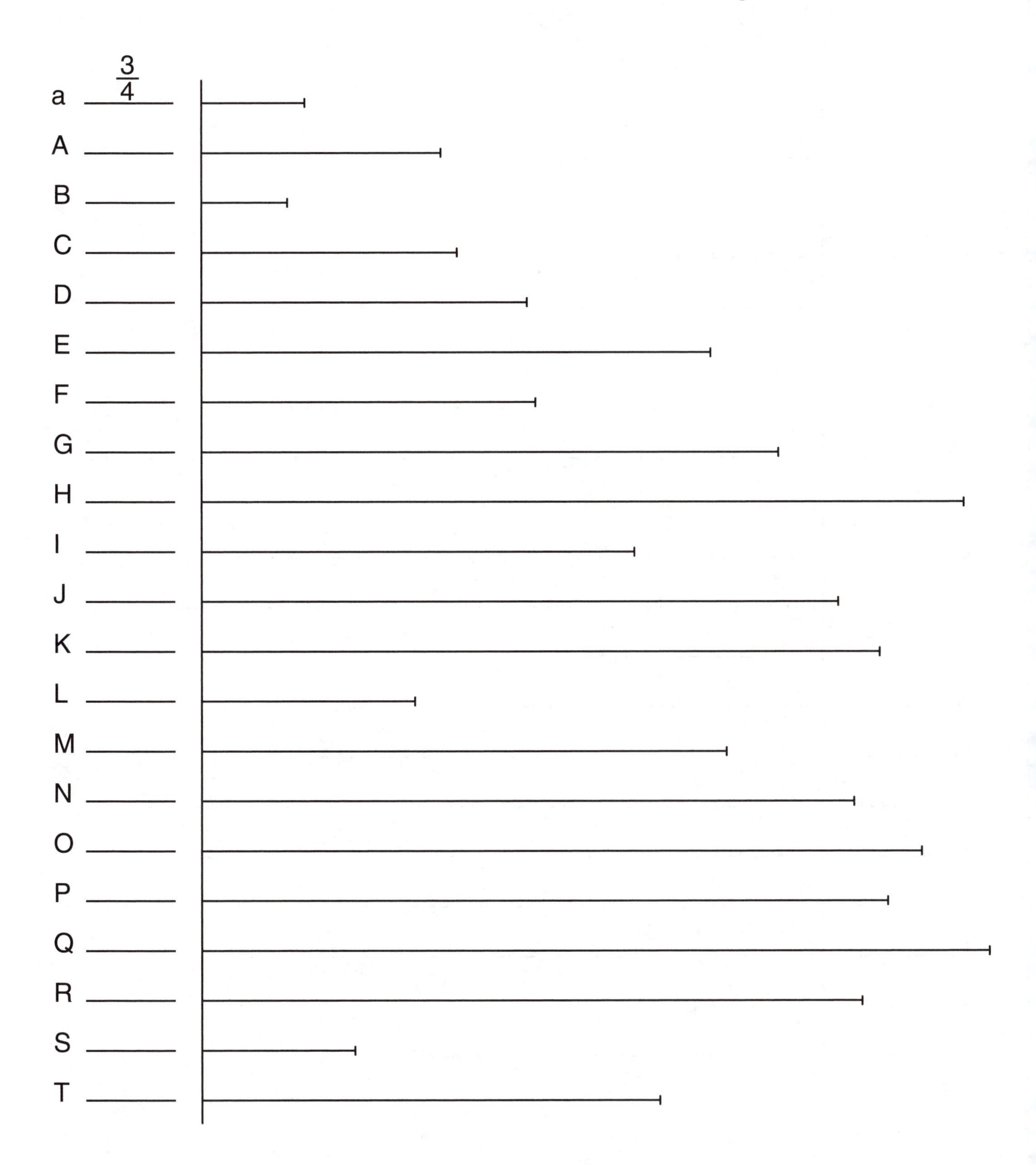

Part V

Use a rule with metric graduations and measure the length of each line. Place your answer in the space to the left of each line.

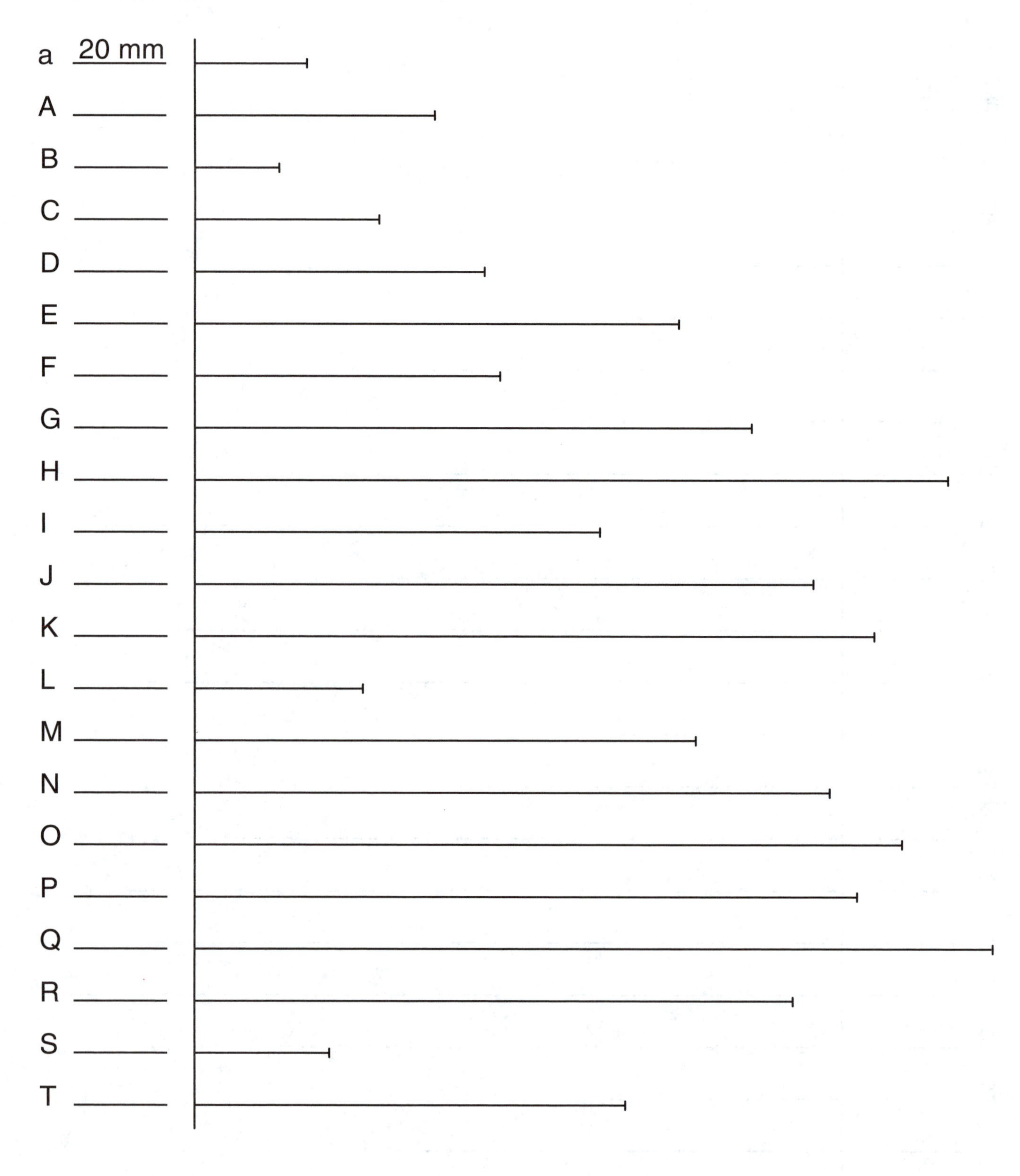

Name ____

Part VI

Use the plate cutout to answer the following questions. Using a rule with 1⁄16″ graduations, measure the distances listed. Find the fractional measurement and then find the decimal inch conversion for each measurement. Remember to change any mixed numbers to improper fractions before changing fractions to decimal inches.

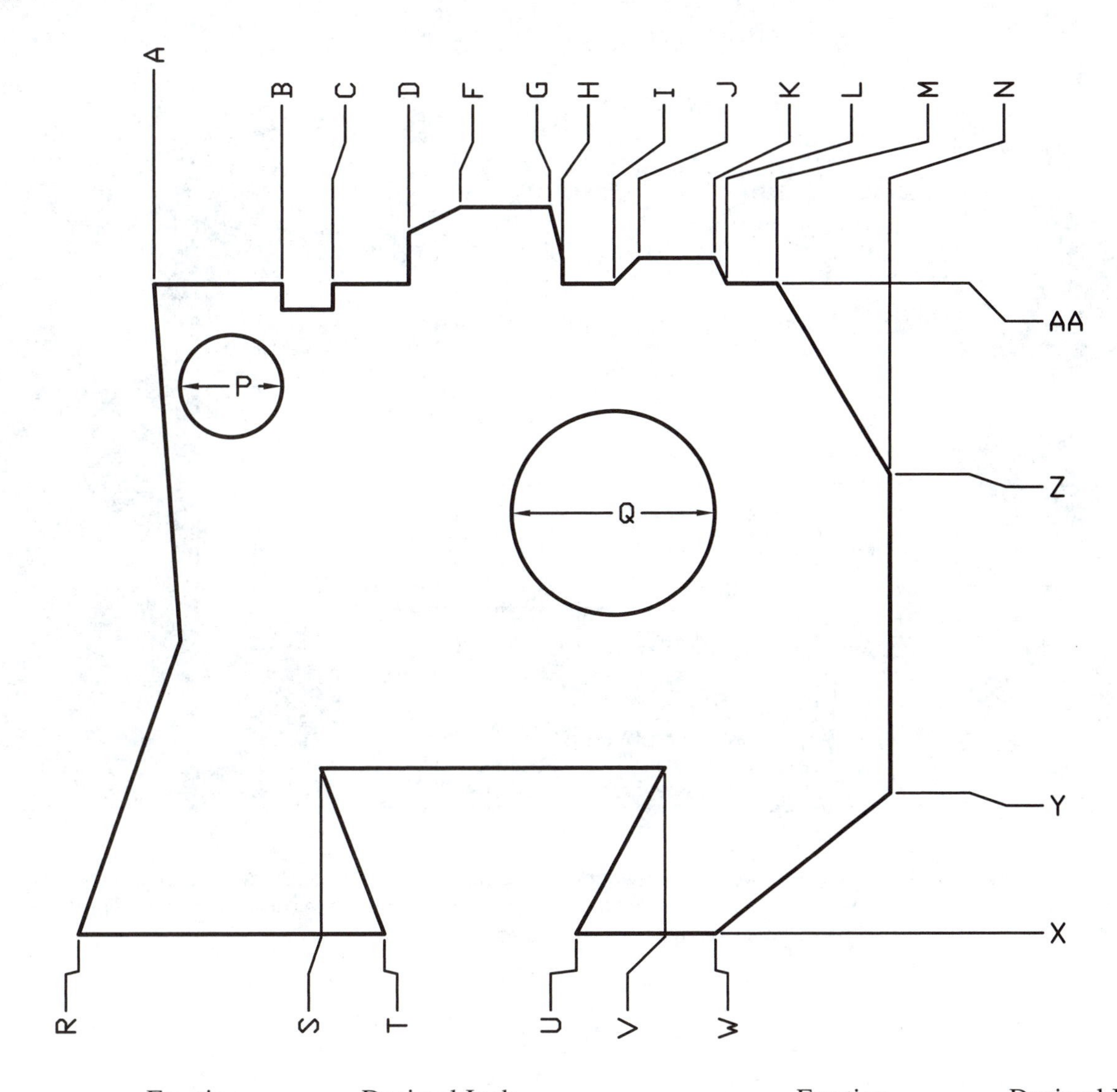

	Fraction Measurement	Decimal Inch Conversion
1. A to D	________	________
2. H to N	________	________
3. Y to Z	________	________
4. A to M	________	________
5. G to H	________	________
6. D to L	________	________
7. S to V	________	________
8. R to T	________	________

	Fraction Measurement	Decimal Inch Conversion
9. R to W	________	________
10. J to K	________	________
11. R to N	________	________
12. X to AA	________	________
13. F to G	________	________
14. Diameter P	________	________
15. Diameter Q	________	________

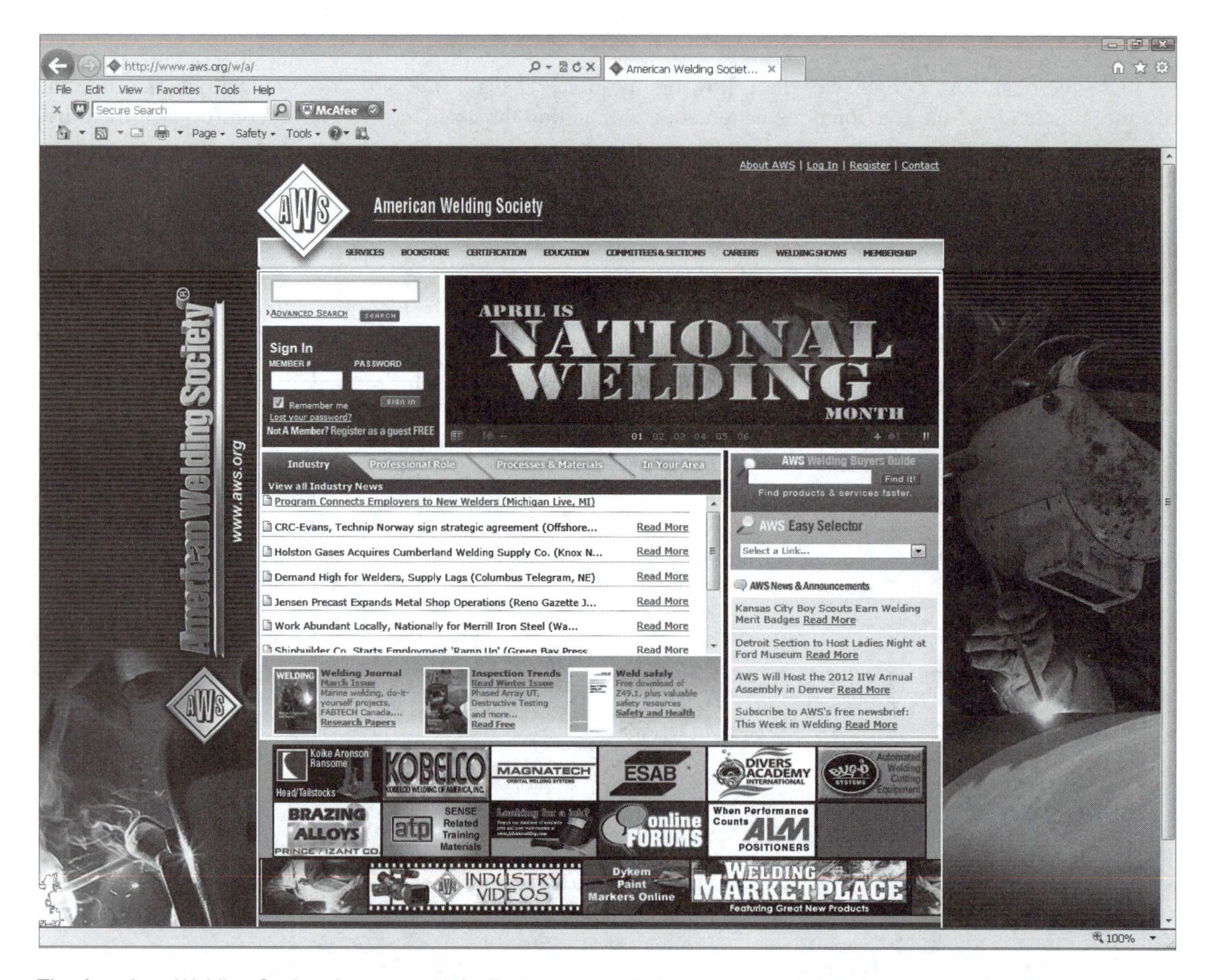

The American Welding Society (www.aws.org) offers a variety of resources for welders. Students can use the website to find information on how to become a certified welder, links to companies offering career and internship opportunities, and free publications on the latest updates in technical standards and safety.

Unit 3
A Review of Fractions and Decimals

After completing Unit 3, you will be able to:

- Define common fractions and explain rules for using them correctly.
- Add, subtract, multiply, and divide fractions.
- Explain various functions of, read, and write decimal fractions.
- Add, subtract, multiply, and divide decimal fractions.
- Explain the metric system as used by welders.
- Add, subtract, multiply, and divide metric decimals.
- Use a conversion chart to convert fractions to decimals, decimals to fractions, U.S. customary units to metric units, and metric units to U.S. customary units.

Key Words

caret mark
common fraction
denominator
dividend
divisor
equal fractions
fraction
higher terms
improper fraction
lowest terms
meter (m)
millimeter (mm)
mixed numbers
numerator
product
proper fraction
quotient
repeating decimal fractions
rounded off
whole numbers

Anyone using welding prints on the job must be able to make calculations using fractions and decimals. For this reason, this unit will provide a quick review of these basic mathematical skills.

Fractions

In math, a ***fraction*** is a number that tells what part of a whole unit is taken. A ***common fraction*** has one number placed above the other. A bar or short line is placed between the numbers. The fraction reading thirteen sixteenths, for example, would be written $^{13}/_{16}$. The number *below* the bar (16 in this example) is called the ***denominator.*** The denominator tells how many parts a whole unit has been divided into, Figure 3-1. The number on *top* of the bar (13 in this example) is called the ***numerator.*** The numerator indicates how many parts of the whole unit are taken.

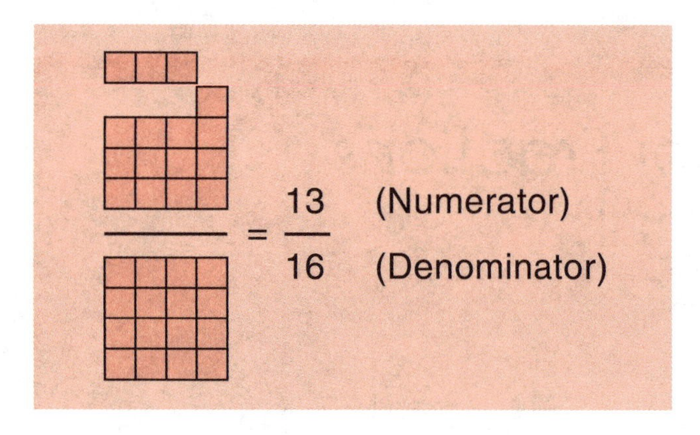

Figure 3-1.
A common fraction consists of a numerator (how many parts of the whole are taken) placed above the denominator (how many parts a whole unit has been divided into). A bar is placed between the numbers.

Numbers can take different forms, including the following:

- ***Whole numbers*** include zero and any positive number containing no fractional parts.
- ***Equal fractions*** have the same value but may have different forms. For instance, $\frac{1}{2} = \frac{2}{4} = \frac{4}{8}$.
- A ***proper fraction*** has a numerator smaller than the denominator. Two examples are $\frac{3}{8}$ and $\frac{7}{8}$.
- An ***improper fraction*** has a numerator equal to, or greater than, its denominator. The fractions $\frac{3}{3}$ and $\frac{8}{7}$ are examples.
- A ***mixed number*** consists of a whole number and a common fraction. Two examples of mixed numbers are $1\frac{1}{3}$ and $2\frac{3}{4}$.

Rules for Using Fractions

- *Whole numbers* are changed to fractions by multiplying the numerator and denominator by the same number.

  For example, this is how to change the whole number 5 into eighths:

  Each unit will contain 8 eighths and 5 units will contain 5 times 8 eighths or 40 eighths.

  $$\frac{5}{1} \times \frac{8}{8} = \frac{40}{8}$$

- *Mixed numbers* are changed to fractions by converting the whole number to a fraction with the same denominator as the fractional part of the mixed number. Then add the two fractions.

  For example, follow these steps to change $4\frac{3}{4}$ to a fraction:

  1. Change the whole number 4 to a fraction with the same denominator as the fraction $\frac{3}{4}$. Since each unit will contain 4 fourths, 4 units will contain 4 times 4 fourths or 16 fourths.

     $$\frac{4}{4} \times \frac{4}{1} = \frac{16}{4}$$

  2. Add the $\frac{3}{4}$ part of the mixed number to $\frac{16}{4}$.

     $$\frac{3}{4} + \frac{16}{4} = \frac{19}{4}$$

- *Improper fractions* are reduced to a whole or mixed number by dividing the numerator by the denominator.

  For example, this is how to convert $\frac{35}{16}$ to a whole or mixed number:

  $$\frac{35}{16} = 35 \div 16 = 2\frac{3}{16}$$

- Reduce fractions to their ***lowest terms*** by dividing the numerator and denominator by the largest number common to both. The value of a fraction is not changed if the same number divides the numerator and denominator.

  For example, to reduce $\frac{12}{16}$ to its lowest terms, divide by 4:

  $$\frac{12 \div 4}{16 \div 4} = \frac{3}{4}$$

- Change fractions to ***higher terms*** (equal fraction) by multiplying the numerator and denominator by the same number. The value of a fraction is not changed if the same number multiplies the numerator and denominator.

  For example, multiply by 4 to convert $\frac{3}{4}$ to sixteenths:

  $$\frac{3 \times 4}{4 \times 4} = \frac{12}{16}$$

Adding Fractions

Fractions can be added only if the denominators are the same. Then add the numerators.

For example, to add $\frac{2}{5} + \frac{1}{2} + \frac{1}{3}$, follow these steps:

1. Convert the fractions to ones that have the same denominator. The common denominator is found by multiplying their present denominators ($5 \times 2 \times 3$). The answer is 30. Next, convert each fraction to an equal fraction with a denominator of 30.

   $$\frac{2}{5} = \frac{?}{30} = \frac{2 \times 6}{5 \times 6} = \frac{12}{30}$$

   $$\frac{1}{2} = \frac{?}{30} = \frac{1 \times 15}{2 \times 15} = \frac{15}{30}$$

   $$\frac{1}{3} = \frac{?}{30} = \frac{1 \times 10}{3 \times 10} = \frac{10}{30}$$

2. Add only the numerators. The denominator remains the same.

$$\frac{12}{30} + \frac{15}{30} + \frac{10}{30} = \frac{37}{30} = 37 \div 30 = 1\frac{7}{30}$$

Addition Problems (Fractions)

Solve the following problems in the addition of fractions. Reduce answers to lowest terms.

1. $\frac{1}{2} + \frac{2}{3} + \frac{3}{4} =$ ____________
2. $\frac{3}{8} + \frac{15}{16} + \frac{5}{8} =$ ____________
3. $\frac{7}{16} + \frac{7}{8} + \frac{5}{6} =$ ____________
4. $\frac{5}{32} + \frac{1}{8} + \frac{9}{16} =$ ____________
5. $\frac{9}{64} + \frac{3}{32} + \frac{5}{16} + \frac{3}{4} =$ ____________
6. $\frac{3}{20} + \frac{7}{10} + \frac{3}{5} =$ ____________
7. $3\frac{3}{8} + \frac{7}{10} + 1\frac{5}{16} =$ ____________
8. $5\frac{7}{8} + 3\frac{1}{4} + \frac{3}{8} + 3 =$ ____________
9. $\frac{3}{64} + \frac{15}{32} + \frac{1}{2} + 4 =$ ____________
10. $1\frac{3}{4} + 3\frac{1}{8} + 1\frac{1}{4} + \frac{3}{2} =$ ____________
11. $\frac{5}{16} + \frac{5}{32} + \frac{5}{64} + \frac{5}{16} =$ ____________
12. $\frac{3}{8} + \frac{6}{16} + \frac{12}{32} + \frac{1}{8} =$ ____________
13. $2\frac{3}{8} + 1\frac{1}{8} + 5\frac{1}{16} =$ ____________
14. $1\frac{2}{3} + \frac{3}{5} + \frac{4}{5} + 2\frac{3}{10} + 1\frac{5}{6} =$ ____________
15. $\frac{1}{16} + \frac{3}{32} + \frac{5}{64} + \frac{17}{16} + 1\frac{1}{8} =$ ____________

Subtracting Fractions

Fractions cannot be subtracted unless the denominators are the same. Then subtract the numerators. Use the following steps:

1. For example, to subtract ⅜ from ¾, the fractions must be converted so they have the same denominator. In this case, the common denominator would be 8.

$$\frac{3}{4} = \frac{?}{8} = \frac{3 \times 2}{4 \times 2} = \frac{6}{8}$$

$$\frac{3}{8} = \frac{?}{8} = \frac{3 \times 1}{8 \times 1} = \frac{3}{8}$$

2. Subtract only the numerators. The denominator remains the same.

$$\frac{6}{8} - \frac{3}{8} = \frac{3}{8}$$

Subtraction Problems (Fractions)

Solve the following problems in the subtraction of fractions. Reduce your answers to their lowest terms.

1. $\frac{5}{8} - \frac{1}{4} =$ ____________
2. $\frac{1}{3} - \frac{1}{4} =$ ____________
3. $\frac{4}{5} - \frac{1}{6} =$ ____________
4. $\frac{11}{16} - \frac{15}{32} =$ ____________
5. $1\frac{3}{4} - \frac{15}{16} =$ ____________
6. $12\frac{3}{8} - 12\frac{3}{16} =$ ____________
7. $9\frac{13}{16} - 5\frac{5}{32} =$ ____________
8. $4\frac{5}{8} - 1\frac{1}{3} =$ ____________
9. $17\frac{1}{16} - \frac{15}{32} =$ ____________
10. $11\frac{3}{64} - \frac{7}{8} =$ ____________
11. $13\frac{7}{32} - 3\frac{7}{16} =$ ____________
12. $5\frac{1}{10} - \frac{1}{8} =$ ____________
13. $1\frac{5}{8} - \frac{11}{16} =$ ____________
14. $12\frac{1}{2} - \frac{15}{32} =$ ____________
15. $20\frac{9}{16} - 7\frac{3}{32} =$ ____________

Multiplying Fractions

Multiply fractions after the following steps are done:

1. Change all mixed numbers to improper fractions.
2. Multiply the numerators to get the numerator part of the answer.
3. Multiply the denominators to get the denominator part of the answer.
4. Reduce the fraction answer to lowest terms.

For example, 1½ × 2⅛ × 2=?

$$\frac{3}{2} \times \frac{17}{8} \times \frac{2}{1} = \frac{102}{16}$$

$$\frac{102}{16} = 6\frac{6}{16} = 6\frac{3}{8}$$

Multiplication Problems (Fractions)

Solve the following problems in the multiplication of fractions. Reduce your answers to lowest terms.

1. $\frac{1}{2} \times \frac{1}{2} \times \frac{1}{2} \times \frac{1}{2} =$ ______
2. $\frac{3}{14} \times \frac{1}{2} \times 2\frac{3}{4} =$ ______
3. $15 \times \frac{7}{8} =$ ______
4. $10 \times \frac{3}{5} \times 5\frac{1}{2} =$ ______
5. $2\frac{3}{4} \times \frac{3}{4} \times 1\frac{1}{3} =$ ______
6. $4\frac{3}{8} \times 5\frac{3}{4} \times \frac{1}{2} =$ ______
7. $2 \times 198 \times \frac{3}{3} =$ ______
8. $9\frac{3}{5} \times \frac{1}{5} \times \frac{4}{5} \times \frac{2}{5} =$ ______
9. $\frac{1}{3} \times 1\frac{2}{3} \times 1\frac{5}{6} =$ ______
10. $15 \times \frac{3}{5} =$ ______

Dividing Fractions

The number being divided is called the ***dividend.*** The number by which the dividend is divided is the ***divisor.***

Divide fractions after the following steps are done:

1. All mixed numbers must be changed to improper fractions.
2. To divide by a divisor that is a fraction, invert (turn upside down) the fraction and multiply.

For example, divide 4¾ by ¾.

$$\frac{19}{4} \div \frac{3}{4}$$

$$\frac{19}{4} \times \frac{4}{3} = \frac{76}{12} = 6\frac{4}{12} = 6\frac{1}{3}$$

Division Problems (Fractions)

Solve the following problems in the division of fractions. Reduce your answers to lowest terms.

1. $2\frac{1}{2} \div \frac{1}{4} =$ ______
2. $16 \div \frac{3}{8} =$ ______
3. $8\frac{1}{2} \div 4\frac{1}{16} =$ ______
4. $5\frac{1}{8} \div 5\frac{1}{16} =$ ______
5. $15 \div 5\frac{1}{5} =$ ______

Decimal Fractions

Decimal fractions are fractions using the base 10 for the denominator (1/10, 1/100, 1/1000, etc.). Decimal fractions, however, are always written *without* a denominator.

A decimal point (.) is placed to the left of a decimal fraction.

Decimal place is the position of a digit to the right of a decimal point. It indicates the value of the digit, Figure 3-2.

For example:

$\frac{1}{10}$ is written .1 (one tenth)

$\frac{1}{100}$ is written .01 (one hundredth)

$\frac{1}{1000}$ is written .001 (one thousandth)

$\frac{7}{10}$ is written .7 (seven tenths)

$\frac{93}{100}$ is written .93 (ninety-three hundredths)

$\frac{625}{1000}$ is written .625 (six hundred twenty-five thousandths)

$3\frac{753}{1000}$ is written 3.753 (three and seven hundred fifty-three thousandths)

Note that a zero (s) is used as a placeholder (.01 = one hundredth, .001 = one thousandth, etc.).

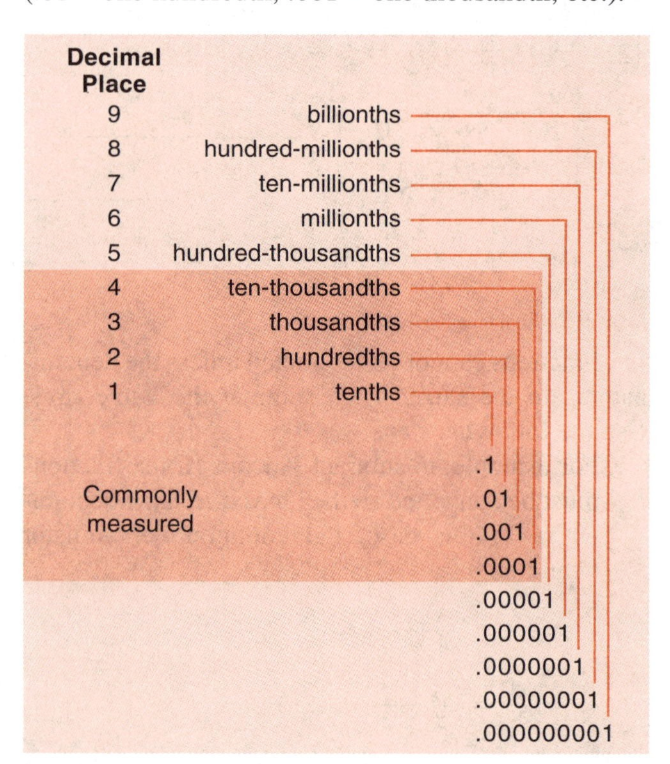

Figure 3-2.
Decimal place is the position of a digit to the right of a decimal point. It indicates the value of the digit.

Writing Problems (Decimal Fractions)

Write decimal fractions for the following:

1. $\frac{3}{10}$ = ______
2. $\frac{5}{100}$ = ______
3. $\frac{9}{1000}$ = ______
4. $2\frac{53}{100}$ = ______
5. $7\frac{625}{1000}$ = ______

Adding and Subtracting Decimal Fractions

Decimal fractions are added and subtracted the same way as whole numbers. The decimal points, however, must line up vertically.

For example, what does 1.317 plus 5.81 plus 3.1 plus 7.01 equal?

```
  1.317
  5.81
  3.1
 +7.01
 17.237
```

Another example: 10.001 minus 9.3 equals what?

```
 10.001
 -9.300
   .701
```

The decimal point in the answer is placed directly below its position in the columns of numbers being added or subtracted.

Addition Problems (Decimal Fractions)

Solve the following problems in the addition of decimal fractions. Double-check your answers and make sure you align the decimal points.

1. 2.123 + 3.234 + 4.345 = ______
2. 5 + 1.125 + 3.250 + 5.500 = ______
3. 10.534 + 7.3 + 5.152 + 3.304 = ______
4. 1 + .005 + .825 + .0234 = ______
5. 3.217 + .01 + 9.5 + 4.05 + 2.005 = ______

Subtraction Problems (Decimal Fractions)

Solve the following problems in the subtraction of decimal fractions.

1. 9.750 – 5.650 = ______
2. 7.897 – 4.925 = ______
3. 5.650 – 3.750 = ______
4. 15.345 – 2.456 = ______
5. 30.7 – 19.875 = ______

Multiplying Decimal Fractions

Decimal fractions are multiplied in the same way as whole numbers. Do not be concerned with the placement of the decimal point until multiplication is completed. Then the number of decimal places in the ***product*** (the answer to a multiplication problem) is the sum of the decimal places in the two numbers multiplied. The decimal point in the answer is set off from the right.

For example, what does 23.53 times 2.1 equal?

```
   23.53        2 decimal places
 ×   2.1      + 1 decimal place
   23 53
  4706
  49.4 13       3 decimal places
```

Multiplication Problems (Decimal Fractions)

Solve the following multiplication problems for decimal fractions. Make sure the decimal point is correctly located.

1. 15.1 × .25 = ____________
2. 6.3756 × 4 = ____________
3. 18.312 × 2.001 = ____________
4. 6.250 × .123 = ____________
5. 3.1416 × 8.2 = ____________
6. 9 × .125 = ____________
7. 8.125 × .25 = ____________
8. 9.373 × 10.2 = ____________
9. 93.73 × 1.02 = ____________
10. 17.33 × .333 = ____________

Dividing Decimal Fractions

Divide decimal fractions in the same way you divide whole numbers. The decimal point, however, must be considered and properly placed in the ***quotient*** (the answer to a division problem). To aid in the proper placement of the decimal point, a device called a ***caret mark*** (^) is used.

The decimal point in the quotient will be located directly above the caret mark in the dividend (the number being divided).

To locate the position of the decimal point in the answer do the following steps:

1. Count the number of decimal places in the divisor (the dividing number). Indicate the last position with a caret mark.
2. The number of places in the divisor is counted to the right of the decimal point in the dividend and indicated with a caret mark.
3. The decimal point in the answer (quotient) is placed directly above the caret mark in the dividend.

For example: What does 4.21875 divided by 3.75 equal? (Work to 3 decimal places in the answer.)

```
         1.125
3.75‸)4.21‸875
      3 75
        468
        375
         937
         750
         1875
         1875
```

Division Problems (Decimal Fractions)

Solve the following decimal fraction division problems to three decimal places.

1. 8 ÷ 2.5 = ____________
2. 4.25 ÷ .875 = ____________
3. 5.123 ÷ 5.1 = ____________
4. 567.765 ÷ 12.37 = ____________
5. 24 ÷ .8 = ____________
6. 123.786 ÷ 12.3786 = ____________
7. 8.875 ÷ .125 = ____________
8. 7.50 ÷ 1.25 = ____________
9. 750.5 ÷ 125.375 = ____________
10. 19.875 ÷ .025 = ____________

Additional Information on Decimal Fractions

Follow these steps to change a common fraction to a decimal fraction:

1. Divide the numerator by the denominator. For example, change ¾ to a decimal fraction.

```
   .75
4)3.00
  2 8
   20
   20
```

2. To change a decimal fraction to a common fraction, remove the decimal point and write in the proper denominator.

   For example, change .875 to a common fraction.

   .875 = eight hundred seventy-five thousandths

   $\frac{875}{1000} = \frac{7}{8}$

3. Decimal fractions must often be ***rounded off*** (the value of the number is closely approximated). After deciding which decimal place the number is to be rounded, the digit to the right of the rounded digit determines the rounding. A number less than 5 is rounded off to the *lesser* value. For example, the digit 2 causes 3.812 to round off to 3.81.

   If the number to the right is greater than 5, the number is rounded off to the next *greater* value. For example, the digit 5 causes 3.815 to round off to 3.82.

4. Some common fractions when converted to a decimal fraction produce ***repeating decimal fractions.*** For practical purposes, this means you cannot complete the division. In this situation, use as many decimal places as necessary for the particular problem and round off.

   For example, performing these division problems results in repeating decimal fractions:

   $\frac{1}{3}$ = .333333333 . . . Round off to .333

   $\frac{2}{3}$ = .666666666 . . . Round off to .667

Metrics

In welding, as in other occupations using drawings and prints, metric dimensioning may be encountered. Figure 3-3 is a conversion chart that can be used to change from U.S. customary units to metric values. It also gives approximate equivalents.

The ***meter (m)*** is the basic unit of the metric system. Units that are multiples or fractional parts of the meter are shown by prefixes to the word "meter." The ***millimeter (mm),*** a very small unit of measure, is commonly used on metric blueprints.

SI Units & Conversions

Property	Unit	Symbol	Exact Conversion			Approximate Equivalency
			From	To	Multiply by	
length	meter centimeter millimeter	m cm mm	inch inch foot	mm cm mm	2.540×10 2.540 3.048×10^2	25 mm = 1 in. 300 mm = 1 ft.
mass	kilogram gram tonne (megagram)	kg g t	ounce pound ton (2000 lb)	g kg kg	2.835×10 4.536×10^{-1} 9.072×10^2	2.8 g = 1 oz. kg = 2.2 lbs. = 35 oz. 1 t = 2200 lbs.
density	kilogram per cu. meter	kg/m^3	pounds per cu. ft.	kg/m^3	1.602×10	16 kg/m^3 = 1 lb./ft.^{3}
temperature	deg. Celsius	°C	deg. Fahr.	°C	(°F – 32) × 5/9	0°C = 32°F 100°C = 212°F
area	square meter square millimeter	m^2 mm^2	sq. inch sq. ft.	mm^2 m^2	6.452×10^2 9.290×10^{-2}	645 mm^2 = 1 in.^{2} 1 m^2 = 11 ft.^{2}
volume	cubic meter cubic centimeter cubic millimeter	m^3 cm^3 mm^3	cu. in. cu. ft. cu. yd.	mm^3 m^3 m^3	1.639×10^4 2.832×10^{-2} 7.645×10^{-1}	16 400 mm^3 = 1 in.^{3} 1 m^3 = 35 ft.^{3} 1 m^3 = 1.3 yd.^{3}
force	newton kilonewton meganewton	N kN MN	ounce (Force) pound (Force) Kip	N kN MN	2.780×10^{-1} 4.448×10^{-3} 4.448	1 N = 3.6 oz. 4.4 N = 1 lb. 1 kN = 225 lb.
stress	megapascal	MPa	pound/in^2 (psi) Kip/in^2 (ksi)	MPa MPa	6.895×10^{-3} 6.895	1 MPa = 145 psi 7 MPa = 1 ksi
torque	newton-meters	N-m	in-ounce in-pound ft-pound	N-m N-m N-m	7.062×10^3 1.130×10^{-1} 1.356	1 N-m = 140 in-oz. 1 N-m = 9 in-lb. 1 N-m = .75 ft-lb. 1.4 N-m = 1 ft-lb.

Figure 3-3.
Study the SI units Conversion Chart. Multiply by a given factor to convert from a U.S. customary value to an SI metric value.

For example:

1 millimeter (mm) = 0.001 meter or 1/1000 meter

1 centimeter (cm) = 0.01 meter or 1/100 meter

1 decimeter (dm) = 0.1 meter or 1/10 meter

1 meter (m) = 1.0 meter or 10/10 meter

1 decameter (dkm) = 10 meters

1 hectometer (hm) = 100 meters

1 kilometer (km) = 1000 meters

These prefixes may be applied to any unit of length, weight, volume, etc. The meter is adopted as the basic unit of length, the gram for mass, and the liter for volume.

In the metric system, area is measured in square millimeters (mm^2), square centimeters (cm^2), etc. Volume is commonly measured in cubic millimeters (mm^3), centimeters (cm^3), etc. One liter (L) is equal to 1000 centimeters.

The following are some examples of length equivalents.

10 millimeters = 1 centimeter

10 centimeters = 1 decimeter

10 decimeters = 1 meter

1000 meters = 1 kilometer

When working with metrics in welding, you should *not* convert metric dimensions to inches, feet, yards, etc. Calculations involving metrics are done in much the same manner as calculations made with decimal fractions. The major difference is *all units* must be the same (millimeters, centimeters, meters, etc.), before they can be added, subtracted, multiplied, or divided.

For example, add the following values with these different units:

25.8 mm + 0.82 m + 15.7 mm + 30.52 mm =

Before the problem can be worked, the 0.82 m must be converted to millimeters (mm). Then all units will be the same.

Since there are 1000.0 mm in each m, multiply 0.82 × 1000.0 to get 820.0 mm. Now addition can be performed.

```
   25.80
  820.00
   15.70
+  30.52
--------
  892.02 mm
```

Metric Addition Problems

Solve the following problems. Double-check your answers. Remember to convert to equal units.

1. 12.5 mm + 17.5 mm + 0.85 mm + 0.15 mm = _____ mm
2. 1.25 mm + 1.25 m + 1.25 cm = _____ mm
3. 987.98 cm + 456.7 m + 456.7 cm + 987.98 mm = _____ m
4. 0.98 cm + 0.89 cm + 89.0 cm + 10.0 mm = _____ cm
5. 1.000 mm + 1000.0 mm + 1.0 cm + 10.10 cm = _____ mm

Metric Subtraction Problems

Solve the following metric subtraction problems. Convert to equal values before subtracting.

1. 12.5 cm – 12.5 mm = _____ mm
2. 125.0 mm – 12.5 cm = _____ mm
3. 87.5 mm – 37.5 mm = _____ mm
4. 625.0 mm – 2.5 cm = _____ mm
5. 375.5 mm + 37.55 cm – 562.5 mm = _____ mm

Metric Multiplication Problems

Solve the following metric multiplication problems.

1. 12.5 mm × 12.5 mm = _____ mm^2
2. 56.3 mm × 5.63 cm = _____ cm^2
3. 875.25 mm × 12.3 cm = _____ cm^2
4. 438.2 cm × 2.15 cm = _____ cm^2
5. 75.0 mm × 6.5 cm = _____ cm^2

Metric Division Problems

Solve the following division problems to two (2) decimal places.

1. 8.0 mm ÷ 2.5 mm = _____
2. 8.0 cm ÷ 2.5 mm = _____
3. 88.75 mm ÷ 0.25 cm = _____
4. 19.99 mm ÷ 0.25 cm = _____
5. 25 mm ÷ 2.5 cm = _____

Conversions

Sometimes it is necessary to convert a decimal fraction to a fraction or a fraction to a decimal fraction. Sometimes a U.S. customary unit is changed to a metric unit or a metric unit to a customary measurement. Many of the prints in use today are dimensioned in decimal inches. However, many of the rules and tape measures used are graduated in fractions of an inch. You should become familiar with the common fractions and their decimal fraction equivalents at least in quarters of an inch and possibly eighths of an inch.

Decimal Fraction to Fraction Conversion

The most common conversion you will find is the decimal fraction to fraction conversion. Often for welding and fabrication, the same precision needed for machining and forming is not possible or required. But, many mechanical drawings show all measurements as decimal fractions.

You need to convert the more precise three-place decimal fraction to a common fraction. Although it is possible to mathematically determine the fraction (refer to Chapter 2), it is more common and easier to use a conversion table.

Figure 3-4 shows part of the Conversion Chart found in the reference section. Notice the first column shows drill sizes, the second column fractions, the third column decimal fractions (expressed in inches) and the last column millimeters. To convert a decimal fraction measurement to a fraction, first find the decimal fraction in the third column. Then find the closest fraction to that measurement.

It is often helpful to know the graduations on your rule or tape measure so you can find the correct fraction from the table. Here is an example. If your tape measure is graduated in eighths of an inch and the print dimension is 0.123, first find 0.123 on the Conversion Chart. Next look for the closest fraction with a denominator of 8. In Figure 3-4 you will see that when you find 0.123, ⅛ is the closest fraction with an 8 in the denominator.

Decimal Fraction to Fraction Conversion Problems

Use the Conversion Chart found in the reference section to complete the following problems.

Convert the decimal fraction (in inches) to the closest ⅛ of an inch fraction.

Decimal Fraction	Fraction
1. .120 =	______________
2. .256 =	______________
3. .515 =	______________
4. .630 =	______________
5. .880 =	______________

Metric to Decimal Fraction or Fraction Conversion

The same method used for converting decimal fractions to fractions is used for converting millimeters to decimal fractions or fractions. Although it is best to stay in the same unit of measure for welding and fabrication, sometimes you may not have a metric rule or tape measure.

To convert to a decimal fraction or fraction, first find the millimeter measurement from the print measurement in the last column of the conversion chart. Then find the decimal fraction or fraction in the third or second column. You may have to find the nearest fraction if the conversion is not direct.

For example, if your tape measure is graduated in sixteenths of an inch and the print dimension is 4.7 mm, first find 4.7 mm on the Conversion Chart. Next look for the closest fraction with a denominator of 16.

In Figure 3-4 you will see that when you find 4.7 mm, $3/16$ is the closest fraction with a 16 in the denominator and the direct conversion to an inch decimal fraction is 0.186″. Notice that without a second decimal place in the millimeter measurement, only an approximation of the decimal fraction is possible.

Metric to Decimal Fraction and Fraction Conversion Problems

Use the Conversion Chart found in the reference section to complete the following problems.

Convert the millimeters to the closest direct decimal fraction (in inches) and fractions to the closest $1/16$ of an inch. Reduce if necessary.

Millimeter	Decimal Fraction	Fraction
1. 1.57 =	__________	__________
2. 3.12 =	__________	__________
3. 6.30 =	__________	__________
4. 11.0 =	__________	__________
5. 11.2 =	__________	__________

Conversion Chart Inch/mm

Drill No. or Letter		Inch	mm
		.001	0.0254
		.002	0.0508
		.003	0.0762
		.004	0.1016
		.005	0.1270
		.006	0.1524
		.007	0.1778
		.008	0.2032
		.009	0.2286
		.010	0.2540
		.011	0.2794
		.012	0.3048
		.013	0.3302
80	.0135	.014	0.3556
79	.0145	.015	0.3810
	1/64	.0156	0.3969
78		.016	0.4064
		.017	0.4318
77		.018	0.4572
		.019	0.4826
76		.020	0.5080
75		.021	0.5334
		.022	0.5588
74	.0225	.023	0.5842
73		.024	0.6096
72		.025	0.6350
71		.026	0.6604
		.027	0.6858
70		.028	0.7112
		.029	0.7366
69	.0292	.030	0.7620
68		.031	0.7874
	1/32	.0312	0.7937
67		.032	0.8128
66		.033	0.8382
		.034	0.8636
65		.035	0.8890
64		.036	0.9144
63		.037	0.9398
62		.038	0.9652
61		.039	0.9906
		.0394	1.0000
60		.040	1.0160
59		.041	1.0414
58		.042	1.0668
57		.043	1.0922
		.044	1.1176
		.045	1.1430
56	.0465	.046	1.1684
	3/64	.0469	1.1906
		.047	1.1938
		.048	1.2192
		.049	1.2446
		.050	1.2700
		.051	1.2954
55		.052	1.3208
		.053	1.3462
		.054	1.3716
54		.055	1.3970
		.056	1.4224
		.057	1.4478
		.058	1.4732
		.059	1.4986
53	.0595	.060	1.5240
		.061	1.5494
		.062	1.5748
	1/16	.0625	1.5875
52	.0635	.063	1.6002

Drill No. or Letter		Inch	mm
		.064	1.6256
		.065	1.6510
		.066	1.6764
51		.067	1.7018
		.068	1.7272
		.069	1.7526
50		.070	1.7780
		.071	1.8034
		.072	1.8288
49		.073	1.8542
		.074	1.8796
		.075	1.9050
48		.076	1.9304
		.077	1.9558
47	.0785	.078	1.9812
	5/64	.0781	1.9844
		.0787	2.0000
		.079	2.0066
		.080	2.0320
46		.081	2.0574
45		.082	2.0828
		.083	2.1082
		.084	2.1336
		.085	2.1590
44		.086	2.1844
		.087	2.2098
		.088	2.2352
43		.089	2.2606
		.090	2.2860
		.091	2.3114
		.092	2.3368
42	.0935	.093	2.3622
	3/32	.0937	2.3812
		.094	2.3876
		.095	2.4130
41		.096	2.4384
		.097	2.4638
40		.098	2.4892
		.099	2.5146
39	.0995	.100	2.5400
		.101	2.5654
38	.1015	.102	2.5908
		.103	2.6162
37		.104	2.6416
		.105	2.6670
36	.1065	.106	2.6924
		.107	2.7178
		.108	2.7432
		.109	2.7686
	7/64	.1094	2.7781
35		.110	2.7490
34		.111	2.8194
		.112	2.8448
33		.113	2.8702
		.114	2.8956
		.115	2.9210
32		.116	2.9464
		.117	2.9718
		.118	2.9972
		.1181	3.0000
		.119	3.0226
31		.120	3.0480
		.121	3.0734
		.122	3.0988
		.123	3.1242
		.124	3.1496
	1/8	.125	3.1750
		.126	3.2004

Drill No. or Letter		Inch	mm
		.127	3.2258
		.128	3.2512
30	.1285	.129	3.2766
		.130	3.3020
		.131	3.3274
		.132	3.3528
		.133	3.3782
		.134	3.4036
		.135	3.4290
29		.136	3.4544
		.137	3.4798
		.138	3.5052
		.139	3.5306
28	.1405	.140	3.5560
	9/64	.1406	3.5719
		.141	3.5814
		.142	3.6068
		.143	3.6322
27		.144	3.6576
		.145	3.6830
		.146	3.7484
26		.147	3.7338
		.148	3.7592
25	.1495	.149	3.7846
		.150	3.8100
		.151	3.8354
24		.152	3.8608
		.153	3.8862
23		.154	3.9116
		.155	3.9370
		.156	3.9624
	5/32	.1562	3.9687
22		.157	3.9878
		.1575	4.0000
		.158	4.0132
21		.159	4.0386
		.160	4.0640
20		.161	4.0894
		.162	4.1148
		.163	4.1402
		.164	4.1656
		.165	4.1910
19		.166	4.2164
		.167	4.2418
		.168	4.2672
		.169	4.2926
18	.1635	.170	4.3180
		.171	4.3434
	11/64	.1719	4.3656
		.172	4.3688
17		.173	4.3942
		.174	4.4196
		.175	4.4450
		.176	4.4704
16		.177	4.4958
		.178	4.5212
		.179	4.5466
15		.180	4.5720
		.181	4.5974
14		.182	4.6228
		.183	4.6482
		.184	4.6736
13		.185	4.6990
		.186	4.7244
		.187	4.7498
	3/16	.1875	4.7625
		.188	4.7752

Drill No. or Letter		Inch	mm
12		.189	4.8006
		.190	4.8260
11		.191	4.8514
		.192	4.8768
		.193	4.9022
10	.1935	.194	4.9276
		.195	4.9530
9		.196	4.9784
		.1969	5.0000
		.197	5.0038
		.198	5.0292
		.199	5.0546
8		.200	5.0800
7		.201	5.1054
		.202	5.1308
		.203	5.1562
	13/64	.2031	5.1594
6		.204	5.1816
5	.2055	.205	5.2070
		.206	5.2324
		.207	5.2578
		.208	5.2832
4		.209	5.3086
		.210	5.3340
		.211	5.3594
		.212	5.3848
3		.213	5.4102
		.214	5.4356
		.215	5.4610
		.216	5.4864
		.217	5.5118
		.218	5.5372
	7/32	.2187	5.5562
		.219	5.5626
		.220	5.5880
2		.221	5.6134
		.222	5.6388
		.223	5.6642
		.224	5.6896
		.225	5.7150
		.226	5.7404
		.227	5.7658
1		.228	5.7912
		.229	5.8166
		.230	5.8410
		.231	5.8674
		.232	5.8928
		.233	5.9182
A		.234	5.9436
	15/64	.2344	5.9531
		.235	5.9690
		.236	5.9944
		.2362	6.0000
		.237	6.0198
B		.238	6.0452
		.239	6.0706
		.240	6.0960
		.241	6.1214
C		.242	6.1468
		.243	6.1722
		.244	6.1976
		.245	6.2230
D		.246	6.2484
		.247	6.2738
		.248	6.2992
		.249	6.3246
E	1/4	.250	6.3500

Figure 3-4.
Using a conversion chart is an easy way to convert an inch decimal fraction measurement to a fraction. The same chart is used for converting millimeters to inch decimal fractions or fractions.

Name ______________________ Date ____________ Class ____________

Review Questions

Part I

1. The number above the bar in a common fraction is called the _____. The number below the bar is called the _____.

2. When dividing fractions, the number being divided is called the _____.

3. What is the difference between a proper fraction and an improper fraction?

4. A mixed number consists of a(n) _____ number and a common fraction.

Part II

Solve the following problems. Reduce fractions to lowest terms.

1. $1/2 + 3/4 + 3/5 =$ _____
2. $1/16 + 5/8 + 3/64 =$ _____
3. $5/6 - 1/3 =$ _____
4. $6\,1/2 - 3\,1/4 =$ _____
5. $1/2 \times 5/8 =$ _____
6. $3\,1/2 \times 5\,2/3 \times 1\,1/8 =$ _____
7. $4\,1/2 \div 3/4 =$ _____
8. $6\,3/8 \div 2\,3/4 =$ _____
9. $12.632 \times 2.25 =$ _____
10. $22.43 \div .044 =$ _____

Part III

Use Figure 3-4 to make the following conversions. Find the closest 1/16″ fraction to the measurement given.

1. .063″ into fractional inches _____
2. .122″ into fractional inches _____
3. .185″ into fractional inches _____
4. 3.15 mm into fractional inches _____
5. 6.33 mm into fractional inches _____

Section 2 Print Reading Basics

Unit 4

Alphabet of Lines

After completing Unit 4, you will be able to:

- Identify the types of lines found on a print.
- Describe the proper use of types of lines found on a print.

Key Words

alphabet of lines
break line
center line
cutting-plane line
dimension line
extension line
hidden line
leader
phantom line
section line
visible line

As you study a print, you will see that lines of different weight (thicknesses) are used to describe the shape and size of the object, Figure 4-1. Each type of line has a particular meaning. Differences between the line weights also help make a print easier to read or understand.

Types of Lines

To interpret a print, it is essential to know the characteristics of the various lines and how they are used on a drawing. These lines, known as the ***alphabet of lines,*** are universally used throughout industry, Figure 4-2.

Visible Line

The ***visible line*** (also called ***visible object line*** or ***object line***) is a thick, continuous line. It is used to outline the visible edges or contours of the object, Figure 4-3.

Hidden Line

A ***hidden line*** (also called ***hidden object line***), Figure 4-4, represents a hidden edge or internal feature of the object. It is medium weight and composed of short dashes about ¼″ (6.0 mm) long. The dashes are typically separated by 1⁄16″ (1.6 mm) spaces. However, hidden object line size may vary according to the size of the drawing.

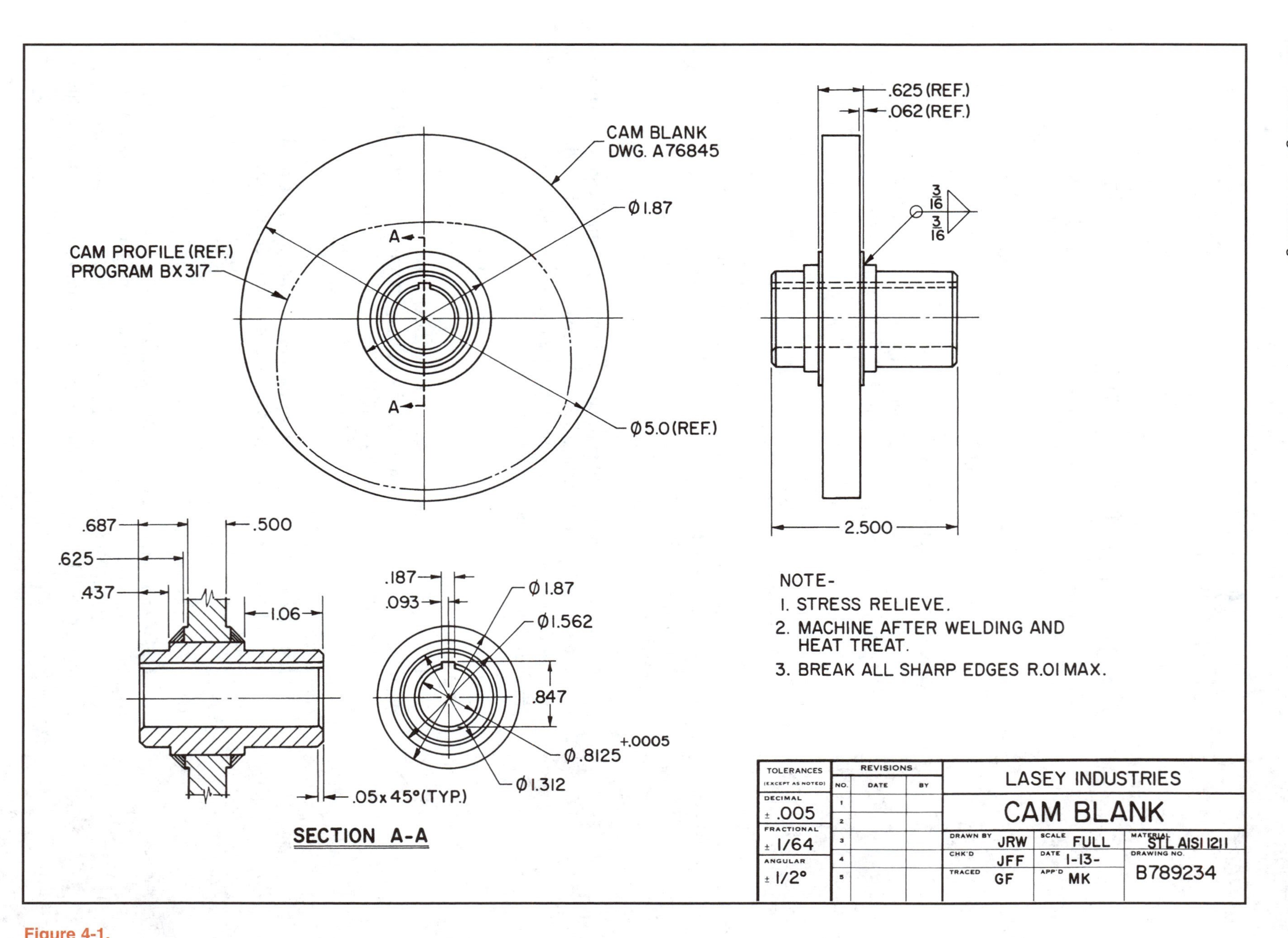

Figure 4-1.
Lines of different weights or thickness describe the shape and size of the object shown on a drawing.

Alphabet of Lines				
Name	**Use**	**Line Conventions**	**Line Weight**	**Example**
Visible (object) line	Define shape. Outline and detail objects	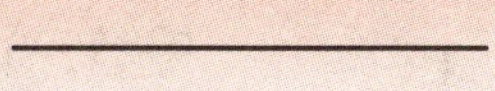	Thick	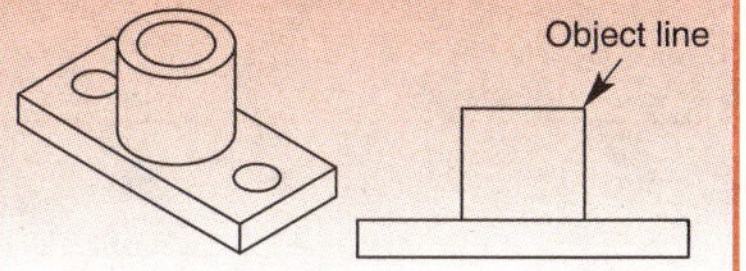 Object line
Hidden (object) line	Show hidden features		Thin	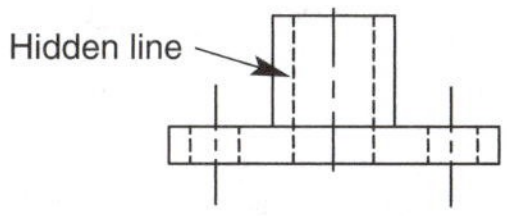 Hidden line
Center line	Locate center points of arcs and circles		Thin	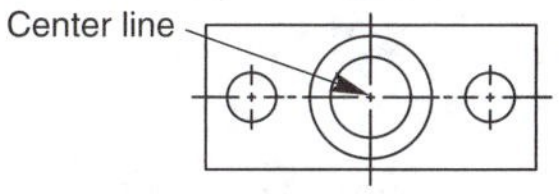 Center line
Dimension line	Show size or location	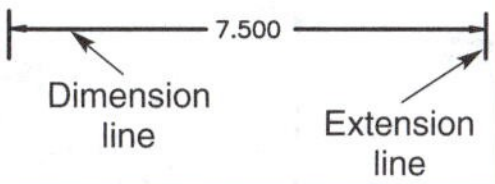 7.500 Dimension line	Thin	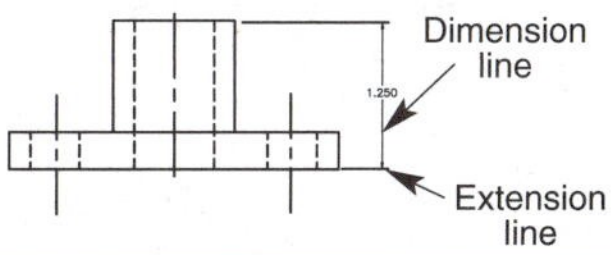 Dimension line 1.250
Extension line	Define size or location	Extension line	Thin	Extension line
Leader	Call out specific features	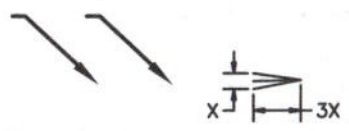 X 3X	Thin	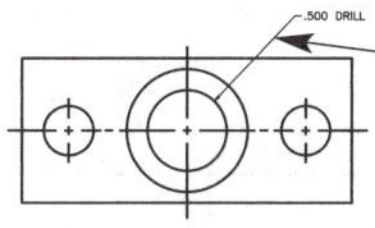 .500 DRILL Leader line
Cutting-plane	Show internal features	A A	Thick	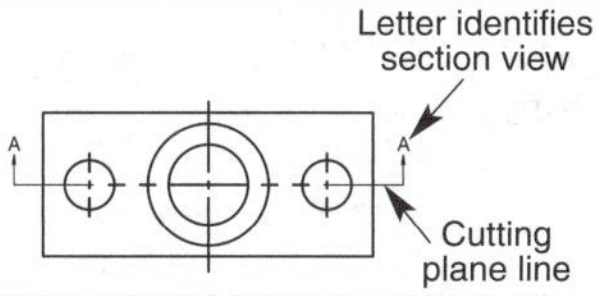 Letter identifies section view A A Cutting plane line
Section line	Identify internal features	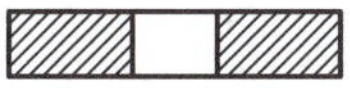	Thin	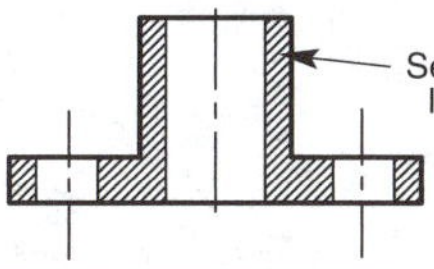 Section lines
Break line	Show long breaks	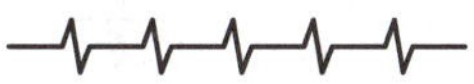	Thin	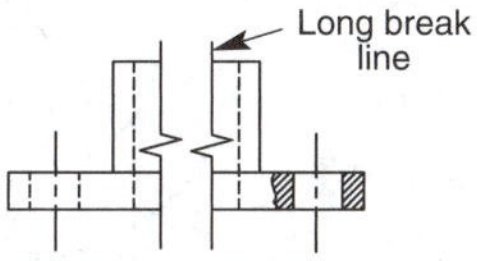 Long break line
Break line	Show short breaks		Thick	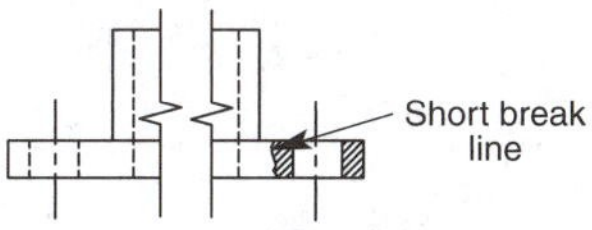 Short break line
Phantom line	Show different position or movement	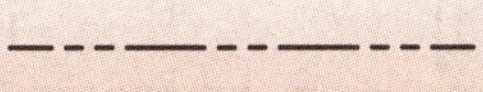	Thin	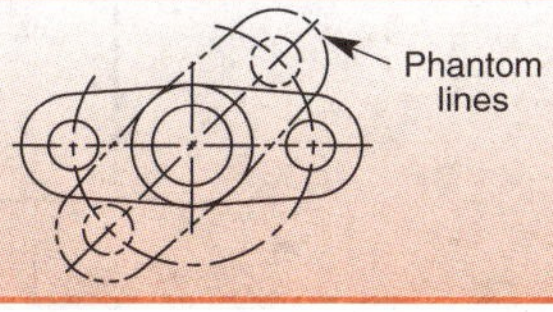 Phantom lines

Figure 4-2.
Interpreting prints accurately requires the welder to understand the universal meaning of the lines and line weights on the drawing.

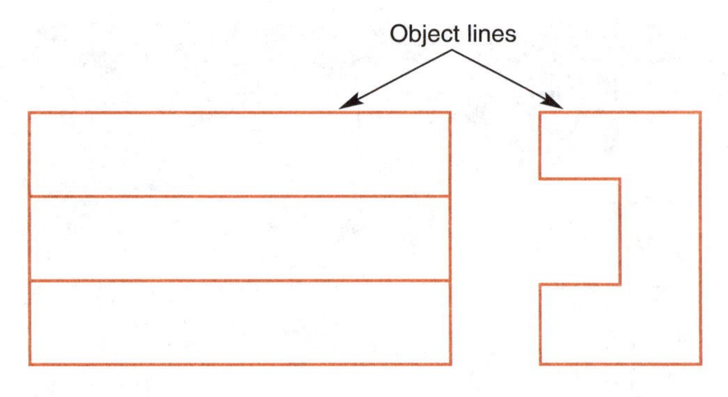

Figure 4-3.
A visible line (also called an object line) outlines the visible edges of a drawn object.

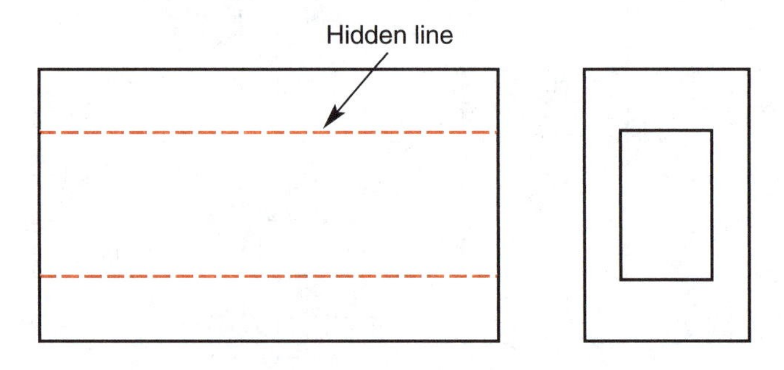

Figure 4-4.
A hidden object line (or invisible line) shows hidden features of an object.

Center Line

The ***center line*** is used to indicate the center of symmetrical objects. They are fine, dark lines composed of alternate long and short dashes with spaces between the dashes, Figure 4-5. On welding prints, center lines are used to show the center of features such as holes, round parts, plug welds, or intermittent fillet welds.

On a same size drawing, the long section of the center line is about ¾″ (19.0 mm) long. The shorter dashes are about ¼″ (6.0 mm) long. The spaces are 1/16″ (1.6 mm) long.

Section Line

Section lines are used to show the cut surfaces of the object in section views. They are fine, dark lines, Figure 4-6. Various types of section lines may also indicate the type of material cut by the cutting plane line. However, a general purpose section line (symbol for cast iron) is often used on a drawing when the material specifications ("specs") are shown elsewhere.

Cutting-Plane Line

The ***cutting-plane line*** indicates an imaginary cut made through the object to reveal its interior characteristics, Figure 4-7. It is a heavy line used with sectional views.

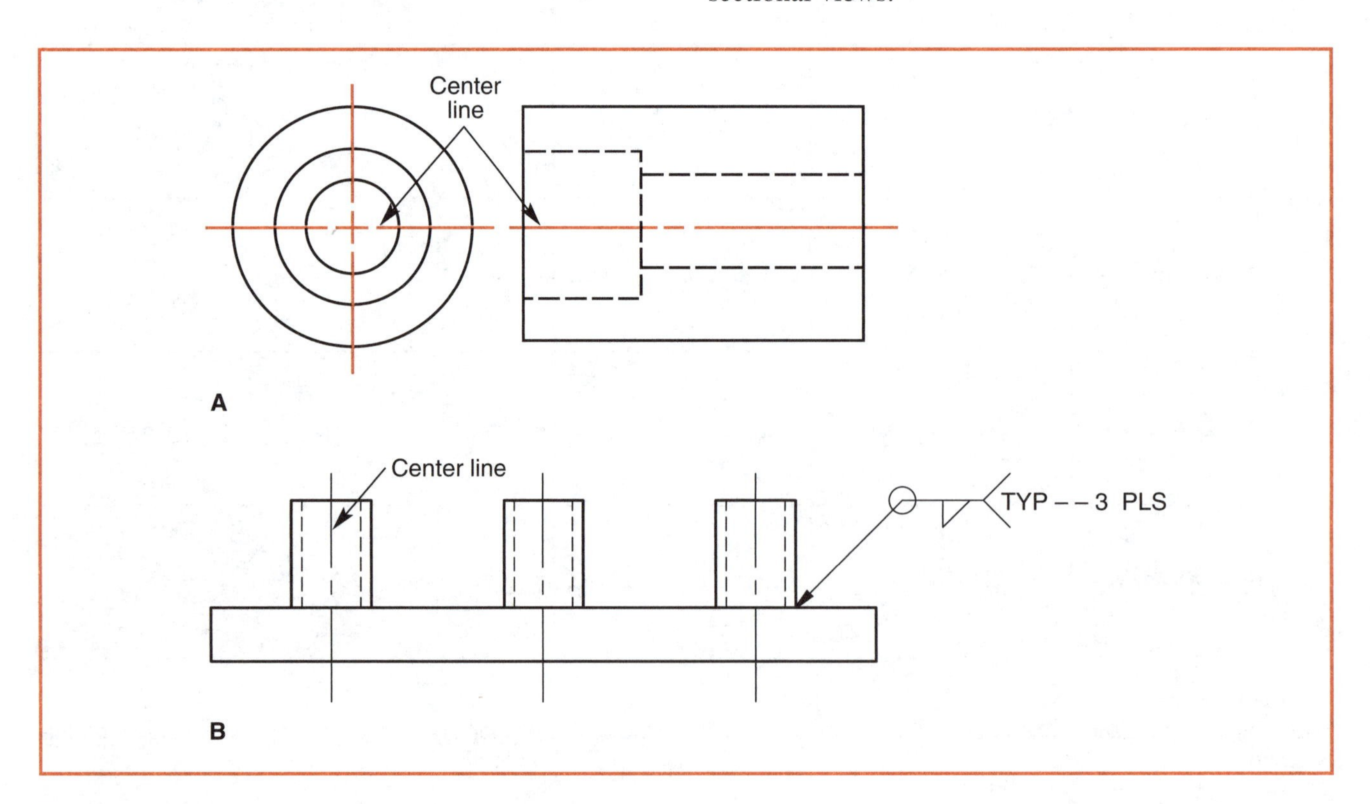

Figure 4-5.
A center line indicates the center of symmetrical objects.

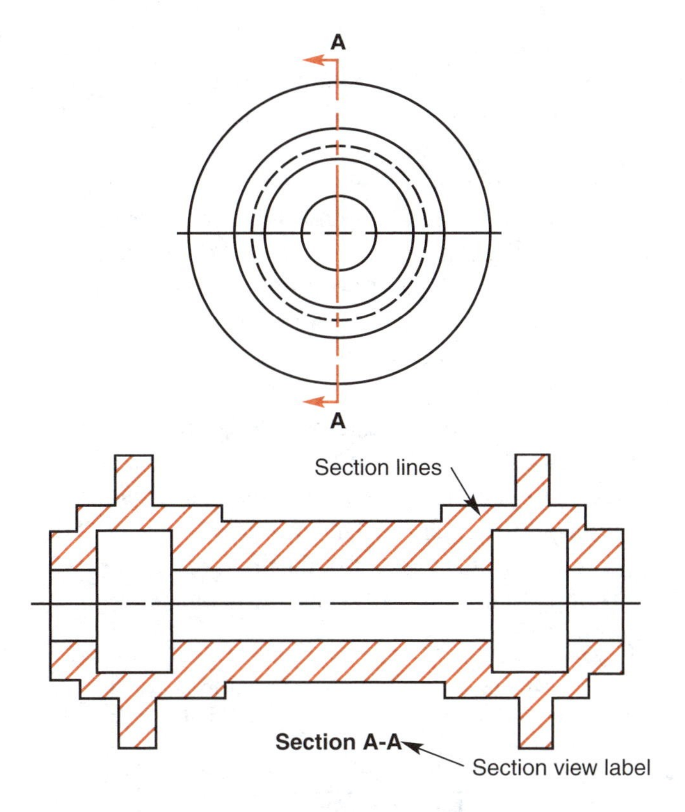

Figure 4-6.
Section lines are used when interior features of an object are shown. They may also indicate the material of an object where it has been cut by the cutting-plane line.

The ends of the lines are at 90° and capped with arrowheads that indicate from what direction the sectional view is taken. Three forms are recommended for general use. Refer to Figure 4-1 and Figure 4-2:

- Evenly spaced dashes
- Alternating long dashes and pairs of short dashes
- Dashes between the line ends are omitted

Letters A-A, B-B, etc., identify the section if it is moved to another position on the print or if several sections are used on a single print.

Break Line

A ***break line*** "breaks out" or removes sections for clarity. It provides clearer detail in viewing parts that lie directly below a removed part. Also, when the object is uniform in cross section for its entire length, break lines are used to shorten long parts and reduce the size of the drawing.

Figure 4-8 shows three types of break lines:

- A heavy, irregular freehand line indicates a short break
- A thin, light-ruled line with freehand zigzags shows long breaks
- A thick "S" break indicates the break in round stock

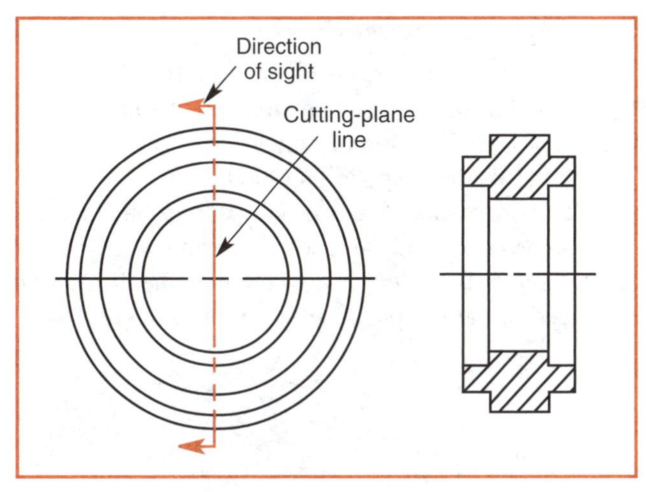

Figure 4-7.
A cutting-plane line shows the location of an imaginary cut made through the object to reveal its interior details.

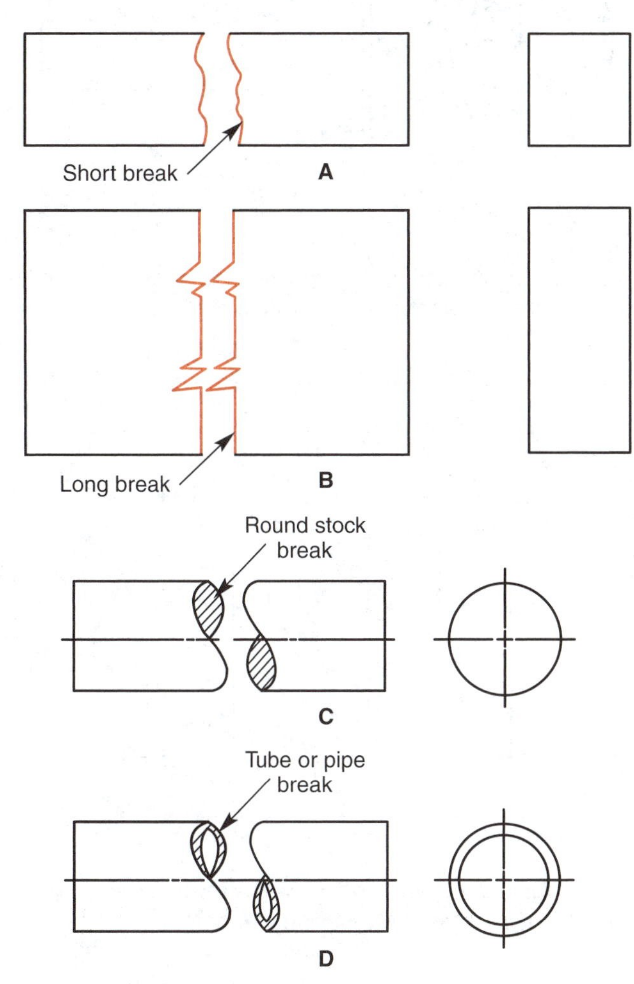

Figure 4-8.
Break lines are used to limit a partial view of a broken section. A—Short break. B—Long break. C—Round stock break. D—Tube or pipe break.

Dimension and Extension Lines

Dimension and extension lines are fine, solid lines used together on a print. The ***dimension line*** indicates the extent and direction of dimensions. The ***extension line*** indicates the termination of a dimension.

The dimension line is usually capped at each end with arrowheads and is placed between two extension lines, Figure 4-9. With few exceptions, the dimension is placed at the midpoint between the arrowheads.

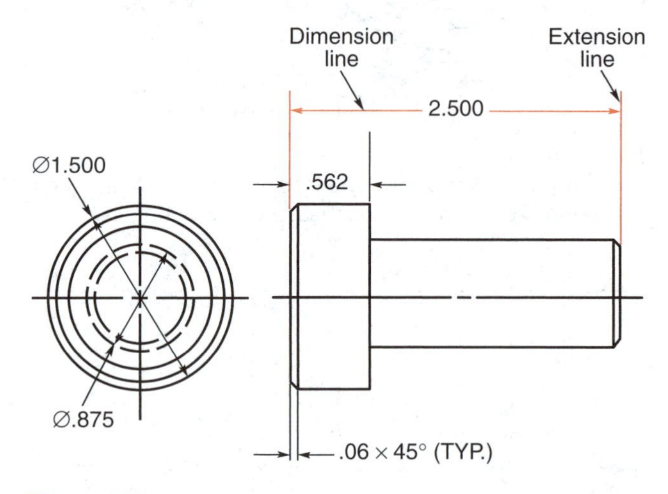

Figure 4-9.
A dimension line usually has arrowheads at each end. It is used to show direction and the extent of dimensions.

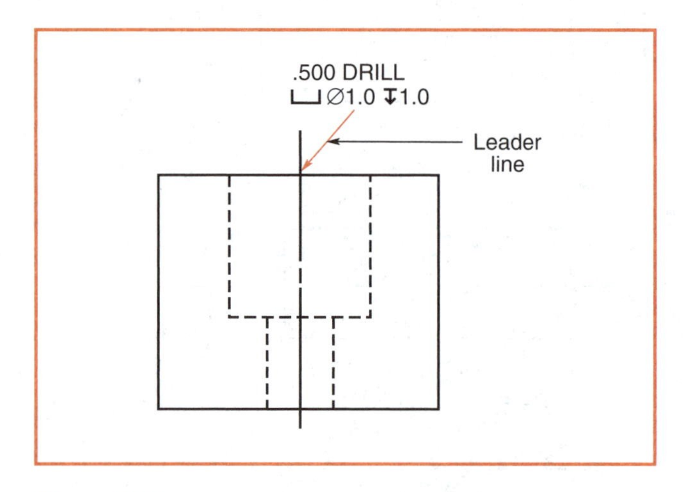

Figure 4-10.
A leader is a dimension line with an arrowhead on only one end. It indicates a dimension or a note.

A dimension line with an arrowhead on only one end is used as a ***leader,*** Figure 4-10. It is used to indicate a dimension or for adding a note.

Phantom Line

A ***phantom line,*** Figure 4-11, indicates several different situations:

- Shows adjacent positions of related parts
- Shows alternate positions of moving parts
- Shows repeated details, like threads and springs
- Shows filleted and rounded corners

It is a thin, dark line made of long dashes alternated with pairs of short dashes. On a same size drawing, the long dashes are ¾ – 1½″ (19.0 – 37.0 mm) long. The short dashes are typically ¼″ (6.0 mm) long with 1/16″ (1.6 mm) spaces.

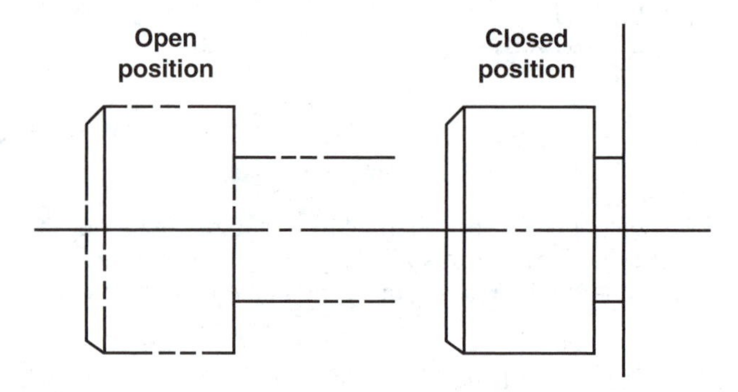

Figure 4-11.
A phantom line shows adjacent parts and alternate positions of a moving part.

How the Alphabet of Lines Is Used

The sample print, shown in Figure 4-12, illustrates how the alphabet of lines is used on a drawing. Study it closely!

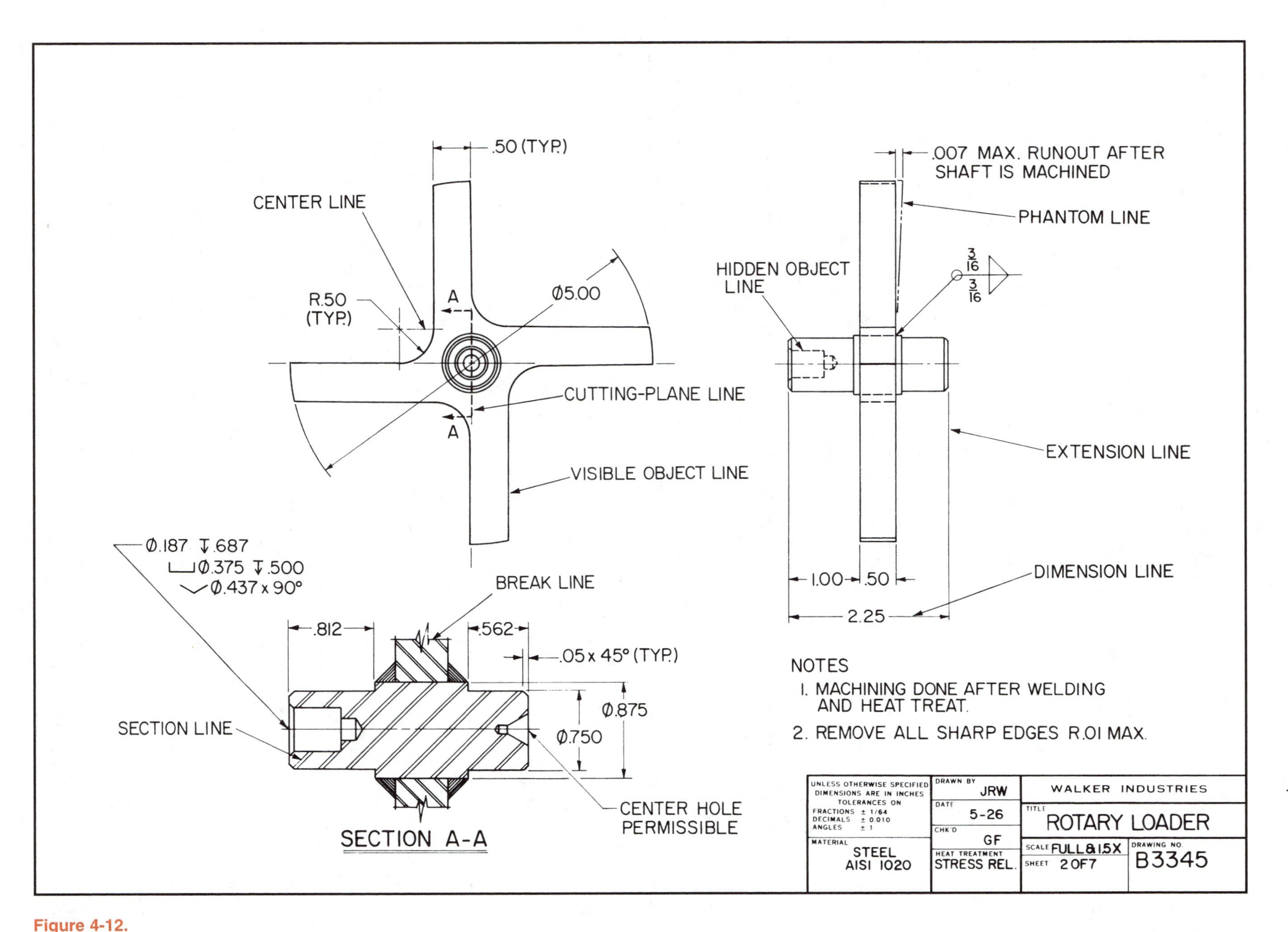

Figure 4-12.
Note how the alphabet of lines is used on a drawing.

Notes

Name ____________________ Date ____________ Class ____________

Print Reading Activities

Part I

Use Figure 4-13 and identify the type of line shown. Write the name of the line in the appropriate space.

A. ____________________

B. ____________________

C. ____________________

D. ____________________

E. ____________________

F. ____________________

G. ____________________

H. ____________________

I. ____________________

Part II

Use Figure 4-14 and identify the type of line shown. Write the name of the line in the appropriate space.

A. ____________________

B. ____________________

C. ____________________

D. ____________________

E. ____________________

F. ____________________

G. ____________________

H. ____________________

I. ____________________

J. ____________________

K. ____________________

L. ____________________

M. ____________________

N. ____________________

O. ____________________

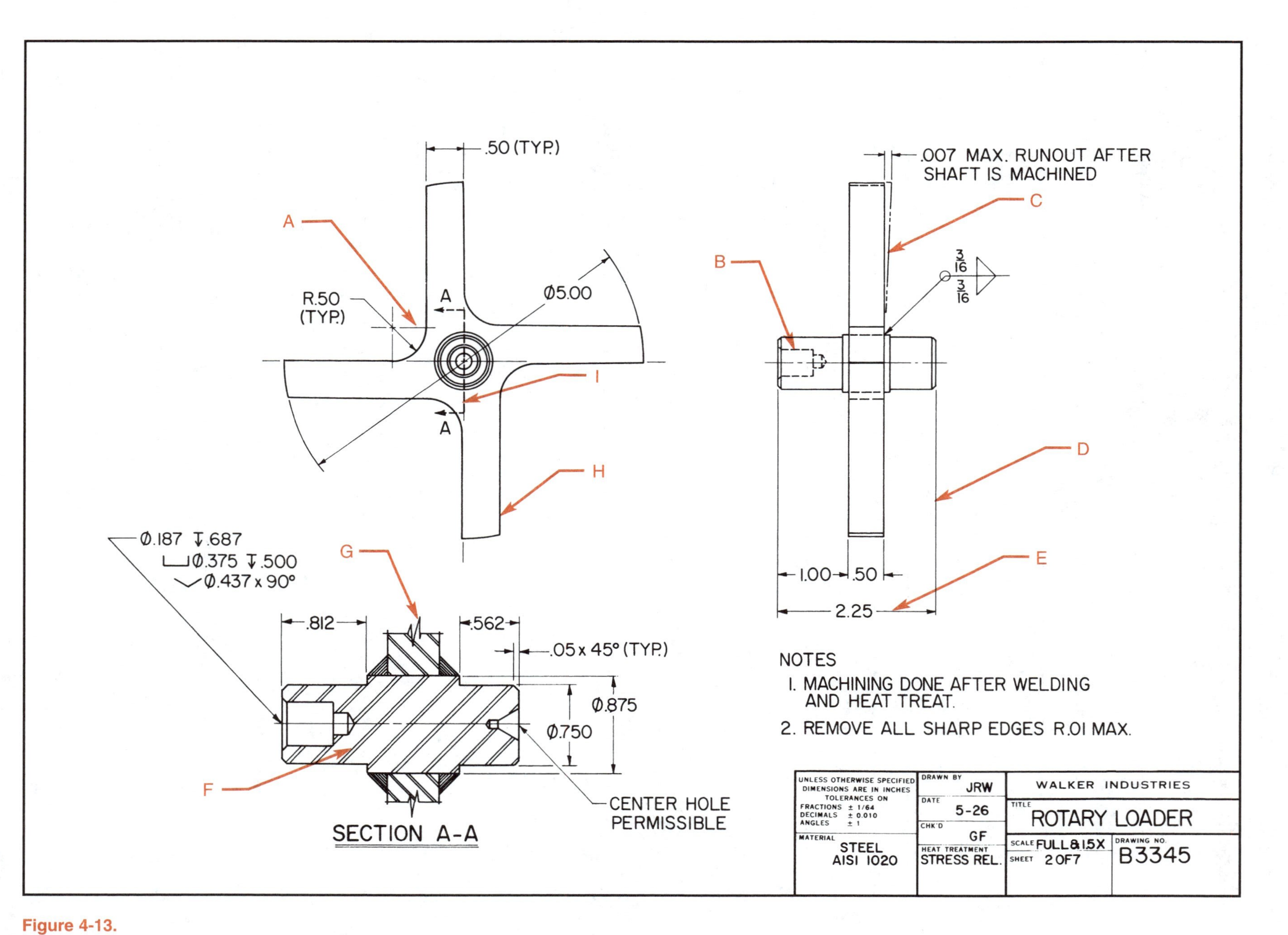

Figure 4-13.
Use this drawing to complete Part I of the Print Reading Activities.

Name ____

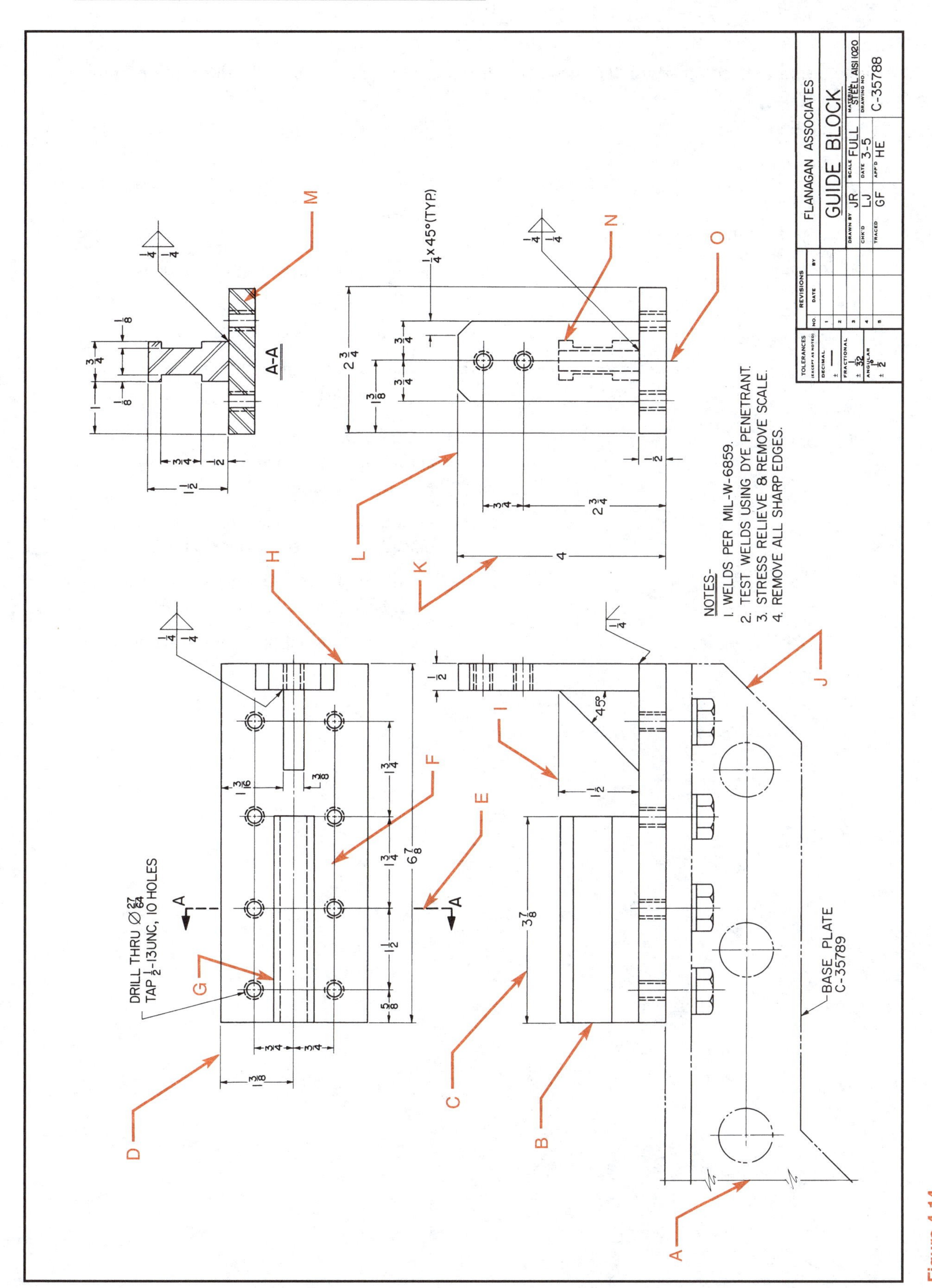

Figure 4-14.
Use this drawing to complete Part II of the Print Reading Activities.

Part III

Match the term in the column with the correct description. Place the letter of the description in the appropriate blank.

1. _____ Visible line
2. _____ Hidden line
3. _____ Center line
4. _____ Cutting-plane line
5. _____ Section lines
6. _____ Break lines
7. _____ Dimension line
8. _____ Extension line
9. _____ Leader
10. _____ Phantom line

(A) Shows adjacent parts and alternate positions of moving parts.

(B) Used when interior features of an object are shown.

(C) Indicates location of an imaginary cut made through object to reveal its interior characteristics.

(D) Used to outline visible edges of object.

(E) Indicates center of symmetrical object.

(F) Shows hidden feature of object.

(G) Used for purpose of breaking out sections for clarity or when object is uniform in cross section for its entire length.

(H) Usually capped at each end with arrowheads and placed between two extension lines.

(I) Used to indicate a dimension or for adding a note. Has an arrowhead on one end only.

(J) Indicates termination of a dimension.

Unit 5

Understanding Prints

After completing Unit 5, you will be able to:

- Identify and explain the significance of the principal views on a multi-view drawing.
- Recognize and explain uses of an auxiliary view and the principal view from which it is projected.
- Explain how the major types of section views are read and used on a print.
- Identify methods to read prints.

Key Words

auxiliary view
broken-out section
conventional breaks
crosshatching
cutting-plane line
first angle projection
full section
half section
multiview drawings
offset section
orthographic projection
partial auxiliary view
principal views
removed section
revolved section
section lining
sectional view
thin section
third angle projection
title block

A print shows a series of views that give the welder an exact shape and size description of an object, Figure 5-1. Additional information necessary to make or assemble the product is also included on the print.

The best way to read a print is to mentally break it into smaller parts. First, try to look at the shape of the part with the dimensions and notes removed. Second, try to determine the overall size of the part so you have some understanding of how big or small the part is. Third, look at the ***title block*** for information about the title part, material, scale, tolerance requirements, and other general pieces of information. Finally, read all of the notes on the drawing.

Multiviews

Most prints are in the form of multiviews. ***Multiview drawings*** are needed when more than one view is required to give an accurate shape description of the object. On most prints, the object is drawn in the operating position.

The views are arranged on the print in a systematic manner accepted as standard throughout industry. For example, in Figure 5-1 the top view appears above the front view. The right side view appears to the right of the front view.

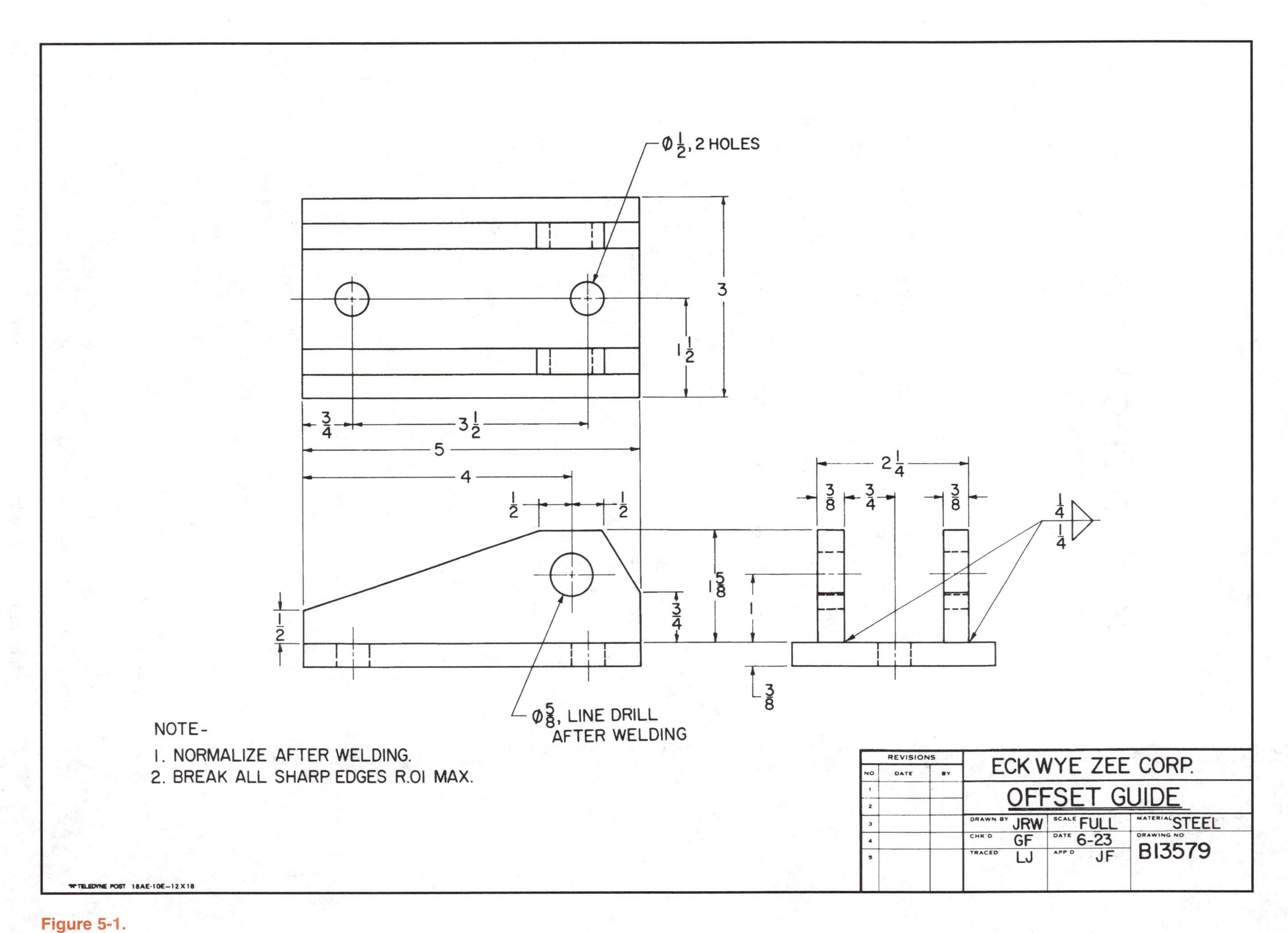

Figure 5-1.
A print shows a series of views giving an exact shape and size description of a part.

This makes it easier for the welder to merge the views in his or her mind and to form a mental picture of the object, Figure 5-2.

In developing the needed views, the entire object is normally viewed from six directions: front, top, left side, right side, rear, and bottom views, Figure 5-3. The same object is viewed separately from these various directions of sight, Figure 5-4.

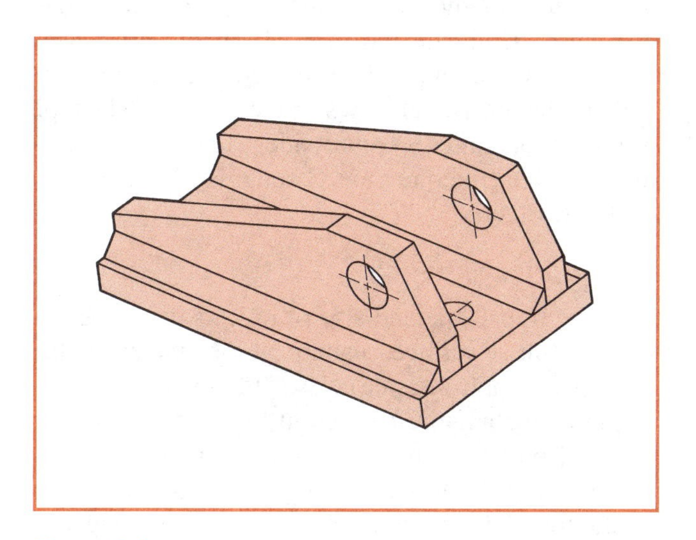

Figure 5-2.
This is the same part as in Figure 5-1 but the views in Figure 5-1 are arranged in a systematic manner. This makes it easier to form a mental image of the weldment. Shown is a pictorial view.

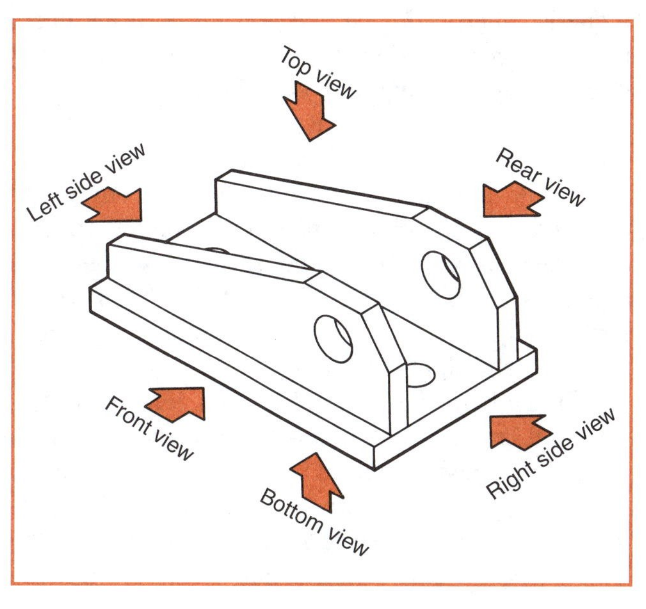

Figure 5-3.
Six directions are normally considered when developing views for an object on a print.

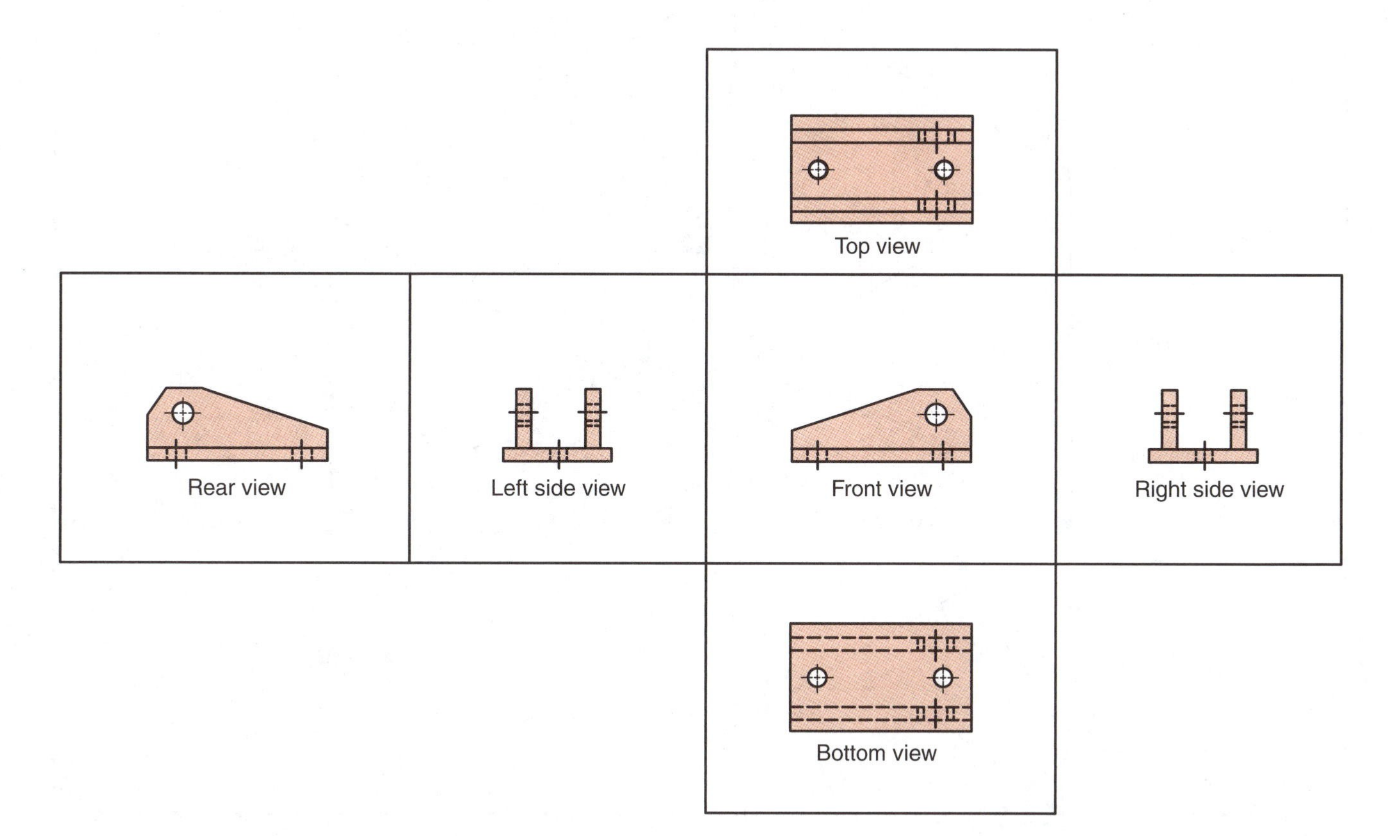

Figure 5-4.
If an object is viewed from the six directions of sight, the views would look like the ones shown.

To obtain the views, think of the object as being enclosed in a hinged glass box. Study Figure 5-5 carefully. Imagine that the views are projected on the sides of the box. The top view of the object is seen on the top of the box, the front view on the front of the box, and so on for the remaining views. This technique is called ***orthographic projection.*** It permits a three-dimensional object to be described on a flat sheet of paper having only two dimensions.

As can be seen, at least six views will be developed. Not all of them, however, are needed. Only those views required to give an accurate shape description of the object are included on the print. A view that repeats the same shape description as another view is *not* used, Figure 5-6. The ***principal views*** commonly shown on a print are the front view, top view, and right side view.

In the United States and Canada, all engineering drawings are drawn in ***third angle projection,*** with the object drawn as viewed in the glass box and the views projected to the six sides of the box. The projected views are drawn to resemble views when the box is opened out. The top view is always directly *above* the front view. The right side view is to the *right* of the front view and in line with it. The left side view is to the *left* of the front view.

Drawings used in European countries are sometimes drawn in ***first angle projection,*** with the object drawn as if it were placed on each side of the glass box. Figure 5-7 shows both third angle and first angle projection. Compare the two. The main differences include how the object is projected and the arrangement of the individual views on the drawing. The type of projection angle is usually identified on the drawing by ISO (International Standards Organization) symbols, Figure 5-8.

Auxiliary Views

The true shape and size of objects having angular or slanted surfaces cannot be shown using the principal (top, front, side) views, Figure 5-9. The true length of the angular surface in Figure 5-9 is shown on the front view, but this view does not show its width. The true width of the angular surface is shown on the top and side views but neither view shows its true length.

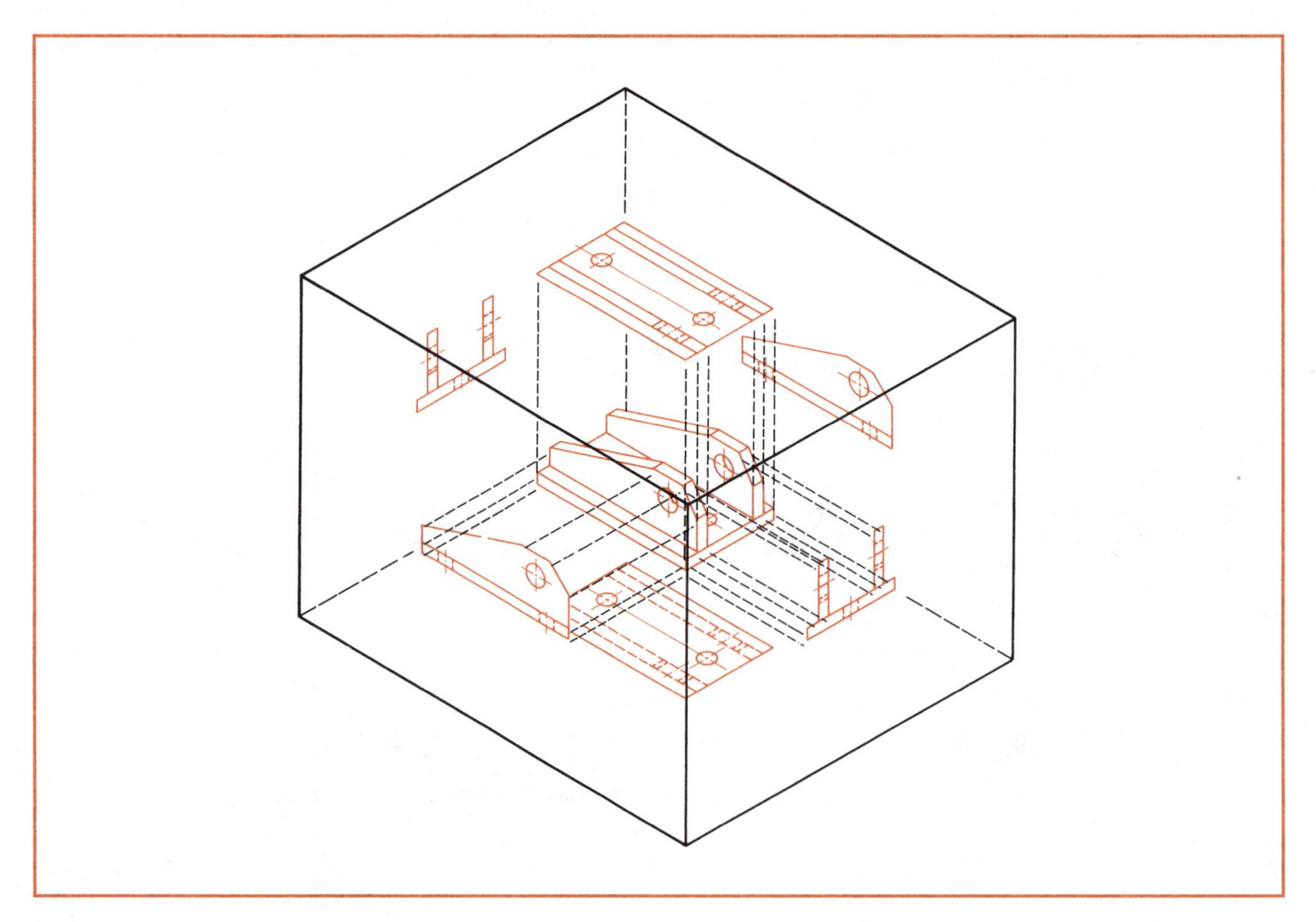

Figure 5-5.
Note relationships among views by thinking of the object as enclosed in a hinged glass box.

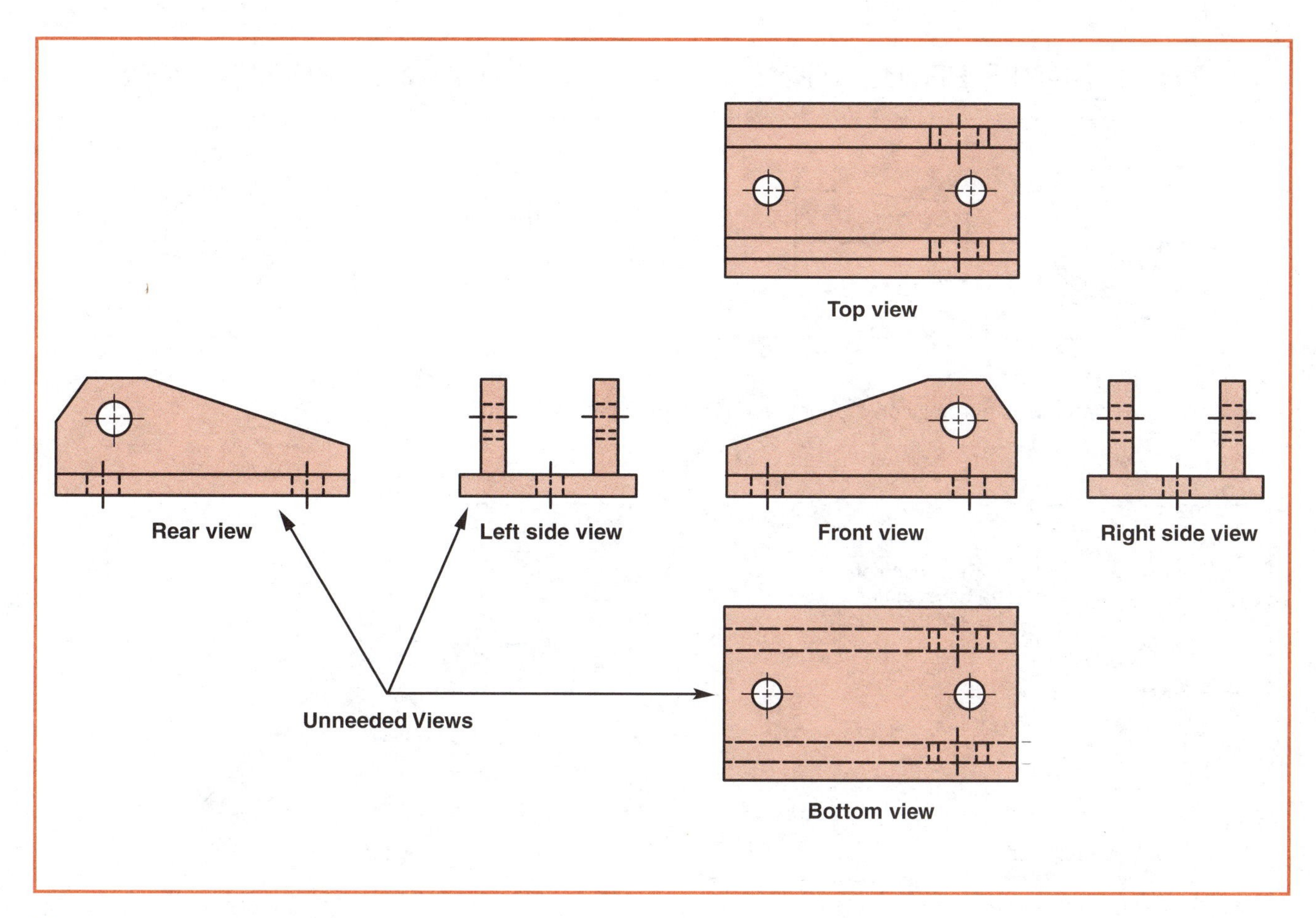

Figure 5-6.
Some views are seldom used on a print since they repeat the same shape description as another view.

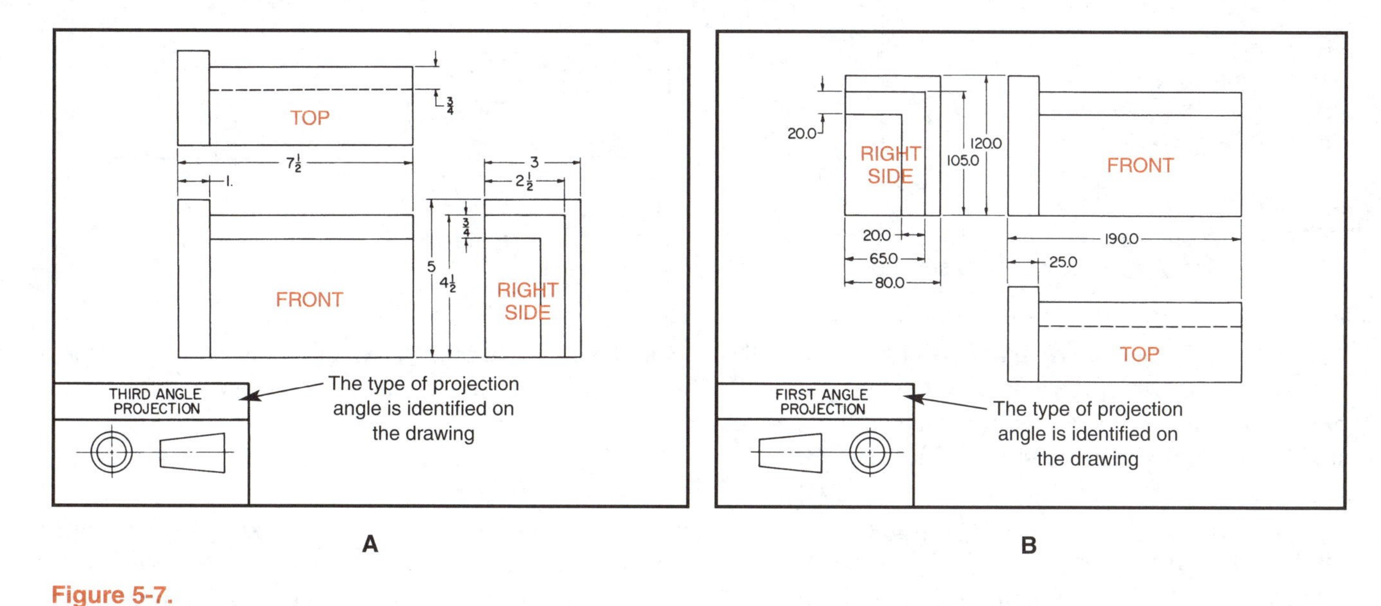

Figure 5-7.
Study third and first angle projection. A—Drawings made in the United States use third angle projection, which has views arranged as shown. B—Drawings used in European countries use first angle projection.

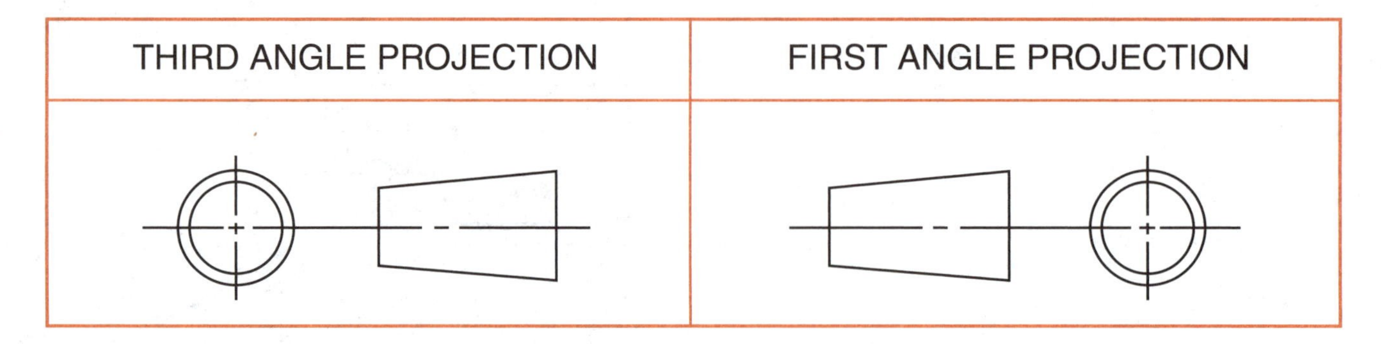

Figure 5-8.
A block with ISO symbols is normally on a print to identify the angle of projection.

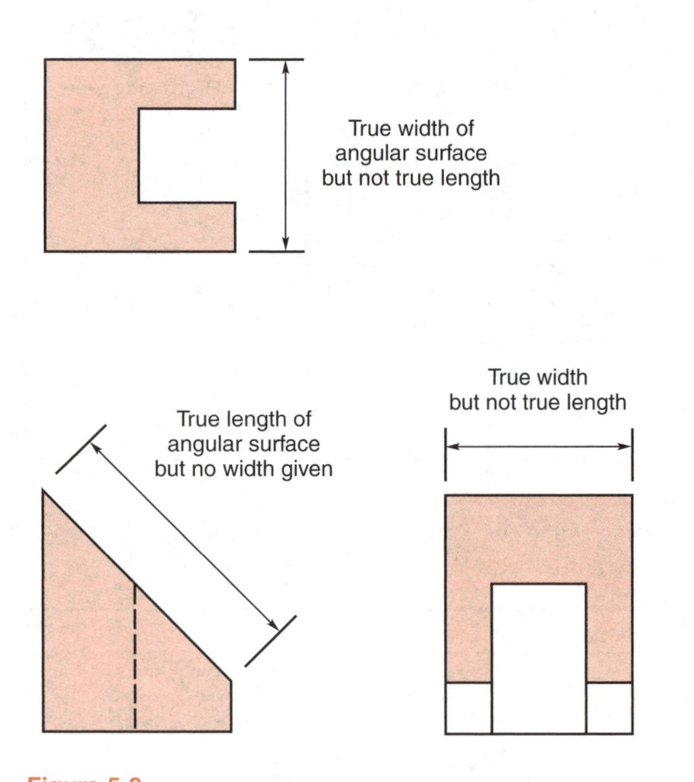

Figure 5-9.
The regular top, front, and side views do not show true shape and size of angular or slanted surfaces.

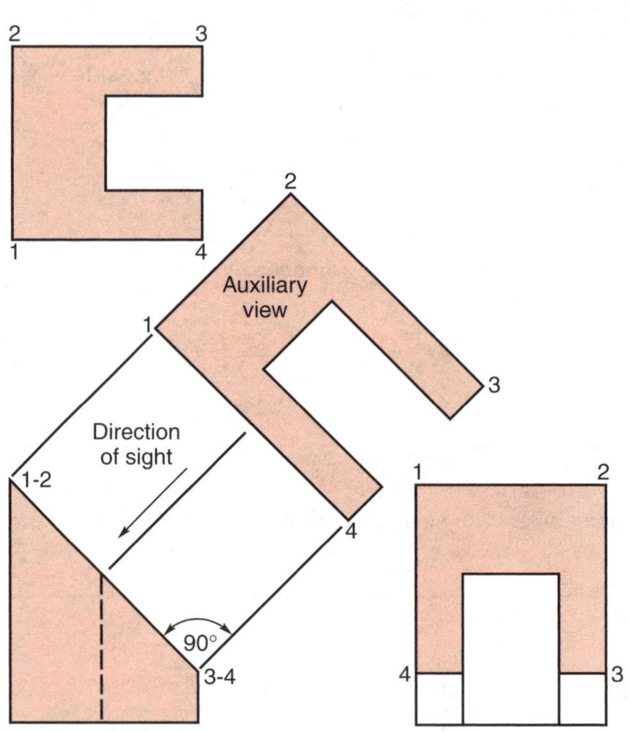

Figure 5-10.
Dimensions of an angular surface are given by an auxiliary view.

An additional or ***auxiliary view*** is used to show the true shape and size of the angular surface, Figure 5-10. Because of the extensive use of angled surfaces, sheet metal and press-break drawings often use auxiliary views.

The auxiliary view is always projected at right angles (90°) from the principal view on which the angular surface appears as a line. Quite often with auxiliary views, it is possible to eliminate one of the principal views, as in Figure 5-11. To avoid confusion and unnecessary sections of the part, some auxiliary views are drawn as ***partial auxiliary views.*** Only the angular surface is shown in the auxiliary view, and break lines indicate where part of the view was removed.

Sectional Views

When an object is relatively simple in design, its shape can be described on a drawing without difficulty, Figure 5-12. For a complex object with many features obscured from view, however, it is often not easy to show its internal structure without a "jumble" of hidden lines, Figure 5-13. The drawing is hard to understand and interpret.

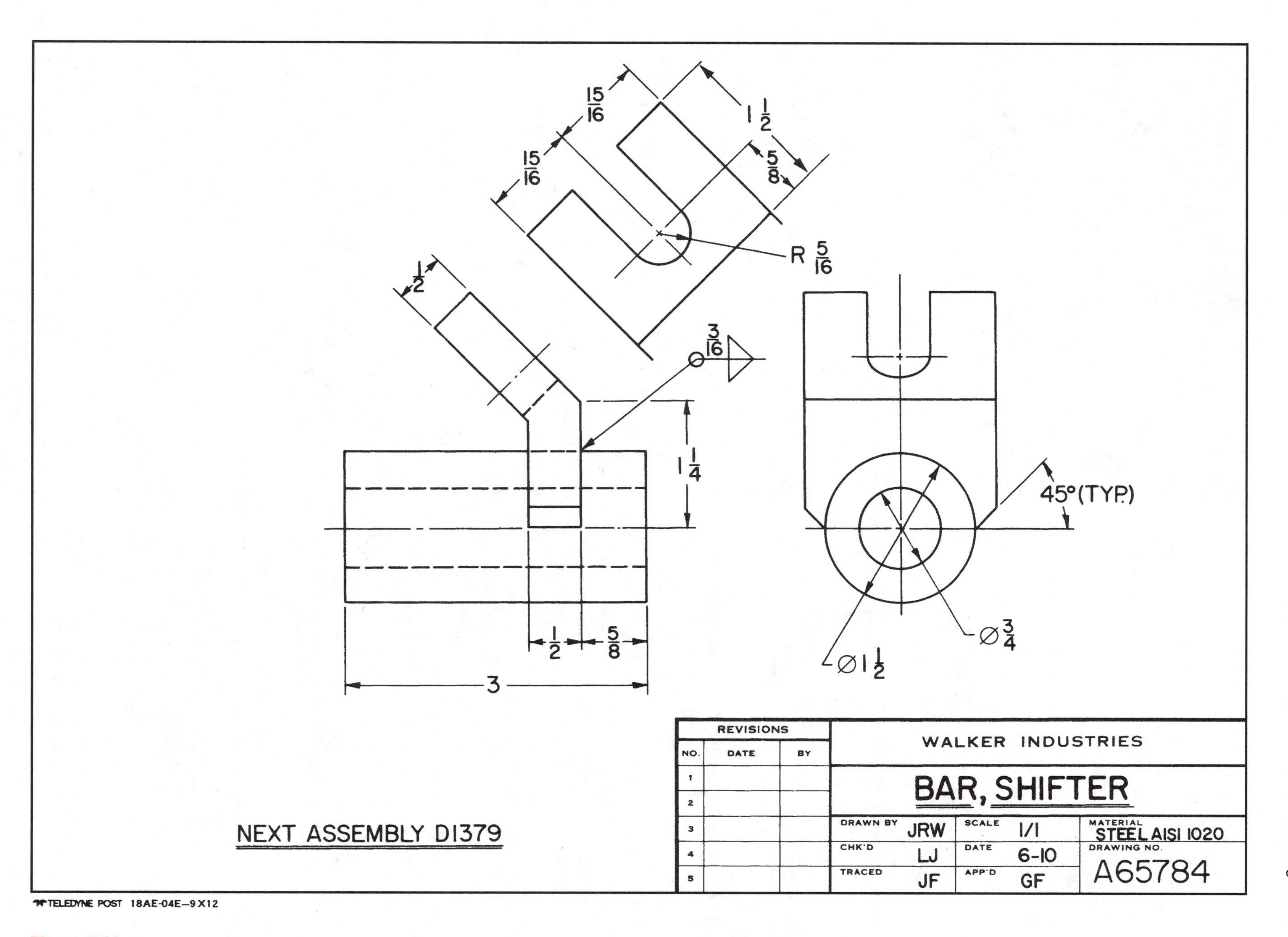

Figure 5-11.
Notice how a print may not have one of the regular views when an auxiliary view is used.

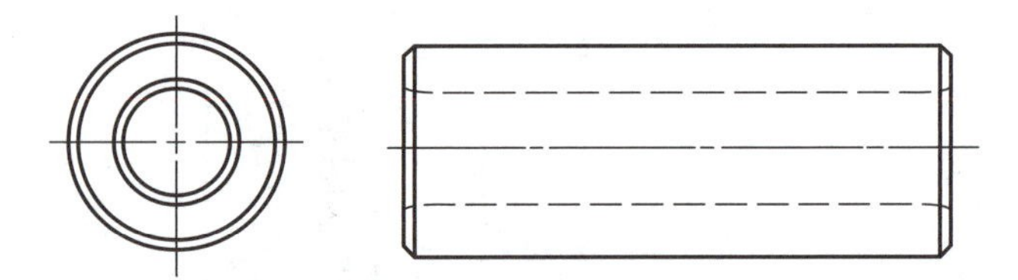

Figure 5-12.
Only two views are needed on simple objects like this piston pin, which can be fully illustrated on a drawing with little difficulty.

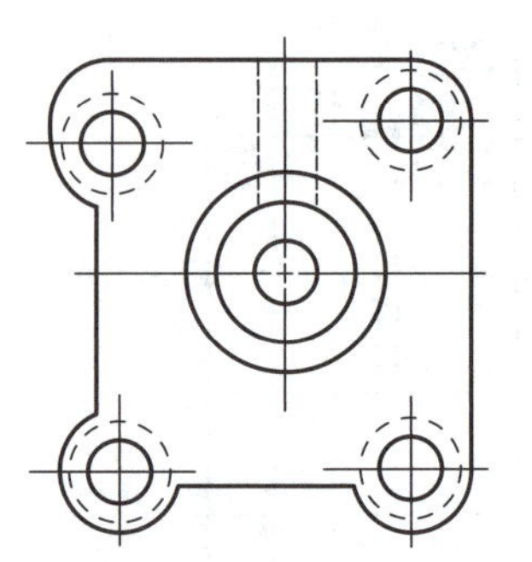

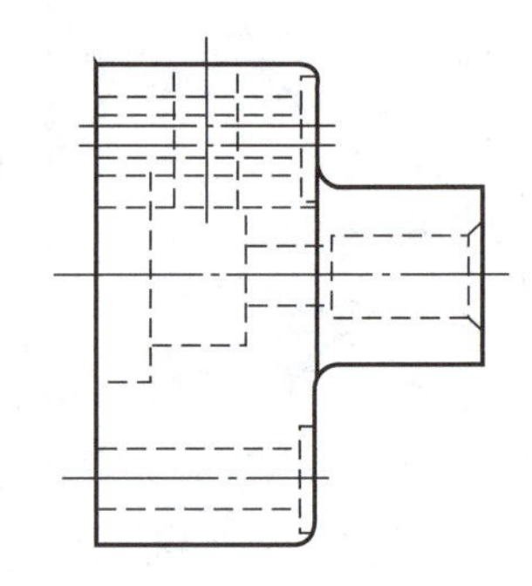

Figure 5-13.
A complex object with many interior features may result in a drawing with a maze of confusing hidden lines.

Sectional views (or sections) permit the true internal shape of a complex object to be shown without the confusion caused by a myriad of hidden lines. A sectional view shows how the object would appear if an imaginary cut (known as the cutting plane) were made through the object perpendicular to the direction of sight. Shown in Figure 5-14, the section or portion of the object between the eye and the cutting plane is removed or broken away to reveal the interior features of the object. This makes the shape of the object more understandable.

Figure 5-15A shows the exterior surface of a pin. Notice that the interior features shown in Figure 5-15B are easier to visualize if a full section view is provided.

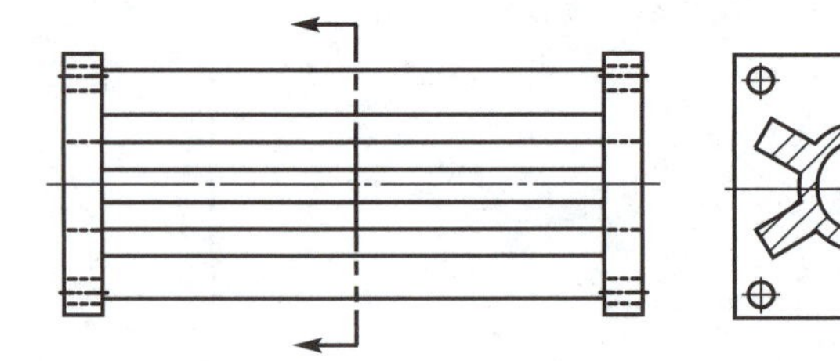

Figure 5-14.
A sectional view shows how an object would appear if an imaginary cut were made through the object perpendicular to the direction of sight. This allows interior features to be seen without the confusion of many hidden object lines.

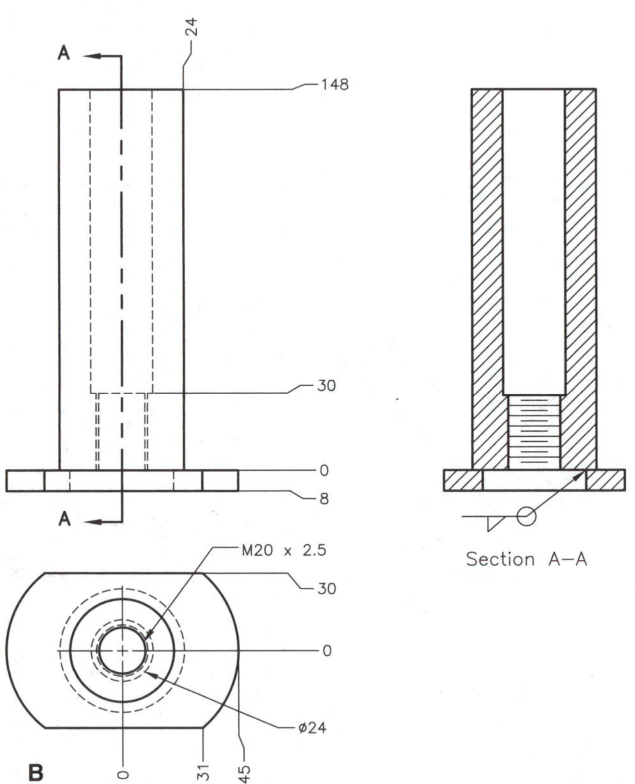

Figure 5-15.
A—The exterior surfaces of a pin are shown. B—The interior features of a pin are shown.

Cutting-Plane Line

The ***cutting-plane line*** indicates the point from which the imaginary cut of the section is taken from the part, Figure 5-16. The arrows at the end of the cutting-plane line show the direction of sight for viewing the section.

Three forms of cutting-plane lines are accepted for general use, Figure 5-17. Sections are usually identified with bold capital letters (A-A, B-B, etc.) if they are moved to another position on the drawing.

Section Lining

Section lining, sometimes called ***crosshatching,*** represents the type of exposed cut surface of a section. The American National Standards Institute (ANSI) has recommended the symbols for section lining shown in Figure 5-18. The symbols depict the different types of material specifications. General-purpose section lining (cast iron) is usually used on drawings when exact material specifications are located elsewhere on the print.

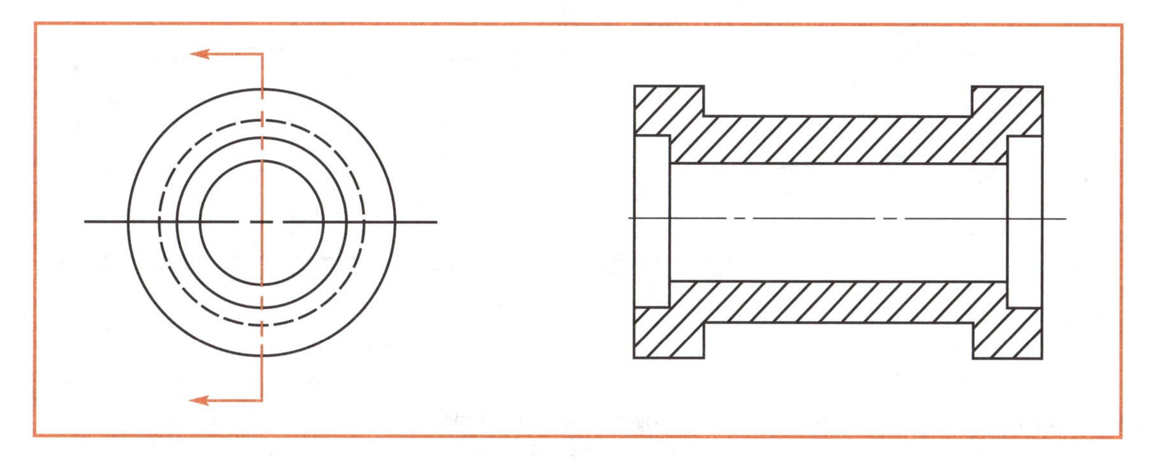

Figure 5-16.
A cutting-plane line shows the point from which a section is removed from the part.

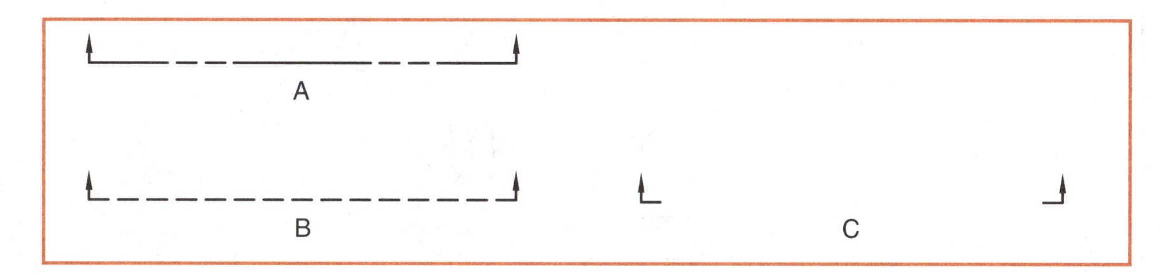

Figure 5-17.
Any of three forms of cutting-plane lines can be found on a drawing. A—Cutting-plane line with long dashes and pairs of short dashes. B—Cutting-plane line with equal length short dashes. C—Cutting-plane line with only the ends and arrowheads.

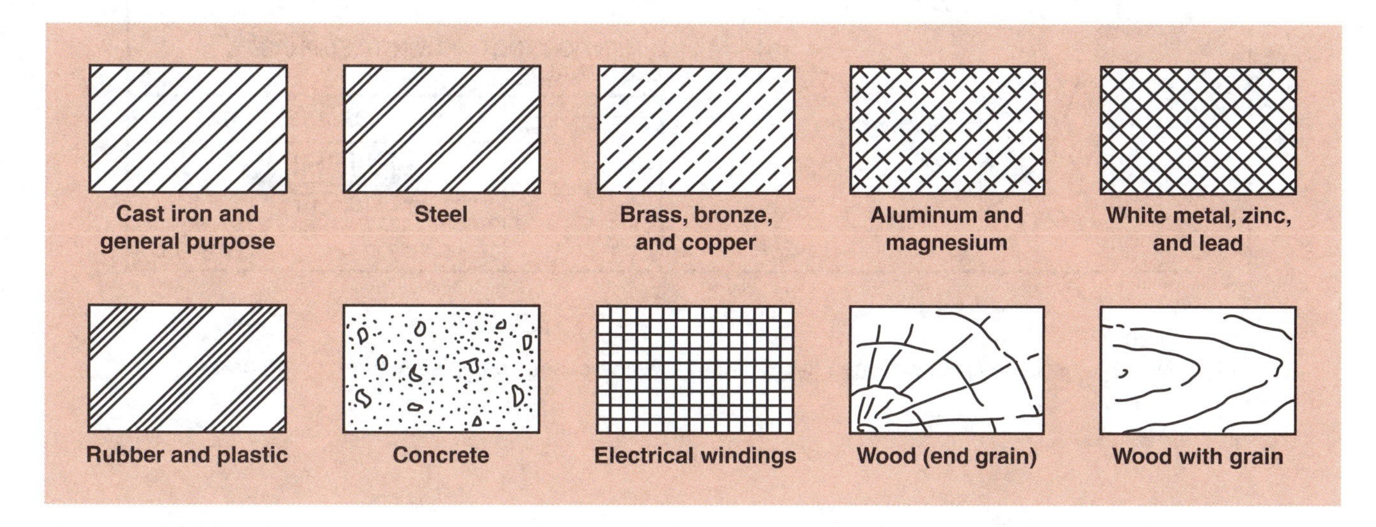

Figure 5-18.
Standard code symbols exist for various materials in a section.

Thin Section

A ***thin section*** or section not thick enough for conventional section lining (sheet metal, gaskets, etc.) is shown as solid black lines, Figure 5-19.

Full Section

A ***full section*** is shown when the cutting-plane line passes entirely through the object, Figure 5-20. The interior features of the object are revealed.

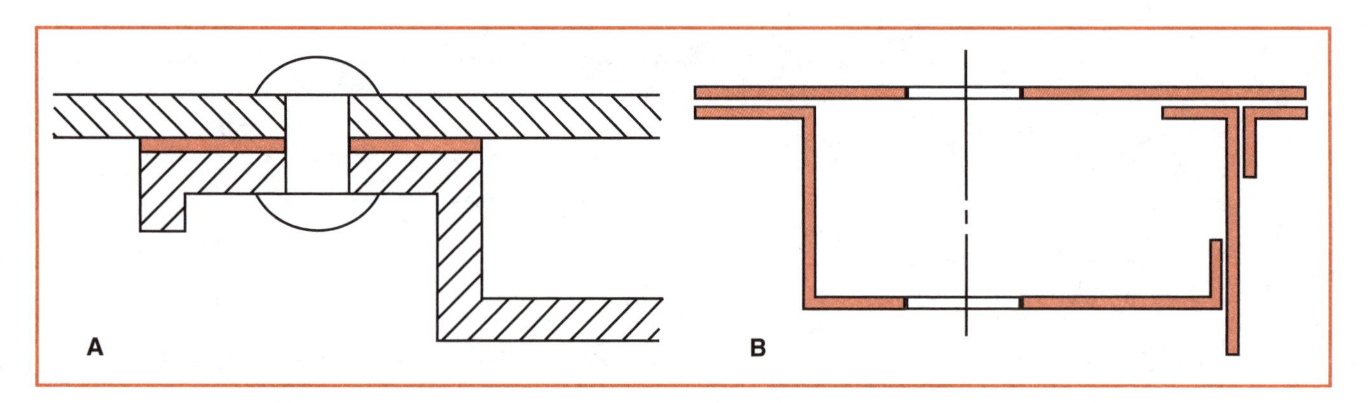

Figure 5-19.
This shows a thin section. A—Sections too thin for conventional crosshatching are shown as solid black lines. B—Adjacent thin sections are separated by very narrow white spaces.

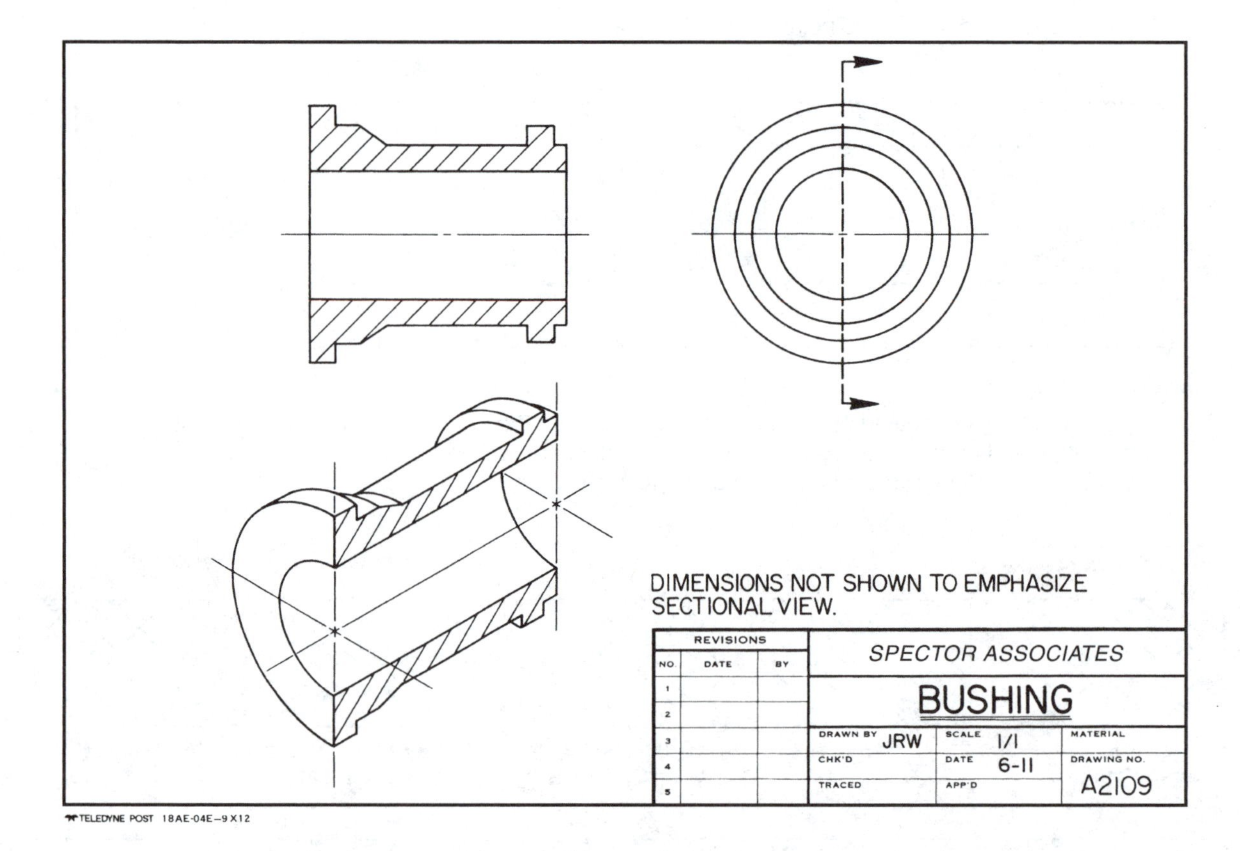

Figure 5-20.
Note how the full section is shown when the cutting-plane line passes completely through the part.

Half Section

Half sections are primarily limited to symmetrical objects. The shape of one-half of the interior features and one-half of the exterior features of the object are shown in the ***half section,*** Figure 5-21. Cutting-plane lines are passed at right angles to each other. One-quarter of the object is "removed" and a half section view is exposed.

Revolved Section

Revolved sections rotate or turn the cut section 90°. They are primarily used to show the shape of such things as spokes, ribs, and stock metal shapes, Figure 5-22.

Removed Section

A ***removed section*** is used when it is not possible to show the sectional views on one of the principal views, Figure 5-23. The section is moved to a new location on the print, allowing it to be enlarged for clarity.

Offset Section

The ***offset section*** is used when a single, straight cutting plane cannot show the needed information, Figure 5-24. In this case, the cutting-plane line is stepped or offset to pass through these features that lie in more than one plane.

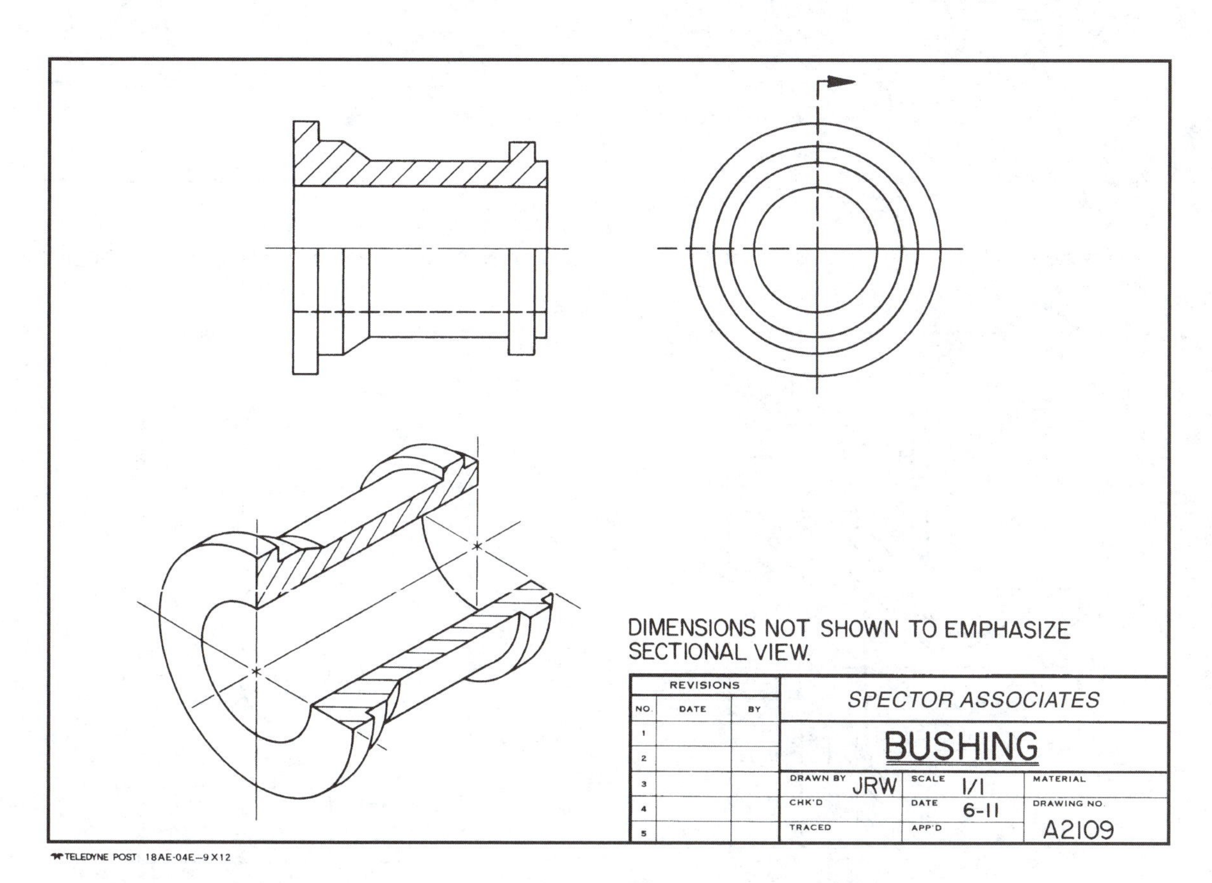

Figure 5-21.
A half section shows the shape of one-half of the interior and exterior of the object.

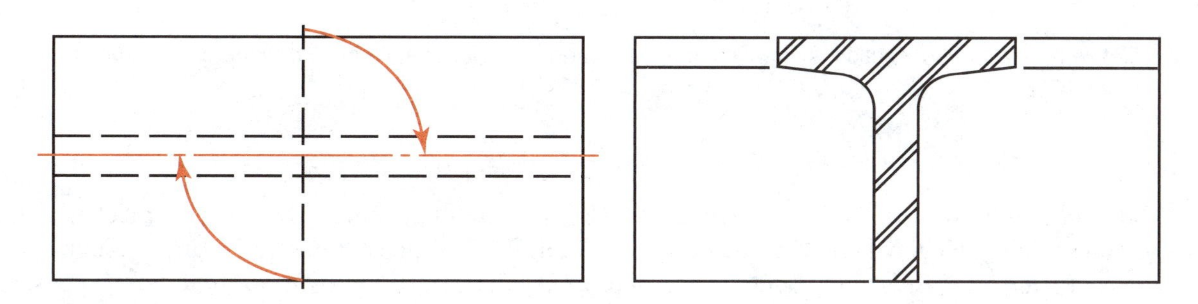

Figure 5-22.
A revolved section helps show the shape of such things as spokes, ribs, and stock metal shapes.

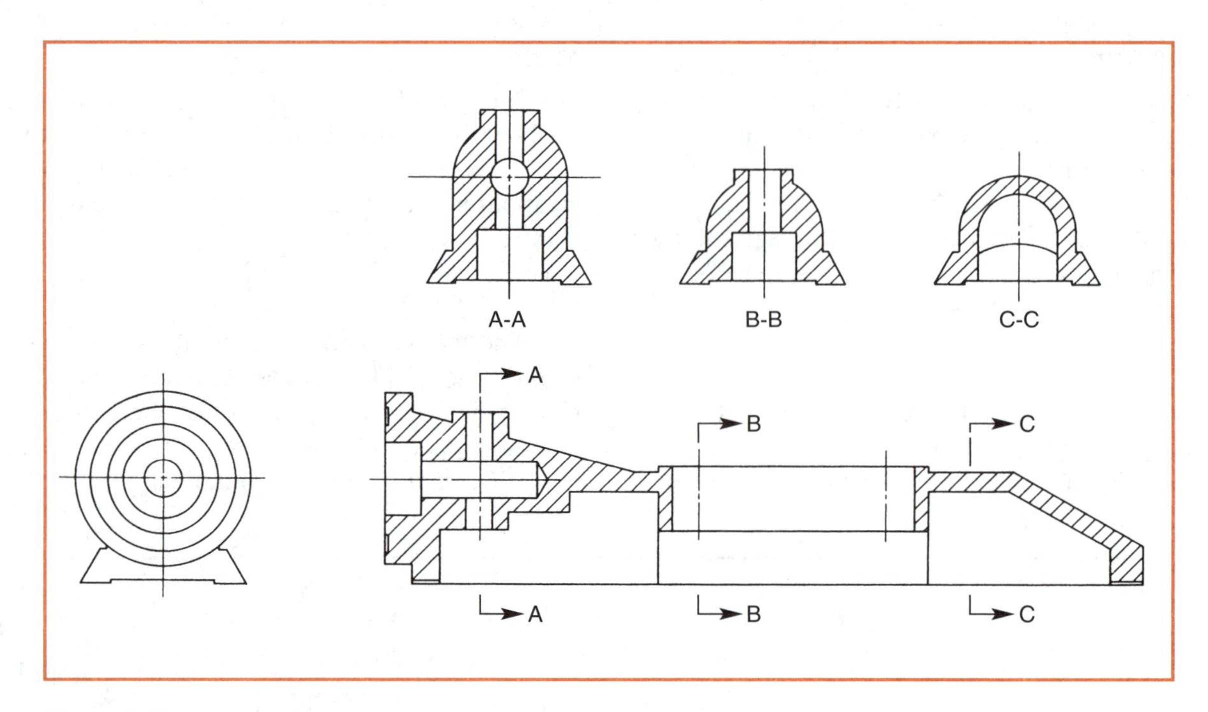

Figure 5-23.
A removed section shows a sectional view taken from the object and shows it on another part of the print.

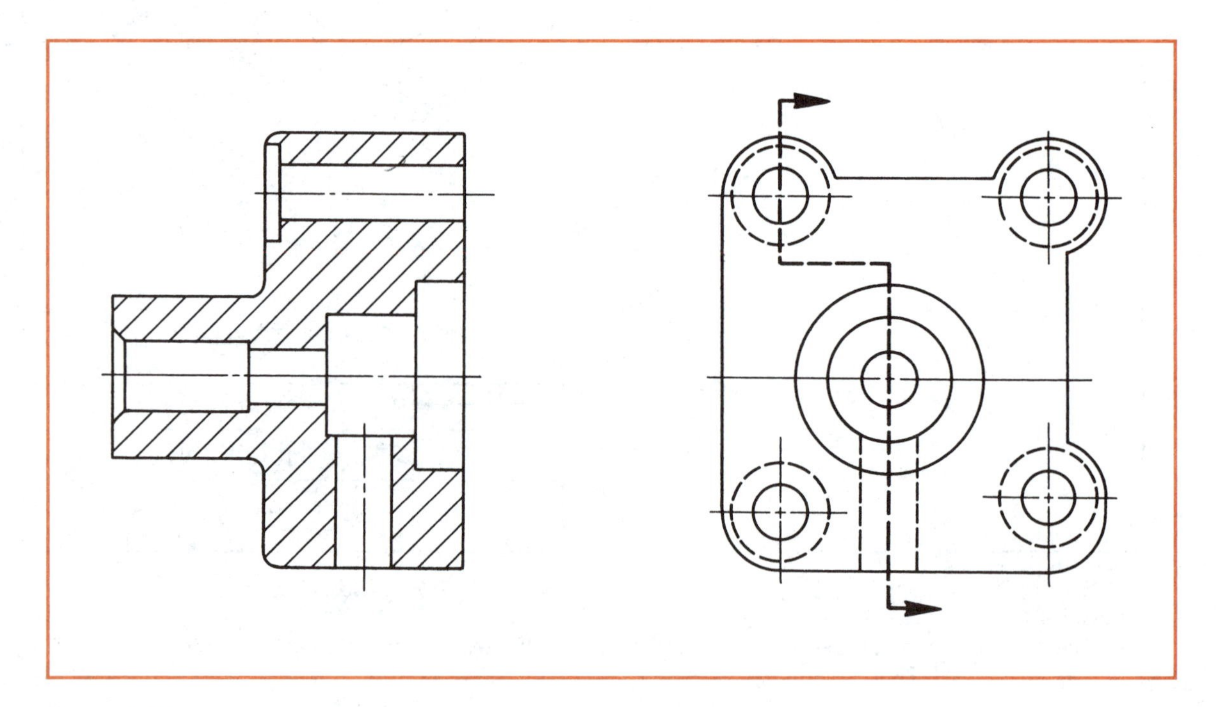

Figure 5-24.
Notice how the offset section is used. It is needed when a single, straight cutting plane cannot show information.

Broken-Out Section

The ***broken-out section*** is used when a small portion of a sectional view will provide the necessary information, Figure 5-25. Break lines define the section.

Conventional Break

A long, uniformly shaped object is sometimes difficult to present in a scale large enough to show its details clearly. A ***conventional break*** permits elongated objects to be shortened so a large enough scale can be used to present details with clarity, Figure 5-26.

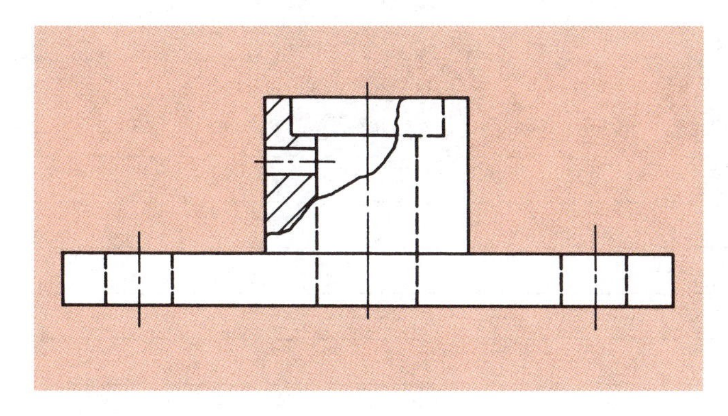

Figure 5-25.
With a broken-out section, only a small portion of a sectional view will adequately show essential information.

Sections through Webs and Ribs

Webs and ribs are added to some objects to increase the part's strength and rigidity. Figure 5-27 shows how ribs and webs are represented in sectional views.

Rules on Reading Prints

There is no one best way to read a print or drawing. Most welders come up with their own method. The following rules are suggested to help you get started. Eventually, you will develop a method best suited to your own needs.

- Carefully review the print.
- Study one view at a time. Identify surface limits and lines that describe the intersection of surfaces. This will help you to visualize the shape of the object.
- Establish sizes from the dimensions.
- Review the other information (notes, title block, revisions, etc.) on the print.
- Determine what your responsibilities will be in producing the object and the sequence you will follow in performing the operations.
- Do not be afraid to ask for help if you do not understand something on a print. A mistake caused by not understanding something can be very costly to your employer and may cause injury to someone using the product at a later date.
- Practice print reading until it becomes second nature to you.

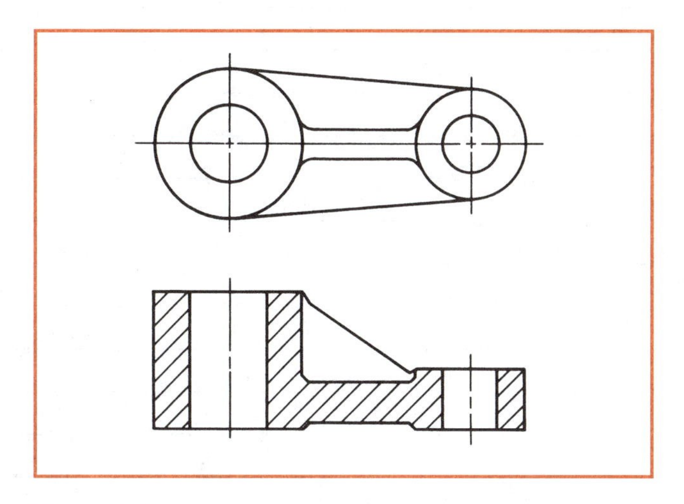

Figure 5-27.
Note this section through a web or rib. Section lines are not drawn through webbed or ribbed areas.

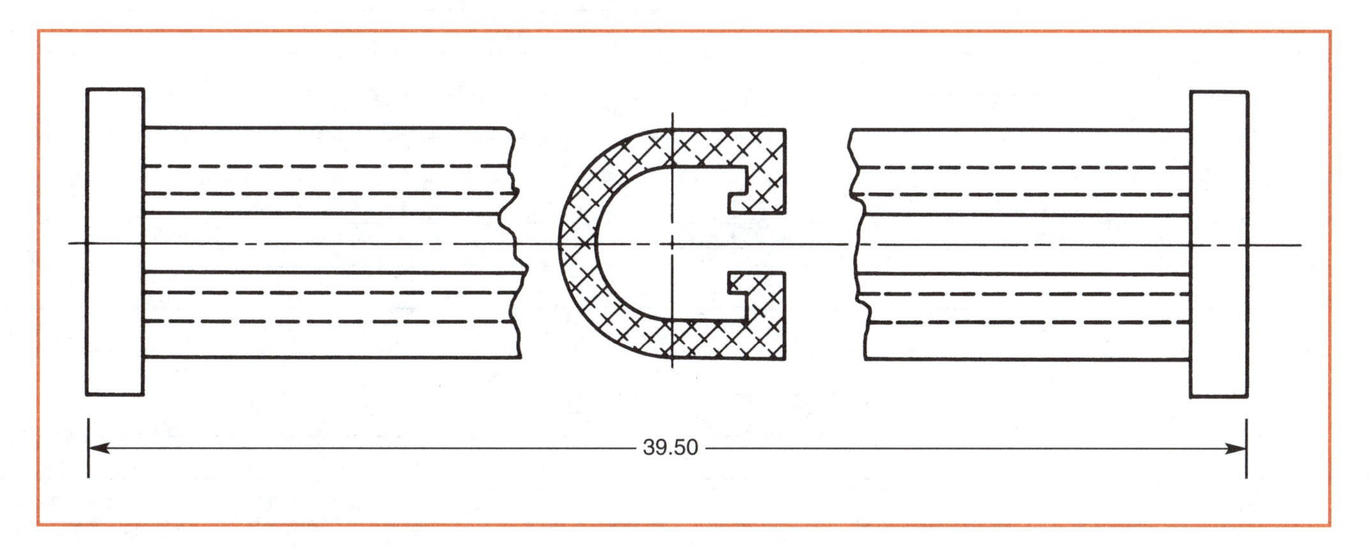

Figure 5-26.
A conventional break allows elongated objects to be shortened so a larger scale can be used to show details more clearly.

Notes

Name ______________________ Date ____________ Class ____________

Print Reading Activities

Study Figure 5-28 and answer the questions in Parts I, II, and III.

Part I

Use the pictorial view of the bracket and write in the dimensions indicated by each of the following letters.

1. A. ______________________
2. B. ______________________
3. C. ______________________
4. D. ______________________
5. E. ______________________
6. F. ______________________
7. G. ______________________
8. H. ______________________
9. I. ______________________
10. How many holes are specified? ______

Part II

Identify the type of lines indicated by the following letters.

11. J. ______________________
12. K. ______________________
13. L. ______________________
14. M. ______________________
15. N. ______________________

Part III

Answer the following questions.

16. Surface X indicates which view of the bracket?

____

17. Surface Y indicates which view of the bracket?

____

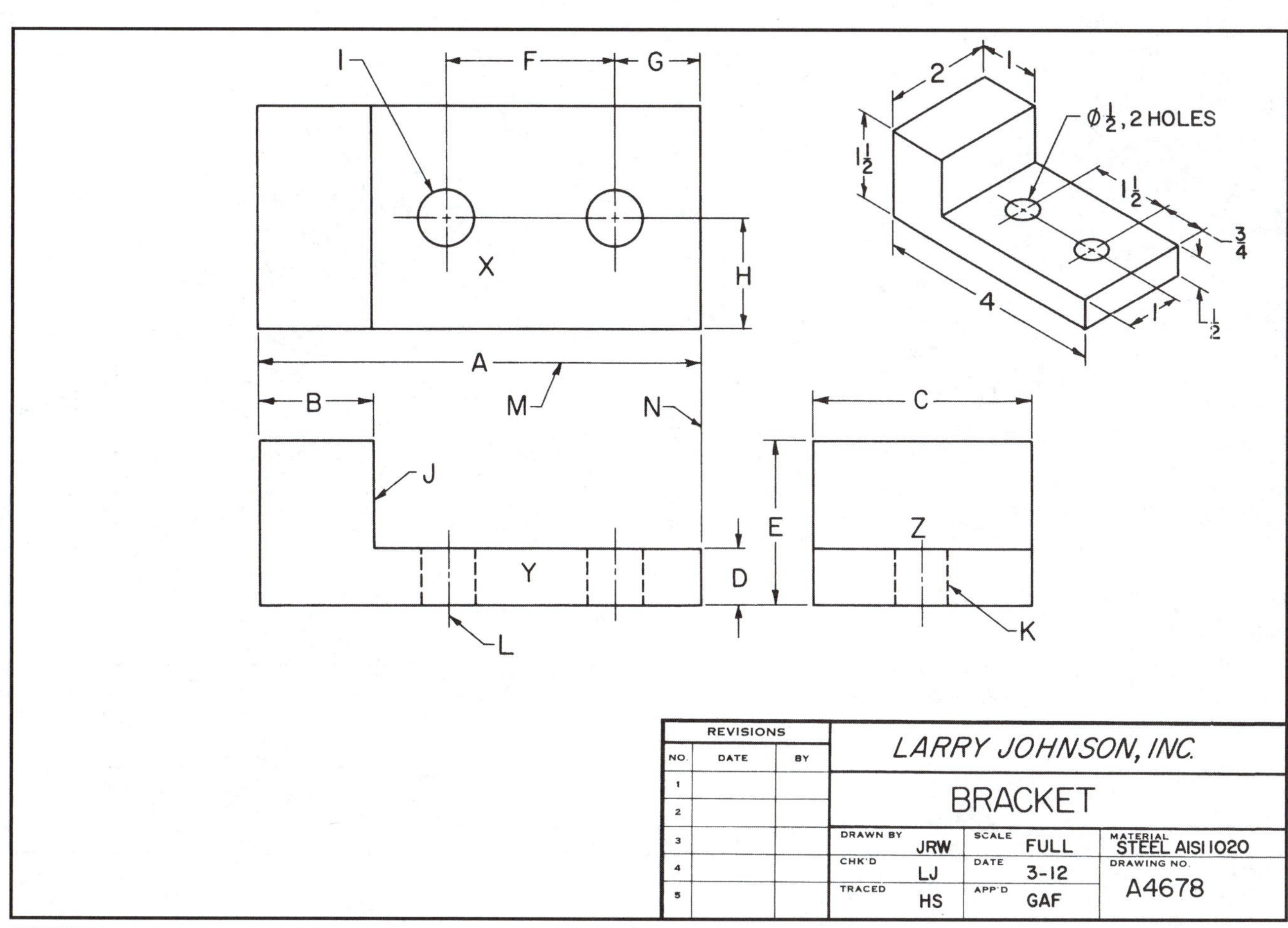

Figure 5-28.
Use this print of a bracket to answer questions in Parts I, II, and III of the Print Reading Activities.

18. Surface Z indicates which view of the bracket?

____

19. What is the print number?

____

20. What material is specified?

____

Part IV

Study the pictorial drawing of the shifter shown in Figure 5-29 and write in the dimensions indicated by the following letters on the orthographic views.

1. A. _______________
2. B. _______________
3. C. _______________
4. D. _______________
5. E. _______________
6. F. _______________
7. G. _______________
8. H. _______________
9. I. _______________
10. J. _______________
11. Surface L indicates which view?

____

12. Surface K indicates which view?

____

13. What is the scale of the original drawing?

____

14. What is the drawing number?

____

15. What material is specified?

____

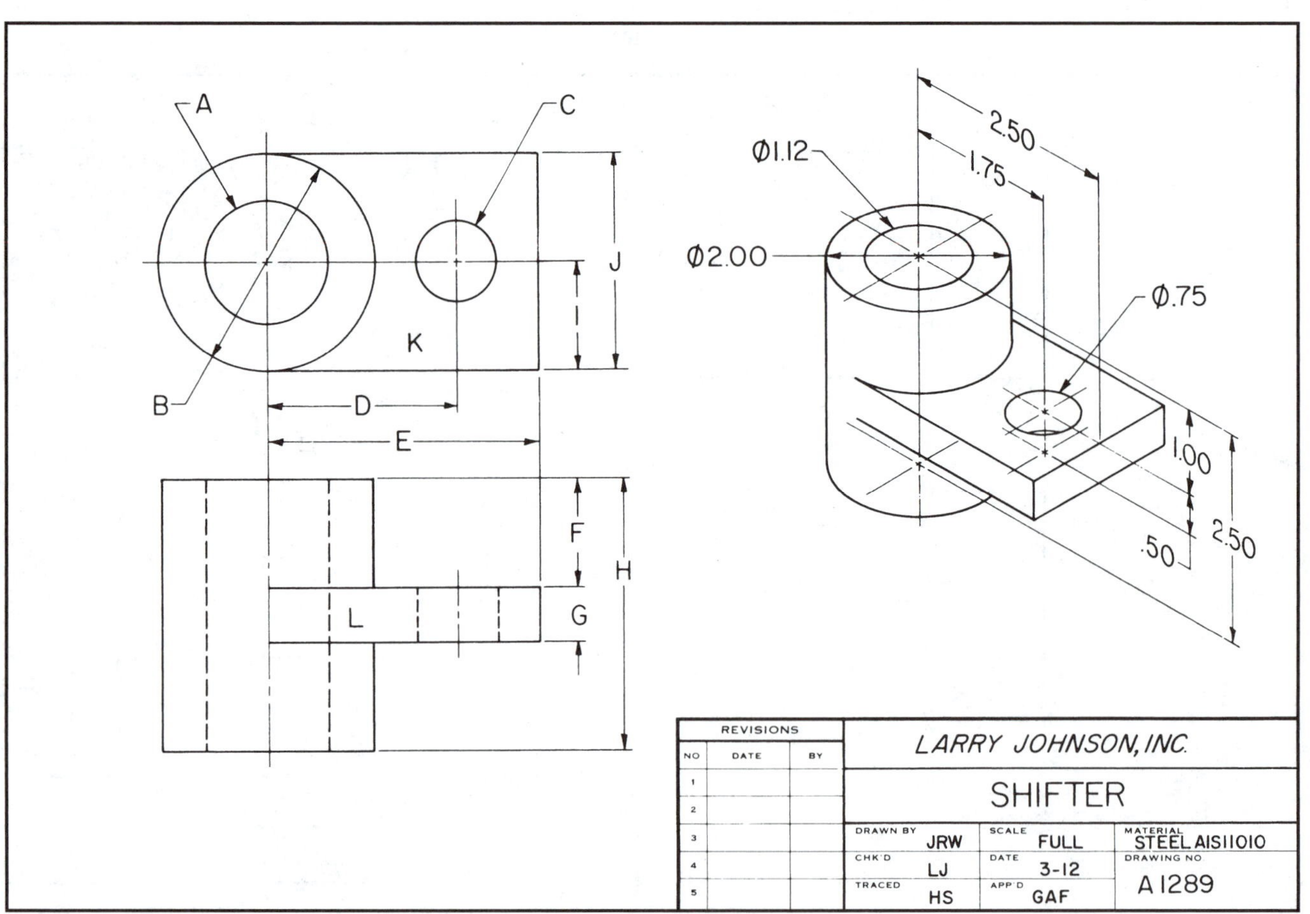

Figure 5-29.
Use this print of a shifter to answer questions in Part IV of the Print Reading Activities.

Name ____

Part V

Study the pictorial views and match each orthographic drawing with its pictorial drawing. Place the correct letter in the space provided.

A.

B.

C.

D.

E.

F.

1. ____

2. ____

3. ____

4. ____

5. ____

6. ____

Part V (continued)

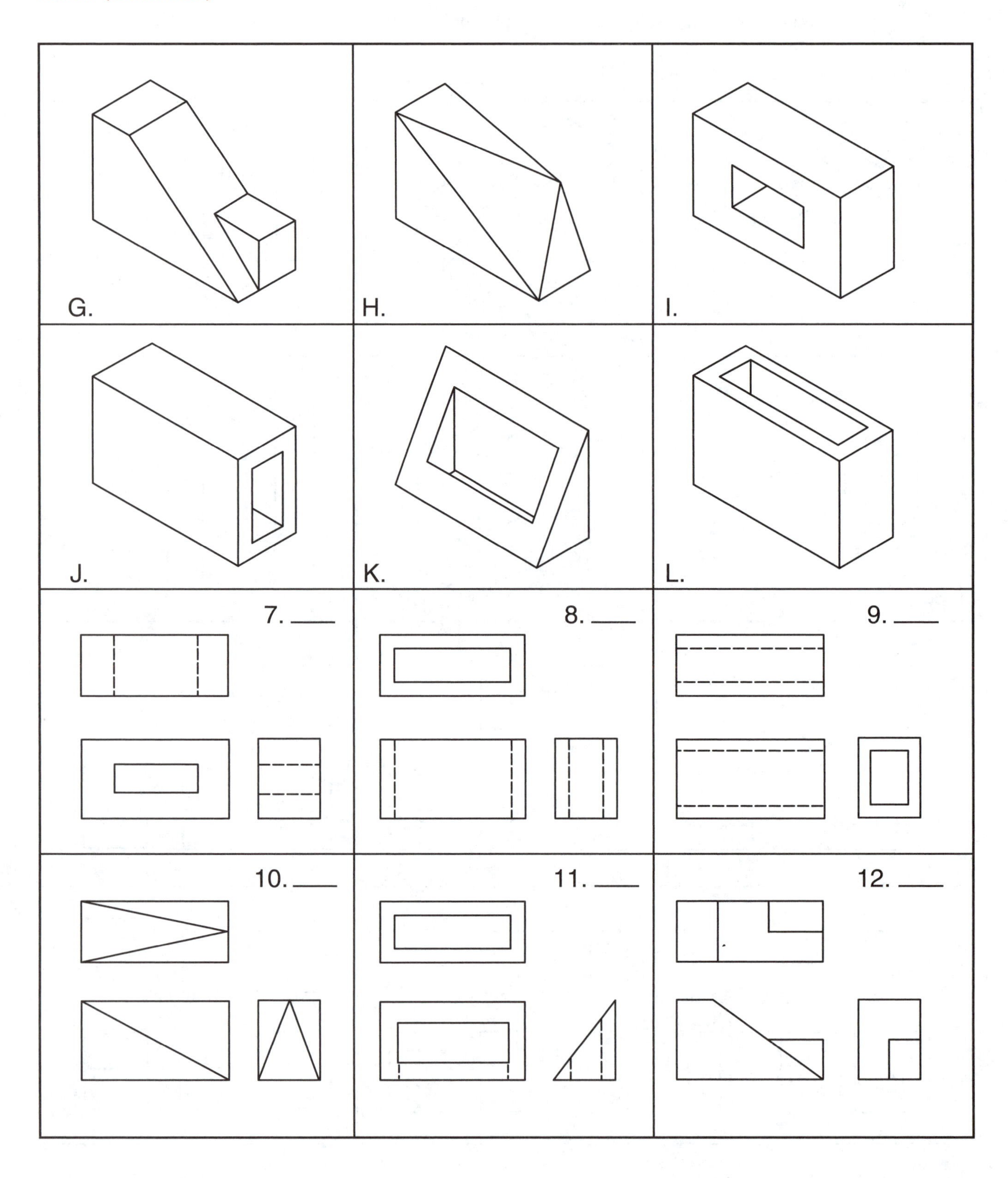

Name ______________________________

Part VI

Sketch in the correct missing lines in the orthographic projection drawings.

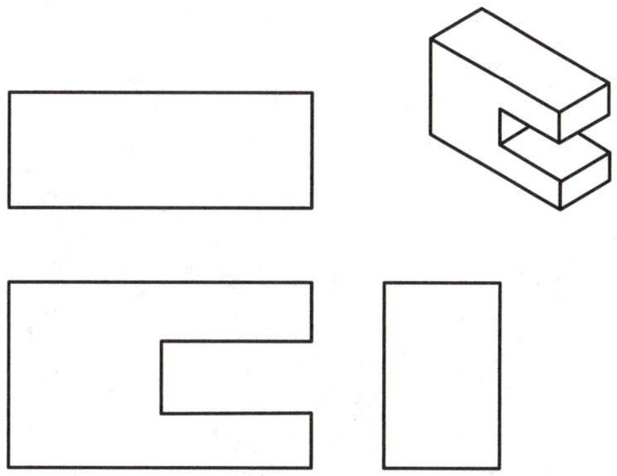

13.

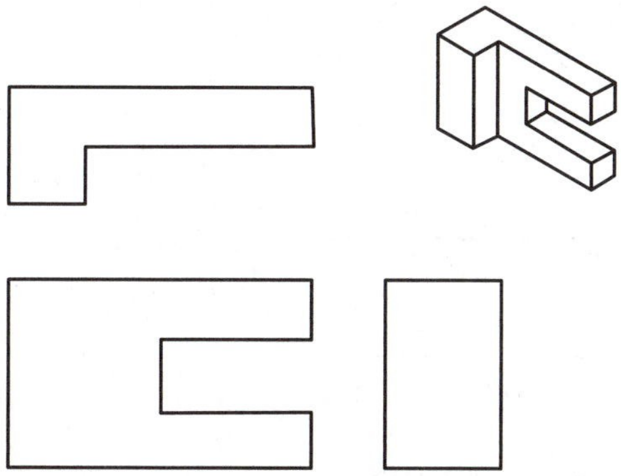

14.

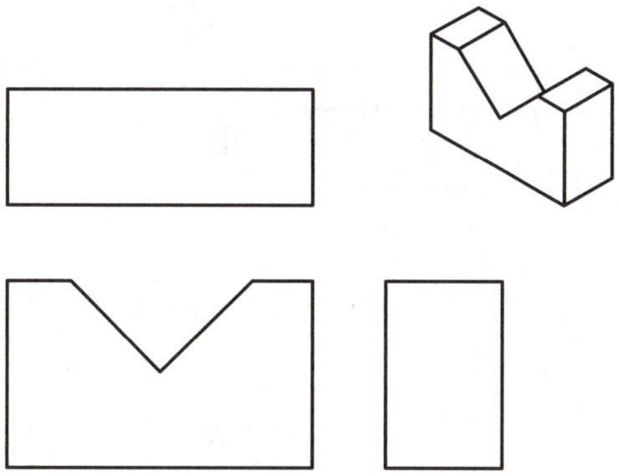

15.

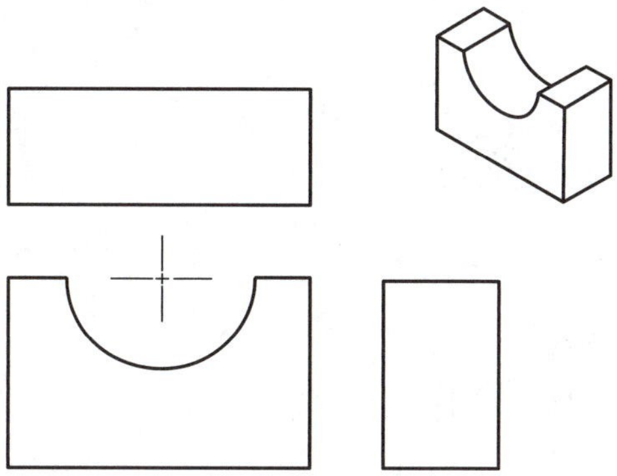

16.

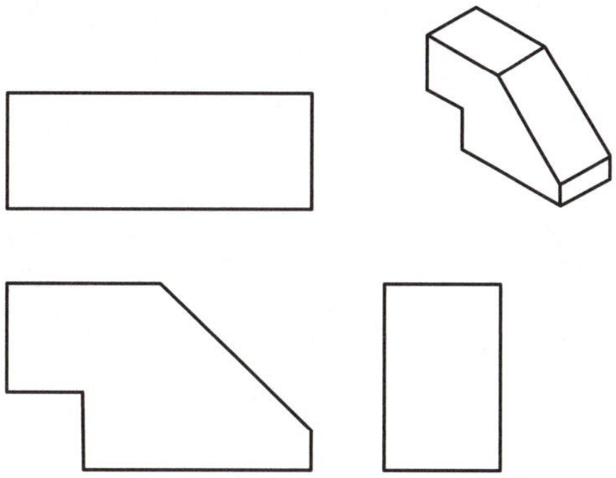

17.

18.

Part VII

Complete the missing view for each.

19.

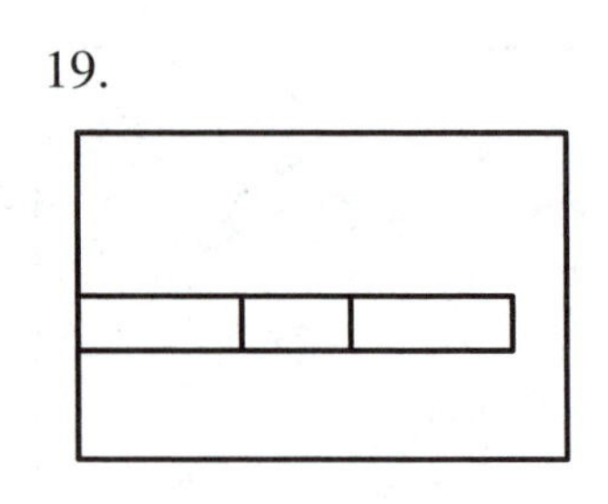

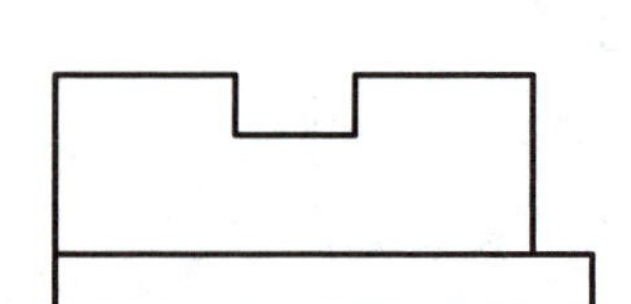

20.

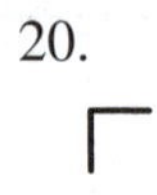

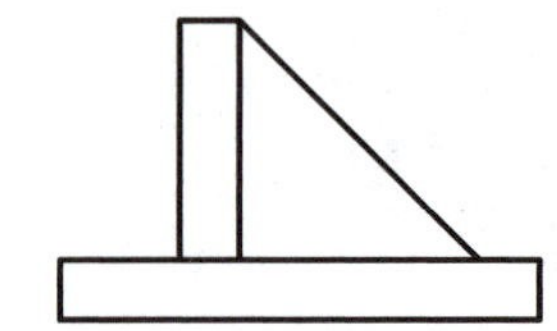

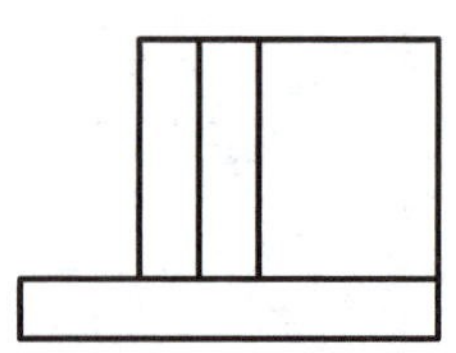

21.

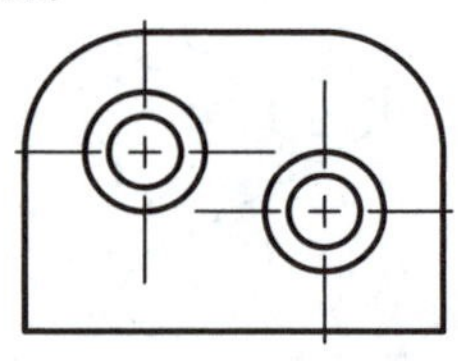

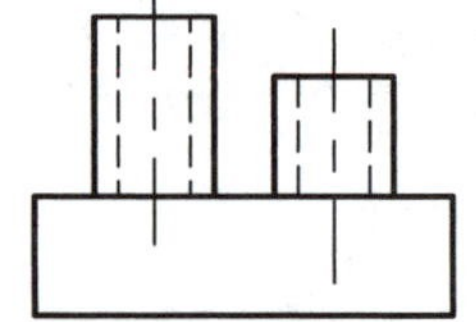

22.

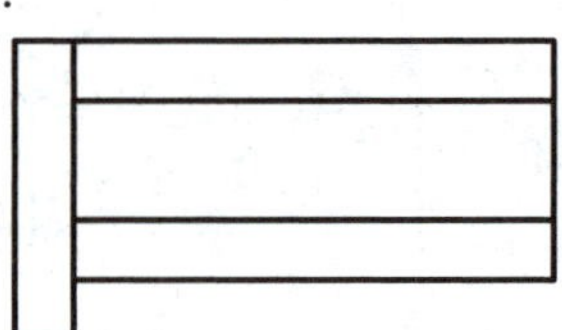

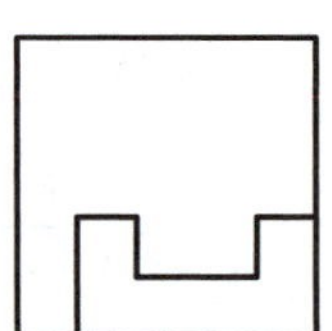

Name ____

Part VIII

Study the pictorial drawings and sketch an orthographic projection (three views) for each.

23.

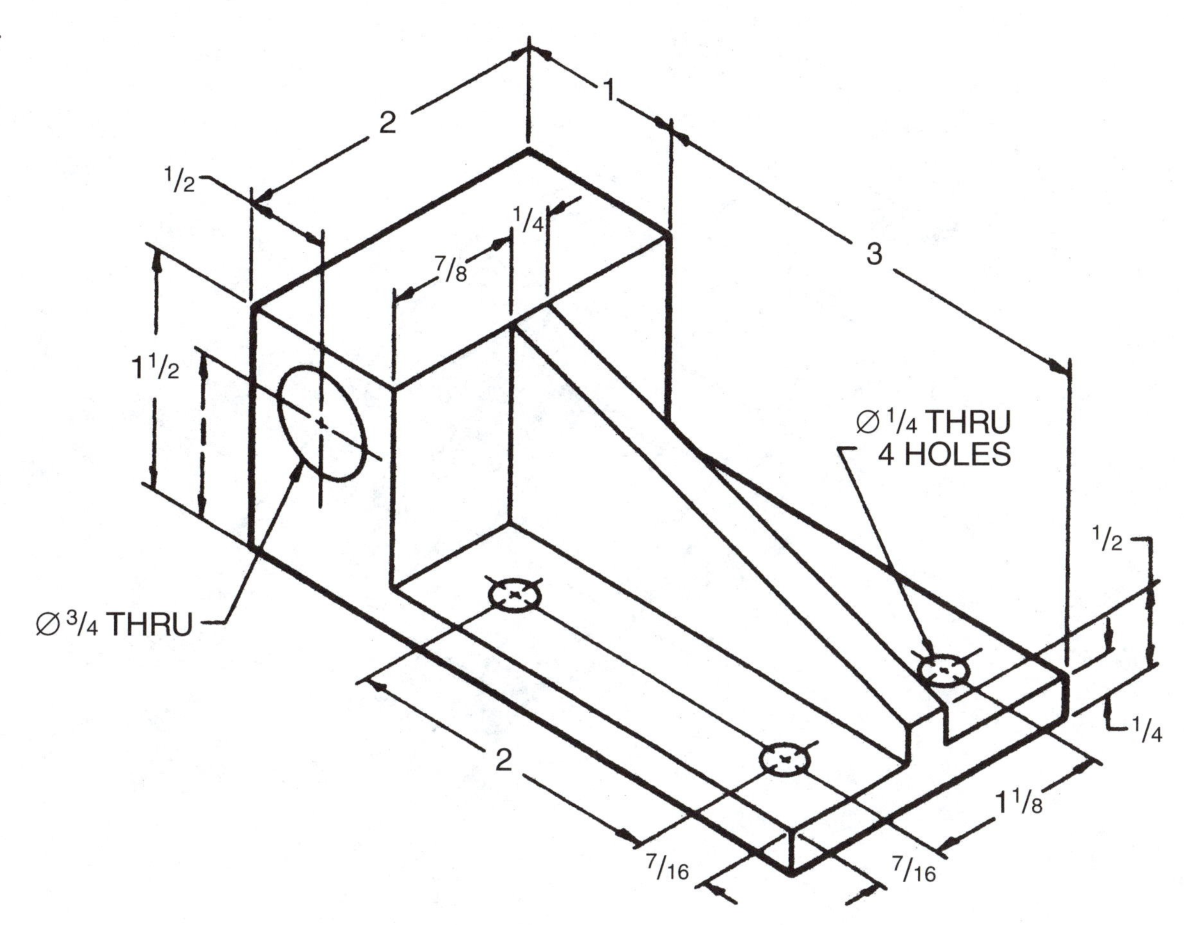

Part VIII (continued)

24.

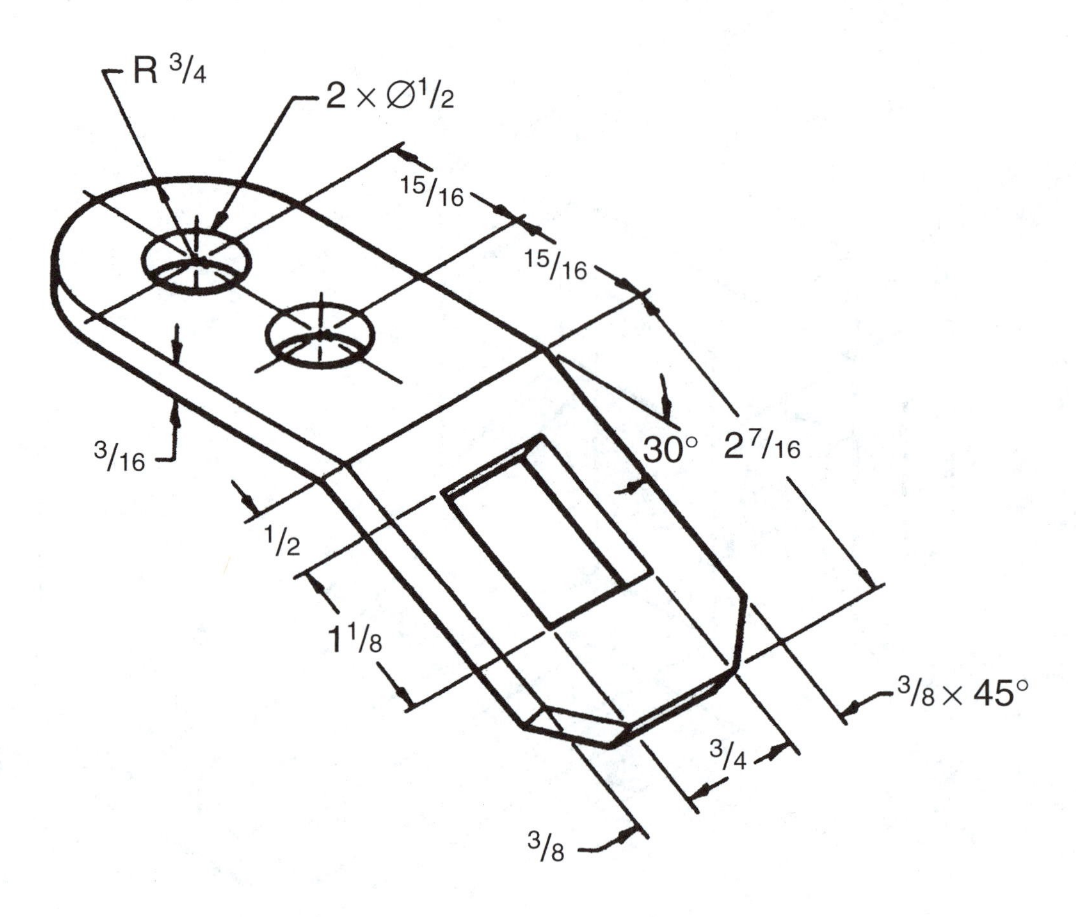

Unit 6

Types of Prints

After completing Unit 6, you will be able to:

- Identify several types of skilled welders.
- Describe two types of working drawings.
- Explain the similarities and differences among types of working drawings and specialized drawings.
- Identify information found on a parts list.

Key Words

assembly drawing
balloons
bill of materials
detail assembly drawing
detail drawing
erection drawing
job shop welder
maintenance welder
material list
parts list
process drawing
production welder
schedule of parts
specialized drawing
subassembly drawing
working drawings

It would not be practical for industry to manufacture a product without using prints or drawings that provide complete manufacturing details. Each part in a product, even the smallest rivet, requires a print.

Skilled welders, whether they work as a production welder, a job shop welder, or a maintenance welder must be able to read and interpret drawings.

Welders Who Read Prints

A ***production welder*** usually does the same welding procedure over and over. Mass production of a weldment requires this type of welder. The ***job shop welder*** seldom does the same job twice. He or she commonly makes single, special order weldments for individual jobs. A ***maintenance welder*** repairs machine parts, fabricates pipelines, builds structural components for manufacturing facilities, and modifies existing equipment. See Figure 6-1.

The drawings and prints welders use vary from a freehand sketch for a simple job, Figure 6-2, to highly detailed prints for a complex job or product, Figure 6-3.

Working Drawings

Working drawings supply the information needed to make and assemble the many pieces and parts that make up a product. Welders and related workers rely on these prints when doing their assigned tasks. There are two types of working drawings: detail drawings and assembly drawings.

Figure 6-1.
This maintenance welder must be able to interpret prints. (Miller Electric Mfg. Co.)

Detail Drawing

A ***detail drawing*** includes a print of the part (one or more views) with dimensions and other information needed to make the part, Figure 6-4.

Assembly Drawing

An ***assembly drawing*** shows where and how the parts described in the detail drawings fit into the complete assembly of the product, Figure 6-5.

Subassembly Drawings

A ***subassembly drawing*** is frequently used on large or complicated products, Figure 6-6. Each drawing shows the assembly of a small portion or section of the complete product.

Detail Assembly Drawing

The detail drawing, in most instances, gives information on only one item. If the mechanism, however, is small in size or if it is composed of only a few parts, the details and assembly may appear on the same print, Figure 6-7. This type of drawing is often referred to as a ***detail assembly drawing.***

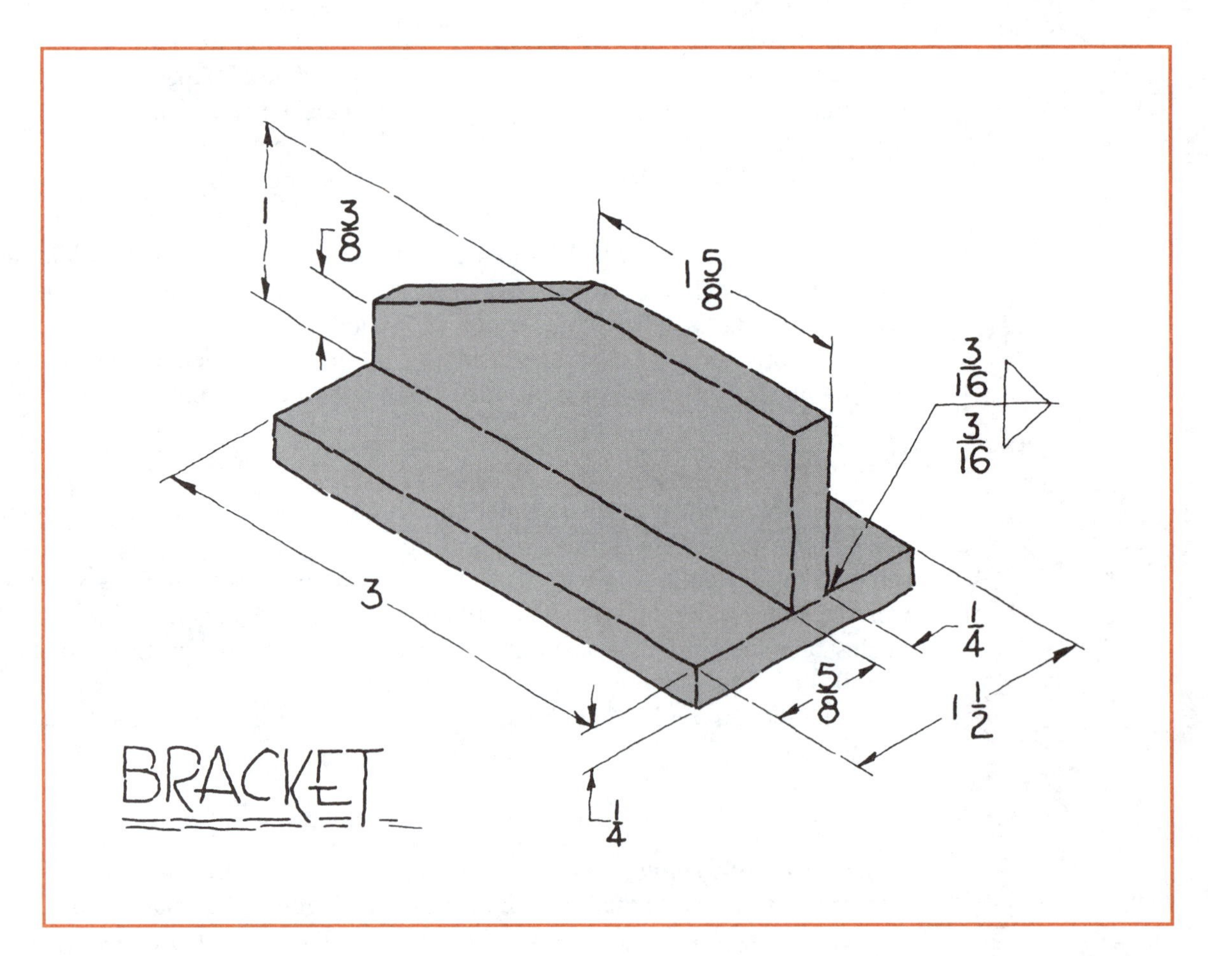

Figure 6-2.
A welder may work from a sketch when the weldment is relatively simple.

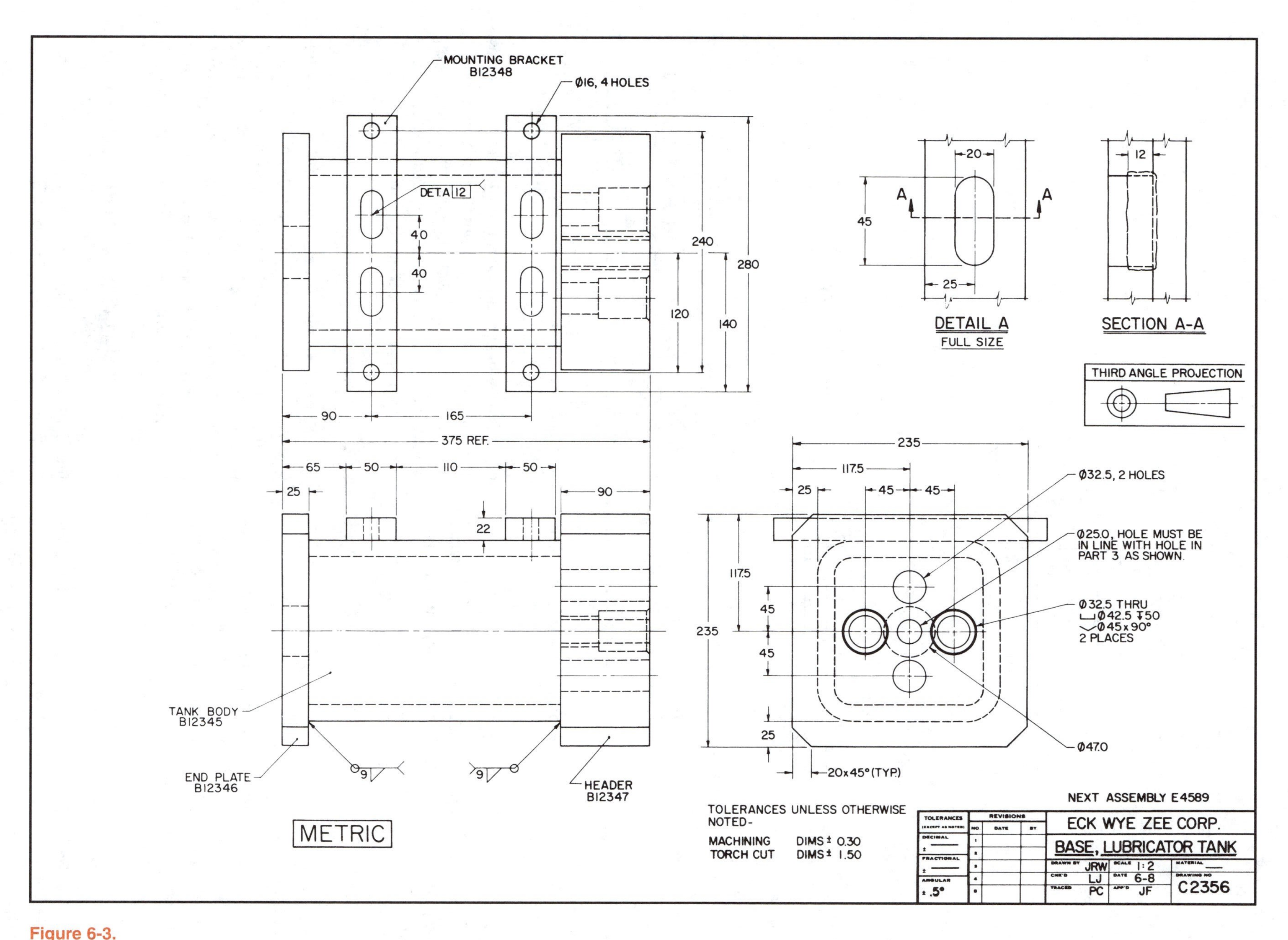

Figure 6-3.
A welder must be able to read and interpret highly complex detailed prints like this one.

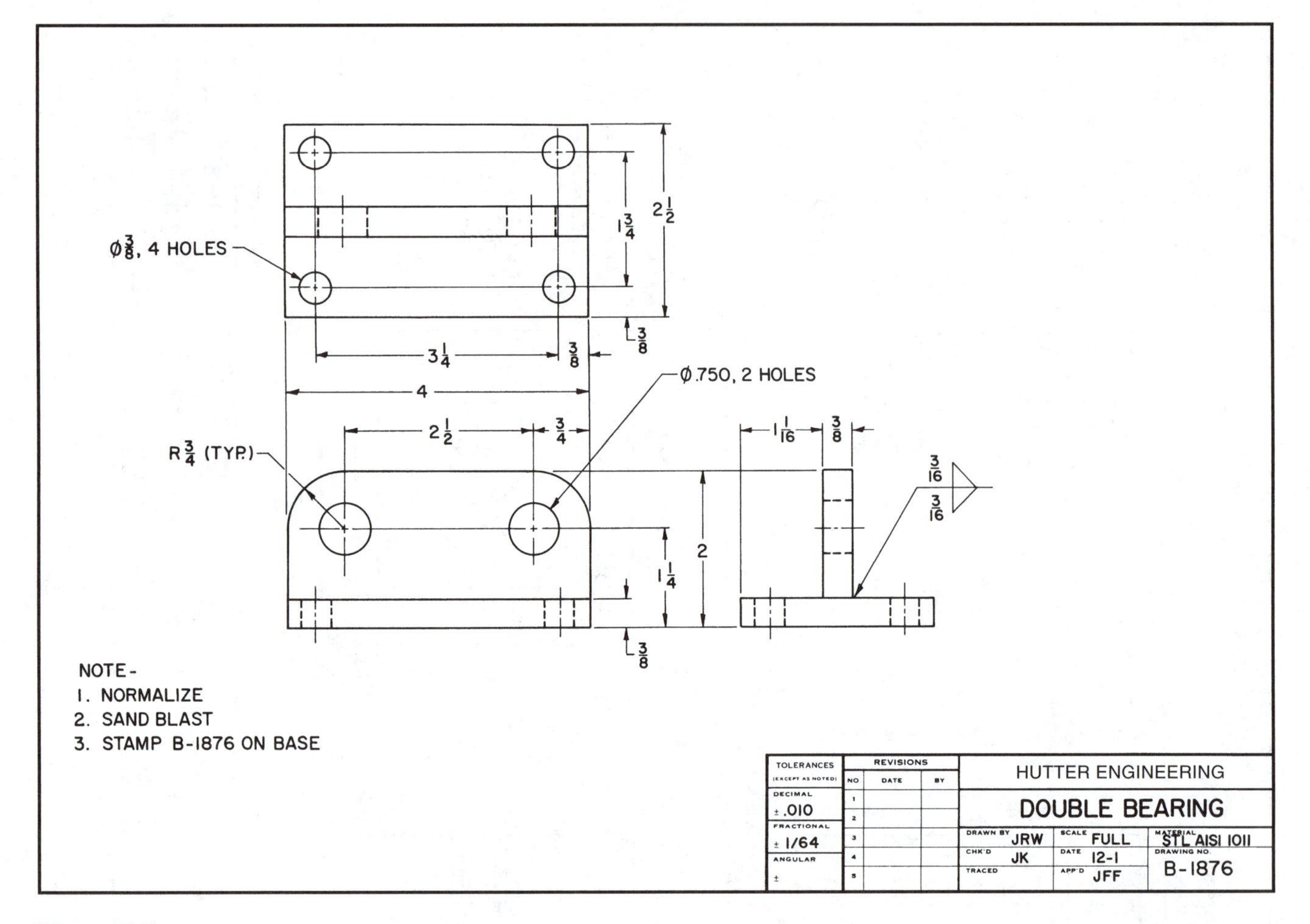

Figure 6-4.
A detail drawing includes all information needed to manufacture or fabricate one part.

Specialized Drawings

When a product is made in quantity, ***specialized drawings*** show separate detail drawings usually prepared for each specific manufacturing process. Figures 6-8, 6-9, and 6-10 show individual detail drawings of the same object for cutting stock, welding, and machining.

Erection Drawing

An erection drawing is a type of assembly drawing. For the welder, an ***erection drawing*** provides the information needed to fabricate (usually in the field) and erect structures such as buildings, bridges, transmission line towers, etc., Figure 6-11.

Process Drawing

A ***process drawing*** includes all of the detail information needed to complete a specific process or group of processes on a part. For example, a machining process drawing would include the dimensions, notes, and tolerances needed to complete machine operations on a part. For welding, the process drawing would include the symbols, dimensions, and notes needed to prepare and weld the part.

Parts List

A ***parts list*** (also called a ***material list, schedule of parts,*** or ***bill of materials***) includes all of the parts required in the manufacture of a product. It usually appears above the title block (which is in the lower right-hand corner). Also included is a description of each part, the quantity of each part needed per assembly, part number, and the number of the drawing used to manufacture each part.

If the product is fabricated from parts provided by a number of different manufacturers, the name of the manufacturer of each component is also included. See Figure 6-12 for a typical parts list.

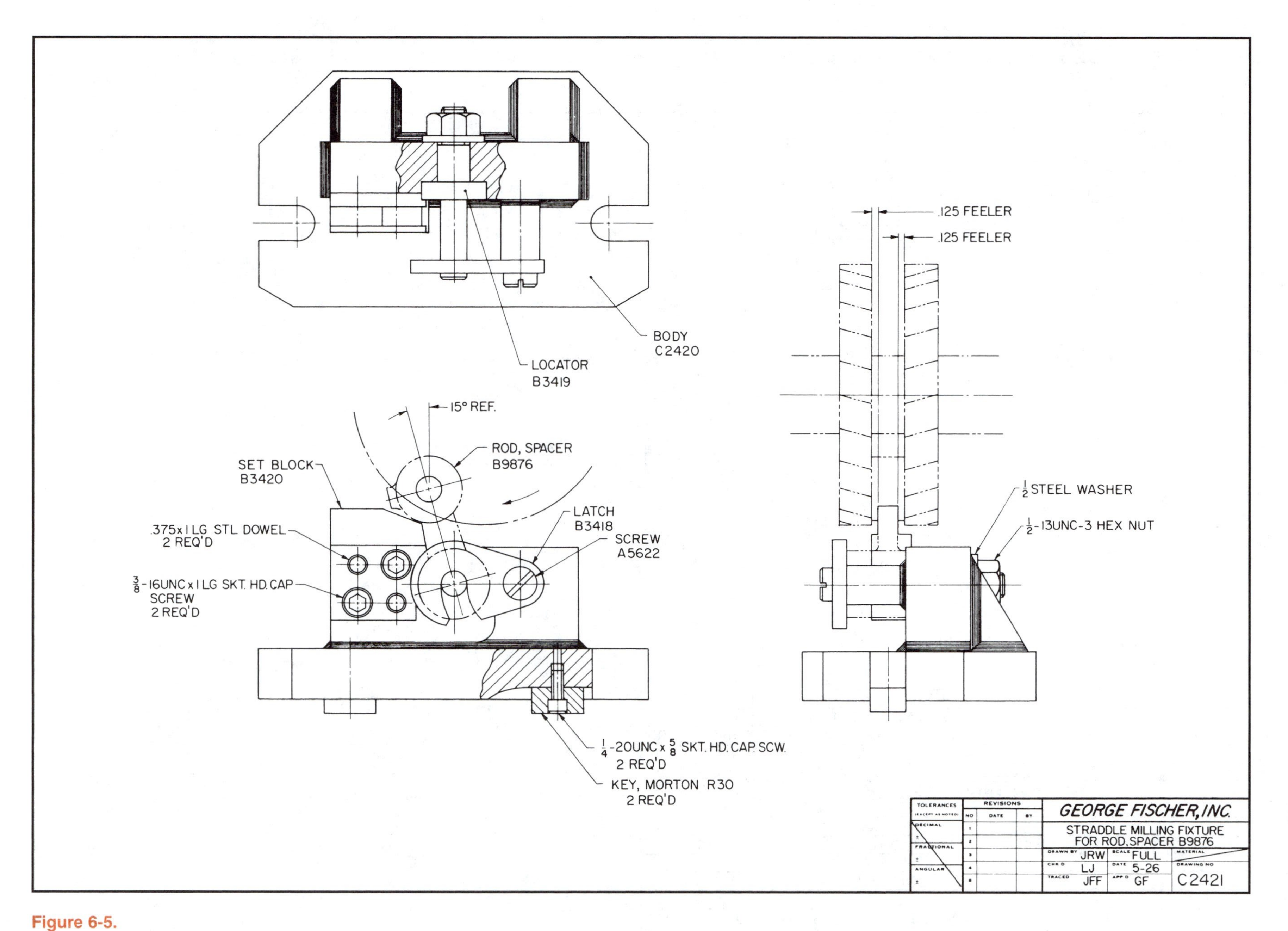

Figure 6-5.
An assembly drawing shows where and how parts described on a detail drawing fit into a complete assembly.

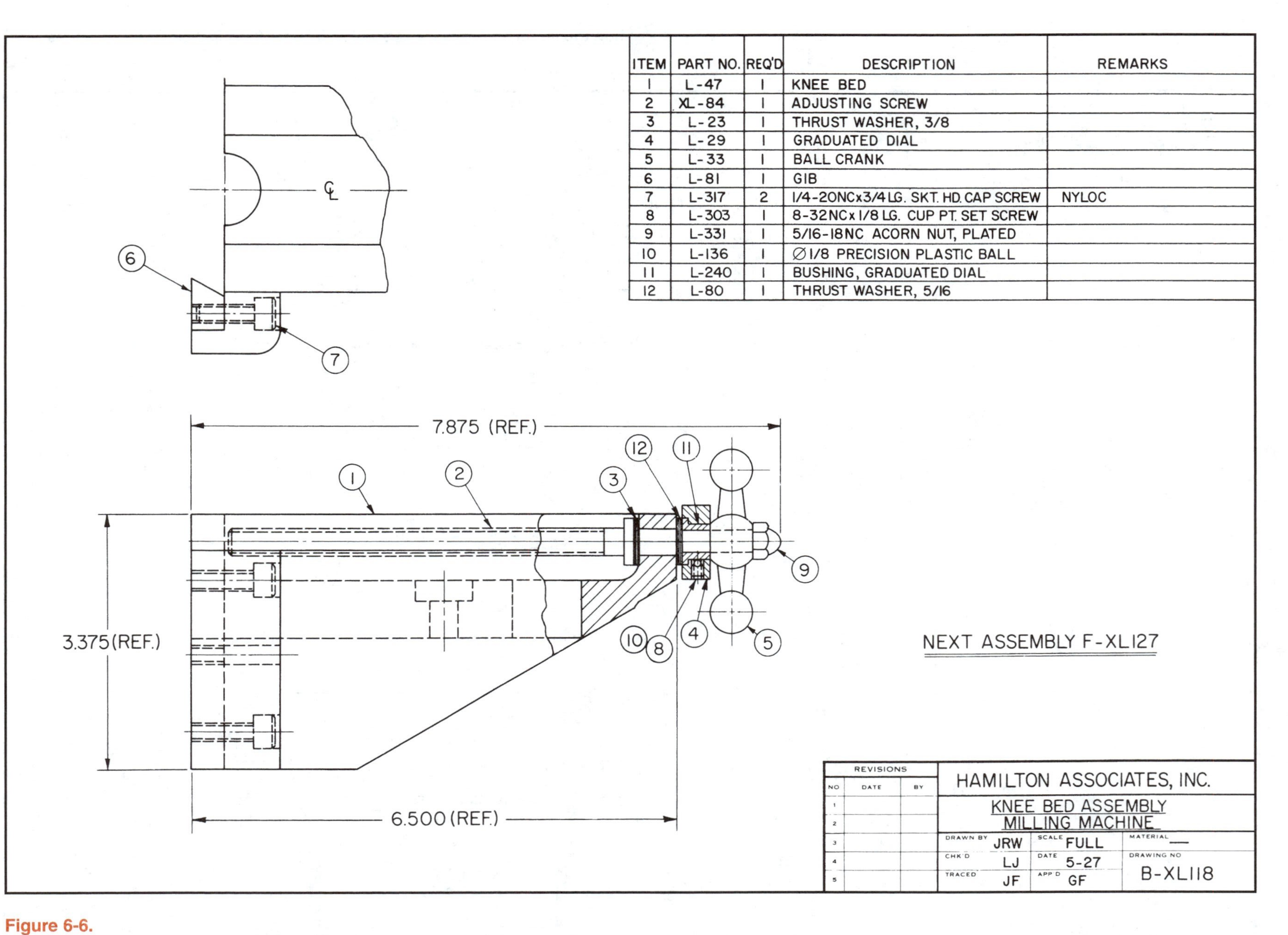

Figure 6-6.
Subassembly drawings are often required for large or complex products. Each subassembly drawing shows a small portion or section of a completed product. Note how numbers show the names of parts.

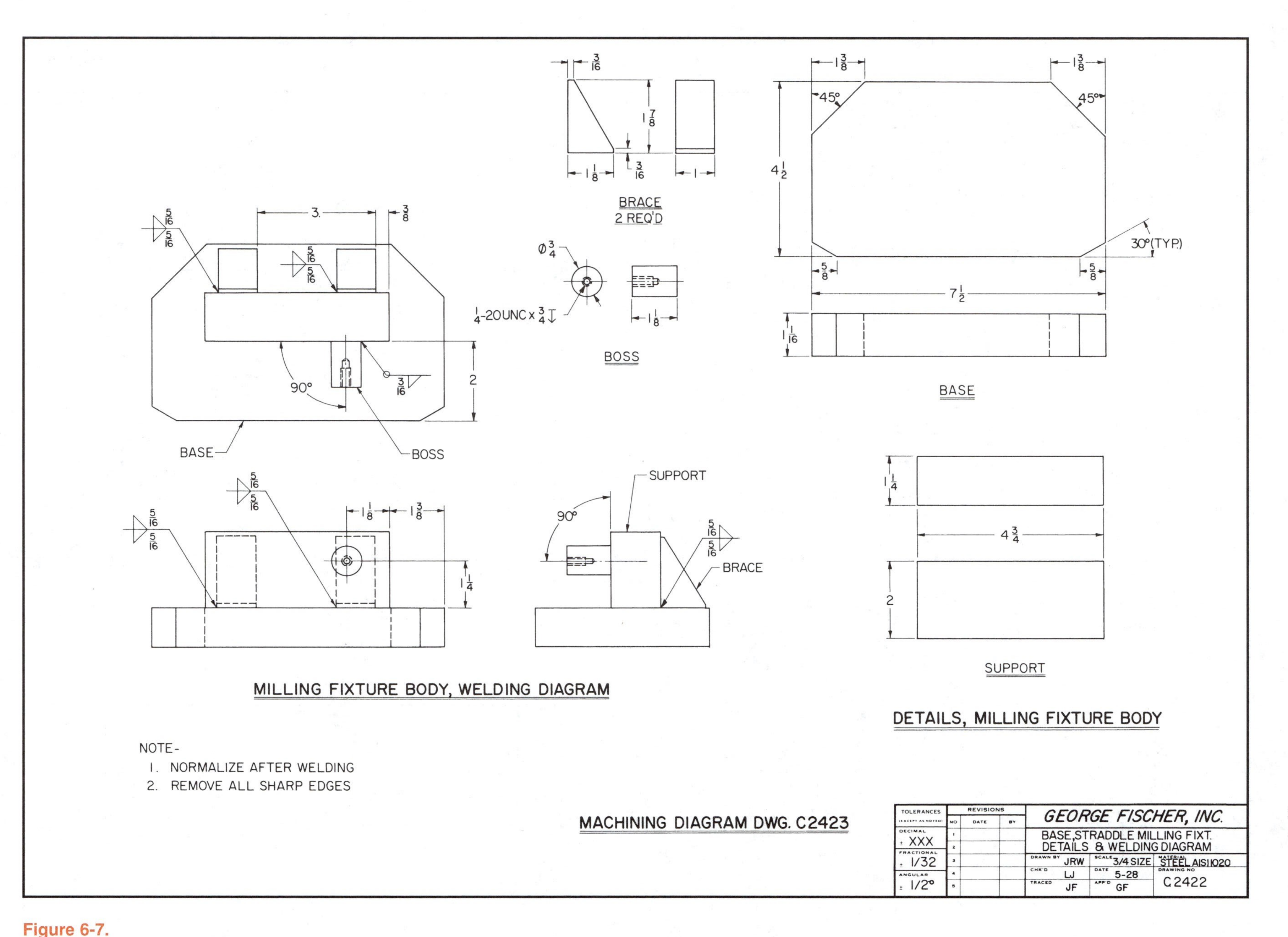

Figure 6-7. A detail assembly drawing includes details and assembly of a simple part on the same print.

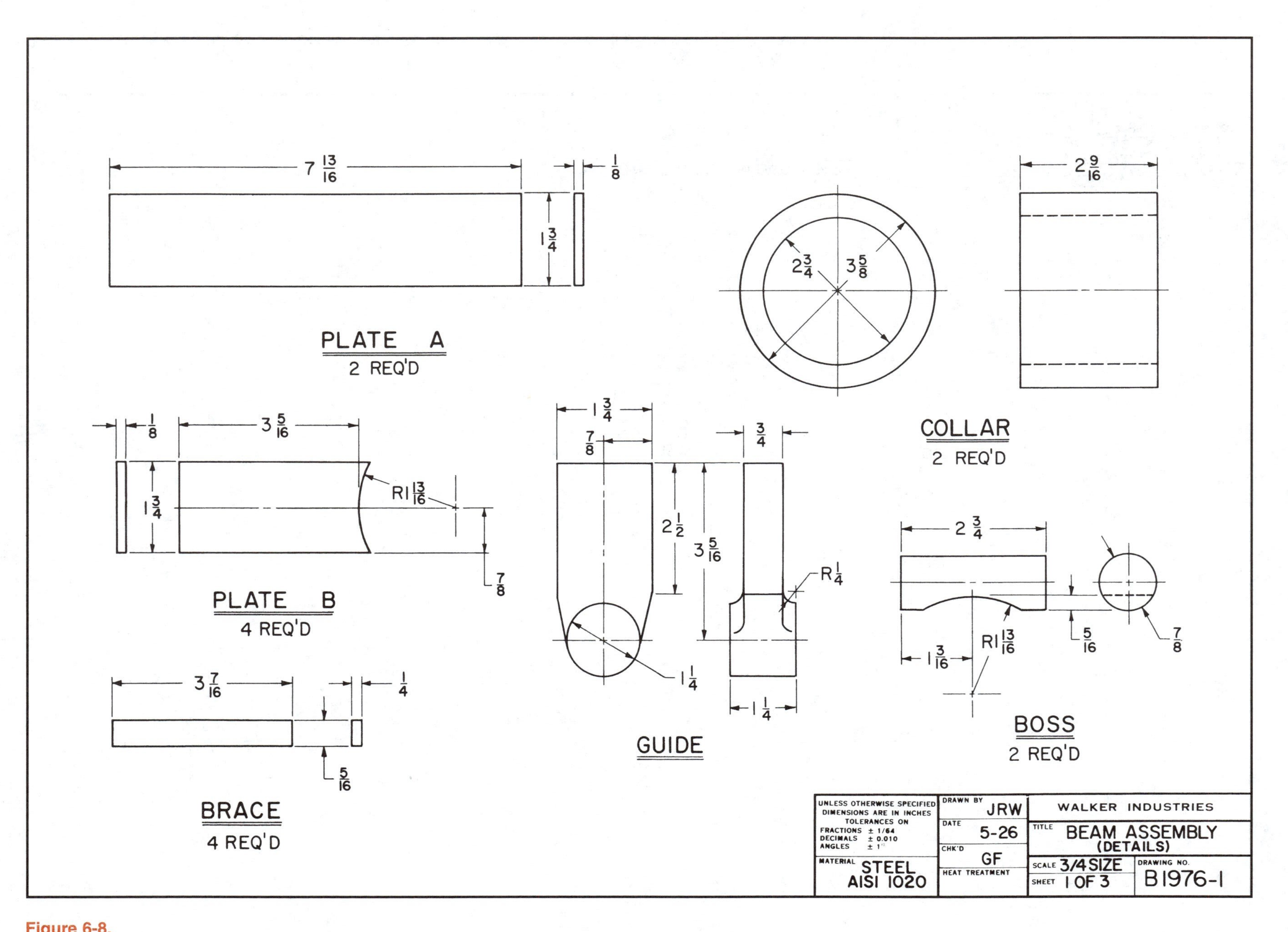

Figure 6-8.
This detail drawing shows how parts should be cut to manufacture a beam assembly. Note how the drawing is numbered.

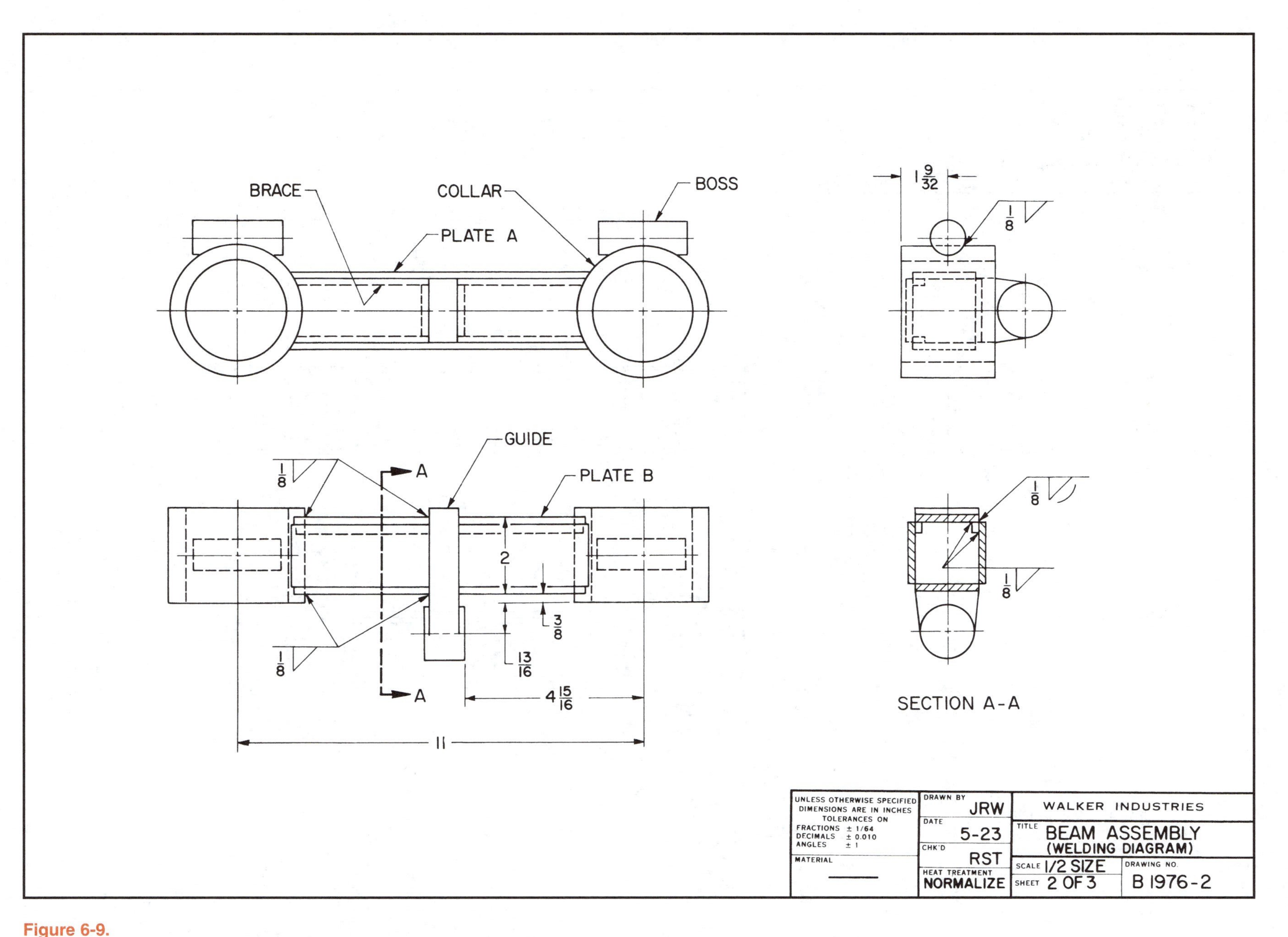

Figure 6-9.
This welding diagram shows how parts described in a detail drawing of a beam assembly are to be fabricated.

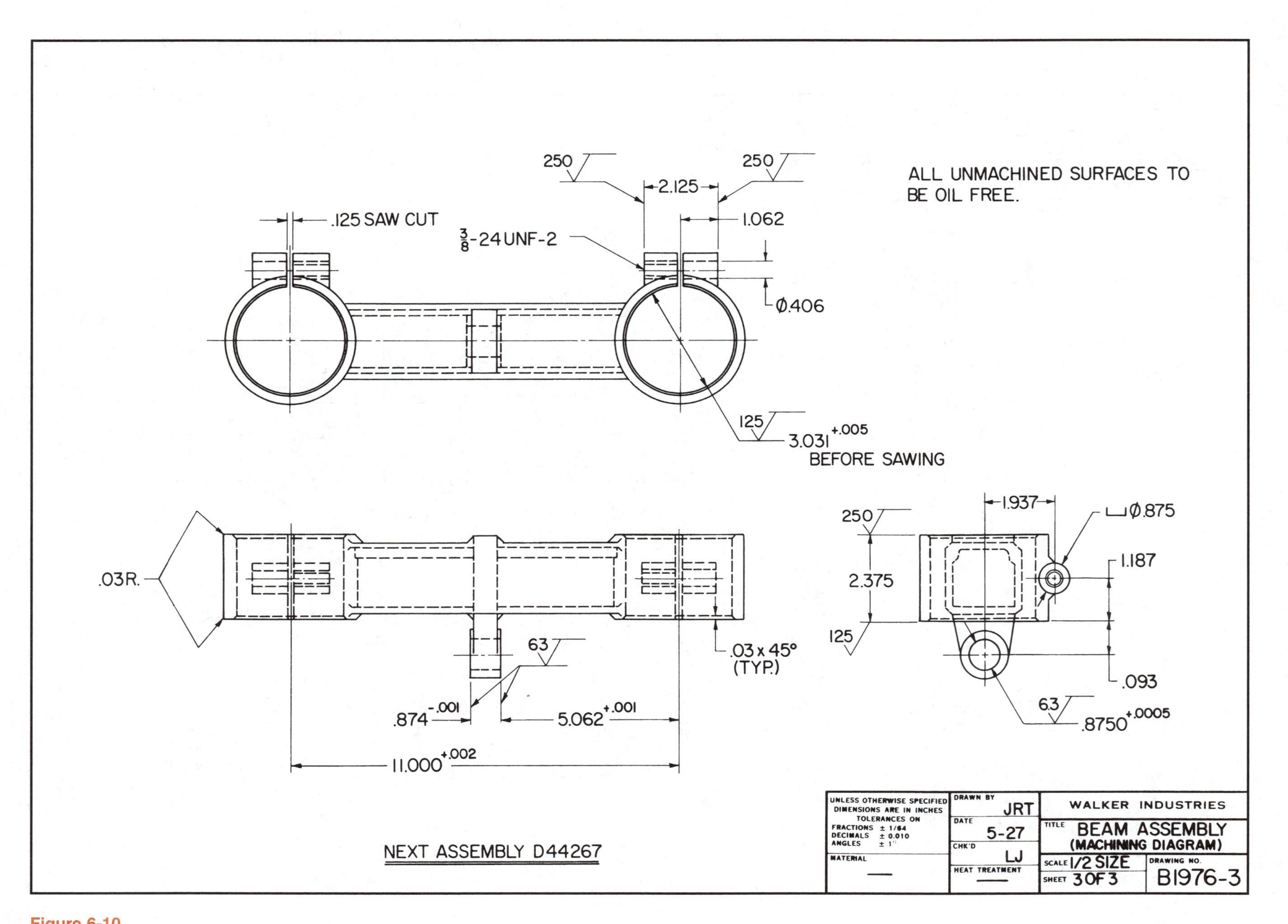

Figure 6-10.
A machining diagram of a beam assembly describes work to be done by a machinist. The welder may not see this drawing.

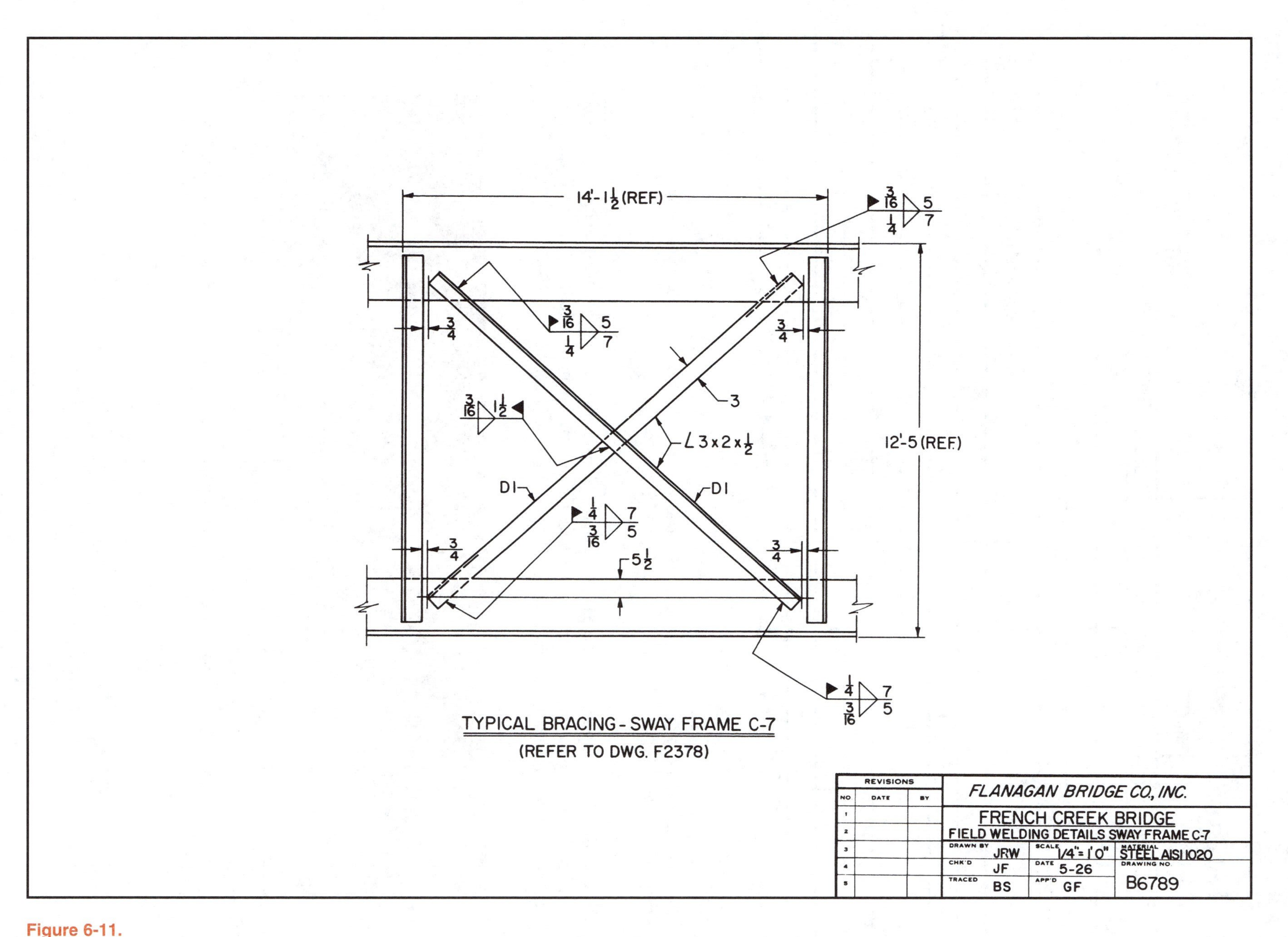

Figure 6-11.
An erection drawing provides the welder with information needed to fabricate and erect a structure, usually in the field.

PARTS LIST	AMERICAN UNCIAL	CODE 81361	DRAWING NUMBER	
	DATE		SHEET OF SHEETS	APP.

	SPECIFICATION	DESIGNATION	USED ON	CODE	UNIT DESCRIPTION
GOV'T					
CONTR					
CONTRACT NO.					

DWG. SIZE	DRAWING NUMBER	PART NUMBER	REV.	PART DESCRIPTION	LINE NO.
					1
					2
					3
					4
					5
					6
					7
					8

Figure 6-12.
A parts list indicates all parts required to make a product. It also includes a description of parts, quantity of parts, part numbers, and the number of drawings.

To identify each component of the part, the print may have ***balloons.*** A balloon is a divided circle with an attached line leading to the component. The top half of the circle holds the item number from the parts list and the bottom half indicates the quantity needed to build the part, Figure 6-13.

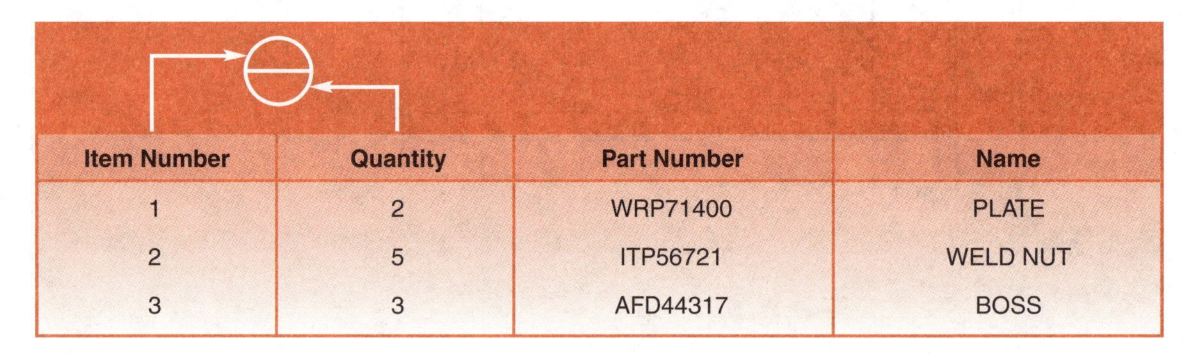

Item Number	Quantity	Part Number	Name
1	2	WRP71400	PLATE
2	5	ITP56721	WELD NUT
3	3	AFD44317	BOSS

Figure 6-13.
Adding a balloon to a print identifies the item number from the parts list and the quantity needed to build the part.

Name ______________________________ Date ______________ Class ______________

Review Questions

Each word in the left column matches one of the sentences. Place the letter of the sentence in the appropriate blank space.

1. _____ Production welder
2. _____ Job shop welder
3. _____ Working drawing
4. _____ Detail drawing
5. _____ Assembly drawing
6. _____ Subassembly drawing
7. _____ Detail assembly drawing
8. _____ Specialized drawing
9. _____ Erection drawing
10. _____ Process drawing

(A) Shows assembly of a small portion of completed product.

(B) Drawing of part that includes dimensions and information needed to make it.

(C) Seldom does same welding job twice.

(D) Shows where and how parts fit into complete unit.

(E) Includes all detail information needed to complete a specific process on a part.

(F) Drawing used to show specific manufacturing process for a product that will be produced in quantity.

(G) Does same welding job over and over.

(H) Supplies information needed to make and assemble many pieces and parts that make up a product.

(I) Contains details and assembly information on a single print.

(J) Provides information needed to fabricate and set up structures such as bridges, transmission line towers, etc.

11. List three other names for a parts list. ______________________________

____

____

____

12. What is a parts list and what is included on it? ______________________________

____

____

____

13. The top half of a callout balloon generally indicates _____.

____

____

____

Print Reading Activities

Use the weld assembly drawing shown in Figure 6-14 to answer questions 1–3.

1. The quantity required for part number A71400-2 on the print is _____.
2. The part number of item 4 on the print is _____.
3. The quantity of item 1 required to build 5 parts is _____.

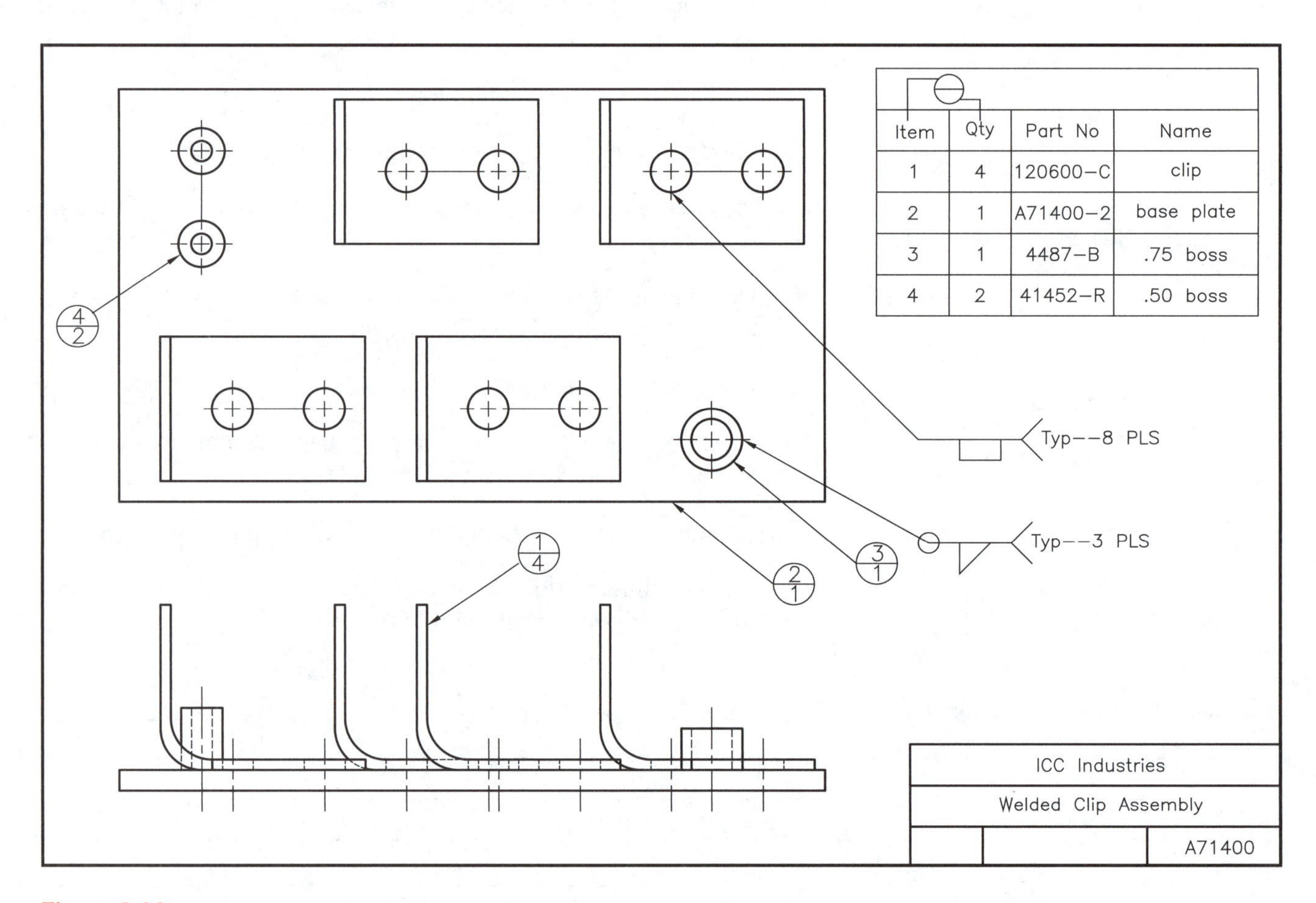

Figure 6-14.
Use this weld assembly print to answer the Print Reading Activities questions.

Unit 7

Print Format

After completing Unit 7, you will be able to:

- Interpret a variety of print formats that use various units of measure for dimensions.
- Locate and explain the purpose of information found in the title block.
- Determine tolerance variances of a part, feature, and weld from the print dimension and specified tolerance.
- Recognize and explain other formats appearing on prints, including sheet size, zoning, and security classification.

Key Words

anneal
application
bilateral tolerance
drawing title
finish
heat treatment
identifying number
material specification
next assembly
print number
revision block
routing sheet
scale drawings
security classification
shape
shop traveler
size
tolerance
unilateral tolerance
work order
zoning

In order to be useful, a drawing or print must include a "picture" or "map" of the product and all of the information needed to make or assemble the product. When first "reading" a print, the views, lines, dimensions, notes, and amount of detail can be overwhelming.

The best way to read a print is to mentally break it into smaller parts. This unit will help you understand the information found on a print. Remember, try to look at the shape of the part with the dimensions and notes removed. Second, try to determine the overall size of the part so you have some understanding of how big or small it is. Third, look at the title block for information about the scale, tolerance requirements, material, and part name. Finally, read all of the notes on the drawing.

By using this method to read a print, no important detail will be overlooked. You will also have a better understanding about where to find specific information you may need to lay out and fabricate the part.

Shape Description

Visible lines are used to show the visible features of a part. Views are used to show the length, width, and thickness of a part. Features that are not visible on the surface of a view are shown as hidden lines.

The views and features combine to describe the ***shape*** of an object. Only the views required to adequately describe the shape of the object are found on the print.

Size Description

A properly dimensioned drawing includes all dimensions needed to give a complete ***size*** description of the product. The United States is in the process of converting from English measure (inch, pound, quart, etc.) to metric measure (meter, millimeter, kilogram, etc.). Total conversion to the metric system is expected to take many years. But many companies (especially those selling products to the rest of the world) have converted to the metric system. Therefore, during this conversion period, you must be familiar with several print formats. A welder may have to work from prints that have dimensions in:

- Fractions, Figure 7-1.
- Decimal fractions, Figure 7-2.
- Metrics, Figure 7-3.
- Dual dimensions, Figure 7-4.

A *dual dimensioned print* has both English and metric values. Metric dimensions are usually in millimeters (mm). Figure 7-5 shows the various dual dimensioning techniques and how the dimensions are placed on a drawing.

Drawings are usually dimensioned in inches and common fractions when the object does *not* require a high degree of accuracy. Greater precision is indicated when dimensions are in inches and decimal fractions, Figure 7-6.

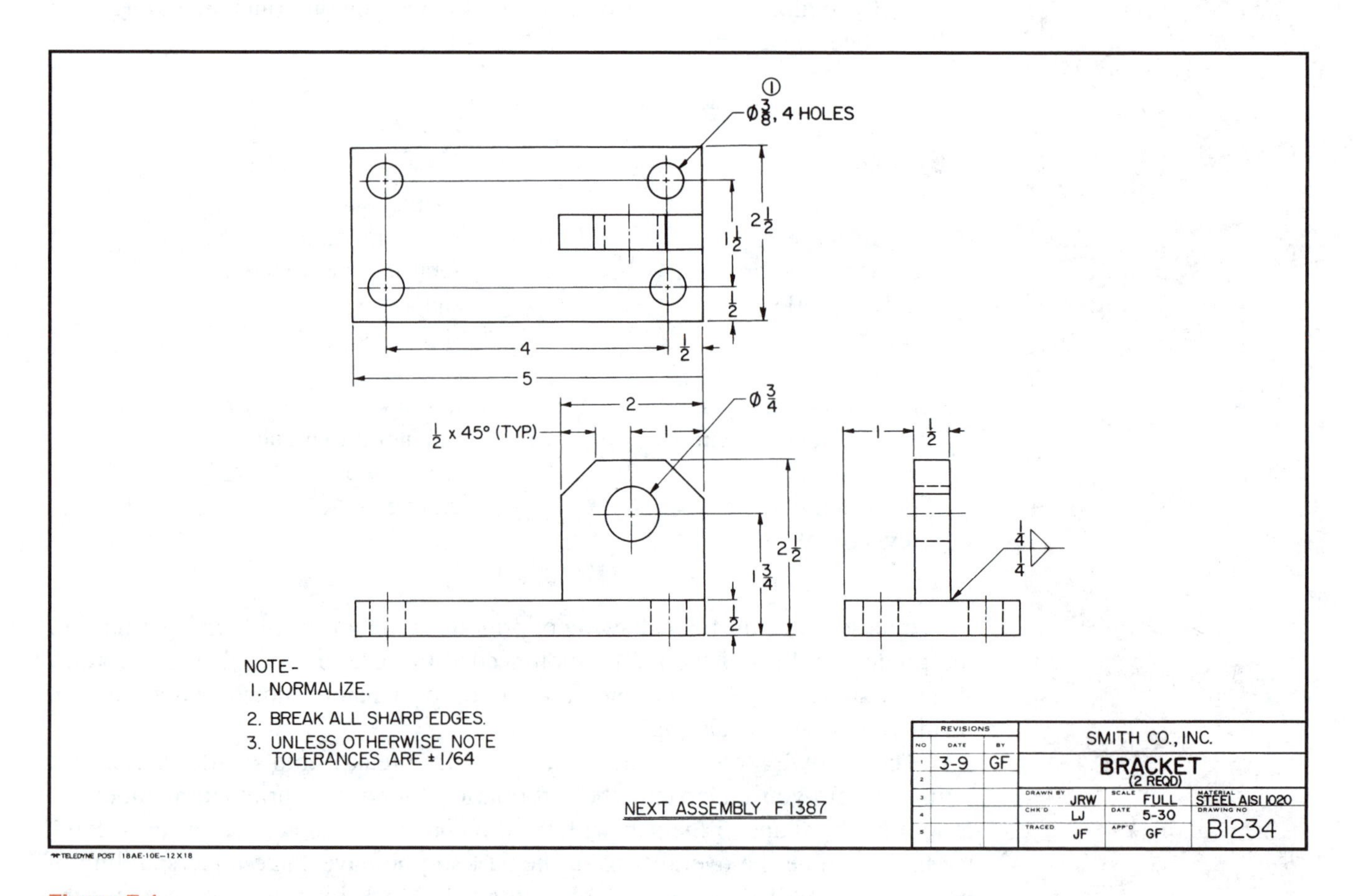

Figure 7-1.
This typical drawing is dimensioned in inches and fractions of an inch.

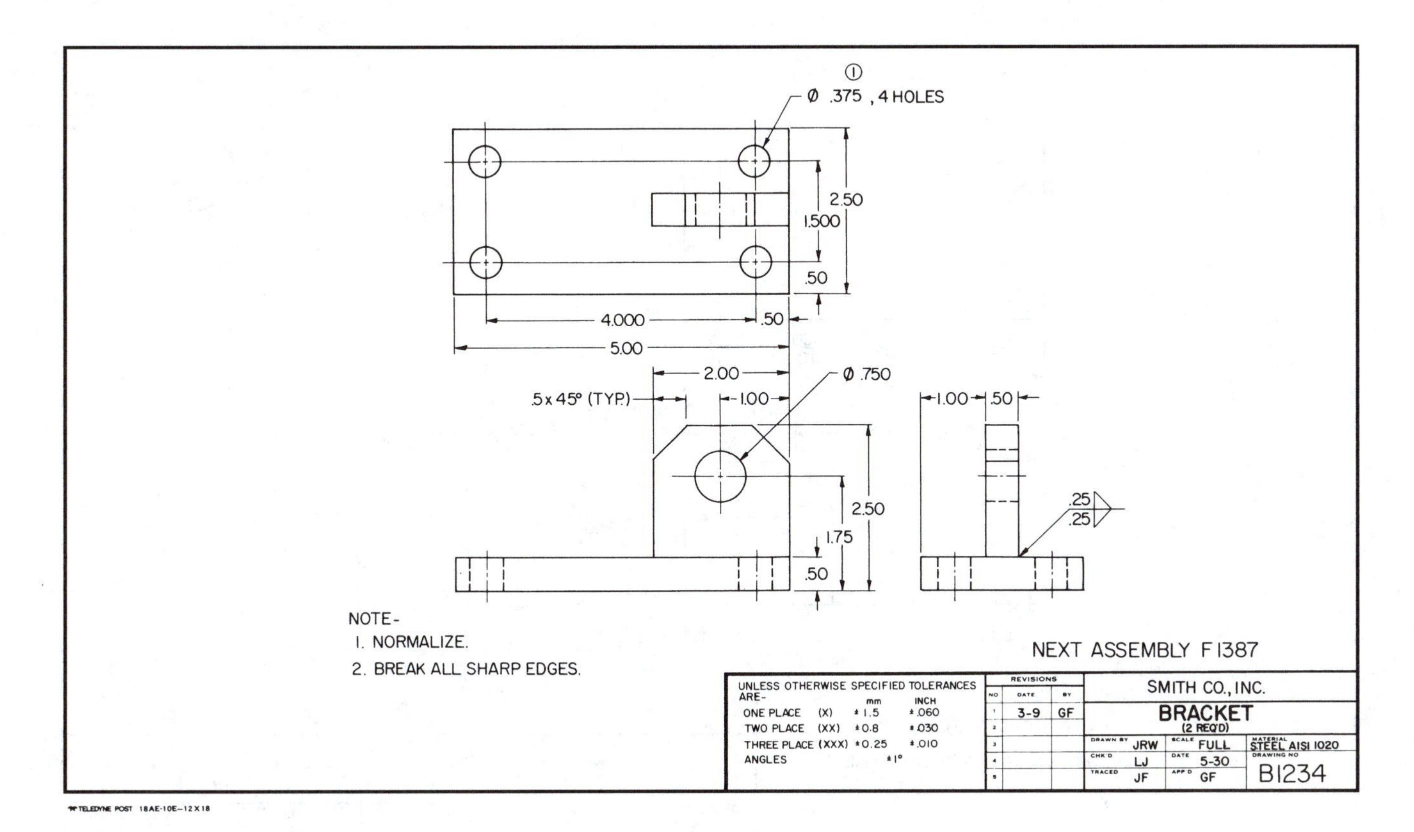

Figure 7-2.
The same drawing is dimensioned in decimal fractions.

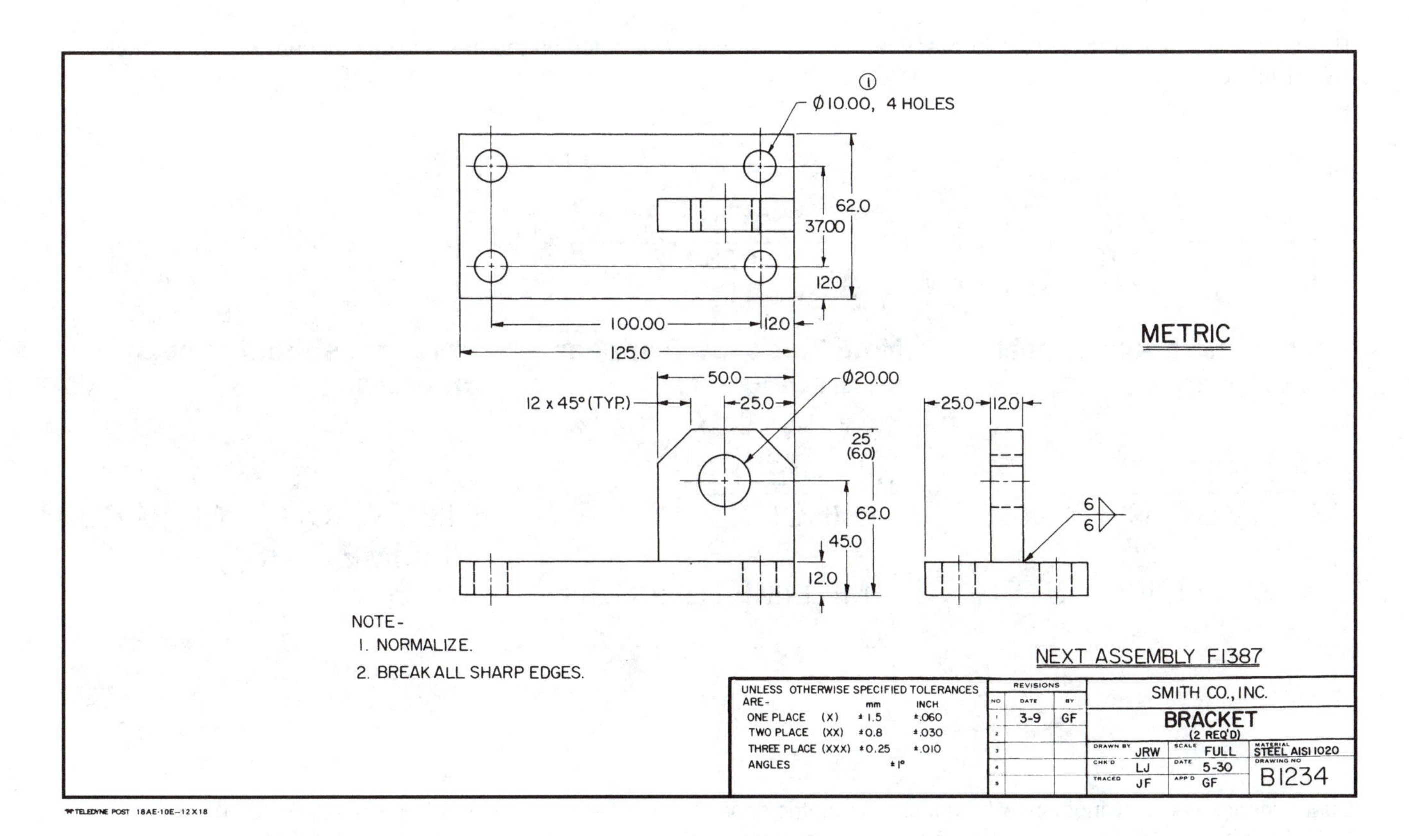

Figure 7-3.
The same drawing is now dimensioned in millimeters.

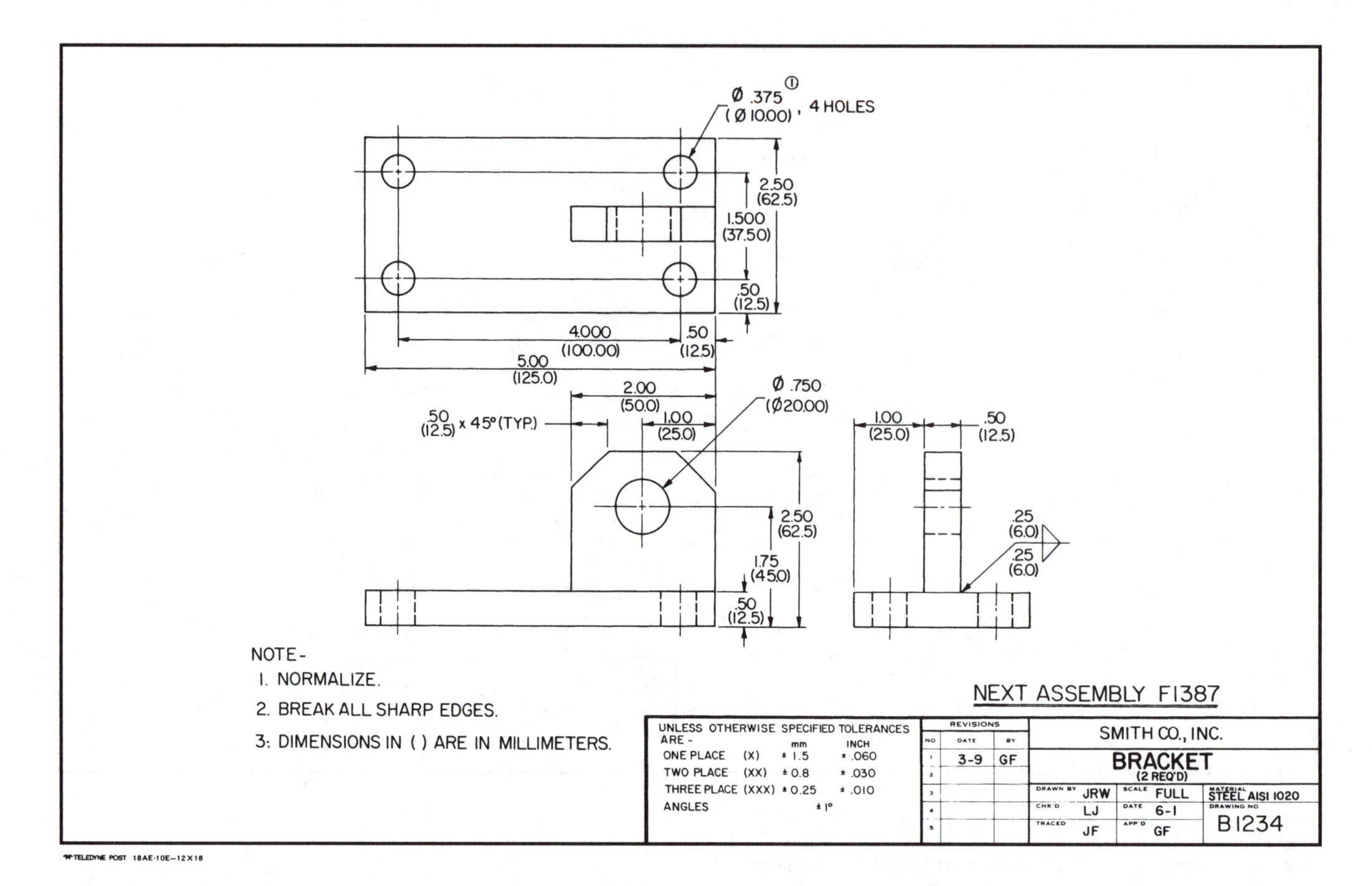

Figure 7-4.
This print uses dual dimensioning to describe the size of a part. The notes on the print indicate dimensions in brackets are millimeters.

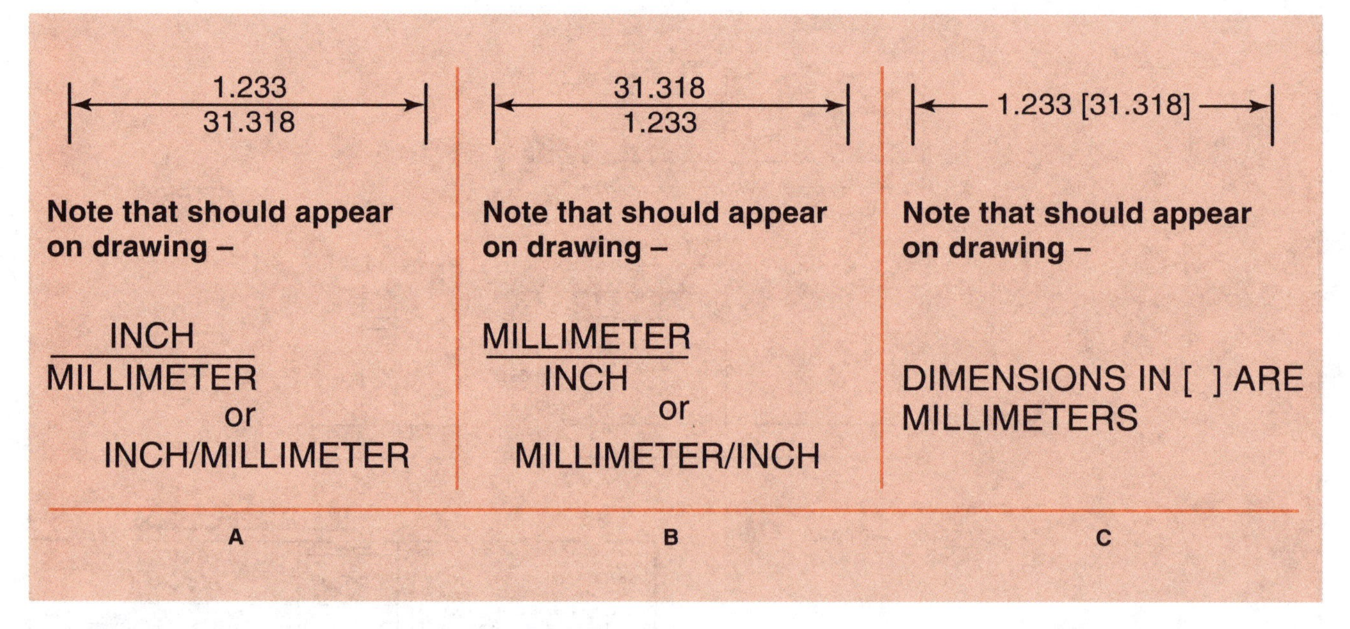

Figure 7-5.
Dual dimensioned drawings show both inch and metric dimensions. A—Method used when drawing is to be used in the United States. B—Method used when the drawing is to be used primarily in a metric country and the United States. C—Brackets sometimes indicate the metric equivalent on a drawing to be used in the United States.

How much more accurate is decimal tolerance?

UNLESS OTHERWISE SPECIFIED DIMENSIONS ARE IN INCHES TOLERANCES ON FRACTIONS ± 1/64 DECIMALS ± 0.010 ANGLES ± 1°	DRAWN BY JRW DATE XX-XX LJ CHK'D GF	WALKER INDUSTRIES	
		TITLE **FLAP SUPPORT**	
MATERIAL ALUM. 6061	HEAT TREATMENT NOTED	SCALE 2:1 SHEET 5 OF 45	DRAWING NO. C9753

Figure 7-6.
When dimensions are in inches and decimal fractions, greater precision is normally required.

Print Title Block

Additional information about the product shown on the print can be found in the title block. Figure 7-7 summarizes most information in a *title block.*

Material Specification

The ***material specification*** on the print gives the exact grade or substance (AISI 1020 steel, 6061-T6 aluminum, etc.) to be used in the weldment. It is usually included in a section of the title block, Figure 7-7A.

Notes can also serve as material specifications *not* given in the title block. If they are used, notes appear elsewhere on the print.

Remember! Under no condition should the welder substitute a different grade or type of material for the material specified on the print.

Tolerances

It would be costly and nearly impossible to make everything to *exact* specified sizes. To keep costs within practical limits, tolerances are permitted.

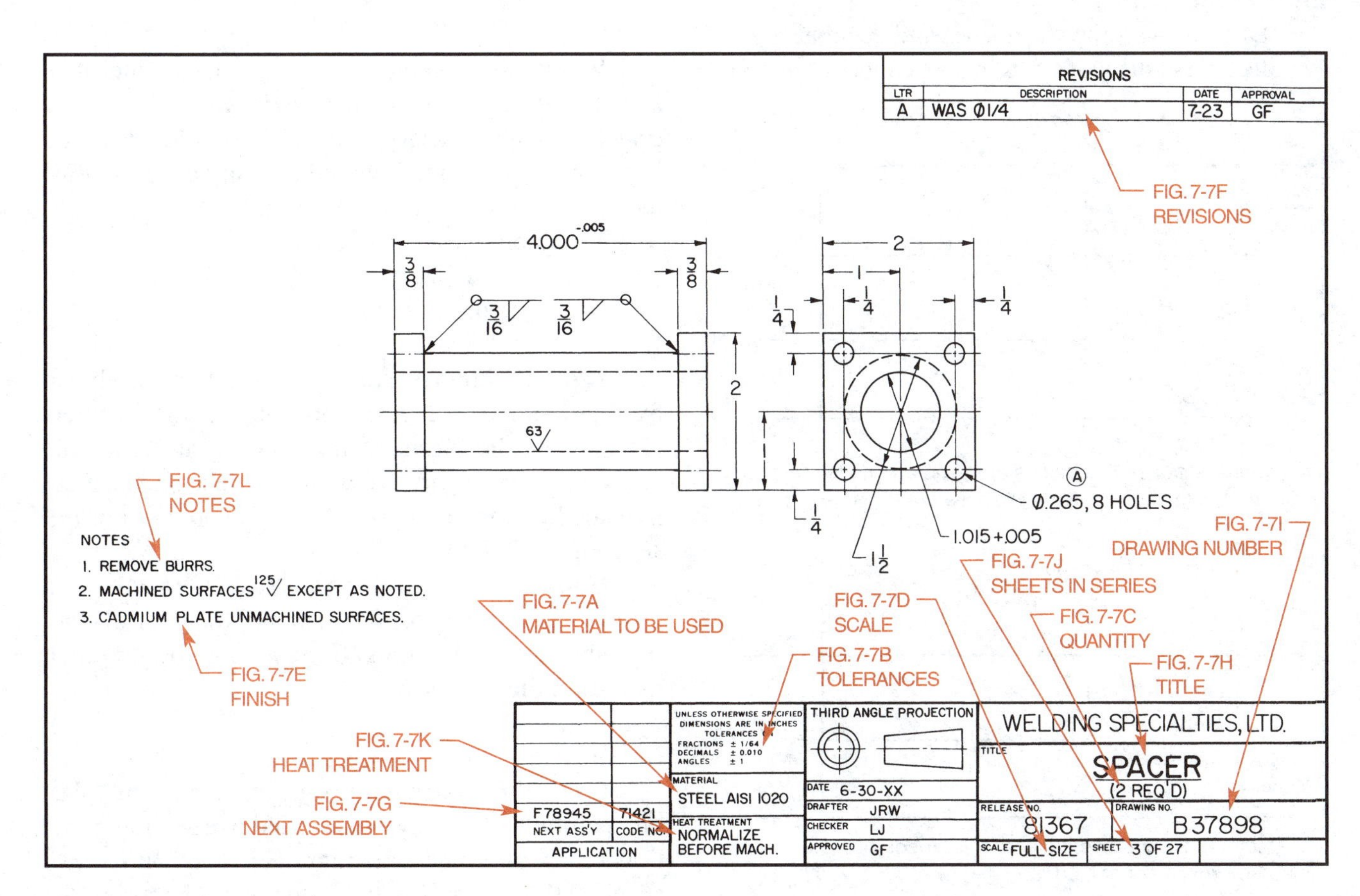

Figure 7-7.
The title block of a drawing contains much information on the part.

A ***tolerance*** is the entire amount of variance allowed in the size of a part or object. It indicates how much larger or smaller a part of a product can be made and still be within specifications. Acceptable tolerances may be shown in the title block or on the drawing in several different ways, Figure 7-7B and Figure 7-8.

In general, when a dimension is given in inches and fractions of an inch, *unless otherwise indicated on the drawing,* permissible tolerances can often be assumed to be $\pm 1/64''$. The symbol ± means the part can be made *plus* (larger) or *minus* (smaller) on that dimension by $1/64''$ and still be acceptable.

When the tolerance is plus (+) *and* minus (–) and the variance is in both directions, it is called a ***bilateral tolerance.*** For example, if it were permissible to make the dimension larger *and* smaller, the dimension would read $2\frac{1}{2}^{\pm 1/64}$ so the part could be as large as $2\frac{33}{64}$ or as small as $2\frac{31}{64}$.

However, if only a plus tolerance is permitted, the dimension would read $2\frac{1}{2}^{+1/64}$, or if only a minus tolerance is permitted, the dimension would read $2\frac{1}{2}^{-1/64}$. When the tolerance permitted is in one direction only, either plus (+) *or* minus (–), it is called a ***unilateral tolerance.***

Dimensions shown as inches and decimals usually indicate that the work must be done more accurately.

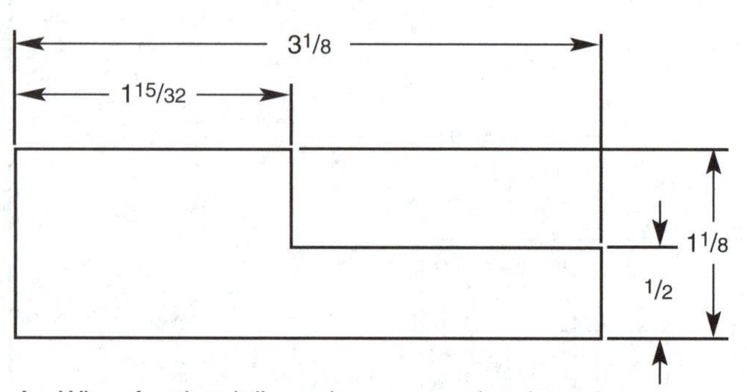

A—When fractional dimensions are used and no tolerances are shown with dimensions, tolerances are usually $\pm\frac{1}{64}$ inch.

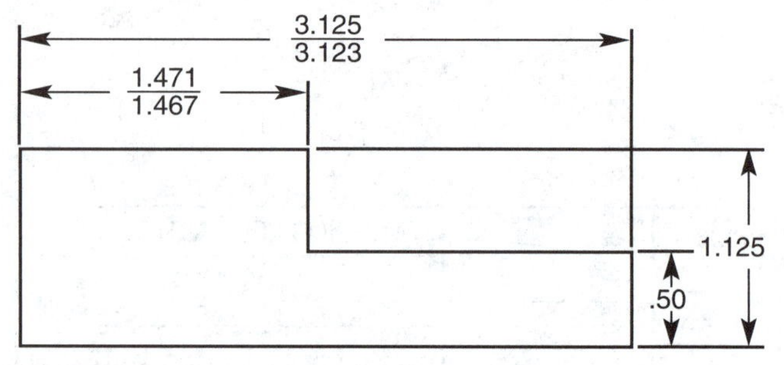

B—Decimal dimensions indicate tolerances of ± 0.010 inch unless otherwise specified.

Figure 7-8.
Two ways of indicating tolerances on a drawing are shown.

Two methods of showing these tolerances are generally used on modern prints.

A *plus* tolerance would be shown as $2.500^{+.010}$ or $^{2.510}/_{2.500}$. A *minus* tolerance would be shown as $2.500^{-.010}$ or $^{2.500}/_{2.490}$.

The dimensions indicate the part can be used as long as that dimension measures within the indicated limits.

Quantity of Units

The print should also show the number of units needed in each assembly, Figure 7-7C. A ***work order*** (included with each job received by the shop) indicates the total number of units to be manufactured. This aids in ordering sufficient material and helps determine the most economical method for manufacturing the part.

A ***routing sheet*** (or ***shop traveler***) may accompany the work order. The routing sheet or shop traveler is used to list the specific processes required to make the part, subassembly, or assembly. The list is in order of operation and may include estimated or standard production times, machine or process numbers, and inspection requirements, Figure 7-9.

Scale of Drawing

When views on the drawing are made other than actual size, they are called ***scale drawings.*** Scale drawings are necessary when the part drawn is too large or too small and it would be impractical to draw the necessary views full size.

Views on a print might be:

- One-half size
- Twice the actual size
- Actual size

A print with views drawn one-half size is shown by the numbers 1:2 in the title block. The numbers 2:1 indicate the views on the drawing are twice the actual size of the part. Drawings made actual size are indicated by the numbers 1:1 or the words *full* or *full size,* Figure 7-7D.

Caution! *Never* measure or scale the views on a print to get a needed dimension. Many prints are reproduced by enlarging or reducing the original; therefore, the print scale changes.

Finish

General ***finish*** requirements (sand blasting, plating, painting, etc.) may be specified in a section of the title block on some drawings. However, full finish specifications are usually listed elsewhere on the drawing, along with the standards the finish must meet, Figure 7-7E.

WRP Industries
Production Routing Sheet

ORDER	# 523463.0	DRAWING	72898-0	These dates are for reference only. On-line system dates supersede.	
PRODUCT#	TAP42100	EC# 2.0/0.0			
DESCRIPTION	PLATE				
QTY-ORDR	17			SCHD-DUE	08-6-XX
COMMENTS				SCHED-STR	08-4-XX
FACTORY	PEORIA FAB	PLANNER	JRP	PRNT-DTE	07-31-XX

OPERATION # DESCRIPTION WORK CNTR

PULL STOCK

1.00 60" × 120" 11GA HPO 247 .456 EA

SHEAR

10.00
INSPECT 1ST PC OPER1 ________ DATE ________
INSPECT LAST PC OPER1 ________ DATE ________

TRIM AND SHEAR BLANK

SHEAR BLANKS SO GRAIN DIRECTION IS IN THE DIRECTION INDICATED ON PRINT 72898-0

PRESS BREAK

20.00
INSPECT 1ST PC OPER1 ________ DATE ________
INSPECT LAST PC OPER1 ________ DATE ________

BREAK 139° AT INDICATED PRINT 72898-0

PRESS BREAK

30.00
INSPECT 1ST PC OPER1 ________ DATE ________
INSPECT LAST PC OPER1 ________ DATE ________

BREAK 120° AT INDICATED PRINT 72898-0

WELD

40.00
INSPECT 1ST PC OPER1 ________ DATE ________
INSPECT LAST PC OPER1 ________ DATE ________

WELD PER PRINT - WRP ENGINEERING STANDARD WRP61384 & WRP4487

Figure 7-9.
A routing sheet is used to list the specific processes required and the order of operation for a part, subassembly, or assembly.

Revisions

The ***revision block*** indicates what changes have been made to the original drawing, Figure 7-7F. The quality assurance system used by your company may require the revisions number on the print be matched to the work order and routing sheet. Always check to make sure the print is the latest revision.

Next Assembly

Next assembly information is necessary to provide the next step in the manufacturing and assembly operations. The term ***application*** is sometimes used in place of next assembly, Figure 7-7G.

Drawing Title

The ***drawing title*** tells the welder the correct name of the part, Figure 7-7H. A section of the title block gives this information.

Print Number

For your convenience in filing and locating drawings, industry provides each master tracing with a ***print number*** (or ***identifying number***), Figure 7-7I.

Each sheet in a series of drawings is also numbered to indicate the consecutive order and the total number of sheets in the series, Figure 7-7J.

The routing sheet generally indicates the print number. To assure the correct print is sent with the routing sheet, always check the print number and revision number on the routing sheet before welding.

Heat Treatment

Heat treatment includes a number of processes involving the controlled heating and cooling of a metal or alloy to obtain desirable changes in its physical characteristics. Heat treatment can improve the metal's toughness, hardness, and resistance to shock. Additionally, heat treatment can also be used to ***anneal*** (soften) metals to make them easier to work.

One heat treatment process that may be specified for metals that have been welded or machined is stress relief. The heat treatment requirements are shown in a section of the title block, Figure 7-7K. A *see note* direction is placed in the block if the heat treatment must conform to specific standards but specifications are placed elsewhere on the print, Figure 7-10.

If no heat treatment is required, the word *none* or a *diagonal line* is sometimes entered in the block.

Notes

Information not included in the title block or dimensioned views, but pertinent to the manufacture or assembly of the part, is included under *notes,* Figure 7-7L.

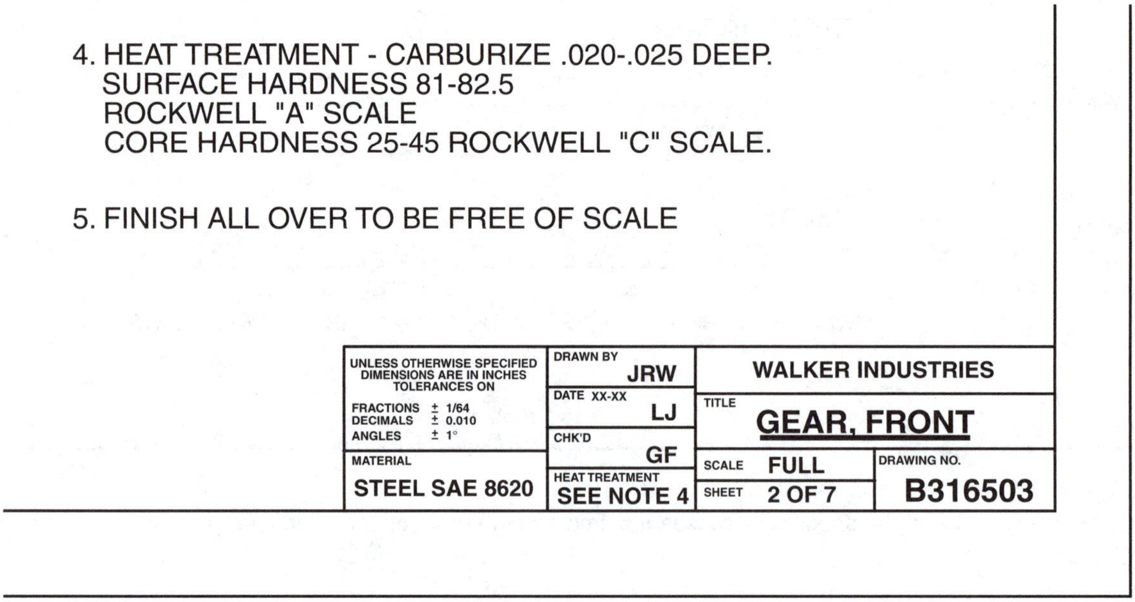

4. HEAT TREATMENT - CARBURIZE .020-.025 DEEP.
   SURFACE HARDNESS 81-82.5
   ROCKWELL "A" SCALE
   CORE HARDNESS 25-45 ROCKWELL "C" SCALE.

5. FINISH ALL OVER TO BE FREE OF SCALE

UNLESS OTHERWISE SPECIFIED DIMENSIONS ARE IN INCHES TOLERANCES ON FRACTIONS ± 1/64 DECIMALS ± 0.010 ANGLES ± 1°	DRAWN BY JRW DATE XX-XX LJ CHK'D GF	WALKER INDUSTRIES TITLE GEAR, FRONT	
MATERIAL STEEL SAE 8620	HEAT TREATMENT SEE NOTE 4	SCALE FULL SHEET 2 OF 7	DRAWING NO. B316503

Figure 7-10.
Heat treatment information is often found in the title block. If heat treatment specifications are given elsewhere on the drawing, *see note* directions may appear in the block.

Print Sheet Size

The bulk of the drawings used by industry are put on standard size drawing sheets. This makes them easier to file and identify. A *title block* often shows sheet size. A prefix *letter* added to the print identification number may also be used to indicate print size.

A listing of standard sheet sizes is shown in Figure 7-11.

Zoning

Zoning is a technique used to aid in locating details on larger size drawings. The zones are indicated outside the border as letters and numbers. Figure 7-12 illustrates how zoning is used on a print.

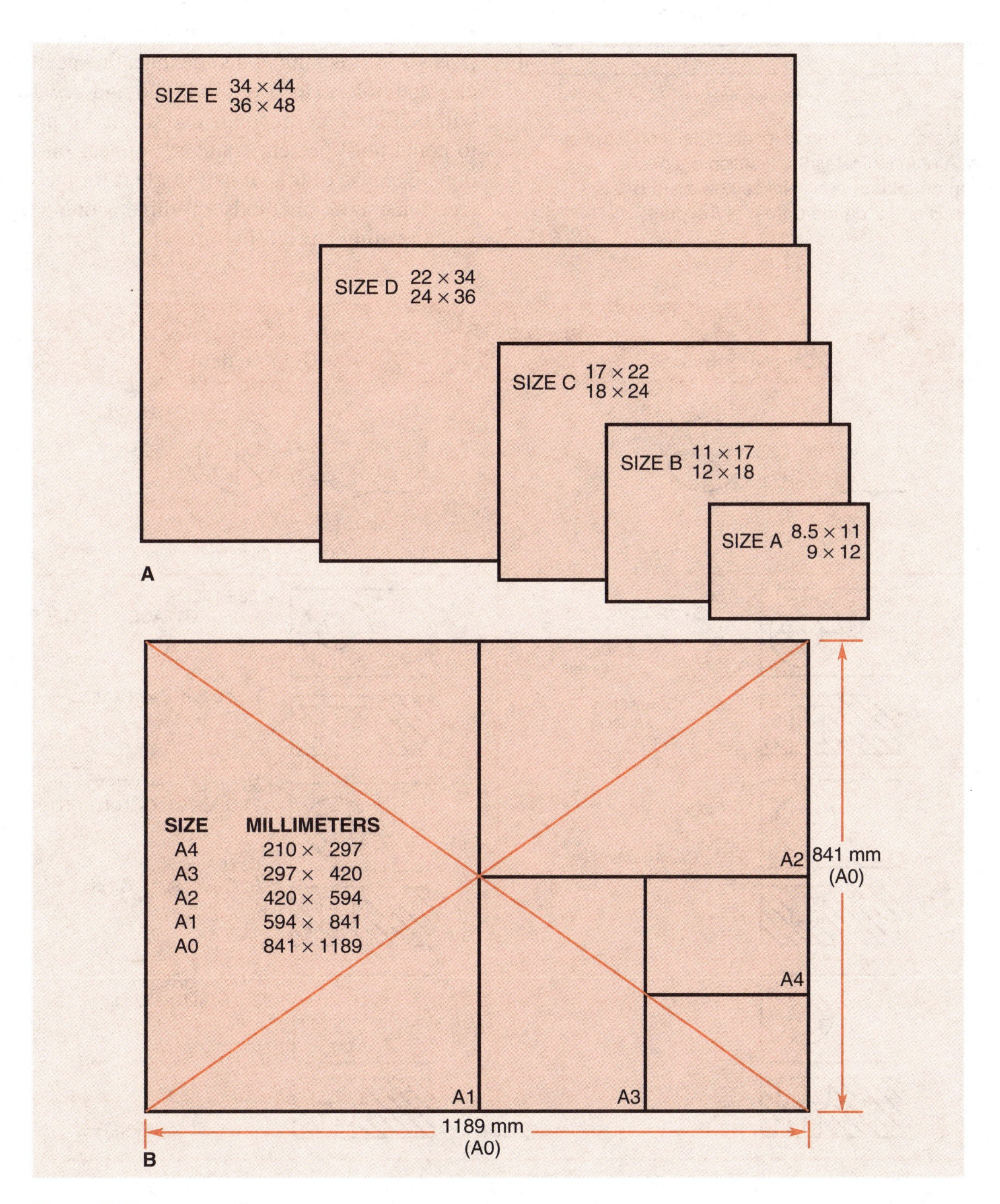

Figure 7-11.
Standard sheet sizes are listed. A—Standard inch print sizes show sheet sizes given as "A", "B", etc. B—Standard metric print sizes show sheet sizes given as A1, A2, etc.

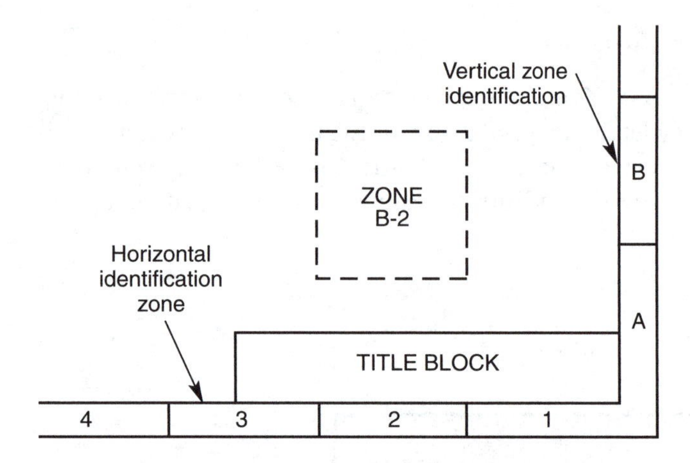

Figure 7-12.
Zoning is a technique used to locate details on large size drawings. A note indicates the location of specific information on a large print. Notice how zone B-2 is adjacent to B and 2 on the border of the print.

Security Classification

Some drawings have a ***security classification*** of top secret, secret, confidential, or restricted. The type of classification is noted on the top of the sheet and below the title block. If your work involves using classified drawings, you will be instructed how to safeguard the security of these prints.

Dimensioning Symbols

The American National Standards Institute (ANSI) has recommended changes in specifying circles and holes. However, both old and new standards will be found on drawings and prints for many years to come until present standards appear on *all* drawings. Because of this, it is important for the welder to recognize both methods of dimensioning symbols when reading a print, Figure 7-13.

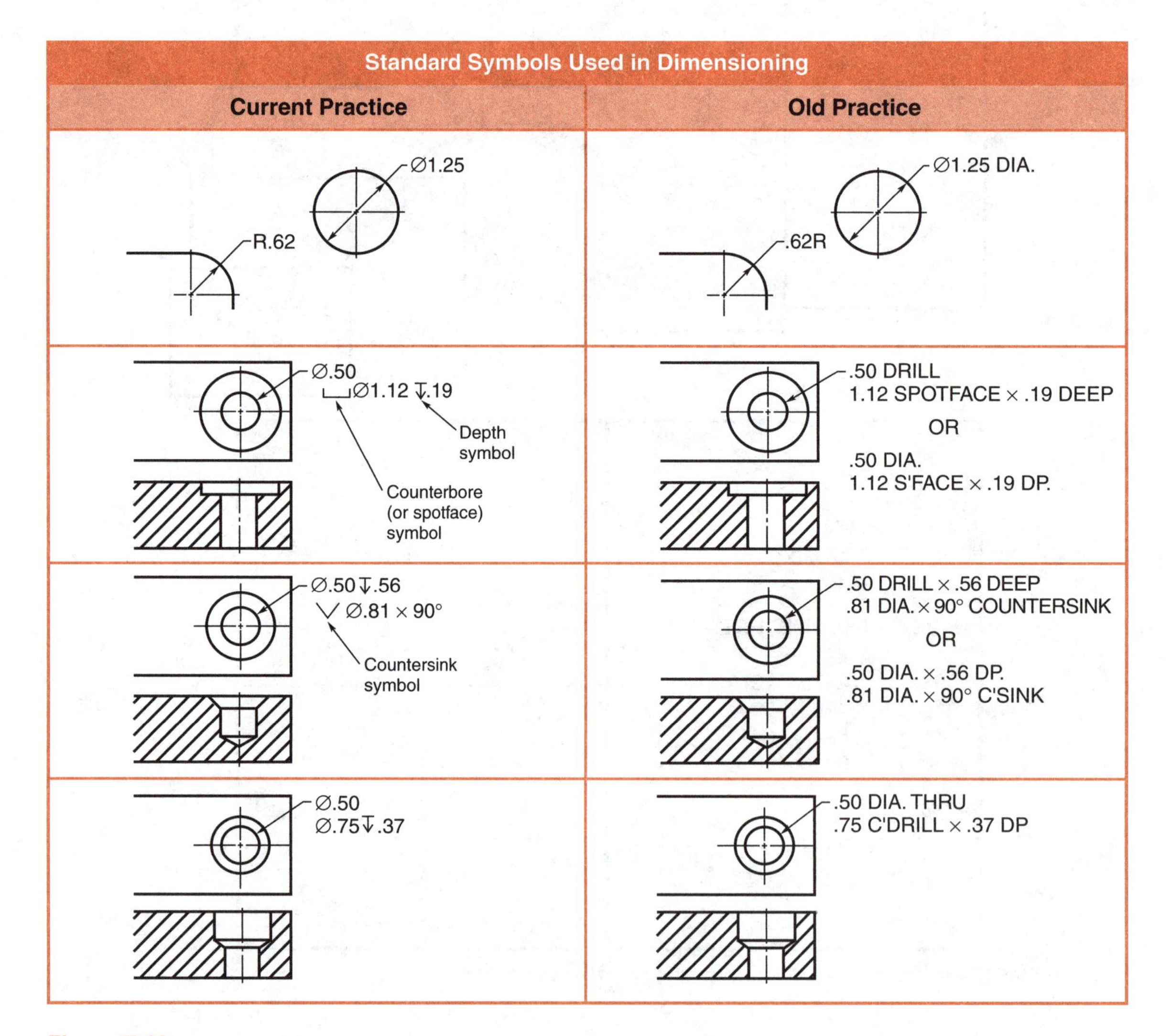

Figure 7-13.
It will take many more years before the most recently recommended standard symbols for dimensioning appear on *all* prints and drawings. It is advisable for a welder to be familiar with both the old and new methods when interpreting a drawing.

Name ______________________ Date ____________ Class ____________

Print Reading Activities

Part I

Use the drawing shown in Figure 7-14 to answer the following questions.

1. What is the name of the object shown on the drawing?
   ____
2. From what material is the object made?
   ____
3. What is the next assembly?
   ____
4. What is the total number of drawings used in the manufacture of this product?
   ____
5. This drawing is number _____.
6. How is this drawing dimensioned?
   ____
7. The views of the object are drawn to what size?
   ____
8. What is the drawing print number?
   ____
9. What revisions have been made to the drawing?
   ____
10. What tolerance is allowed on most dimensions?
    ____
11. What heat treatment is specified?
    ____
12. What finish technique is indicated?
    ____

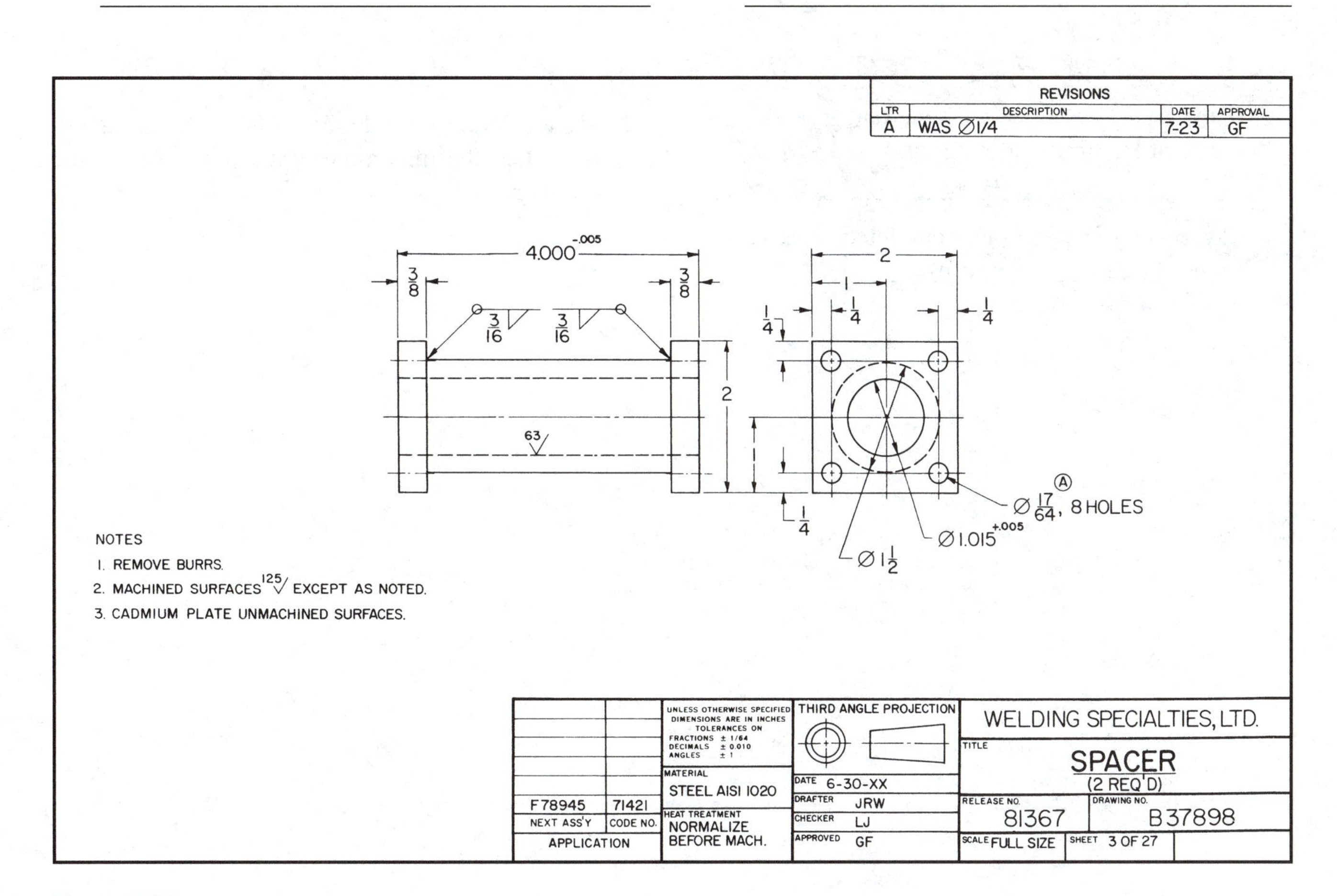

Figure 7-14.
Use this drawing (B37898) to answer the Print Reading Activities questions.

Part II

Use the tolerance table that follows to answer the questions in Part II.

Fillet Weld Tolerance	
Specified Fillet Weld Size in Inches	**Fillet Weld Size Tolerance**
.12 – .25″	– .00 and + .06
Over .25 – .50″	– .03 and + .09
Over .50 – .75″	– .06 and + .12
Over .75 – 1.00″	– .09 and + .19
Over 1.00 – 2.00″	– .12 and + .25

1. What is the maximum leg size for a weld with a print dimension of .625?

____

2. What is the minimum leg size for a weld with a print dimension of .187?

____

3. What is the maximum leg size for a weld with a print dimension of 1.00?

____

4. What is the minimum leg size of a weld with a print dimension of 1.50?

____

5. What is the maximum leg size for a weld with a print dimension of .38?

____

Review Questions

1. If the print scale indicates one-half size, what is the print length of a dimension listed as 12.00″?

____

2. The order of operations completed on a part is found on a(n) _____ or a(n) _____.

3. If the tolerance is stated as ±0.05″ for a dimension, then the maximum length for a dimension given as 2.12″ is _____.

Unit 8

Basic Plane Geometry

After completing Unit 8, you will be able to:

- Identify features of a circle and calculate the circumference and area.
- Identify common types of angles and calculate supplementary and complementary angles.
- Identify features of right triangles, calculate area, and use the Pythagorean Theorem.
- Use angles, triangles, and circles to calculate missing dimensions.
- Use fundamental geometry concepts to solve layout and fabrication problems.

Key Words

acute angle
angle
arc
area
auxiliary line
chord
circle
circumference
complementary angles
diameter
formula
geometry
hypotenuse
legs
obtuse angle
plane figure
Pythagorean Theorem
radius
right angle
right triangle
square root
straight angle
supplementary angles
theorem
triangles

Geometry is the branch of mathematics that deals with points, lines, angles, planes, and shapes. Geometric figures form the basic shapes for parts you will lay out and fabricate. When you lay out a part from a print, a fundamental knowledge of geometry is helpful. Figure 8-1 shows a large welded fabrication with many geometric shapes and angles.

Plane Figures

A ***plane figure*** is a flat figure with no depth. It is composed of straight or curved lines. Circles, triangles, and quadrilaterals are examples of plane figures.

Figure 8-1.
Knowledge of plane geometry is helpful when welding fabrications containing various shapes and angles. (Kress Corporation)

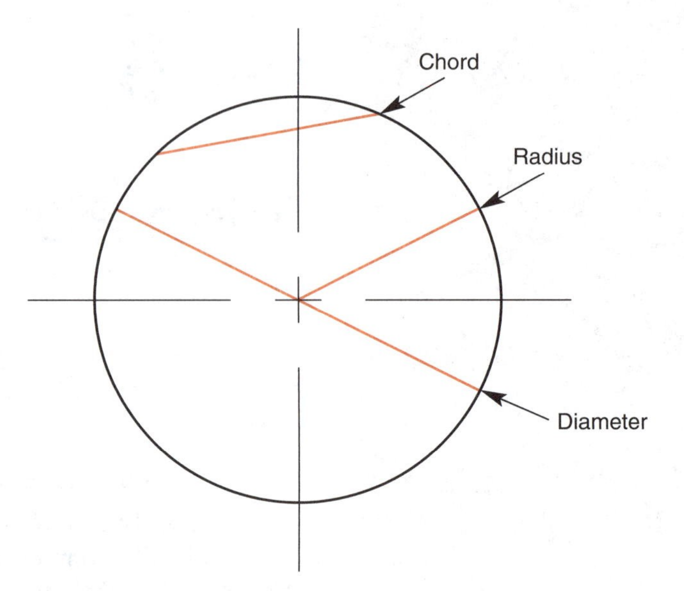

Figure 8-2.
Recognizing a chord, radius, and diameter of a circle helps a welder understand dimensions given on a print.

Circles

A ***circle*** is a geometric figure where all points on the circle are the same distance from the center. The following discussion describes some simple terms that apply to a circle. Figure 8-2 shows an example of the terms.

The ***diameter*** is a line segment through the center of a circle having its endpoints on the circle. The ***radius*** is the distance from the center of a circle to any point on the circle (so its distance is one-half the diameter). A ***chord*** is a line having endpoints on the circle, but the line does not pass through the center of the circle. Often dimensions are given on a print as a diameter or radius rather than as a linear distance, Figure 8-3.

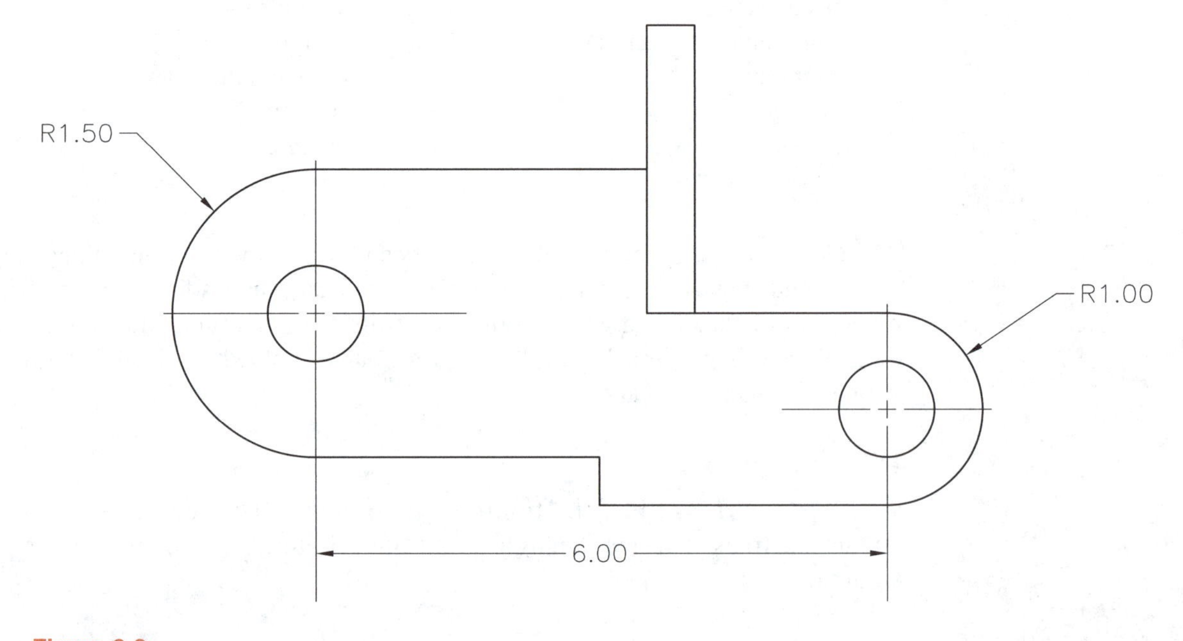

Figure 8-3.
Interpreting dimensions given on a print often requires recognizing diameter or radius information rather than a linear distance.

Circumference

The distance around a circle is called the ***circumference.*** It measures 360°. The distance around a circle can be calculated using the circumference formula. A ***formula*** is a mathematical equation containing a fact, rule, or principle. When the diameter is known, use the simplified formula:

$C = \pi d$
where
C = circumference
π (pi) = 3.14
d = diameter

Look at Figure 8-4 for an example. To calculate the distance around the boss, multiply the diameter of the boss by π. The weld distance is 1.0″ (diameter) times π (pi) or 3.14:

$C = \pi d$
so
$3.14 = 3.14 \times 1.0$

Because the diameter (d) equals twice the radius (2r), another formula for circumference is:

$C = 2\pi r$
where
C = circumference
2 = constant
$\pi = 3.14$
r = radius

Circumference (C), radius (r), and diameter (d) are all measured in the same units.

Arcs

Figure 8-5 shows a round part with multiple welds. Estimating the amount of electrode needed to complete the weld requires determining the circumference.

An ***arc*** is a portion of the circumference. If only a portion of the distance around the boss requires welding, the distance is usually given in degrees. Figure 8-6 shows an example of welding required around only part of the boss.

The arc length is calculated by first determining what part of the 360° circle consists of the 70° arc. To determine this, divide 70° by 360°. The decimal that results is then multiplied by the circumference.

For example:

Arc length decimal = Length of arc ÷ 360°
0.19 = 70° ÷ 360°
Arc length = Arc decimal × Circumference
0.60″ = 0.19 × 3.14

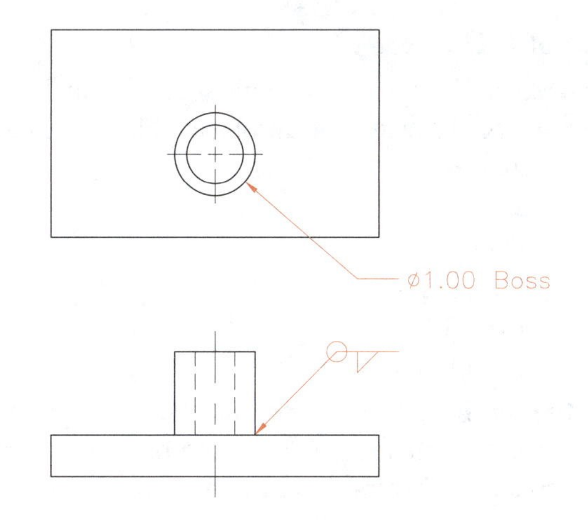

Figure 8-4.
The formula for finding circumference may be used for determining distances.

Figure 8-5.
Welds can be completed once the circumference of rounded parts is determined. (Kress Corporation)

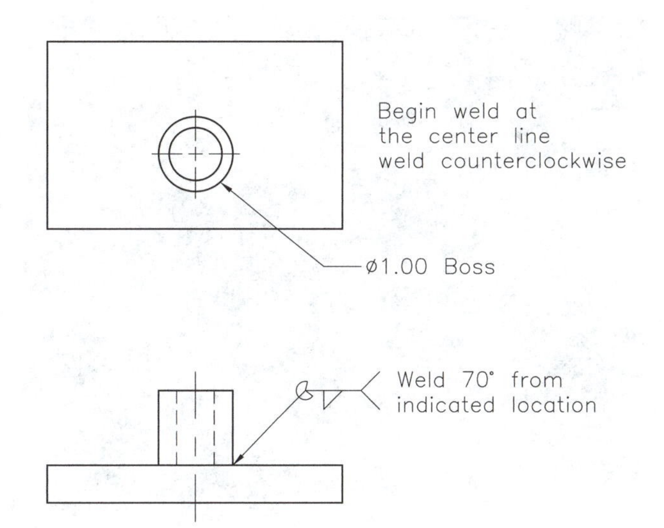

Figure 8-6.
Sometimes only a portion of the distance around the boss requires welding.

Area of a Circle

Area is the number of unit squares equal to the entire surface measure of an object. The formula for circular area is:

$A = \pi r^2$
where
A = area
$\pi = 3.14$
r^2 = radius squared

Remember! The order of mathematics operations is important for this calculation. You need to complete the exponents (powers) first, followed by the multiplication and division, and then addition and subtraction.

For an example, if the circular area of a surface welded section on a wheel loader bucket is needed to determine the amount of required welding electrode, then first the radius of the section needs to be determined. Figure 8-7 shows a wheel loader bucket with a circular section marked. If the diameter of the area is 10″ then the radius is 5″.

Calculate the area by squaring the radius (r^2) and then multiplying the square of the radius by pi (π). The area calculation is used to estimate the amount of electrode needed to surface the area or to determine the extra weight added to the bucket.

For example:

$A = \pi r^2$
$A = \pi \times (5 \times 5)$
$A = 3.14 \times 25$
$A = 78.5 \text{ in.}^2$

Angular Measurement

Did you notice that the distance around the circle might be given in degrees? Angles are also measured in degrees. An ***angle*** (∠) is the intersection of two lines or sides. Many of the parts you will lay out and fabricate are not along straight lines but are at some angle to each other. Figure 8-8 shows a large weldment with many angles.

To lay out angles or to determine missing angular dimensions on a print, you need to know a few common types of angles. Figure 8-9 shows the angular measurement for common angles. A ***straight angle*** forms a straight line and measures 180°. A ***right angle*** is formed when two lines intersect perpendicular to each other. A right angle is always 90°. An ***acute angle*** measures less than 90°. An ***obtuse angle*** measures more than 90°.

The angular measurement for a straight line is 180°. For a right angle the angular measurement is 90°. Notice that two right angles form a straight line. For a right angle, the sum of the angles that form the right angle is always 90°. Figure 8-10 shows the angular measurement of a straight line to a circle. Half of the circle measures 180°, and so does a straight line.

In Figure 8-11, angle *a* is called the complement of the 30° angle. ***Complementary angles*** are two angles formed by three lines with the sum of the angles equal to 90°. For these complementary angles, angle *a* is equal to 60° (90° – 30° = 60°).

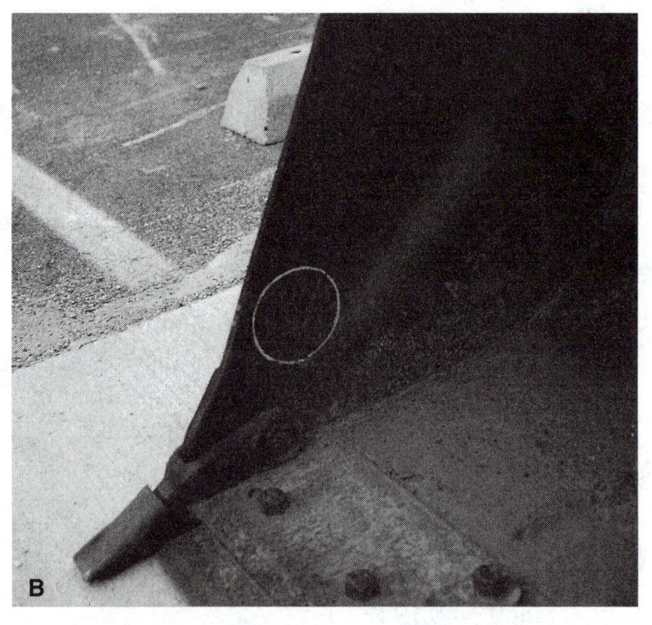

Figure 8-7.
A welder may need to estimate the area of a surface welded section. A—A section of this wheel loader bucket requires welding. B—Use the formula for area of a circle to calculate the amount of electrode needed for the circled area.

If the two angles combine to form a straight line, then the angles are called ***supplementary angles.*** The sum of the angles that form the straight line is 180°. Figure 8-12 shows an example. Angle *b* is equal to 50° (180° – 130° = 50°).

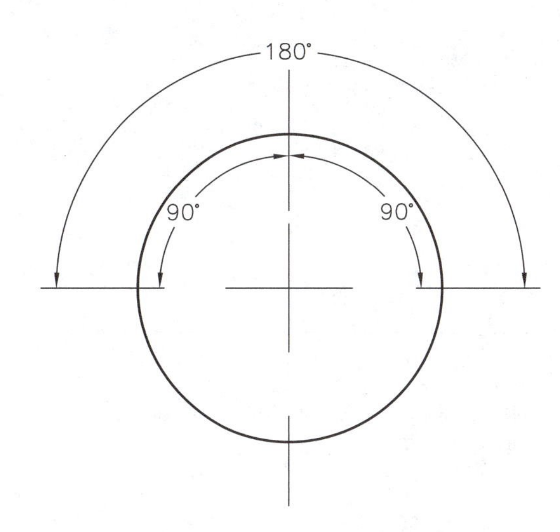

Figure 8-10.
Half of a circle and a straight line both measure 180°.

Figure 8-8.
Since weldments often consist of many angled surfaces, it is important to know how to measure angles. (Kress Corporation)

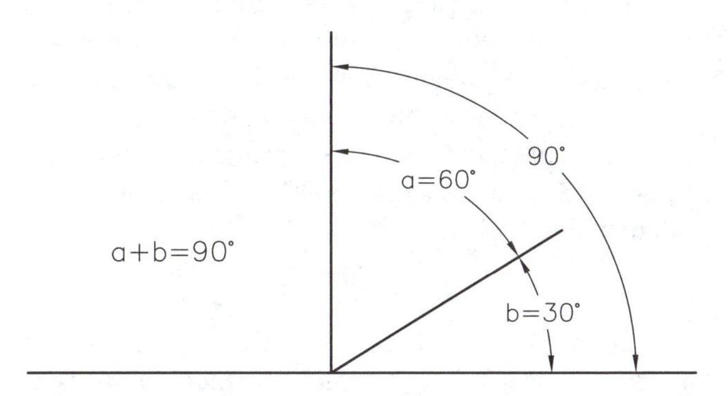

Figure 8-11.
The two angles that form complementary angles have a sum of 90°.

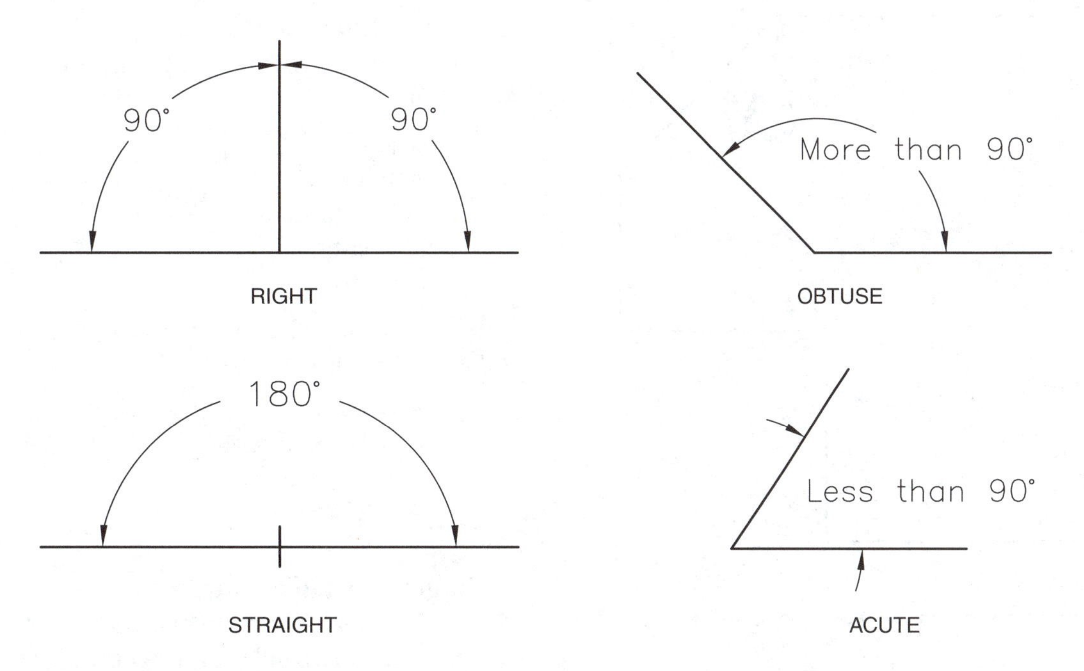

Figure 8-9.
Four types of common angles and their measurements are shown.

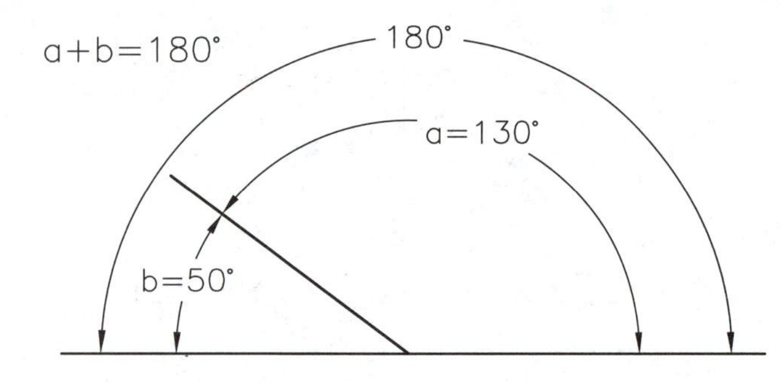

Figure 8-12.
Two angles that form a straight line and equal 180° are called supplementary angles.

Angles on Prints

Look at Figure 8-13. A detailer provided dimensions for this subassembly to the welded component and an angular measurement. However, the boss prevents you from laying out the angle using the 135° angle. You need to calculate the supplement to locate the angled part.

Remember the sum of supplementary angles is 180°. For the part in Figure 8-13, the angle needed is calculated by 180° – 135° = 45°.

Triangles

Triangles are three-sided figures with three interior angles. Triangles can be formed on prints to help calculate missing dimensions. Figure 8-14 shows a ***right triangle,*** which is a triangle with one 90° angle.

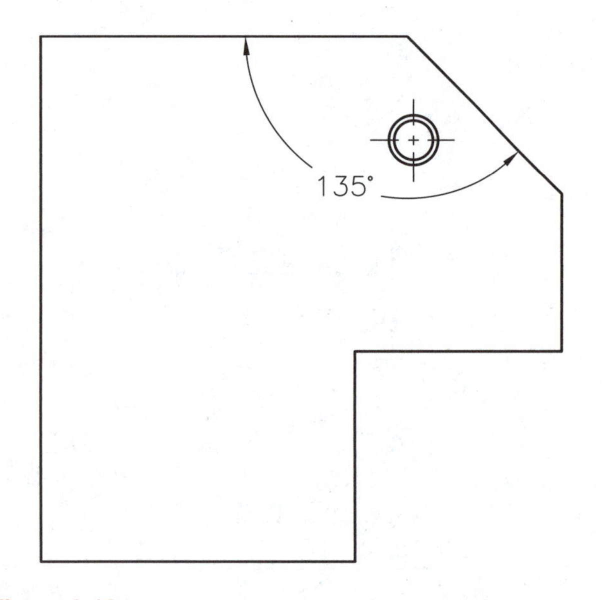

Figure 8-13.
The supplementary angle must be calculated to locate the angled part.

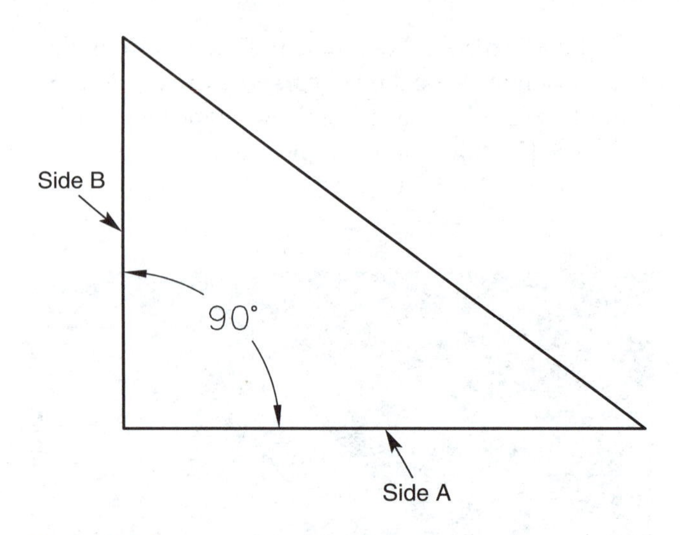

Figure 8-14.
A right triangle contains one 90° angle.

The two sides that form the right angle (side *a* and side *b*) are called ***legs.*** Fillet welds use the same term (leg) for the vertical and horizontal parts of the weld. The side opposite the right angle is called the ***hypotenuse.***

For triangles, the sum of the interior angles is always equal to 180°. For example, in Figure 8-15, the angle formed between leg *a* and hypotenuse *c* is 25°. The angle formed between leg *b* and hypotenuse *c* is calculated by first subtracting the right angle (90°) from the sum of the angles of a triangle (180°), so 180° – 90° = 90°. Then subtract 25° from 90° to determine the angle *bc,* so 90° – 25° = 65°.

Area of a Triangle

The formula for finding the area of a triangle is A = 1/2bh

where
A = area
1/2 = constant
b = base
h = height

For example, the area of a triangle with a 5″ base and a 10″ height would be:

A = 1/2bh
A = 1/2 × (5 × 10)
A = 1/2 × 50
A = 25 in^2

Pythagorean Theorem

A ***theorem*** is a mathematical truth that can be proven. If only two sides of a right triangle are given,

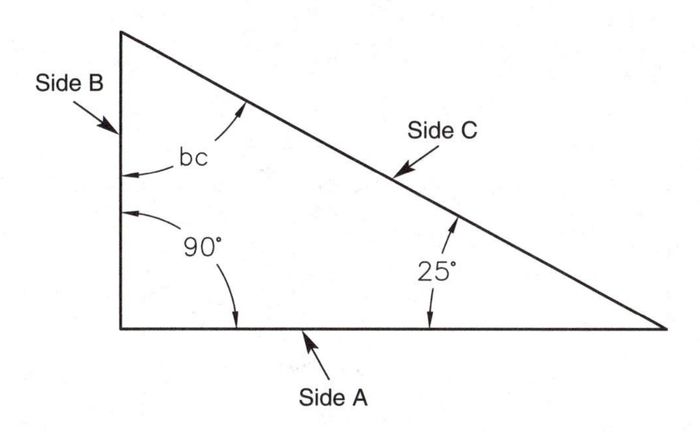

Figure 8-15.
Interior angles of a triangle are equal to 180°.

the third side can be found by using the ***Pythagorean Theorem.*** The theorem states that the square of the hypotenuse is equal to the sum of the squares of the two legs (other sides).

The equation for the Pythagorean Theorem is:

$c^2 = a^2 + b^2$
where
c^2 = length of the hypotenuse squared
a^2 = length of one side squared
b^2 = length of the other side squared
Refer to Figure 8-16.

If the length of the hypotenuse in Figure 8-16 is needed, then the length is calculated by first squaring side *a* and side *b*. To complete the calculation, the squares of side *a* and side *b* are added together and the square root is taken. The ***square root*** of a positive number x ($\sqrt{x}$) is the positive number n such that $n^2 = x$. In equation form, the theorem is:

$c = \sqrt{a^2 + b^2}$
Referring to Figure 8-16, $c^2 = a^2 + b^2$
so
$c^2 = 5.50^2 + 3.00^2$
$c^2 = 30.25 + 9.00$
$c^2 = 39.25$
or
$\sqrt{c^2} = \sqrt{39.25}$
so
$c = \sqrt{39.25}$
$c = 6.26$

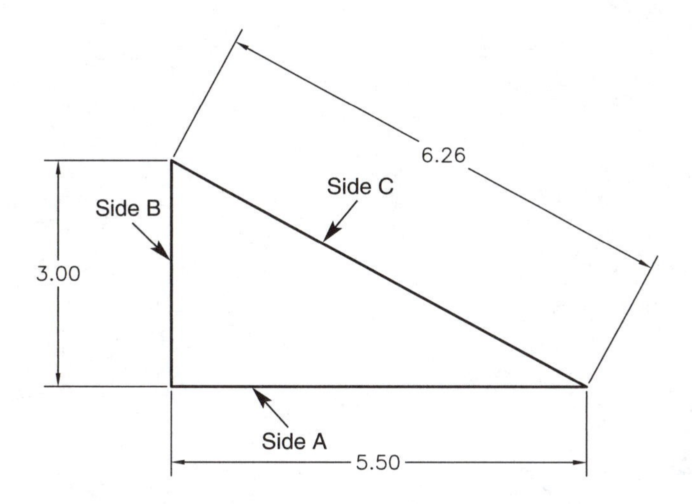

Figure 8-16.
The third side of a right triangle can be found by using the Pythagorean Theorem.

Figure 8-17 shows a gusset (triangular support) on a tee joint. The detailer has provided the length of the hypotenuse and one of the leg (side) lengths. But to fabricate the gusset, it would be easier to use the length of the two legs (sides) positioned at 90°.

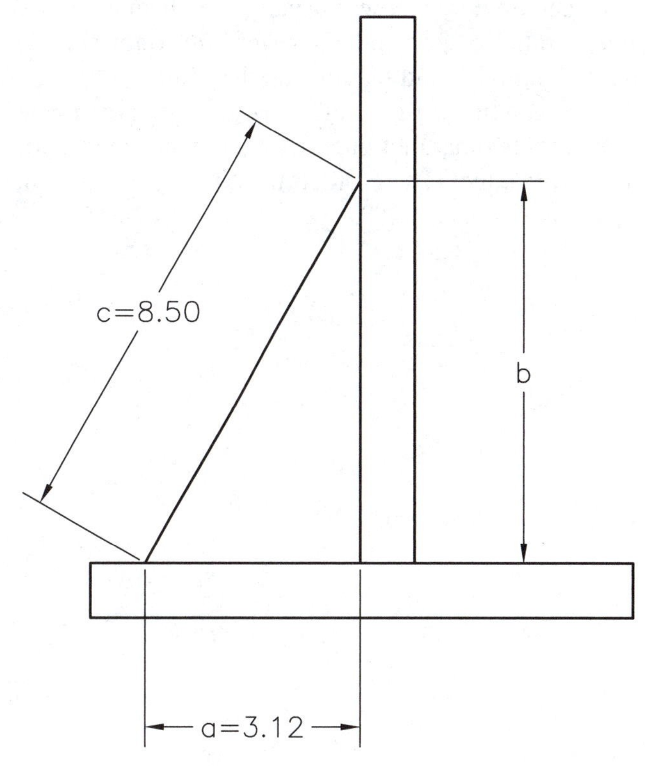

Figure 8-17.
The Pythagorean Theorem is used when fabricating the gusset shown here.

To calculate the missing dimension, the Pythagorean Theorem needs to be algebraically manipulated, beginning with the equation:

$c^2 = a^2 + b^2$
$8.50^2 = 3.12^2 + b^2$
$72.25 - 9.73 = b^2$
$62.52 = b^2$
or
$\sqrt{62.52} = \sqrt{b^2}$
$\sqrt{62.52} = b$
$7.91 = b$

Parallel Lines and Interior Angles

A useful geometry rule is shown in Figure 8-18. Notice that an auxiliary line is drawn parallel (nonintersecting but the same distance apart at any location) to the extension line containing the 20° angular dimension. An ***auxiliary line*** is an additional line used to help you visualize the relationship between lines and angles. You will usually sketch an auxiliary line parallel or perpendicular (forming right angles) to existing lines and features on the print.

A geometry rule states that if a diagonal line intersects two parallel lines, then the opposite interior angles are equal. Therefore, if the 20° angle is known, then angle *a* is also 20°.

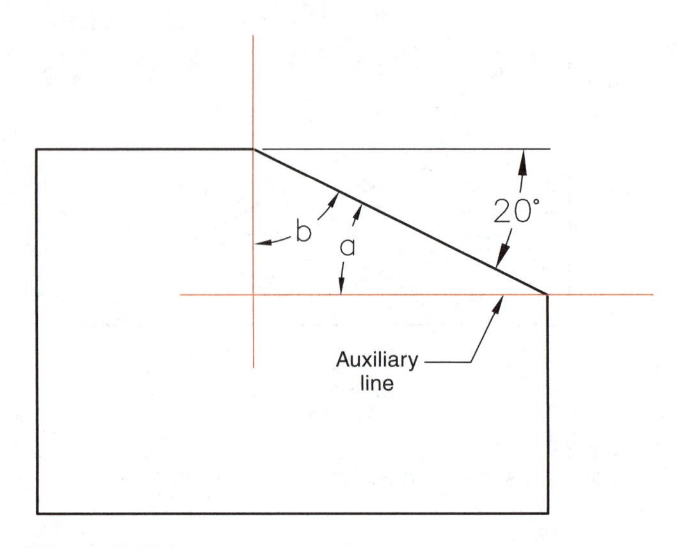

Figure 8-18.
An auxiliary line helps visualize the relationship between lines and angles.

Using the two previous concepts, angle *b* in Figure 8-18 can be calculated. Use a second auxiliary line drawn perpendicular to the first auxiliary line. Connect it to the point where the angled line begins and a right triangle is formed. The sum of the interior angles of a triangle is 180°. The angular measure of a right angle is 90°; therefore, angle *b* is equal to 70°.

Name ______________________ Date ____________ Class ____________

Review Questions

1. The term used to describe a line drawn from one side of a circle to the other through its center is _____.

   ____
2. The distance around a circle is called the _____.

   ____
3. The number of degrees in a circle is _____.
4. The formula for finding the circumference of a circle is _____.

   ____
5. Two angles having a sum of 90° when added together are called _____ angles.

   ____
6. Supplementary angles have a sum of _____.
7. The side opposite the right angle of a right triangle is called the _____.

   ____
8. The sum of the interior angles of a triangle equals _____.
9. An auxiliary line is used _____.

   ____

   ____
10. If a diagonal line intersects two parallel lines, the opposite interior angles are _____.

    ____

Print Reading Activities

Part I

Calculate the angles.

∠ A _____

∠ B _____

∠ C _____

∠ D _____

∠ E _____

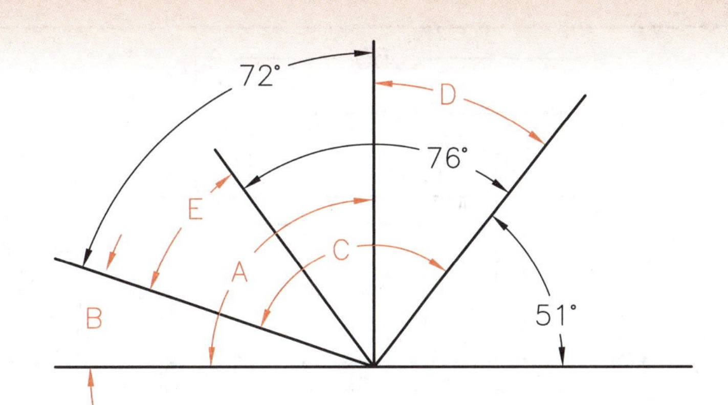

Part II

Calculate the angles.

∠ A _____

∠ B _____

∠ C _____

∠ D _____

∠ E _____

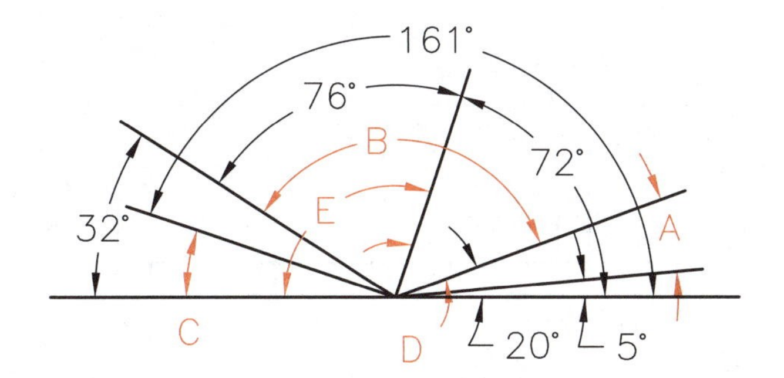

Part III

Use the print below to answer the following questions.

1. What is the maximum overall length of the part?

____

2. What is the radius of the two holes?

____

3. What is the circumference of one of the holes?

____

4. What material is used to fabricate the part?

____

5. What is the complementary angle for the 63° angle?

____

6. Calculate the total length of weld needed if the vertical plates are welded on both sides.

____

7. Calculate the approximate area of the base plate.

____

8. What is the tolerance listed for 3-place decimals?

____

9. How many individual pieces are required to fabricate the part?

____

10. Determine the dimension or angle for the following:

A _____	I _____
B _____	J _____
C _____	K _____
D _____	L _____
E _____	M _____
F _____	N _____
G _____	O _____
H _____	

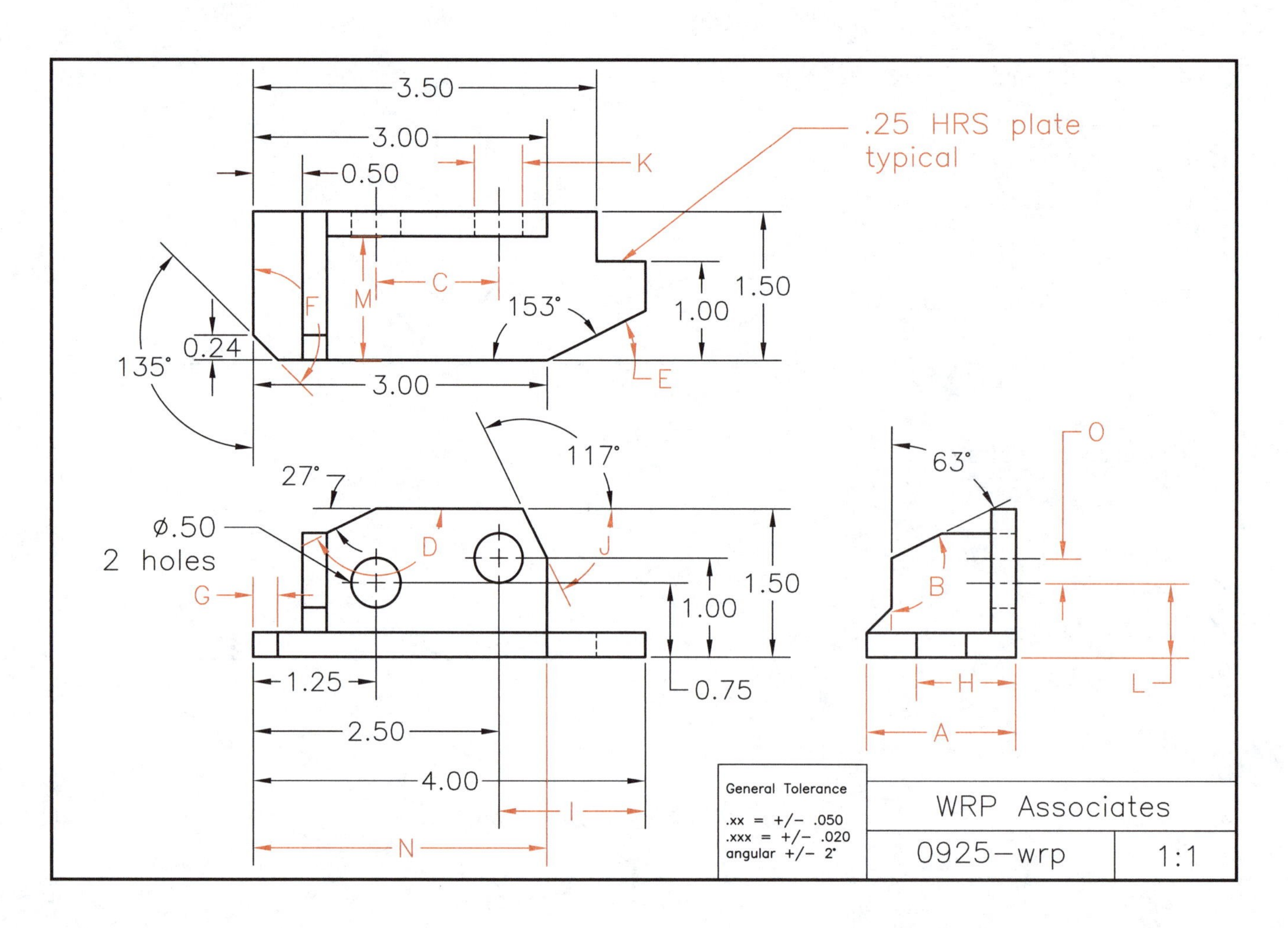

Unit 9 Dimensioning Welding Prints

After completing Unit 9, you will be able to:

- Identify and apply various dimensioning systems used on the print.
- Describe methods of showing fractional, decimal, and angular dimensions on a print.
- Calculate missing dimensions.
- Explain the purpose of linear and angular tolerance.
- Describe the use of geometric tolerance.
- Identify symbols used in geometric dimensioning and tolerancing to indicate the relationship between parts and features.
- Explain applications of geometric dimensioning and tolerancing.

Key Words

angle dimension
broken-chain dimension
calculated dimension
chain dimension method
datum
datum feature
datum identification symbol
datum (or baseline) dimensioning
datum targets
dimensions
direct dimension
explicit tolerance
extension and dimension line method
feature control frame
feature dimension
form geometric tolerances
general tolerance
geometric dimensioning and tolerancing
limits
orientation geometric tolerances
overall dimension
positional (or location) tolerances
primary datum
reference dimension
runout tolerance
unidirectional dimensioning
zero plane (or ordinate) dimensioning

In previous units, objects were described by their shape. Shape description gives an understanding of the proportions of the part but it does not give enough information to build the part. ***Dimensions*** are numerical values describing the size, location, geometric characteristics, or surface texture of a part or its features. Figure 9-1 shows a large weldment held in a fixture. Imagine the fit-up difficulty if the print dimensions were incorrectly interpreted.

The types of print and production methods generally determine which dimensions are found on the prints. The dimensions for a machining print may be different than those used for an assembly print. Likewise, the dimensions given on a welding print may be different than those given on a forming print.

Figure 9-1.
Accurate fit-ups on weldments, such as this frame for a track-type tractor, are impossible if dimensions are interpreted improperly. (Caterpillar Inc.)

Extension and Dimension Lines

Several dimensioning systems have been developed. The most common is the traditional extension-dimension line system. Figure 9-2 shows a print of a simple butt joint. The top view shows the length of the part, or the ***overall dimension***. A feature dimension is used to indicate the length of one of the plates used for the joint. ***Feature dimensions*** are used to show individual features, such as the center of a hole or the location of individual parts welded into an assembly.

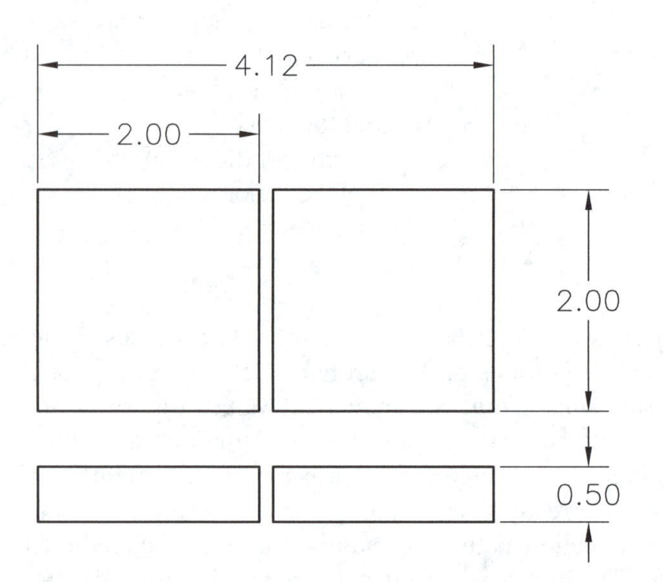

Figure 9-2.
Dimensions for this simple butt joint include an overall dimension and a feature dimension.

Notice the dimension for the length of other individual plates is *not* shown. Only the dimensions required to accurately describe the part are given on the print. Dimensions are *not* duplicated. If the overall length of the part is given in the top view, for example, it is not repeated in the front view.

Usually, the overall length dimension is shown in the front or top view, the overall width is shown in the top or side view, and the overall height is shown in the front or side view. See Figure 9-3. Sometimes, the length dimension is called *width*, the width dimension is called *depth*, and the height dimension is called *thickness*.

Dimensions can be given as linear or angular. Customary English units of measure may be used or metric dimensions may be given. For the customary English system, fractional inches and decimal inches are most common. The millimeter is the most common unit for metric dimensions.

Dimensions are usually read from the bottom of the print. This method of dimension placement is called ***unidirectional dimensioning.*** Today most CAD drawn prints use this method of dimension placement. However, older manually drawn and reproduced prints may have aligned dimensions so the horizontal dimensions are read from the bottom of the print and the vertical dimensions are read from the right side of the print.

The ***extension and dimension line method*** includes the traditional method of an overall dimension located furthest out from the view and ***broken-chain dimensions,*** Figure 9-4, or chain dimensioning as shown in Figure 9-5. The ***chain dimension method*** does not include an overall dimension, and the accumulation of tolerances for each feature often exceeds the tolerance of the overall dimension. When using the broken-chain dimensioning method, the undimensioned feature accumulates the tolerance. More information about tolerance is given later in the unit.

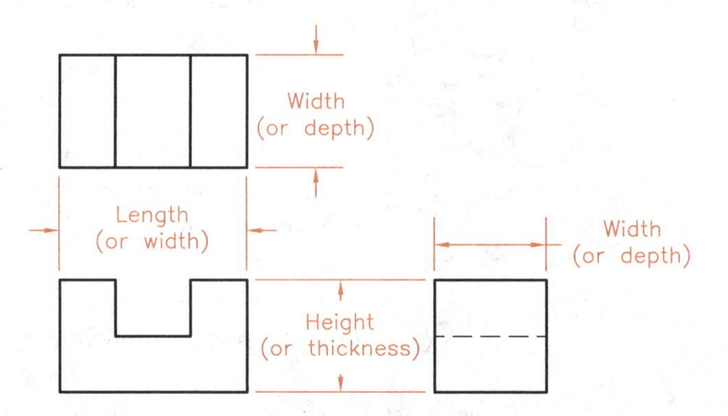

Figure 9-3.
Overall dimensions are used to describe the overall length, width, and height of objects.

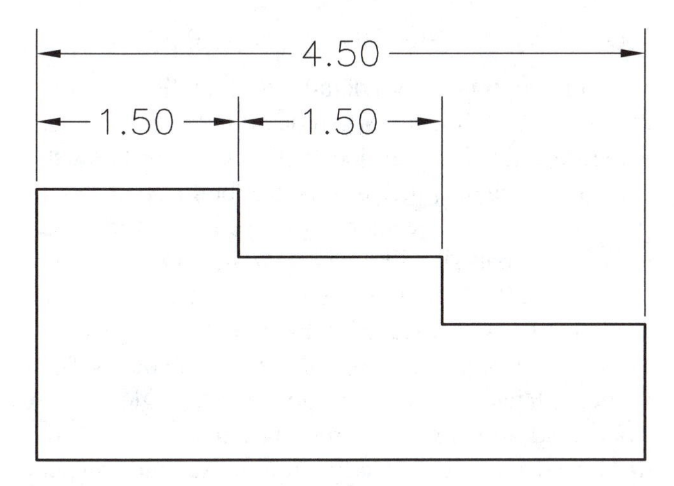

Figure 9-4.
Broken-chain dimensions appear beneath the overall dimension in this illustration.

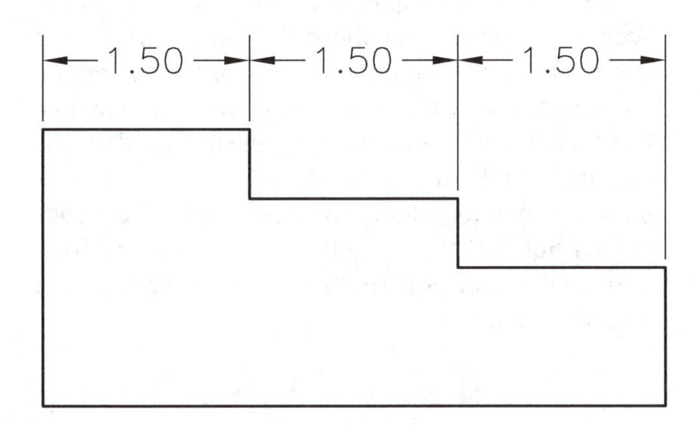

Figure 9-5.
The extension and dimension line method may only include chain dimensioning.

An ***angle dimension*** line is an arc drawn through the center of the angle at its apex (the uppermost point). Angles are dimensioned using degrees (°). For greater precision, they are dimensioned in parts of a degree. Parts of a degree may be shown as minutes and seconds, or decimal degrees. There are 60 minutes (′) to one degree and 60 seconds (″) to one minute. Angle dimension examples are shown in Figure 9-6.

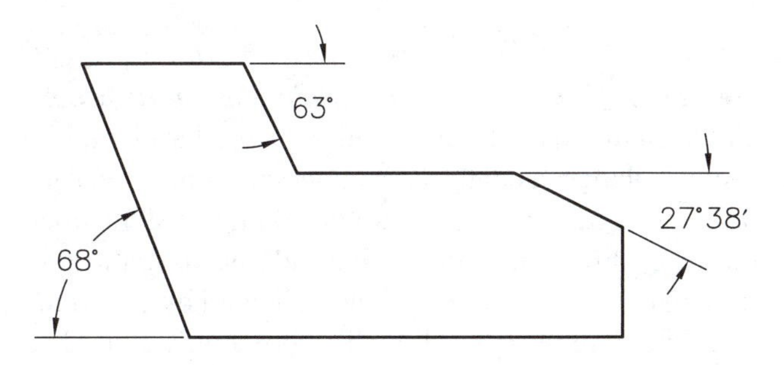

Figure 9-6.
Angle dimensions are shown in degrees. Parts of a degree are used when greater precision is desired.

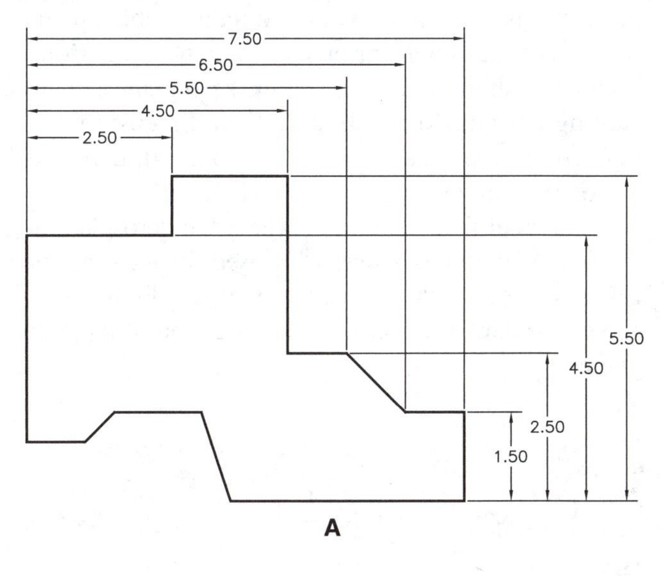

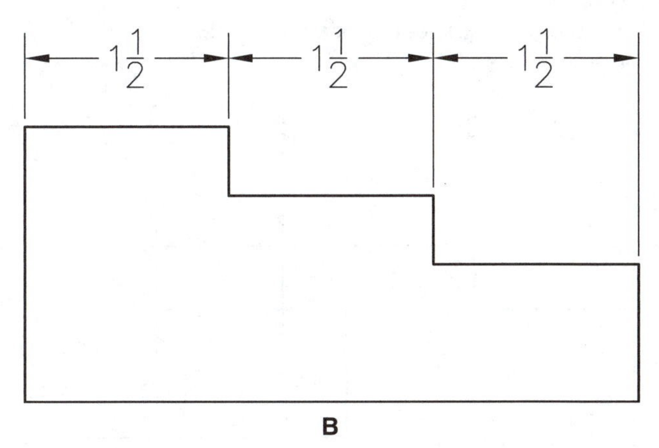

Figure 9-7.
Using the datum dimensioning system minimizes compounding errors when dimensions are given in chains.

Datum Dimensions

When a single side of a part is used to start all of the dimensions, the dimensioning method is referred to as ***datum*** (or ***baseline***) ***dimensioning,*** Figure 9-7. A *datum* or *baseline* is a starting plane or edge. Notice that when laying out a part, the datum dimensioning system minimizes the compounding of error. Refer to Figure 9-7B. If there is a + 1/16″ error in layout for each of the dimensions given in the chain, then the overall error for the part would be 3/16″.

Zero Plane Dimensions

Zero plane (or ***ordinate***) ***dimensioning*** uses a single plane, usually at the edge of the part or intersecting planes, as datums from which dimensions are derived. The zero plane dimensioning system does not use dimension lines. Instead, each feature is dimensioned from a zero plane indicated by an extension line ending with a box. The box holds a number that indicates the zero plane number, Figure 9-8.

The dimensions are shown at the end of extension lines linked to the specific feature requiring a dimension. This method minimizes the chance of introducing error into the part layout and reduces the amount of room needed on the print to contain the dimensions. Figure 9-9 shows a print using the zero plane dimensioning system. Notice that vertical dimensions are read from zero plane 01 and horizontal dimensions are read from zero plane 02.

Be careful to determine where the zero plane is located. Sometimes a feature is used for locating the zero plane rather than the edge of a part. Figure 9-10 shows the center of a hole as the location of the zero planes.

Figure 9-8.
The zero plane number is located in a box.

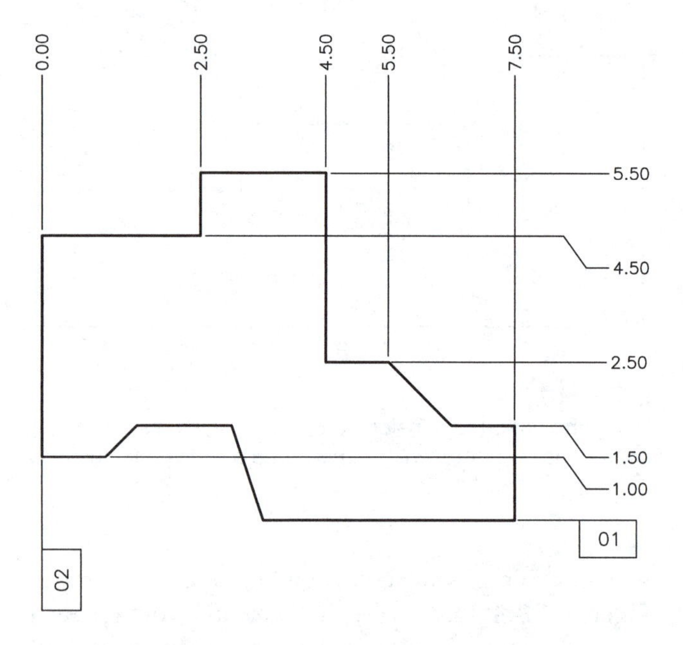

Figure 9-9.
Using the zero plane dimensioning system, the vertical dimensions are read from the zero plane 01. Horizontal dimensions are read from zero plane 02.

Direct and Calculated Dimensions

It is important for a detailer to think like a welder, machinist, or fabricator when detailing a drawing. But sometimes the designer does not consider or know the order in which the processes used to build a part should be completed. Most dimensions given to features used to build the part should be ***direct dimensions.***

Figure 9-11 shows dimensions to the edge of two vertical plates welded to a base plate. However, to minimize error in the layout of the vertical plates, both vertical plates should be laid out from a single edge. A ***calculated dimension*** is required to determine the distance from the left edge to the second vertical plate from the right. Notice the distance to the left edge of the right vertical plate is 2.00″. This type of calculation is common on many welding prints.

For another example, in Figure 9-12, the center-to-center dimension for the holes shown in the top view is missing. Assembly for this part may rely on the accuracy of the center-to-center dimension. Therefore, a center-to-center dimension needs to be calculated. In Figure 9-12, calculate the center-to-center distance by adding the center-to-end distance for both holes (1.00″ + 1.00″) and subtracting from the overall dimension (6.00″). The center-to-center dimension is 4.00″.

Linear and Angular Tolerance

Tolerance is the total permitted amount (in distance or degrees) an exact or basic dimension may vary. Because there is some variation in all manufactured

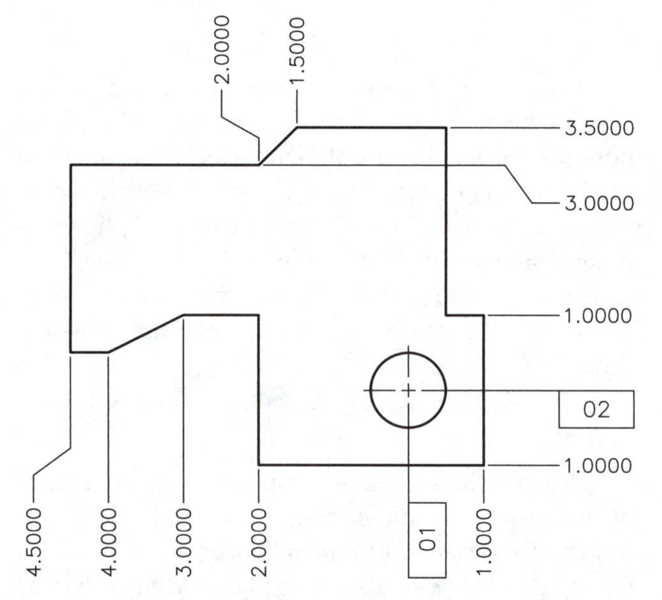

Figure 9-10.
The center of the hole, rather than the edge of a part, is the location of the zero plane in this print.

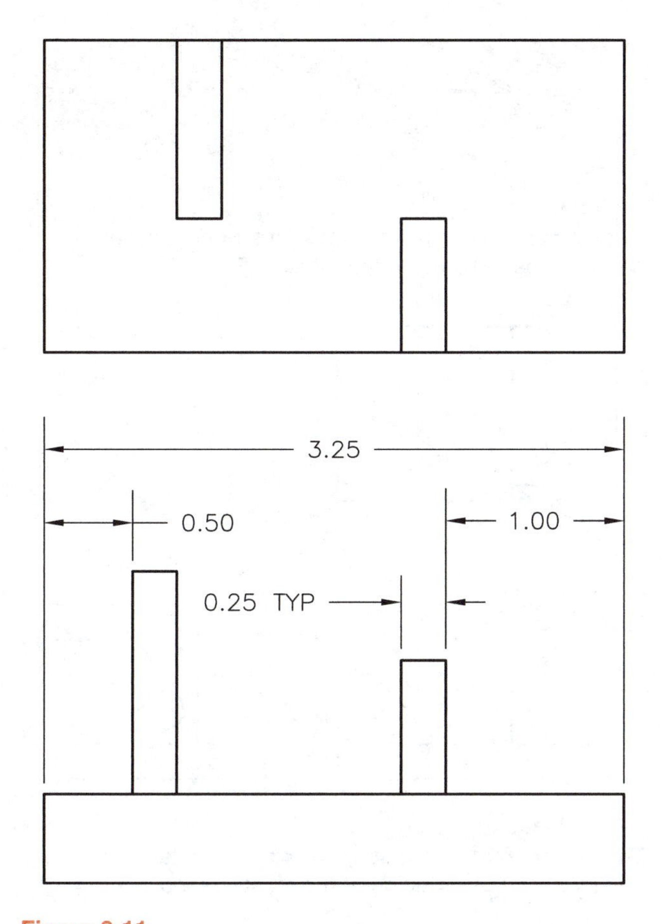

Figure 9-11.
Both vertical plates in this print should be laid out from a single edge so accurate calculations can be determined.

products as a result of the processes and measurement systems, all parts will vary somewhat from the print dimensions. A permissible variation must be determined so the function of the part will not be affected.

Total variation in a part comes from both the manufacturing processes and the measurement system. This is why measurement instruments should be checked against a standard and equipment should be maintained.

Tolerance on a print may be given as a general tolerance or as an explicit tolerance. The ***general tolerance*** is usually found in the tolerance box in the title block or in a note. Figure 9-13 shows the tolerance in the title block. Notice the number of *x*'s represents the number of decimal places to which the tolerance applies. That is, *.xx* indicates a dimension with two decimal places and *.xxx* indicates a dimension with three decimal places. Figure 9-14 shows an example of an ***explicit tolerance*** with the maximum and minimum limits listed.

Tolerance may be given as a fraction, decimal fraction, metric value, or in degrees. Tolerance may also be *unilateral* (one side) or *bilateral* (both

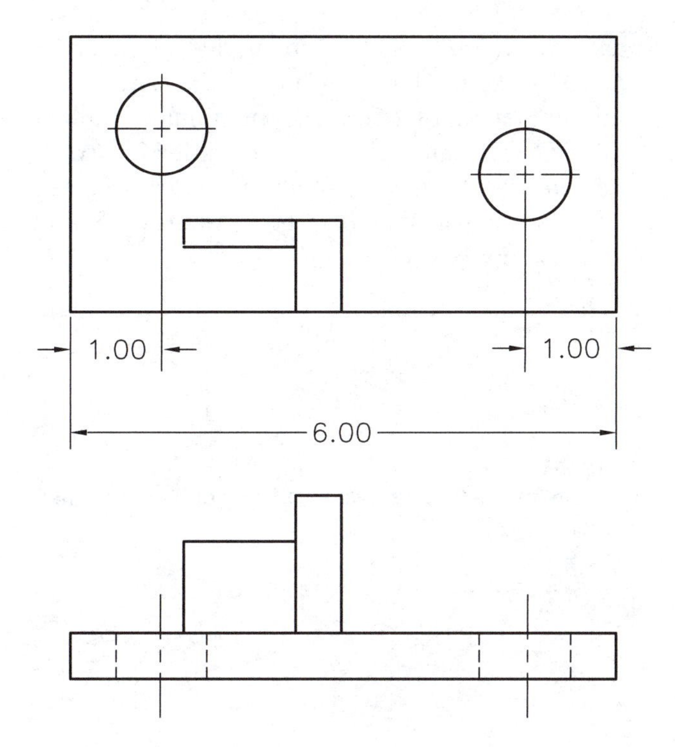

Figure 9-12.
Missing dimensions might need to be calculated by using other stated dimensions on the print.

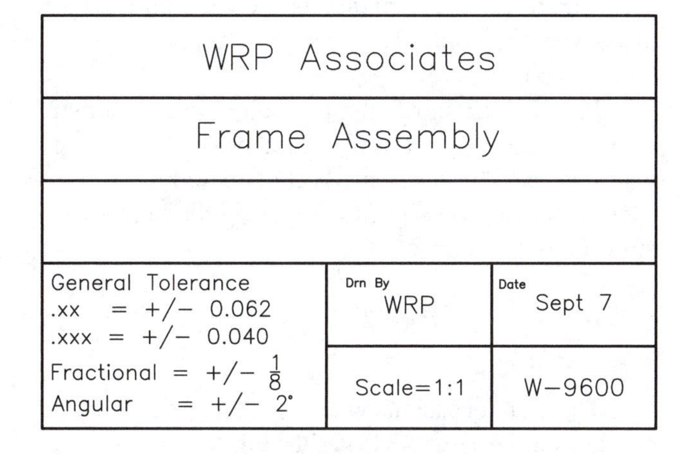

Figure 9-13.
Tolerance is listed in the title block.

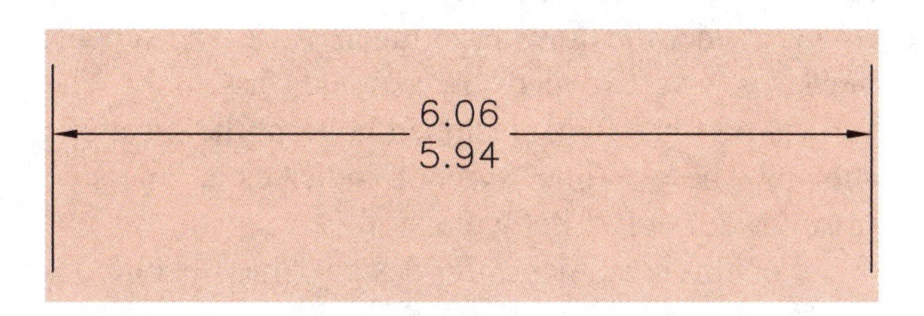

Figure 9-14.
An explicit tolerance lists maximum and minimum allowances.

sides). In Figure 9-15, a bilateral tolerance is shown. Notice the length of the part may vary to a maximum length of 2.56″ or to a minimum length of 2.44″. The maximum distances are called ***limits.***

Figure 9-16 shows a unilateral tolerance. The location of the part may only be more than the print dimension, not less.

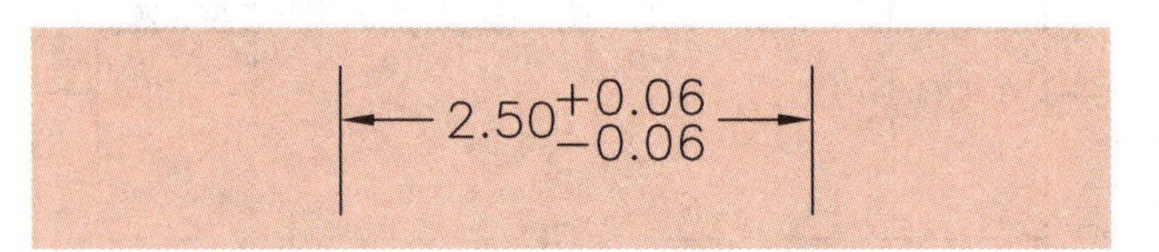

Figure 9-15.
This bilateral tolerance allows a variance of 0.06″ in either direction.

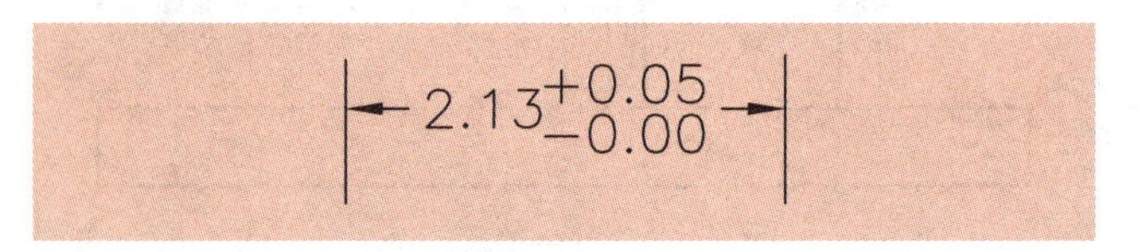

Figure 9-16.
A unilateral tolerance allows a variance in one direction only.

Reference Dimensions

A ***reference dimension*** is used for information of convenience only. No tolerances apply to a reference dimension.

Figure 9-17 shows the use of a reference dimension. Notice the reference dimension is between parentheses. The other method of including a reference dimension on a print is by using the letters REF after the dimension.

Geometric Dimensioning and Tolerancing

There are occasions when linear tolerances do not provide the best method of showing the correct relationship between parts or features on a part. A designer uses ***geometric dimensioning and tolerancing*** to specify the requirements of a part by the actual function and relationship of the features.

Consider the drawing in Figure 9-18. The support bracket shows two parts in a welded assembly. The vertical part has four machined holes at the locations shown. The welding symbol indicates a one-side (arrow side) only fillet weld.

For this weldment, the designer has specified a linear tolerance of + or – 0.062″. When welded, it is likely the fillet weld will shrink and the vertical part will warp and move. When inspected, the position of the bottom of the vertical part is within tolerance. But

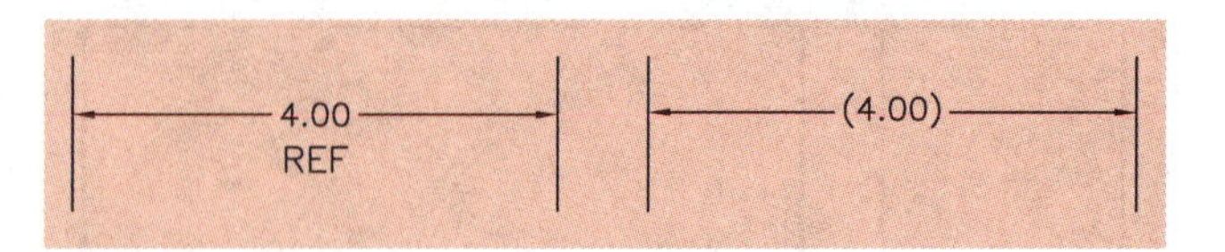

Figure 9-17.
A reference dimension is shown on a print between parentheses or by using the letters REF.

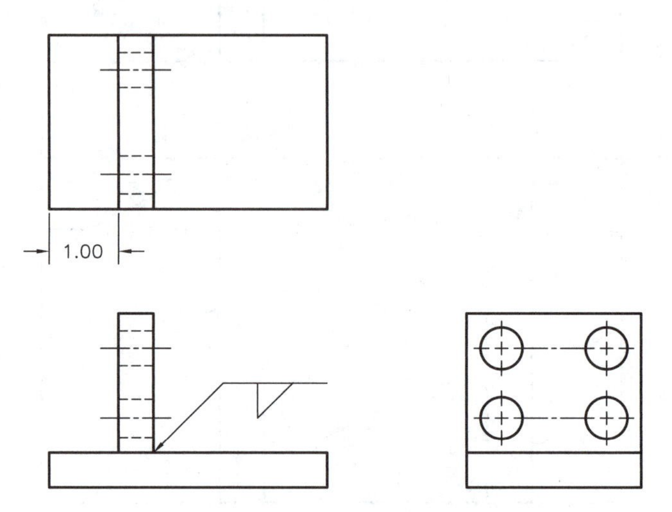

Figure 9-18.
The vertical part of this support bracket contains four holes requiring machining at the locations indicated.

look at the top of the vertical part after the weld is complete. As shown in Figure 9-19, the top of the part has moved so the holes will not be in the position required by the mating part.

The welder can measure the position of the bottom of the vertical part and verify the part is within tolerance. As the part moves to assembly, the part will be difficult or impossible to assemble. The part may require rework to remove the warp, or it might be scrapped.

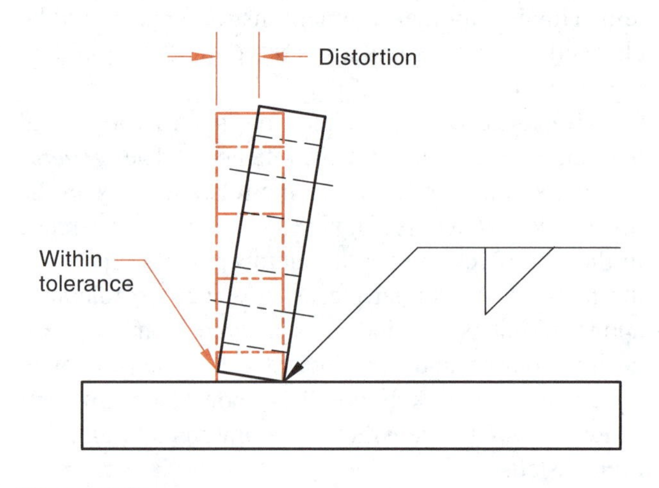

Figure 9-19.
If the vertical part warps or moves after welding, the position of the bottom might be within tolerance, but the top of the vertical part may not be within tolerance.

Definitions and Symbols

Geometric dimensioning and tolerancing (GD&T) symbols are used to indicate the relationship between parts and features. Although the concepts of GD&T are often complicated and confusing, when used in welding and fabricating, you will only need to know how to apply a few of the symbols.

Earlier in the unit, the datum was introduced for dimensioning purposes. In GD&T, the ***datum*** is defined as an exact point, axis, or plane from which the location or geometric features of a part are located. The ***datum identification symbol*** is shown in Figure 9-20. The symbol is relatively new to industry, and many prints still have the older datum symbol shown in Figure 9-21.

A *feature* is often used to describe a physical portion of the part, such as a surface, hole, or edge. Sometimes a feature is used to establish a datum. When an actual feature is used as a datum, it is referred to as a ***datum feature***.

Figure 9-22 shows the symbols used to specify tolerances in geometric dimensioning. The symbols are grouped into five types of characteristics linked to individual features or to other features. The types

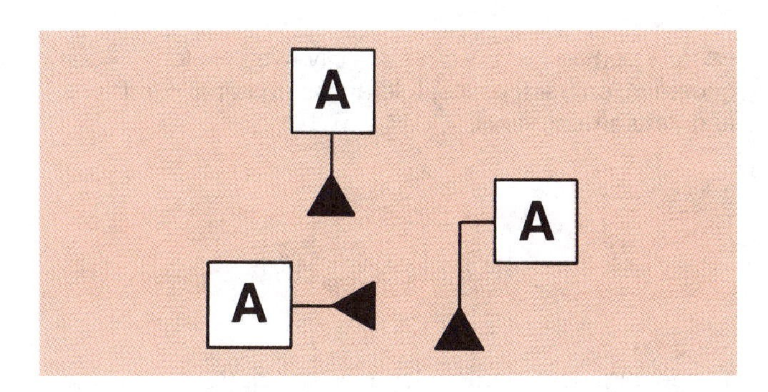

Figure 9-20.
The datum identification symbol shown is used in industry on newer drawings.

Figure 9-21.
This type of datum symbol can still be found on older drawings.

Symbol for:	ASME Y14.5
Straightness	—
Flatness	⏥
Circularity	○
Cylindricity	⌭
Profile of a line	⌒
Profile of a surface	⌓
All-around profile	◄⊖—
Angularity	∠
Perpendicularity	⊥
Parallelism	//
Position	⌖
Concentricity/coaxiality	◎
Symmetry	⌯
Circular runout	*↗
Total runout	*⌰
At maximum material condition	Ⓜ
At least material condition	Ⓛ
Regardless of feature size	NONE
Projected tolerance zone	Ⓟ
Diameter	∅
Basic dimension	[30]
Reference dimension	(30)
Datum feature	[A]◄
Datum target	Ø6/A1 (in circle)
Target point	X
Dimension origin	⌾→
Feature control frame	[⌖ | ∅0.5Ⓜ | A | B | C]
Conical taper	⌲
Slope	⌳
Counterbore/spotface	⌴
Countersink	⌵
Depth/deep	↧
Square (shape)	□
Dimension not to scale	15 (underlined)
Number of times/places	8X
Arc length	⌒105
Radius	R
Spherical radius	SR
Spherical diameter	S∅

* May be filled

Figure 9-22.
Standard symbols are used in industry to specify tolerance information in geometric dimensioning.

include form, profile, orientation, runout, and location. For welded parts, orientation and form are the most important, although some other types may be used.

Application of GD&T

The application of GD&T starts with the identification of a datum. Generally datums are identified by the datum identification symbol and lettered in order of importance starting with the letter *A*. If the part is a rectangular solid, the datum references are established as three mutually perpendicular planes, Figure 9-23. The ***primary datum*** reference is listed first, then the *secondary datum* reference, and finally, the *tertiary datum* reference.

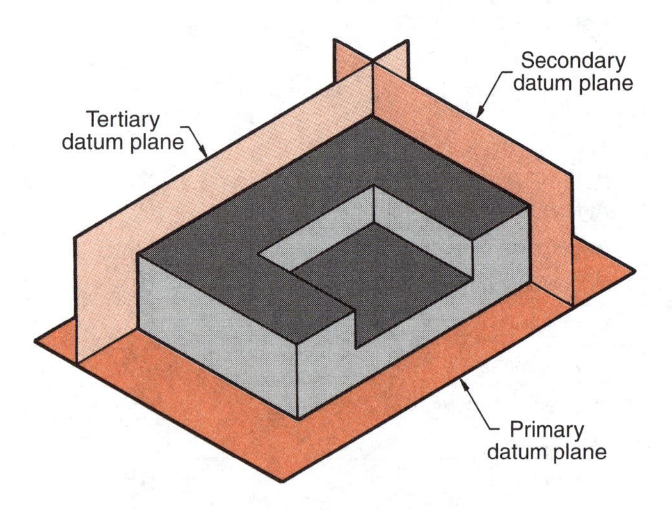

Figure 9-23.
These datum references are perpendicular planes.

Once the datum reference is established, a feature or group of features is linked to the datum by a ***feature control frame***, Figure 9-24. This frame shows the perpendicular symbol, the diameter symbol to establish the tolerance zone, the material condition symbol, and the datum reference.

For a welded part, the feature control frame might look like Figure 9-25. This feature control frame shows the perpendicular symbol, the tolerance, and the datum reference. The feature control frame is read, "The feature must be perpendicular within 0.125″ with respect to datum A."

Notice the tolerance zone is 0.125″ wide, in which the controlled feature is included. Therefore, the feature may be warped, misaligned, or twisted, but the plane established by the edges of the part must *not* break the tolerance zone.

Orientation Geometric Tolerances

Orientation geometric tolerances control the degree of parallelism, perpendicularity, or angularity of a feature with respect to one or more datums. Figure 9-26 shows the orientation geometric symbols.

Figure 9-27 shows a part with two parallel vertical plates welded to a horizontal plate. The left plate is used to establish a datum from which the right plate is only permitted to vary within a tolerance zone of 0.062″.

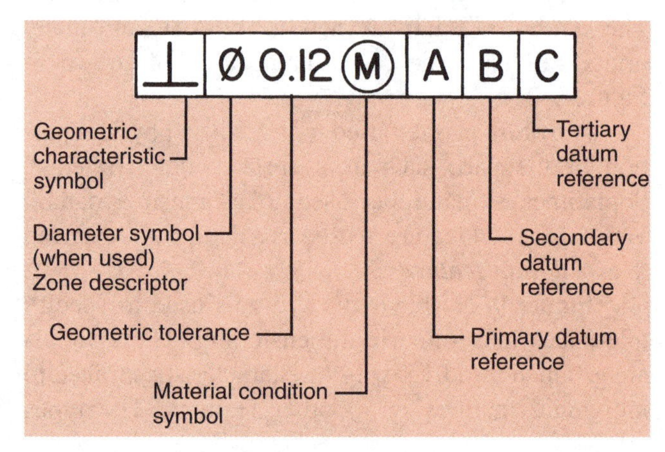

Figure 9-24.
Feature control frames often specify symbols for geometric characteristics, tolerances, material conditions, and datum references.

Figure 9-25.
This feature control frame for a welded part indicates the feature must be perpendicular within 0.125″ with respect to datum A.

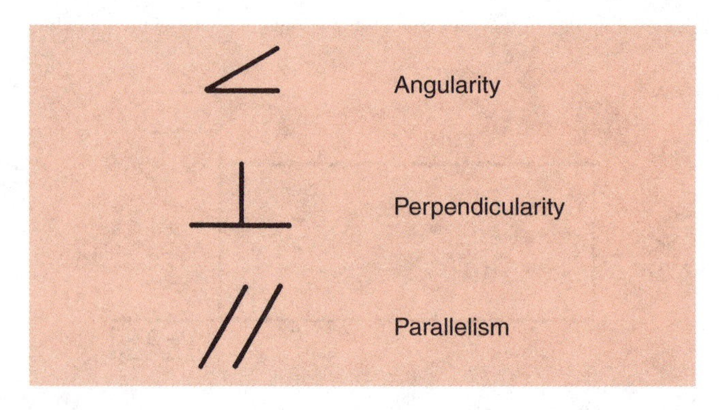

Figure 9-26.
Orientation geometric tolerance symbols control the degree of parallelism, perpendicularity, or angularity of a feature with respect to the datum.

Form Geometric Tolerances

Form geometric tolerances are used to control flatness, straightness, circularity (roundness), and cylindricity. The tolerance controls the feature and is *not* linked to a datum. Figure 9-28 shows the form geometric symbols.

The symbol shown in the feature control frame in Figure 9-29 indicates straightness. The plate surface controlled by the feature control frame is required to be within a 0.005″ tolerance zone.

Datum targets are used to establish consistent locations on parts with rough or irregular surfaces. The datum target is usually circular and indicates where the part should contact the fixture. The use of datum targets helps assure the proper locating of parts in a fixture.

Figure 9-30 shows a datum target symbol. When the datum target is visible on a view, the target appears as shown in Figure 9-30. Hidden targets have hidden lines in the top half of the target. A hidden datum target is shown in Figure 9-31. The targets are usually lettered, starting with the primary target lettered *X*.

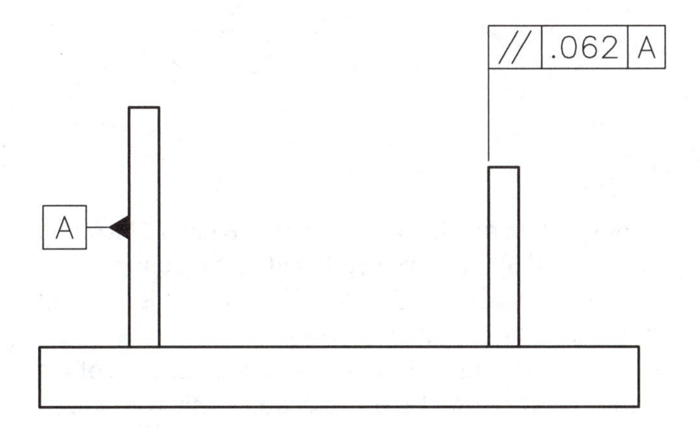

Figure 9-27.
The right plate is only permitted to vary within a tolerance zone of 0.062″.

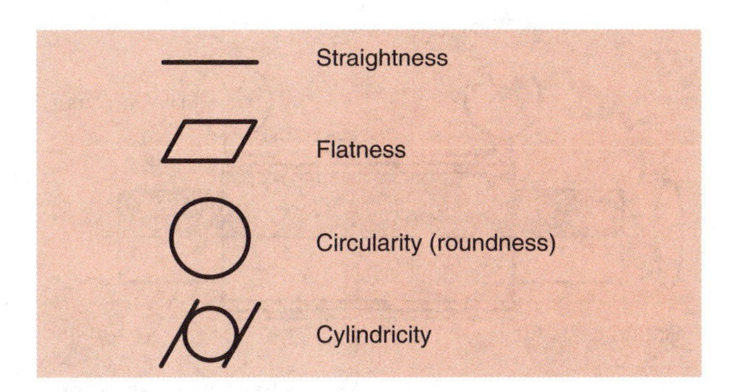

Figure 9-28.
Form geometric symbols are used to control straightness, flatness, circularity (roundness), and cylindricity. (American National Standards Institute)

Other Geometric Tolerances

For welding, the orientation and form geometric tolerances are the most important, although other types of geometric characteristics may be indicated on a print. Some of them may be important, depending on the shape or geometry of the part being welded.

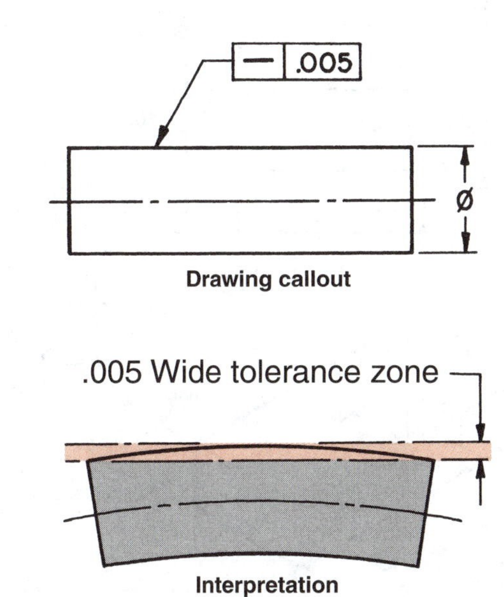

Figure 9-29.
The first symbol in the feature control frame indicates straightness. The tolerance zone for the straightness of the plate is .005.

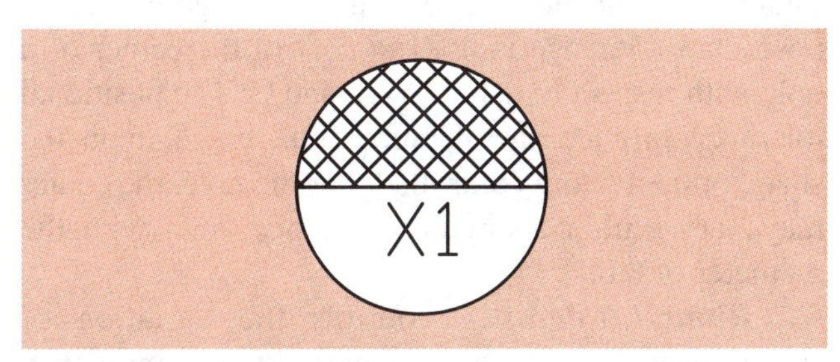

Figure 9-30.
A visible datum target symbol is shown.

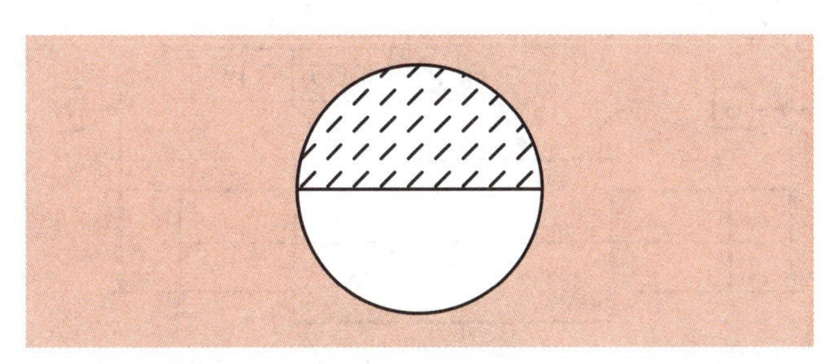

Figure 9-31.
A hidden datum target symbol contains hidden lines in the top half of the target.

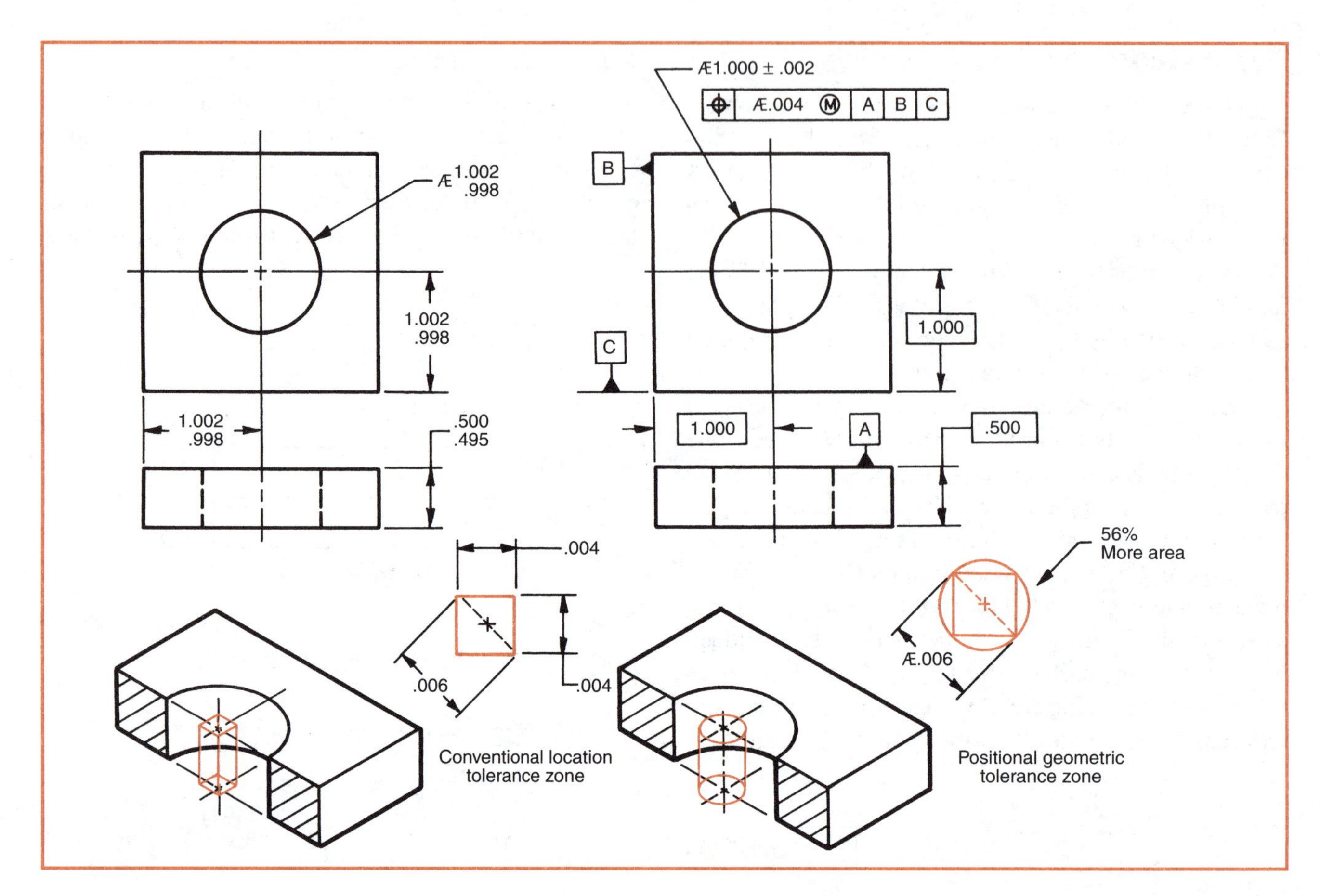

Figure 9-32.
Location of features is established by position tolerances.

Positional (or ***location***) ***tolerances*** are used to establish the location of features. In Figure 9-32, a positional tolerance is used to control the center of a hole with respect to datum A, B, and C. The positional tolerance provides a tolerance zone based upon the shape of the feature controlled. For the hole, the center may vary within a circular tolerance zone with the diameter of 0.004″.

Runout tolerance controls the variation of straightness, circularity, cylindricity, or angularity of surfaces with respect to the center line or axis of cylindrical parts. If two cylindrical parts (like two parts of a shaft) are welded together, the runout tolerance controls the straightness of the shaft. Figure 9-33 shows a welded shaft indicating a total runout tolerance of 0.003. Figure 9-34 shows the application of the runout tolerance and the inspection method to check the runout.

G

.003 G H

H

Drawing callout

Figure 9-33.
The runout tolerance controls the straightness of the shaft.

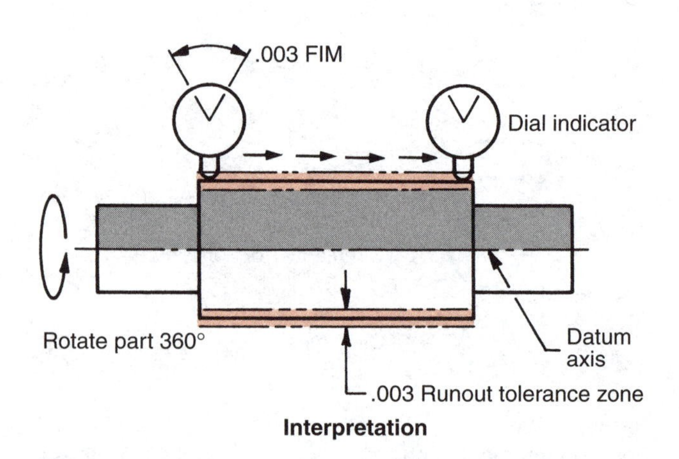

Figure 9-34.
Runout tolerance is applied in the circular cross section.

Name ______ Date______ Class ______

Review Questions

Part I

1. The dimension located furthest from the view in an extension and dimension line method is called the _____ dimension.
2. A dimension that indicates the center of a hole is called the _____ dimension.
3. A dimension system that uses two intersecting planes from which dimensions are located is called _____.
4. A dimension system in which extension lines and dimension lines are used and all dimensions are referenced to an edge of the plane on the part is called _____.
5. A tolerance found in a tolerance box or note is called a(n) _____ tolerance.
6. If a general tolerance and an explicit tolerance next to a specific dimension are used on a print, which tolerance is used for the specific dimension?
_____
7. What indicates datums used for geometric dimensioning and tolerancing? _____
_____
8. Dimensions given directly on a print are called _____ dimensions.
9. Perpendicularity is a geometric characteristic listed as what type of tolerance? _____
10. A symbol that contains a geometric characteristic, tolerance zone, and when needed, a datum reference is a(n) _____.

Part II

General Tolerance	
Decimals	
.XX	±.12
.XXX	±.05
Fractional	± $\frac{3}{16}$
Angular	± 2°

1. What is the maximum size for a feature dimensioned at 4.50?
_____
2. What is the minimum length for an overall length of 11⅜?
_____
_____
3. The angle between two plates is 47°. What is the maximum angle between the parts to remain within tolerance? _____
4. What is the minimum limit for the 9.50 dimension shown below? _____

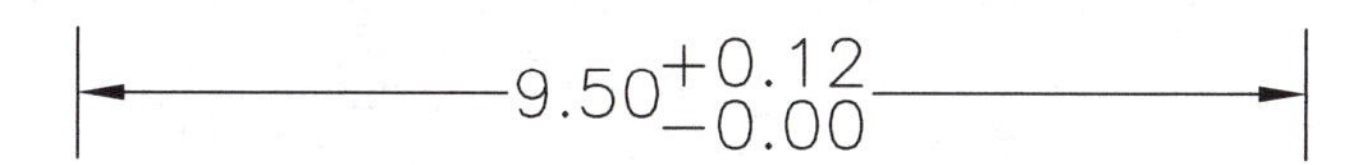

5. Which of the following dimensions has the tightest tolerance? 7.38 or 7.380? _____

Part III

1. What is the tolerance zone listed in the feature control frame shown below?
_____

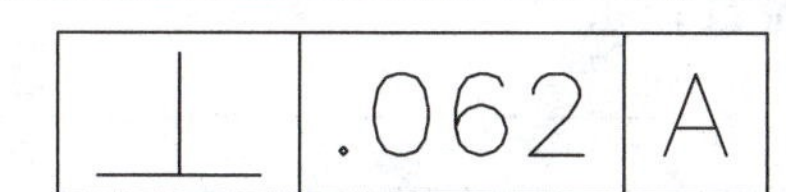

2. Which of the following geometric characteristics is not related to a datum?
_____

3. Which datum shown below is considered the primary datum?
_____

4. Explain the following symbol.

⊥	.06	B

5. Use the circle below and sketch a visible datum target used to fixture a part.

Print Reading Activities

Use the alignment plate in Figure 9-35 to answer the following questions.

1. How many datums are shown on the print? ____
2. List the reference dimension.
3. What is the plate thickness used to fabricate the alignment plate?
4. How many .25 holes are required?
5. What revision was made on 1-12?
6. Explain the feature control frame at F.
7. Explain the feature control frame at G.
8. Explain the feature control frame at H.
9. What is the maximum size of the 5.25 dimension?
10. Determine the dimension for the following letters on the print:

    A.

    B.

    C.

    D.

    E.

Name ______________________________

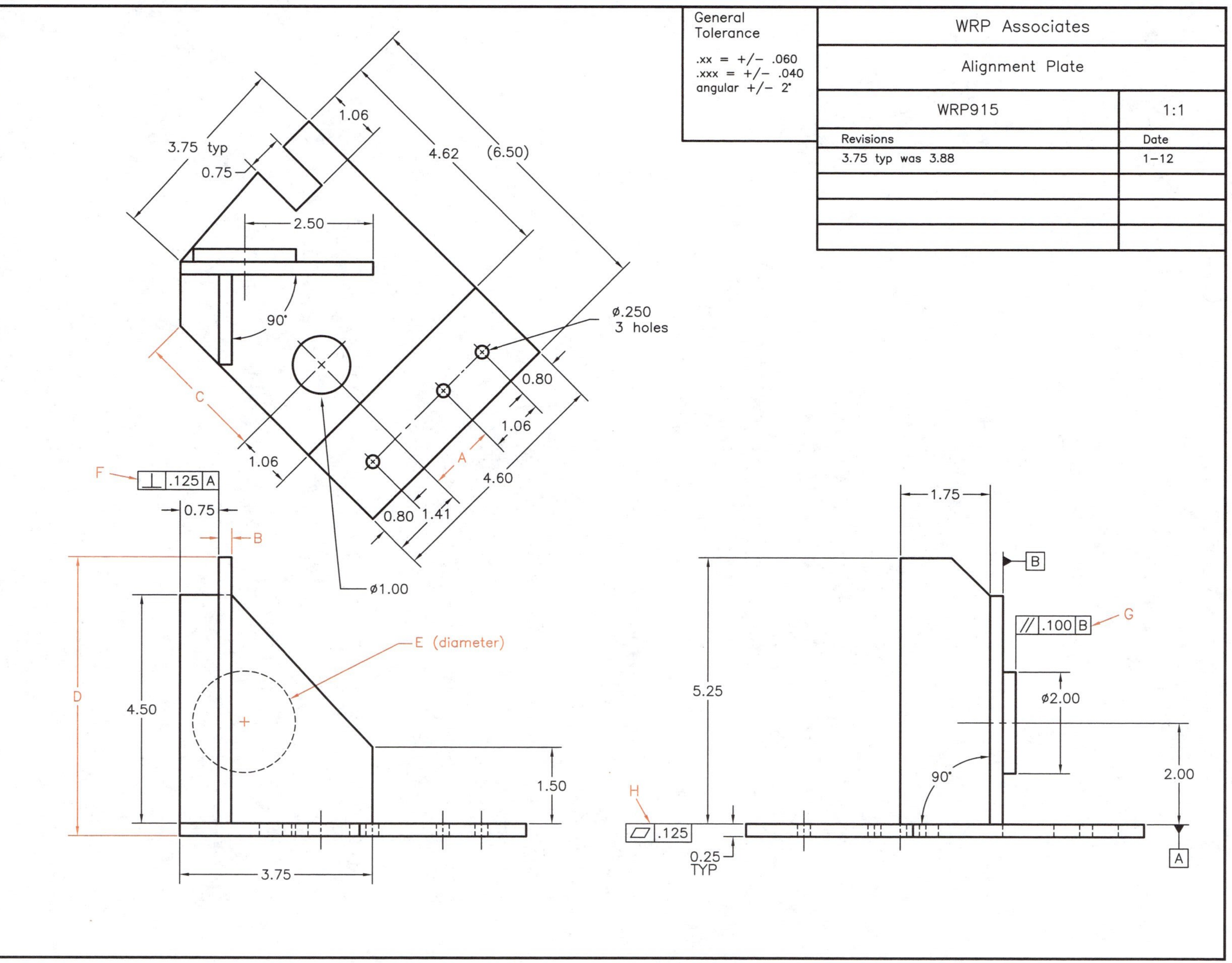

Figure 9-35.
Use this print to answer the Print Reading Activities questions.

Section 3 Welding Fundamentals

Unit **10** Welding Processes

Unit **11** Threaded Fasteners

Unit **12** Structural Metals

Unit **13** Common Types of Joints and Welds

Unit 10

Welding Processes

After completing Unit 10, you will be able to:

- Distinguish among definitions of soldering, brazing, and welding.
- Explain the similarities and differences among the major modern welding processes.
- Recognize and use welding process abbreviations.
- Describe the equipment components of the major welding processes.
- Explain the newest welding advances and processes.

Key Words

arc stud welding (SW)
automatic (robotic) welding
brazing (B)
braze welding (BW)
cold welding (CW)
electron beam welding (EBW)
electroslag welding (ESW)
flame spraying (FLSP)
flash welding (FW)
flux cored arc welding (FCAW)
friction stir welding (FSW)
friction welding (FRW)
gas metal arc welding (GMAW)
gas tungsten arc welding (GTAW)
laser beam welding (LBW)
mechanized welding
metallizing
microwelding
oxyfuel gas welding (OFW)
plasma flame spray
projection welding (PW)
resistance welding (RW)
resistance seam welding (RSEW)
soldering (S)
resistance spot welding (RSW)
seam welding
shielded metal arc welding (SMAW)
spot welding
submerged arc welding (SAW)
thermo spray
ultrasonic welding (USW)
upset welding (UW)
welding

The meaning of the terms soldering, brazing, and welding are sometimes confused since they all use heat to join metals. The differences, however, must be fully understood. The application of the correct welding process is crucial to the success of the repair or fabrication of the component.

The abbreviations shown in parentheses after some terms are what you would see on a welding symbol. The abbreviations are the accepted standards used by the American Welding Society (AWS) to describe a specific welding process.

Soldering (S)

Soldering (sometimes called *soft soldering*) is a method of joining metals with a nonferrous metal filler *without* having to heat them to a point where the base metals melt. Soldering is carried out at temperatures lower than 840°F (450°C). Soft solder is generally a combination of tin and lead made in the form of wire or bar.

The strength of solder is relatively low. It is used for low stress, low pressure, and low temperature applications. Soft soldering is commonly used to join copper tube, copper wire, and sheet metal.

Brazing (B)

Brazing is a group of joining processes using nonferrous alloys as filler metal with melting temperatures above 840°F (450°C). However, the filler metal's melting point is lower than the melting point of the metals being joined. Brazing is usually an automated process that relies on capillary action, while ***braze welding (BW)*** is a manual process of adding the filler metal to the joint.

The strength of brazed joints depends on the type of brazing filler metal and the alloying (resulting metal content in the weld) of the filler metal with the base metal. Most faucet parts are joined through the use of brazing while the repair of some broken steel and cast iron parts is done by braze welding.

Welding

Welding is a method of joining metals by heating them to a suitable temperature, causing them to melt and fuse together. This may be with or without the application of pressure. The welding may occur with or without the use of filler material having similar composition and a melting point of the base metal, or the filler metal may have a different composition than the base metal but the filler will melt above 840°F.

Modern Welding Processes

Many different welding processes have been developed since the days of simple welding done by the blacksmith at a forge (forge welding). Today, there is hardly an industry or business not dependent in some way on welding. Refer to Figure 10-1.

Figure 10-1.
Pipe ranging from 5″ to 16″ outside diameter is produced on this electric resistance weld mill. Pipe running off line at right has been rolled into circular form as it comes down line from left. It is welded together by large rotating circular electrodes in the center of the picture. (Bethlehem Steel Corp.)

Each welding process has its advantages and disadvantages and may be more adaptable to certain applications than other welding processes. Many factors affect the selection of one welding process over another. A few of these factors include:

- Materials to be joined
- Size and/or shape of the part
- Material thickness
- Type of material
- Production requirements

Oxyfuel Gas Welding (OFW)

Oxyfuel gas welding includes a group of welding processes that use burning gases (such as oxy-acetylene welding, known as OAW) or hydrogen mixed with oxygen. The heat produced causes the base metal to melt and fuse. A filler material that has a similar composition and melting temperature as the base metal may or may not be used. Figure 10-2 shows an oxyfuel gas welding setup.

The most common applications for oxyfuel gas welding include thin carbon steel sheet, thin wall tube, and small diameter black iron pipe.

Shielded Metal Arc Welding (SMAW)

Shielded metal arc welding (also called ***stick*** or ***arc welding***) is a joining technique that uses an electric arc to produce the heat necessary to cause the base metals to melt and fuse together, Figure 10-3. Filler metal in the form of an electrode may be added to the joint, Figure 10-4.

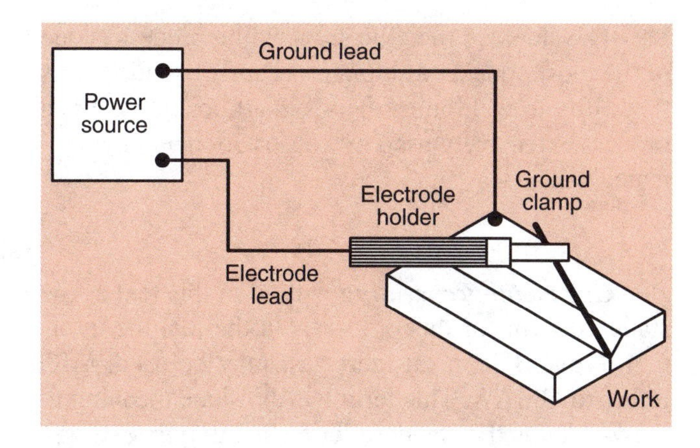

Figure 10-3.
Shielded metal arc welding (SMAW) is a joining technique where electric current makes heat to melt and fuse the base metal.

Figure 10-4.
Filler metal from the electrode is added to the joint.

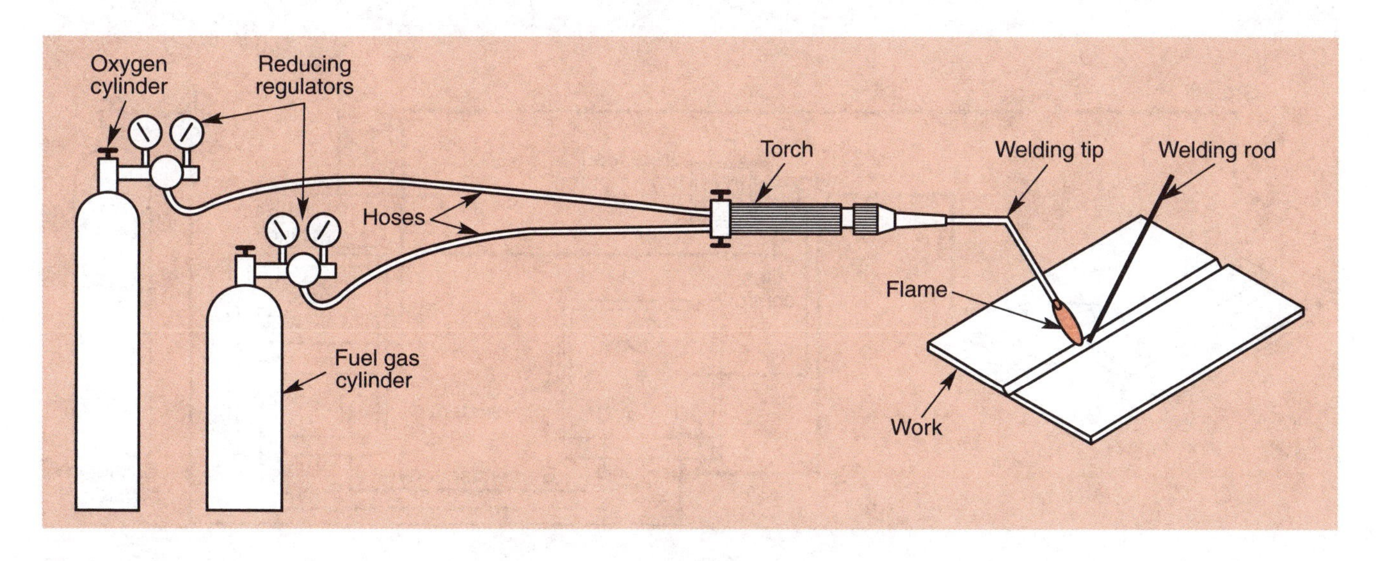

Figure 10-2.
Oxyfuel gas welding (OFW) makes use of burning gases to produce heat to melt and fuse base metal. Study components of gas welding outfit.

The shielded metallic arc welding process is one of the most popular welding processes. Because of its flexibility to weld most metals in all positions, it is a process extensively used by construction and maintenance welders.

Gas Metal Arc Welding (GMAW)

Gas metal arc welding is used in this text as the standard term for this process. On the job, the informal term MIG (metal inert gas) may be used. MIG refers to GMAW. This term has the same meaning to the welder as the standard term used in this text.

GMAW is a welding process that uses an arc formed between continuous filler metal and the weld pool. The filler metal is supplied from a spool or coil and fed into the weld pool by a set of drive rollers found in a wire feeder. The spool may be as small as one pound when held in the welding gun or as large as one thousand pounds when used with a robot welding system. Figure 10-5 shows the equipment used for GMAW welding.

The weld pool is shielded from the atmosphere by an externally supplied gas. The gas may be inert for welding nonferrous metals (like aluminum), or it may be a gas mixture of inert and reactive gases for welding metals (like steel). The most popular gases for welding steel are mixtures of argon and carbon dioxide.

GMAW has become very popular for production welding. Most low carbon steels are welded using the GMAW process, including thin sheet metal, thick plate, structural, and subassembly shapes.

Gas Tungsten Arc Welding (GTAW)

Gas tungsten arc welding is used in this text as the standard term for this process. On the job, the informal term TIG (tungsten inert gas) may be used. TIG refers to GTAW. Gas tungsten arc welding uses an arc produced between an electrode that is *not* consumed in the welding process and the weld pool. The electrode is made of a metal (tungsten), which has a very high melting point. You may also hear the informal term *heliarc* used (referring to inert gas tungsten arc welding).

The filler metal, if needed, is generally fed into the weld pool manually. Like GMAW welding, GTAW uses a shielding gas to protect the weld pool from the atmosphere. The gas is generally argon, but sometimes helium is used. Figure 10-6 illustrates the equipment for GTAW.

Originally designed to weld nonferrous and "hard to weld" materials, GTAW is commonly used today on both ferrous and nonferrous metals. Fabrication of steel and stainless steel pipe, sheet metal, and repair of various metal components are often welded using GTAW.

Flux Cored Arc Welding (FCAW)

Flux cored arc welding is similar to GMAW welding because it uses an arc produced between a continuous wire electrode and the weld pool. The

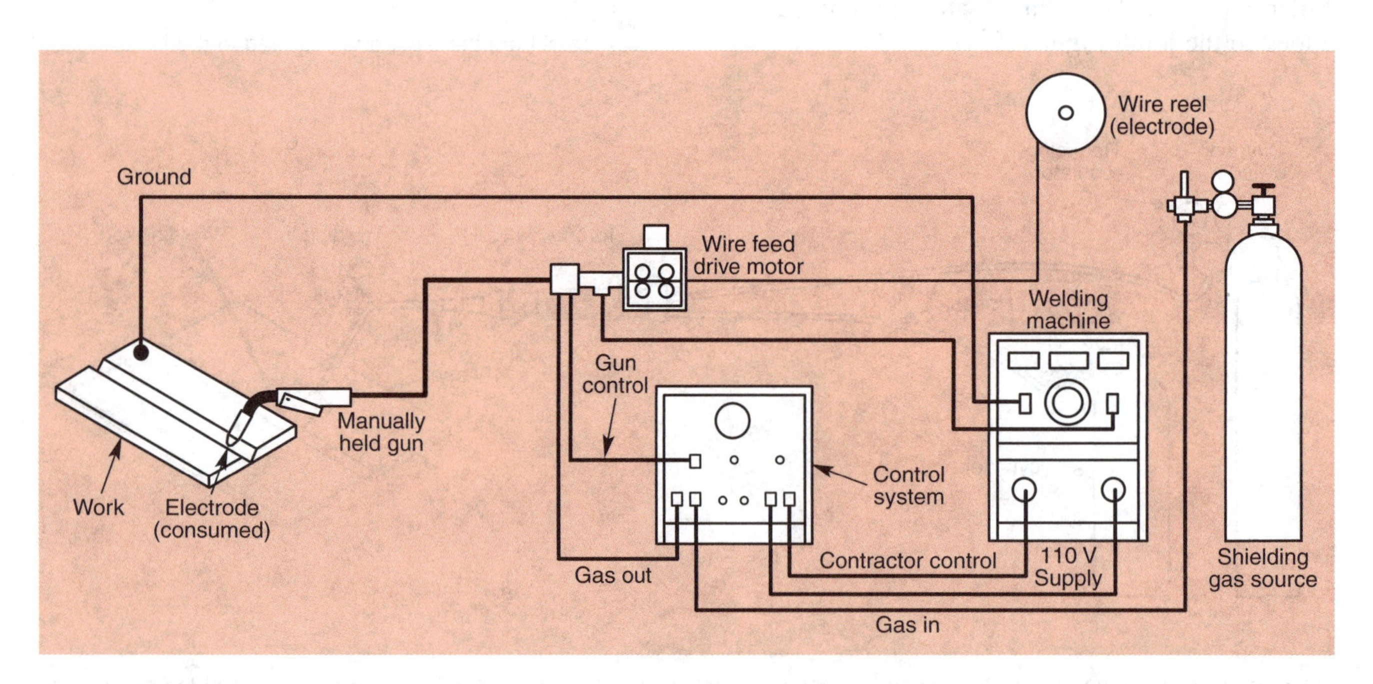

Figure 10-5.
Gas metal arc welding (GMAW) is a gas-shielded arc welding technique that uses an electrode consumed in the welding process. The electrode contributes filler metal to the joint.

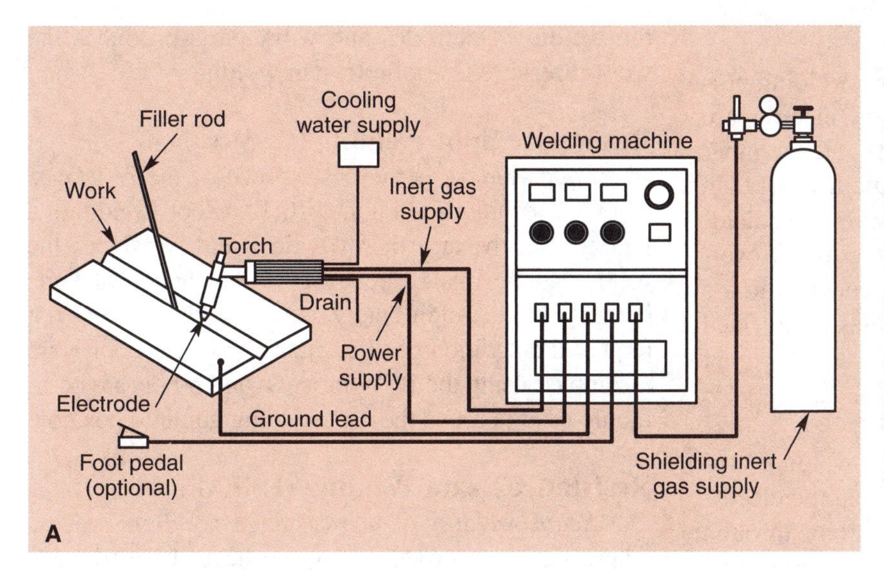

Figure 10-6.
Gas tungsten arc welding (GTAW) is a gas-shielded arc welding technique accomplished with a permanent electrode that is not consumed. A—Study the GTAW welding process. B—The welder may add filler metal into the weld pool manually, as shown.

wire, however, is similar to the SMAW electrode turned "inside out." This means the coating that produces the gas to cover the weld pool and the flux used to clean the metal is found inside the wire.

The FCAW process produces a high-quality weld at high deposit rates. Because of the material inside the wire, welds are less likely to have internal defects than with the similar GMAW process.

The FCAW process has two common variations of wire electrode. One of the wire electrodes needs an externally supplied shielding gas like GMAW welding. Carbon dioxide is the gas frequently used, but many companies use an argon and carbon dioxide mixture like GMAW welding.

The other type of FCAW wire has the shielding gas producing material *and* the cleaning material inside the wire. Use caution when setting up FCAW equipment so the correct electrode polarity is used for the type of wire. Figure 10-5 shows the equipment used for FCAW. The equipment is the same type of power supply and wire feeder as used with GMAW, but sometimes the power supply has a higher output.

Submerged Arc Welding (SAW)

Submerged arc welding is a technique where coalescence (fusion) is produced by heating with an electric arc between a bare metal electrode and the work. A blanket of flux on the work shields the welding arc. Pressure is *not* used, and filler metal is obtained from the electrode and sometimes from a supplementary welding rod or feed into the weld pool from an additional coil of wire.

The process produces little smoke, arc rays, radiant heat, or spatter. To weld, the operator fills the flux cone, points the gun into the joint, allows a pile of flux to accumulate, and then strikes an arc under the flux with the electrode. Because the flux is granular, the process is only used when welding in flat and sometimes horizontal positions.

Once the arc is struck, the electrode automatically feeds into the arc as the gun is moved over the work. The SAW process is often mechanized, using a machine to move the gun over the work, Figure 10-7.

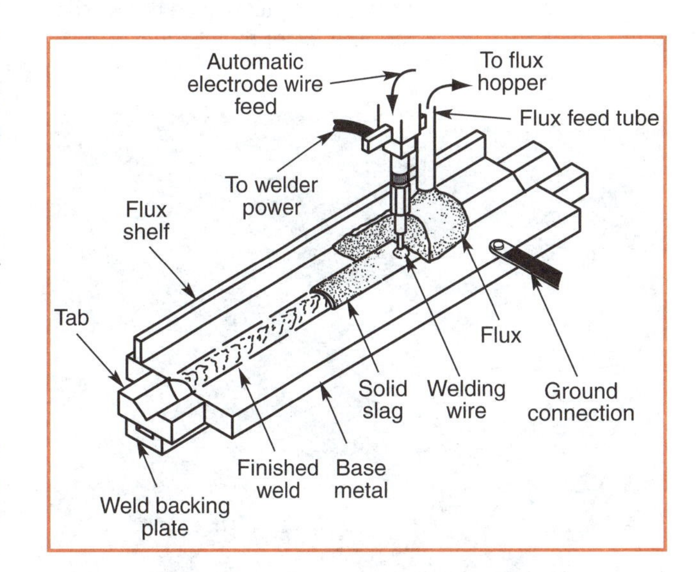

Figure 10-7.
Study the submerged arc welding (SAW) process.

Arc Stud Welding (SW)

Arc stud welding, Figure 10-8, is a welding process where fusion is produced by an electric arc between a metal stud (or similar part) and the other work part. The surfaces to be joined are heated; then they are brought together under pressure. No shielding is used. Studs may be welded to the top of overpass-type bridge members to hold the concrete to the bridge, or screw-threaded fasteners may be welded onto sheet metal, Figure 10-9.

Resistance Welding (RW)

Resistance welding is a group of welding processes where fusion heat is obtained from the electrical resistance of the work. Electric current through the welding electrodes and work parts produces the weld. Pressure is applied when welding.

Resistance Spot Welding (RSW)

Spot welding is the best known of the resistance welding techniques, Figure 10-10. Spot welding is widely used because it saves time and weight as the welds can be made directly between the metal parts being joined. Additionally, spot welding does not require the addition of filler metal. Most car bodies are assembled using the RSW process, and robots are commonly used to move the spot welding gun into position.

Resistance Seam Welding (RSEW)

Seam welding is a resistance welding process where fusion is produced by the heat obtained from resistance to the flow of an electric current through the work parts. The work parts are held together under pressure by circular electrodes. The resulting weld is a series of overlapping spot welds made progressively along the joint by rotating the electrodes. Figure 10-11 pictures this welding process.

Projection Welding (PW)

Projection welding is also a resistance welding technique, as shown in Figure 10-12. Heat is produced by the flow of an electric current through the work parts that are held under pressure by the projection electrodes. The welds are localized at predetermined points by the design of the parts to be welded.

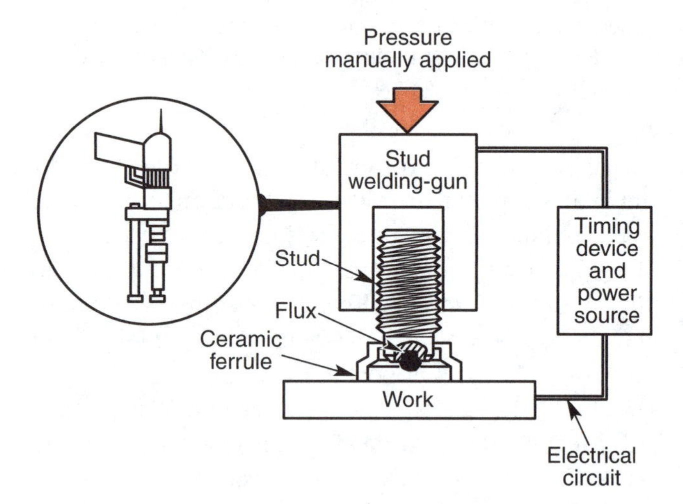

Figure 10-8.
Stud welding (SW) is an arc welding technique that welds a metal stud or similar part to another metal surface.

Figure 10-9.
A stud welding gun and screw-threaded fastener are shown.

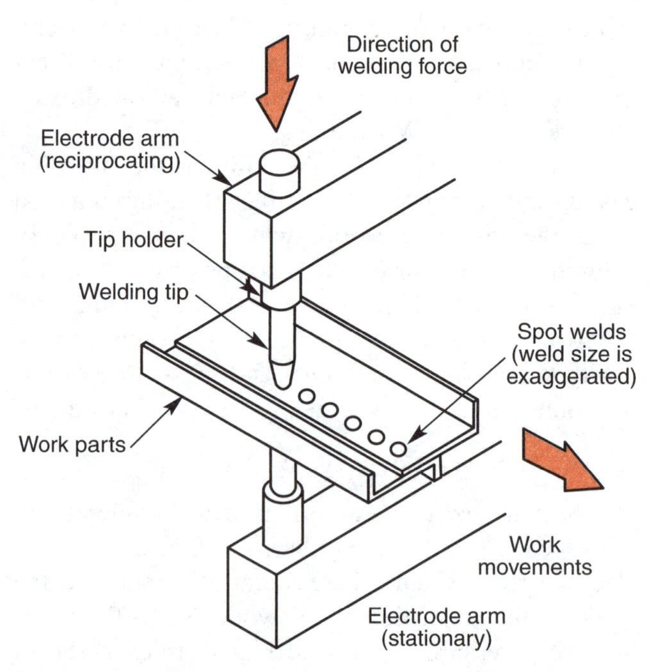

Figure 10-10.
Resistance spot welding (RSW) is a common welding technique.

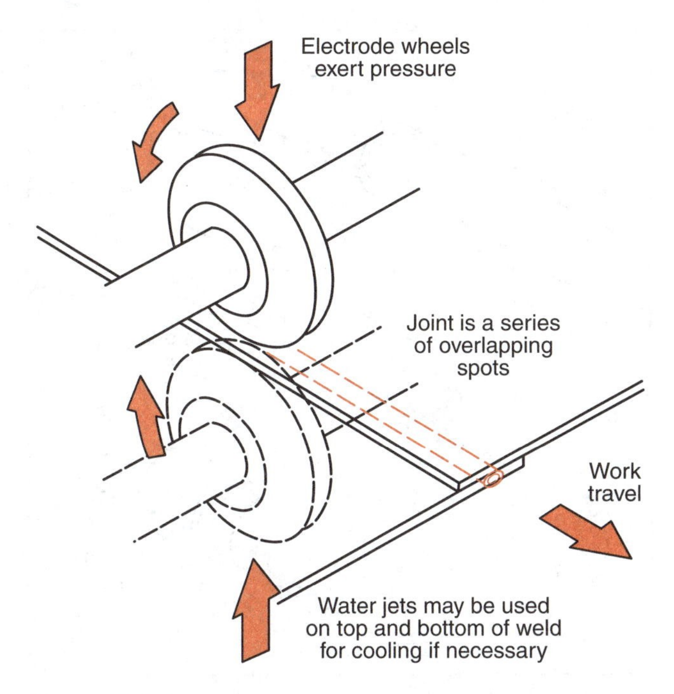

Figure 10-11.
The resistance seam welding (RSEW) process is a technique that produces a series of overlapping spot welds along the joint.

Flash Welding (FW)

In ***flash welding,*** fusion is produced simultaneously over the entire area of abutting surfaces. Figure 10-13 shows a band saw flash welding attachment. Welding heat is obtained from electric current between the two surfaces and by the application of pressure after heating is substantially completed. Flashing is accompanied by the ejection of molten metal from the joint.

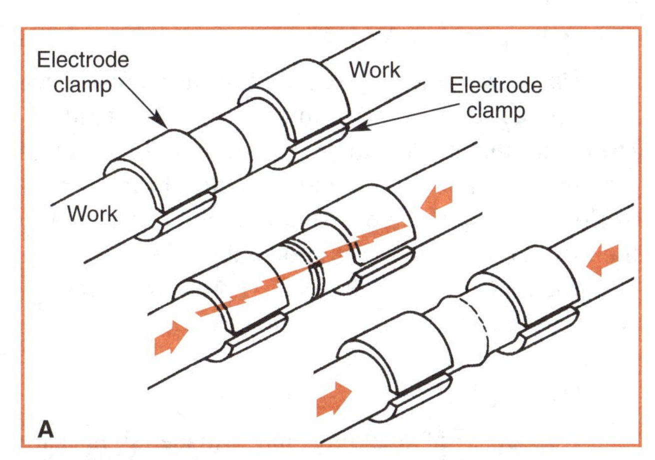

Figure 10-13.
In flash welding (FW), two parts are moved into each other. A—Fusion is produced over the entire area of abutting surfaces by heat obtained from the resistance to the current between the two surfaces and by pressure after heating is almost completed. B—A flash weld machine is shown.

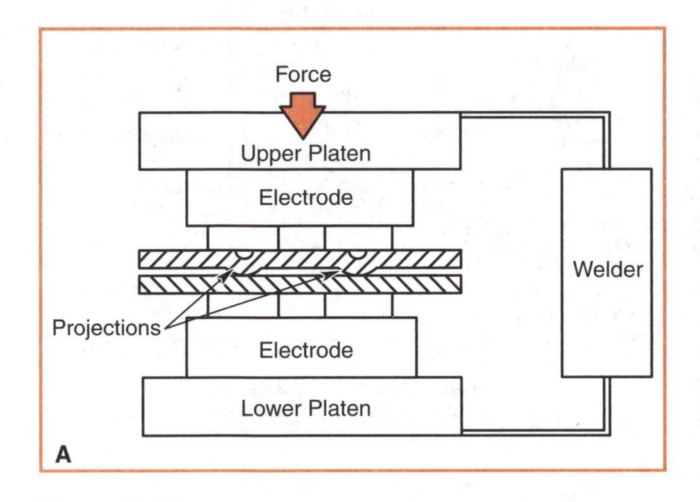

Figure 10-12.
Projection welding (PW) also uses heat produced by current through the work parts held under pressure by electrodes. A—Note the projections on the parts. B—Projection welding equipment is shown.

Upset Welding (UW)

Upset welding, Figure 10-14, is also a resistance welding process very similar to flash welding. However, the pieces to be joined are first butted together under pressure, and then current is allowed to flow through the pieces until the joint is heated to the fusion point. Pressure is continued after the power is turned off. This causes an upsetting (bulging) action at the point where fusion occurs.

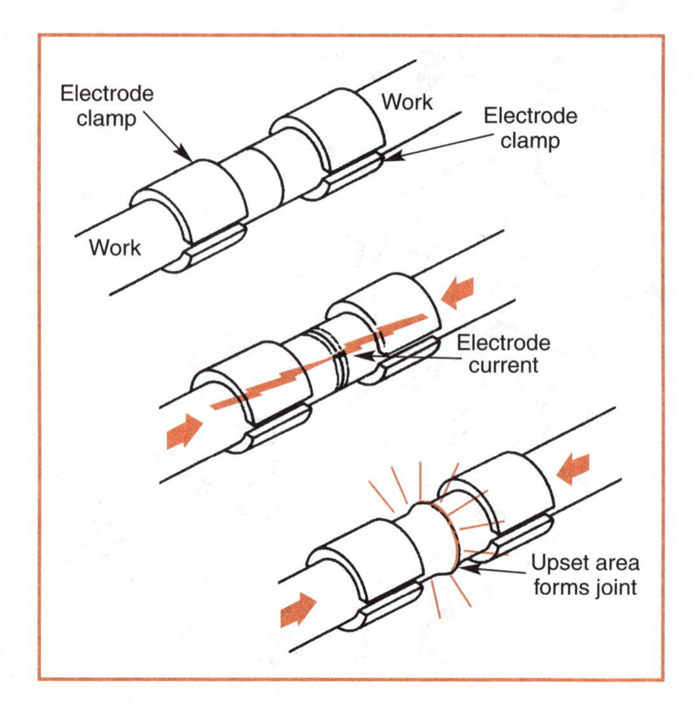

Figure 10-14.
During upset welding (UW), pieces to be joined are butted together under pressure. Then current is allowed to flow through pieces until the joint is heated to fusion point.

Additional Welding Processes

All areas of welding have made significant advances in recent years. Many welding machines and process monitoring systems use computers to improve the quality and speed of making welds. Other advances have come from using new "exotic" metals to improve the properties of the weld.

Electron Beam Welding (EBW)

The ***electron beam welding*** process makes use of a beam of fast-moving electrons to supply the energy needed to melt and fuse the base metals. This is illustrated in Figure 10-15. Welds must be made in a vacuum of 10^{-3} to 10^{-5} mm Hg, which practically eliminates the contamination of weld metal by atmospheric gases. The electron beam is capable of melting any known metal.

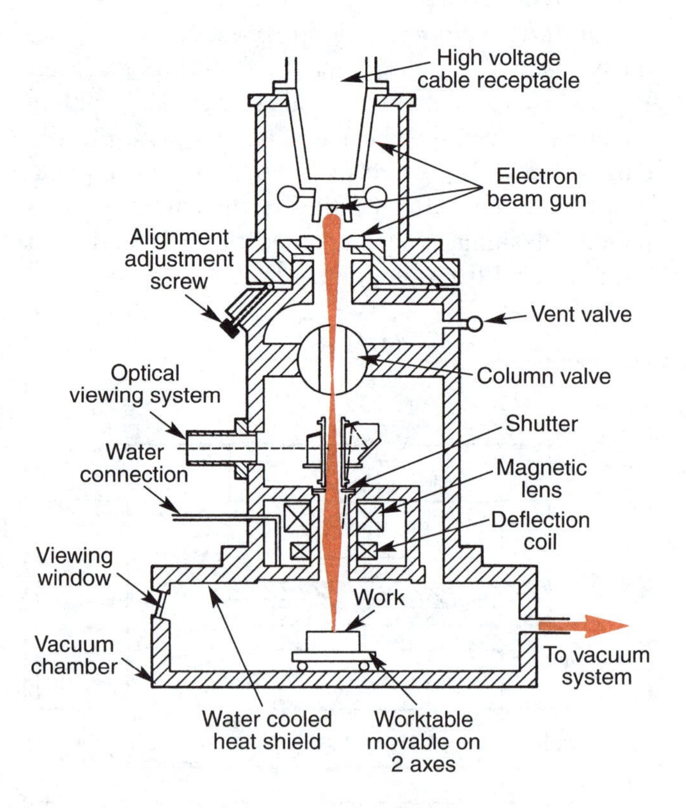

Figure 10-15.
With electron beam welding (EBW), the beam of fast-moving electrons supplies the energy to melt and fuse base metals. Welds must be made in a vacuum.

Friction Welding (FRW)

Friction welding is one of the simplest and most unique of the welding processes. The technique uses frictional heat and pressure to produce full strength welds in a matter of seconds, Figure 10-16.

The parts to be joined may be bar or tubular in shape; however, flat plates or formed shapes can be joined if the interface (portions forming the joint) is generally circular, Figure 10-17.

Friction Stir Welding (FSW)

Friction stir welding is a solid-state joining process. Solid-state means that the metal is not melted during the process. Friction stir welding is often used when the characteristics of the original metal must not be changed. Aluminum is the most common metal joined with the process, and the parts are usually fit together as butt joints, Figure 10-18.

A variation of the friction stir process is laser assisted friction stir welding. A laser is used to heat the metal ahead of the rotating tool. The heating of the metal reduces the wear of the tool and reduces the amount of force required to move the tool through cold metal.

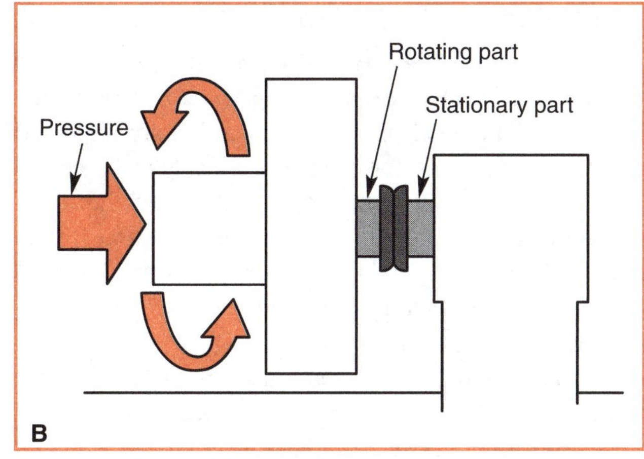

Figure 10-16.
Study the friction welding process. A—In friction welding, the flywheel, chuck, and one of the parts to be welded are accelerated to a preset speed. B—On reaching the required speed, the drive is disengaged and the rotating part is thrust against the stationary part. Energy in the flywheel is discharged into the interface (joint) and makes the weld.

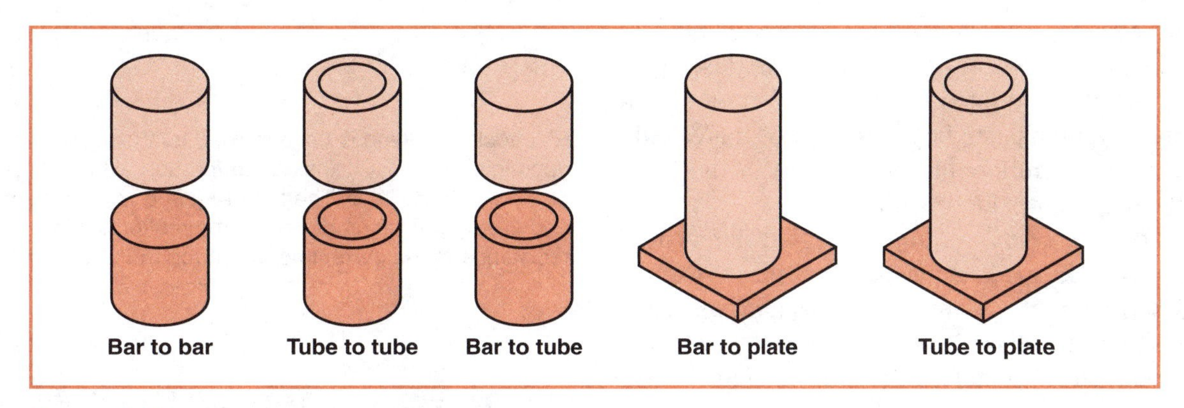

Figure 10-17.
Friction welding requires the joint face of at least one part to be essentially round.

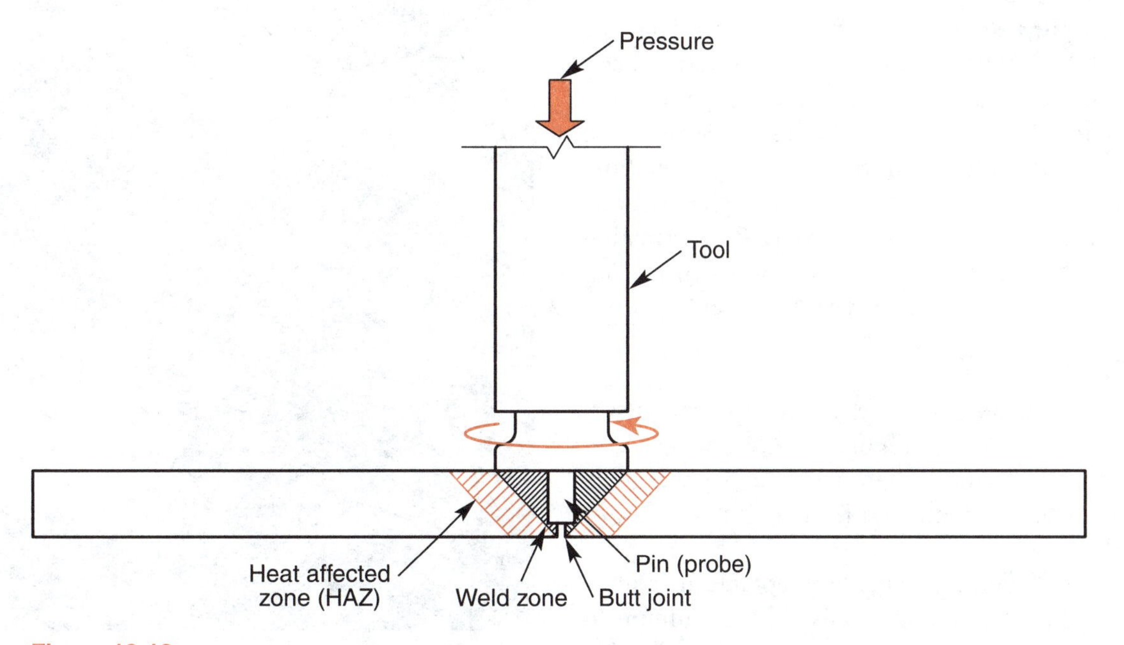

Figure 10-18.
Friction stir welding does not melt the metal but does create heat from the friction between the rotating tool and the metal surface. The RPM and travel speed of the tool are controlled to get the required mixing of the metal.

Electroslag Welding (ESW)

Electroslag welding (ESW) is a high production process used on thick materials, most commonly carbon steel. The steel plates to be welded are arranged as a butt joint. Water-cooled copper plates, called shoes, are clamped tightly on each side of the joint. An arc is struck at the bottom of the joint, and flux is added. Electroslag is not an arc welding process because as the flux produces molten slag, the slag extinguishes the arc, Figure 10-19.

The electroslag process may use one or more wire electrodes, which are consumed during the process. Equipment costs are generally high for the process, but a single pass using the electroslag welding process can complete a weld that normally requires multiple passes.

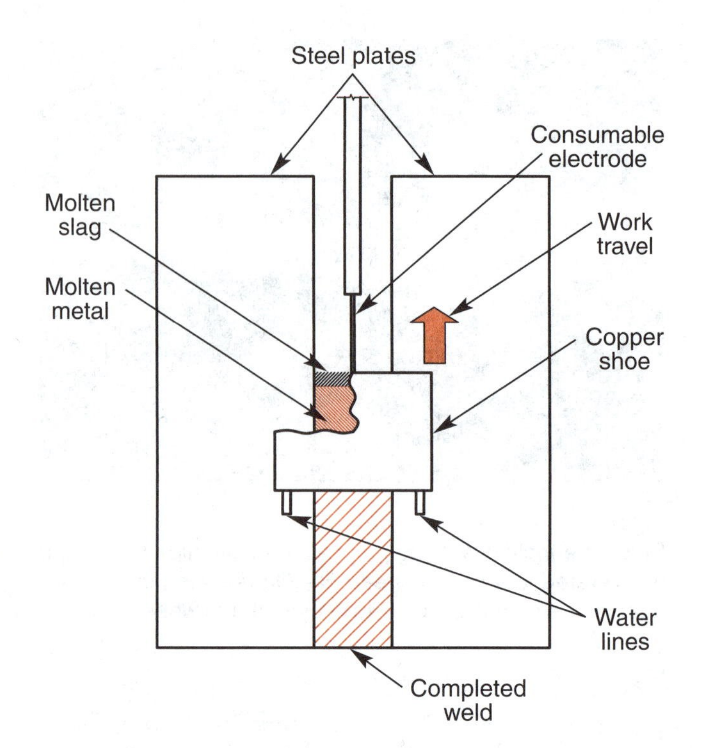

Figure 10-19.
Electroslag welding is used to weld butt joints with large gaps in one pass. Flux is added through either a hopper or the electrode itself. The heat required to melt the base and filler metals is produced by the molten slag's resistance to the current flowing through the electrode.

Laser Beam Welding (LBW)

Laser is the abbreviation for *light amplification by stimulated emission of radiation.* The laser produces a narrow and intense beam of coherent monochromatic light that can be focused onto an area only a few microns (millionths of an inch) in diameter. Figure 10-20 shows laser welding.

With ***laser beam welding,*** the light beam is used to vaporize the work at its point of focus. Molten metal surrounds the point of vaporization when the beam is moved along the path to be welded. A vacuum is not required as with electron beam welding.

Figure 10-20.
CO_2 laser welding of stainless steel is shown. (Laserage Technology Corp.)

Ultrasonic Welding (USW)

Ultrasonic welding is a process for joining metals without the use of solders, fluxes, or filler metals, and usually without the application of external heat. As in Figure 10-21, the metals to be joined are clamped lightly between sonotrodes (welding tips), and ultrasonic energy is introduced for a brief time (usually 1–3 seconds). A strong metallurgical bond is produced. There is little or no external deformation (which is characteristic of pressure welding) and no heat-affected zones found in resistance welding.

Cold Welding (CW)

Pressure alone is used to join the two metal surfaces in ***cold welding,*** Figure 10-22. The process, however, involves more than pressure. Special tools produce the deformation required to direct the flow of metal into a true weld.

Cold welding is especially adaptable to aluminum. However, dissimilar metals (like aluminum to copper, silver, lead, or nickel) can be joined with this process. Thin sheet can be readily bonded to thick sections.

Microwelding

Technology has made it possible to fit tens of thousands of transistors on a chip (microscopic sized electronic circuit). The chip may only be 0.025″ thick (0.64 mm) and 0.20″ square (5.08 mm). An example is given in Figure 10-23.

Microwelding was developed to attach leads to these microcircuits or chips. The first welding stage attaches 0.0015″ (0.04 mm) gold wire to the chip and to heavier leads. This is done in an ultraclean room, Figure 10-24.

Depending on a number of inputs, a microwelding machine can attach approximately 40 wires in 30 to 100 seconds. Very little heat is produced in making the welds. This prevents chip overheating and damage. After checking each weld, chips are enclosed in protective covers.

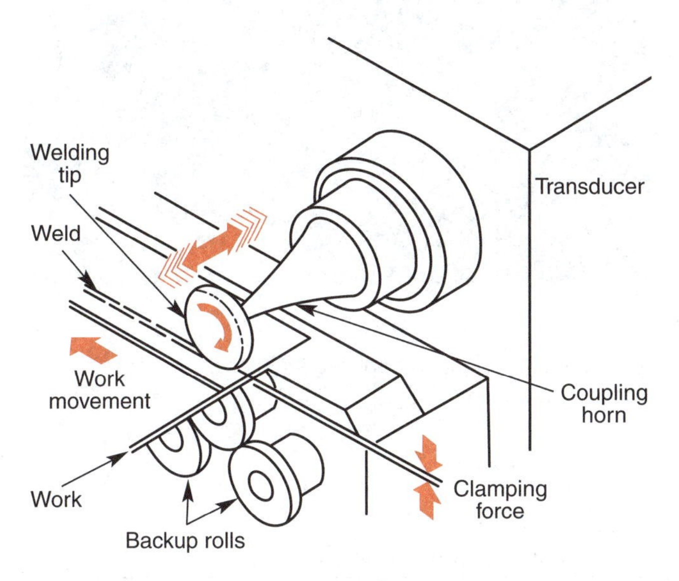

Figure 10-21.
Ultrasonic welding (USW) joins metals without solders, fluxes, or filler metals. Ultrasonic or high-frequency wave energy is used to make a strong metallurgical bond.

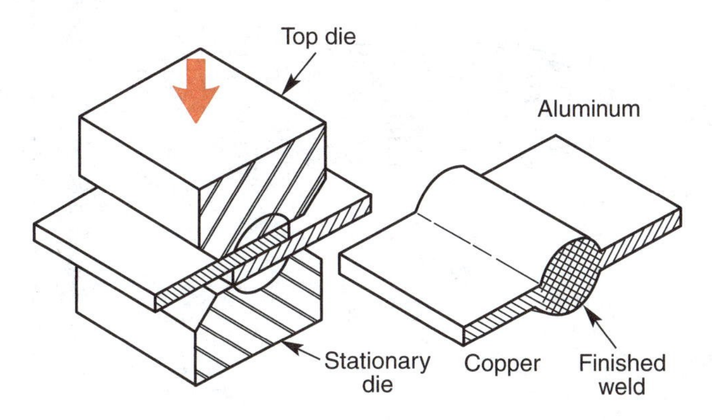

Figure 10-22.
Pressure alone joins two metal surfaces in cold welding (CW). Similar or dissimilar metals can be joined.

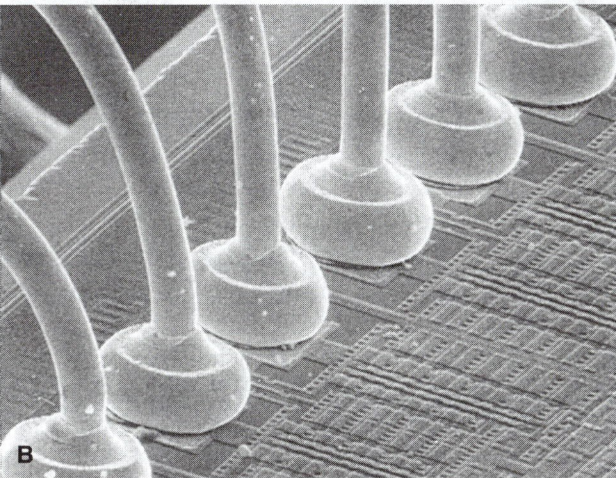

Figure 10-23.
The microwelding process is shown. A—This machine spends less than 30 seconds to "stitch" forty gold threads between the microcircuit and larger external leads. B—Close-up view of tiny gold wires (99.99% pure) used to connect the integrated circuit to the lead frame. (Delco Electronics Div., General Motors Corp.)

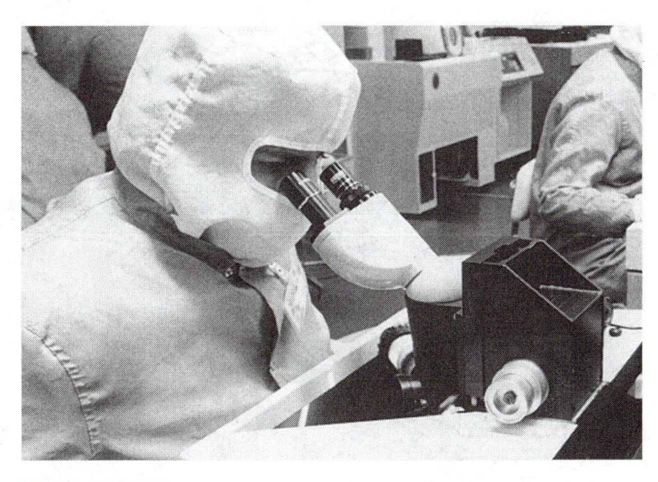

Figure 10-24.
Microwelding must be done in an ultraclean room; otherwise, dust may ruin the microcircuits. (Delco Electronics Div., General Motors Corp.)

Flame Spraying (FLSP)

Flame spraying is the term used to describe the process where a metal is brought to its melting temperature and sprayed onto a surface to produce a coating. The American Welding Society recognizes this technique as a welding process although it is never used to join two metal sections. Flame spraying processes include:

- Metallizing
- Thermo spray
- Plasma flame spray

The term ***metallizing*** describes the flame spraying process that involves the use of metal in wire form. Specially made wire is drawn through a unique spray gun by a pair of powered rolls. The wire is melted in the gas flame and atomized by compressed air. The compressed air forces the molten metal onto the previously prepared work surface, Figure 10-25.

Upon striking the surface, the particles interlock or mesh to produce a coating of the desired metal. Receiving surfaces must be cleaned and roughened before spraying, or the metal will not bond to it. Theoretically, there is no limit to the thickness that can be built up by flame spraying. This process can spray any metal that can be drawn into wire form. Figure 10-26 shows a modern metallizing gun.

Metallizing has been used to apply protective corrosion resistant coatings of zinc and aluminum to steel surfaces. It is also used to build up worn bearings and shafts that might otherwise need to be discarded.

Figure 10-26.
This metallizing gun is applying stainless steel onto a roll surface. (Metco Inc.)

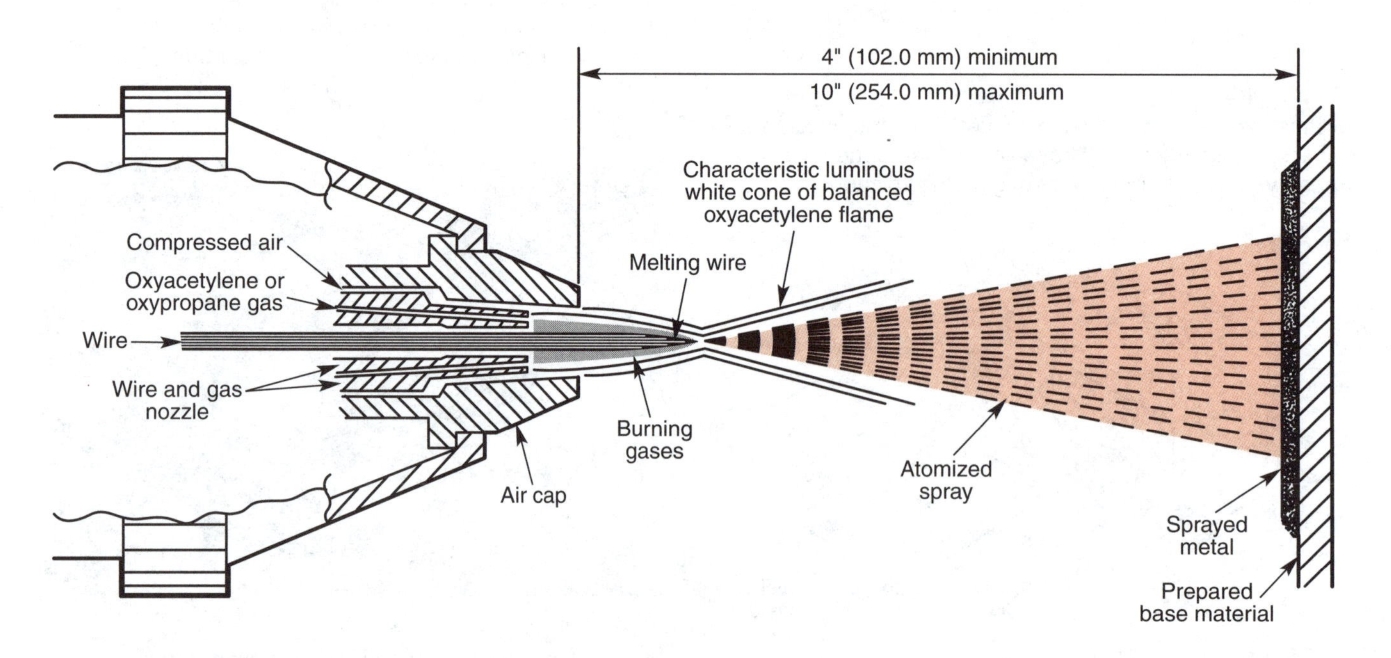

Figure 10-25.
During the metallizing process, wire is fed through a special spray gun, melted by a gas flame, atomized, and sprayed onto the work by compressed air.

The term ***thermo spray*** describes the flame spraying equipment that involves the application of metals and other materials that cannot be drawn into wire. They are used in powder form, Figures 10-27 and 10-28. These special alloys and materials are ideal for hard surfacing critical areas that must operate under severe conditions. High temperature, refractory materials that are also chemically inert can be sprayed.

In the ***plasma flame spray*** process, the spray gun utilizes an electric arc contained within a water-cooled jacket, Figure 10-29. An inert gas, passed through the arc, is "excited" to temperatures up to 30,000°F (16,649°C). In general, most inorganic materials that can be melted without decomposition can be applied, Figure 10-30.

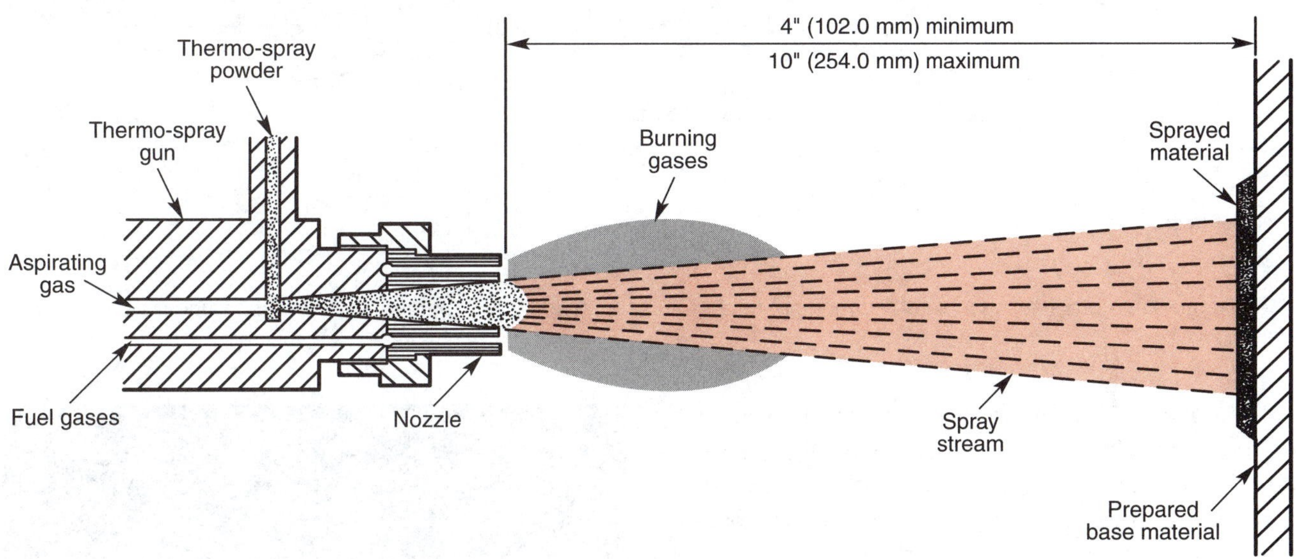

Figure 10-27.
Study the diagram of the thermo spray process and how it operates.

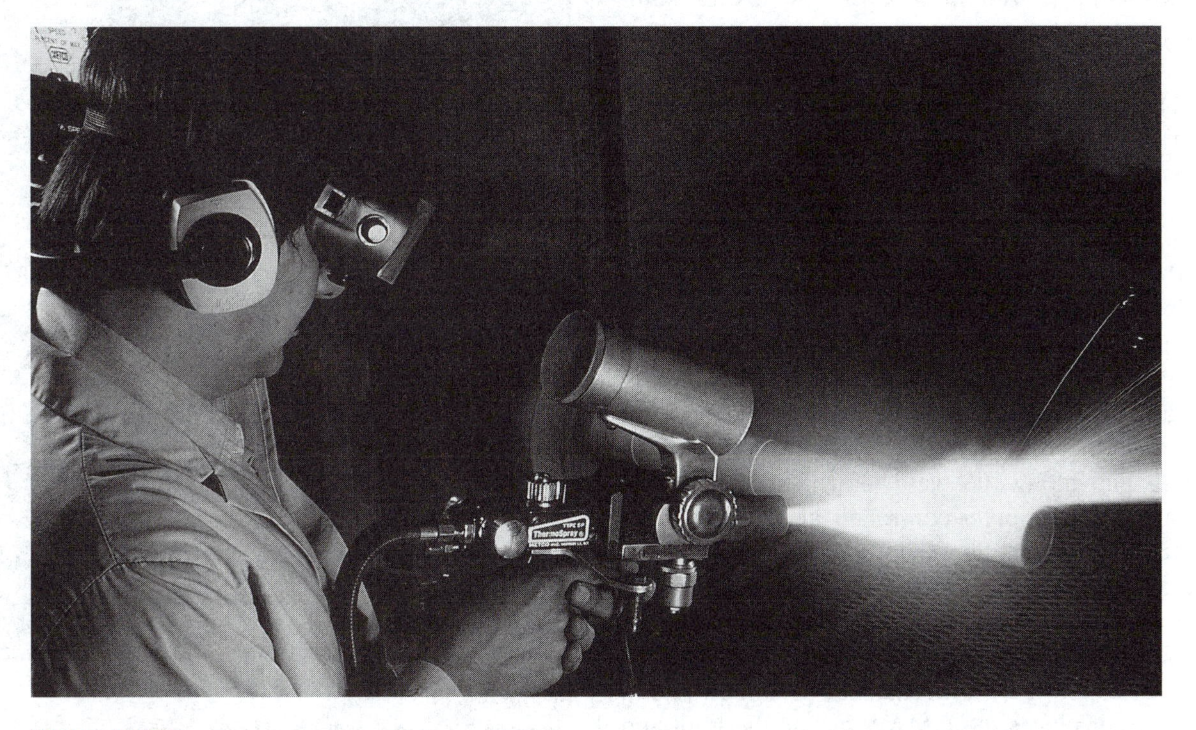

Figure 10-28.
A thermo spray gun uses metal and ceramic materials in powder form and an oxyfuel gas flame to apply coatings. (Metco Inc.)

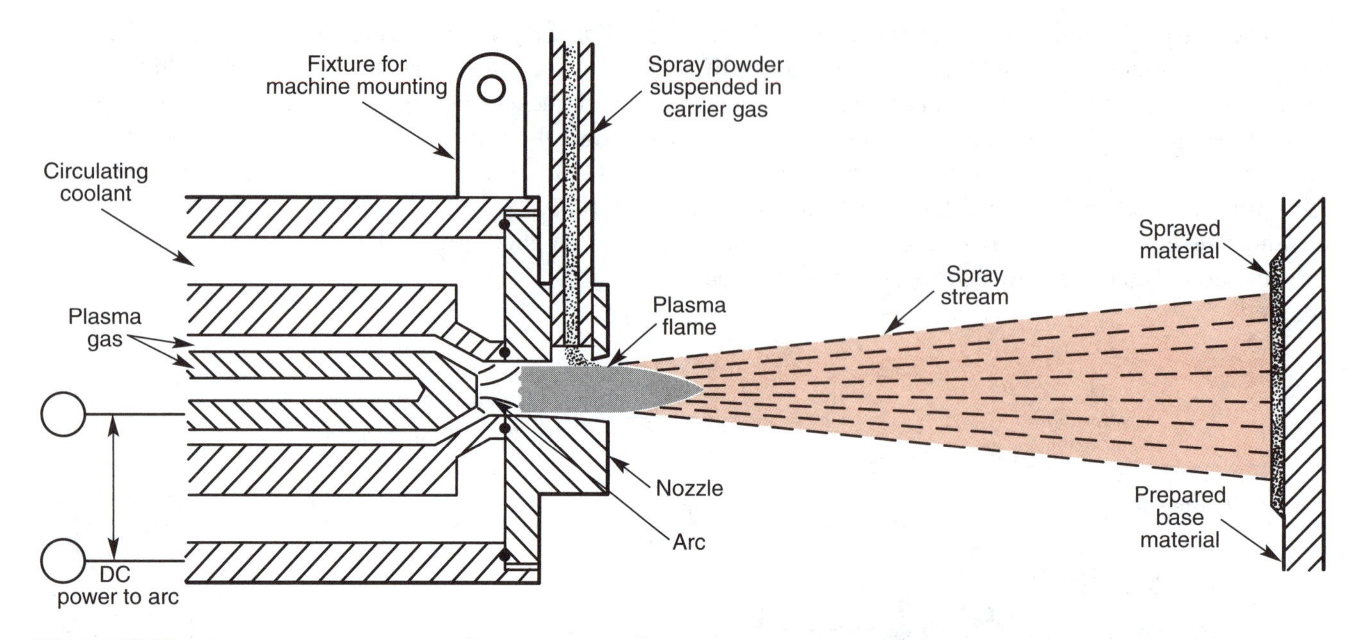

Figure 10-29.
The plasma flame process, capable of "super high" temperatures, can spray any material that will melt without decomposing.

Application of this process includes the spraying of rocket nozzles and nose cones with high melting point materials. Jet engine turbine blades are often given a protective coating by the plasma flame process.

Automatic (Robotic) and Mechanized Welding Systems

Automatic or ***robotic welding*** systems use computer-controlled machines to ensure consistent welds and fits between parts. Figure 10-31 shows an example of robotic welding.

The use of robots is generally limited to production welding. Robots commonly use GMAW or FCAW for welding structural shapes and RSW for welding car bodies.

Robots accurately and consistently place the welding gun in the weld joint. But for maintenance, construction, and some special fabrications, robot systems are impractical due to the cost, programming time, and system maintenance required.

Mechanized welding uses a machine to move the welding gun along the weld joint. The machine is generally not computer-controlled and is under the supervision of a welding operator. This use of mechanized systems improves the productivity and consistency of the welding process, but the equipment is not as expensive as a robot welding system. Figure 10-32 shows an example of a mechanized welding system. The operator places the GMAW gun in the joint and monitors the process. The part is rotated as the filler metal is added to the joint.

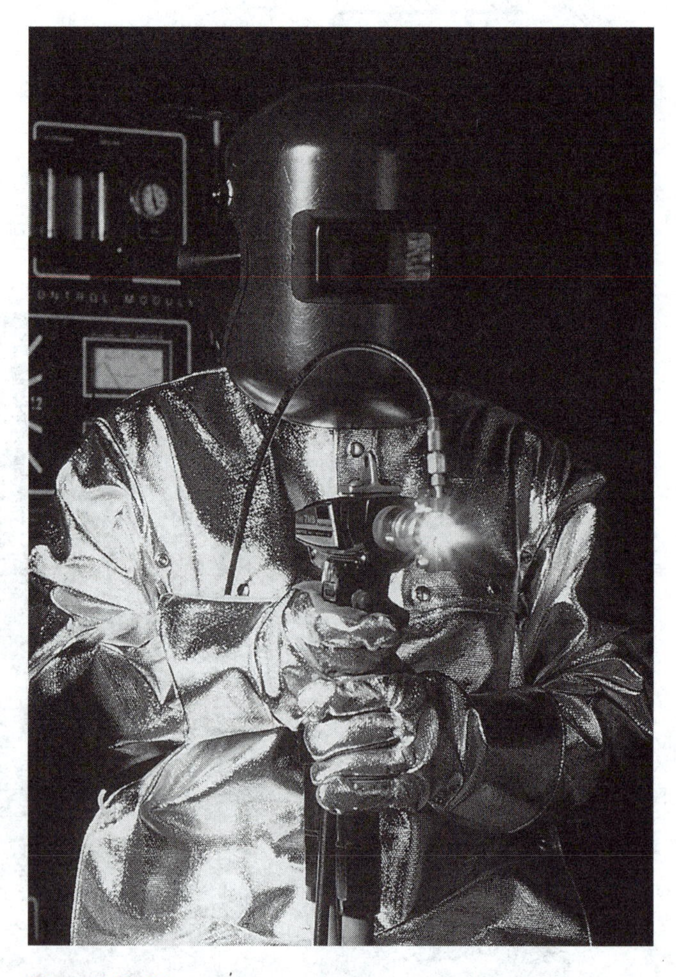

Figure 10-30.
Preparing to use the plasma flame process is shown. Note how the operator is dressed for protection against heat. (Metco Inc.)

Figure 10-31.
Robotic welding, although fast and accurate, is limited to high volume production. A robot welding rollover protection is shown. (Caterpillar Inc.)

Figure 10-32.
A mechanized welding system is supervised by a welding operator. (Kress Corporation)

Standard Welding, Joining, and Allied Processes

As discussed in this unit, the quality of a fabricated part or assembly depends on the correct application of one or more processes. The American Welding Society recognizes the use of 116 welding and allied processes. These are grouped under 11 major headings. See Figure 10-33. As shown in the chart, the different processes are identified with specific abbreviations. These are standard abbreviations used on drawings. When interpreting information on prints, it is important to be able to recognize how the different processes are identified and the operations they describe.

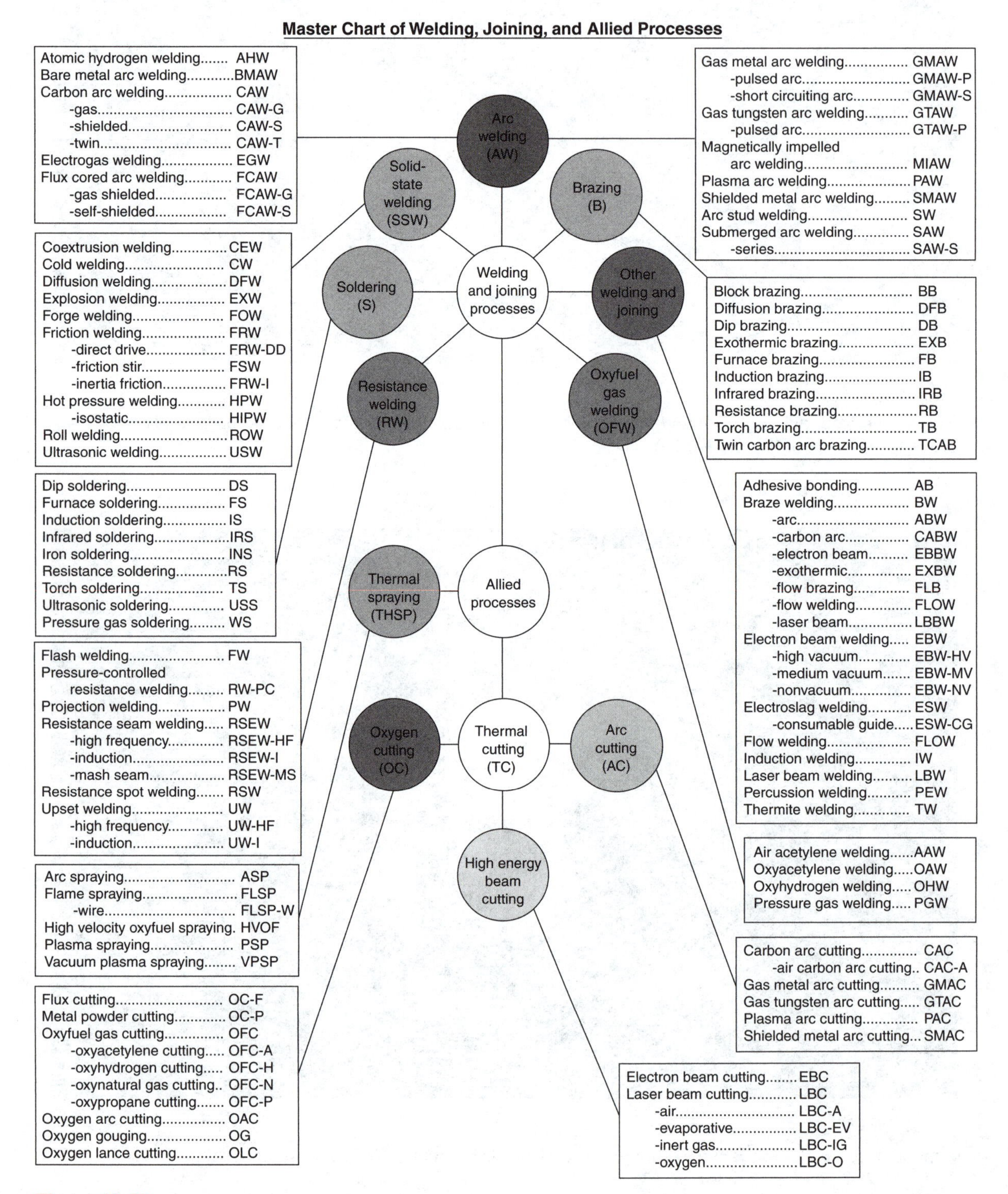

Figure 10-33.
Standard welding processes recognized by the American Welding Society. (Adapted from AWS A3.0:2001)

Name ______________________ Date ____________ Class ____________

Review Questions

Match the following terms to their descriptions by placing the letter of the correct answer in the appropriate blank space next to the number.

1. _____ Microwelding
2. _____ Soldering
3. _____ Metallizing
4. _____ Brazing
5. _____ Welding
6. _____ Gas welding
7. _____ Arc welding
8. _____ Gas tungsten arc welding
9. _____ Gas metal arc welding
10. _____ Thermo spray
11. _____ Submerged arc welding
12. _____ Arc stud welding
13. _____ Plasma flame spray
14. _____ Resistance welding
15. _____ Resistance spot welding
16. _____ Resistance seam welding
17. _____ Projection welding
18. _____ Flash welding
19. _____ Electron beam welding
20. _____ Friction welding
21. _____ Laser beam welding
22. _____ Flux cored arc welding
23. _____ Ultrasonic welding
24. _____ Robotic welding
25. _____ Cold welding

(A) A joining technique that uses an electric arc to produce heat that melts and fuses base metals. Filler metal from an electrode may be added to the joint.
(B) A welding technique that uses a bare metal electrode. A blanket of flux on the work shields the welding arc.
(C) Pressure alone is used to make the joint in this welding process.
(D) Uses frictional heat and pressure to produce a full strength weld in seconds.
(E) Molten metal is often ejected from the joint with this process.
(F) Uses burning gases to produce the heat needed to make the weld.
(G) Uses a gas-shielded permanent electrode that is not consumed in the welding process.
(H) Uses a gas-shielded consumable electrode that melts and contributes filler metal to the joint.
(I) A welding technique used to attach metal studs or similar objects.
(J) A welding process that uses a continuous wire electrode with gas-producing and flux materials within the wire.
(K) Heating joining metals to a suitable temperature to cause them to melt and fuse together. Pressure may or may not be applied, and filler metal may or may not be used.
(L) Uses a nonferrous alloy with melting temperatures above 840°F (450°C) but lower than the melting point of metals being joined.
(M) Widely used process for car bodies. Welds can be made quickly and do not require addition of filler metal.
(N) Must be used in a vacuum to prevent contamination of weld metal by atmospheric gases.
(O) Carried out at a temperature lower than 840°F (450°C) and has a relatively low joint strength.
(P) Group of welding processes where fusion heat is obtained from resistance to flow of electric current through the work and by application of pressure.
(Q) Form of resistance welding where welds are localized at predetermined points by the design of the parts to be welded.
(R) Process that uses a beam of intense light only a few microns in diameter.
(S) Process for joining metals without use of solders, fluxes, or filler metals and usually without application of external heat.
(T) Circular electrodes apply pressure to hold metal pieces being welded together. Resulting weld is a series of overlapping spot welds made by rotating electrodes.
(U) Technique employed to weld very fine wires to microcircuits (chips).
(V) Computer-controlled automatic welding systems.
(W) Flame spraying process where wire is melted in a gas flame and vaporized by compressed air, which carries it to a previously prepared work surface.
(X) Flame spraying equipment used to apply metals in powder form.
(Y) Capable of producing temperatures up to 30,000°F (16,649°C). This flame spraying technique can use any material that will melt without decomposing.

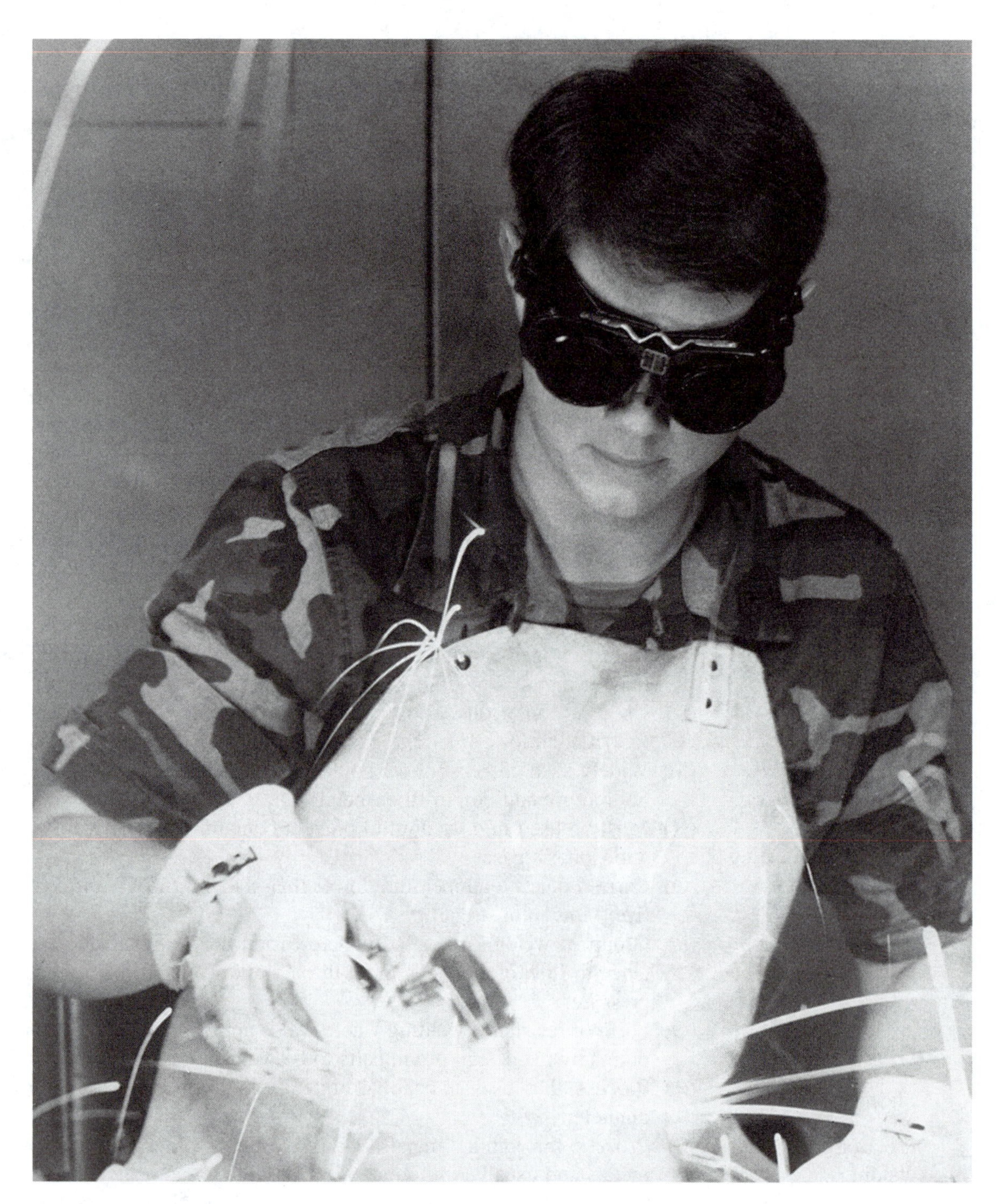

The United States Army and other branches of the Armed Forces offer an excellent opportunity for learning welding skills. Best of all, you get paid while you learn.

Unit 11

Threaded Fasteners

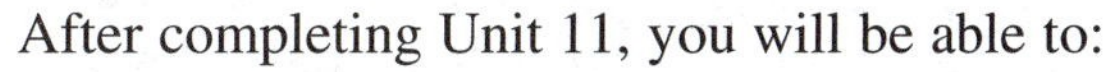

After completing Unit 11, you will be able to:

- Identify the parts of a screw thread.
- Explain why a threaded fastener would be used instead of welding.
- Describe the components of a thread note.
- Use a thread series table to select a tap drill for a specified thread form.
- Explain the way threaded fasteners are represented on a print.
- Identify different types of threaded fasteners.
- Identify threaded fasteners that are commonly welded.

Key Words

bolts
boss
cap screws
class of fit
detailed representation
fastener
helix
helix angle
machine screws
nuts
pitch
schematic representation
screws
set screws
simplified representation
stud
tap drill
thread note
thread series
tolerance
Unified National Coarse (UNC)
Unified National Extra Fine (UNEF)
Unified National Fine (UNF)
weld nut

Welding is a method of permanently joining materials together. But sometimes parts need to be assembled and then disassembled. A ***fastener*** is a device used to hold together parts that may be assembled and disassembled easily. Threaded fasteners, such as bolts, nuts, machine screws, bosses, and studs, hold parts together using threads.

There are many advantages for using threaded fasteners. Because threaded fasteners are manufactured in standard sizes in both English and metric units, they are easily purchased in large quantities (from equipment distributors) or a few at a time (at a local hardware store). Threaded fasteners, unlike welding, can easily join dissimilar materials together—such as steel parts to aluminum or metal parts to plastic. Also, threaded fasteners are easily installed in a shop or in the field (on the job site) using hand or power tools.

The threaded fastener uses the principle of projecting an inclined plane around a cylindrical shape for holding strength. The concept of the threaded fastener and many other simple machines is attributed to the ancient Greeks. The

Roman Legions used threaded fasteners to hold together armor, as did knights in the Middle Ages. However, there was no standard thread form. Each threaded fastener was made by hand, usually by filing the thread into a solid piece of cylindrical material.

Thread Forms

In 1864, near the end of the Civil War, William Sellers described the need for a national screw thread standard at a meeting of machinists and engineers in Philadelphia. Sellers told the audience about a thread shape that could be manufactured easily and checked for accuracy using standard equipment found in a machine shop. This thread form remained the standard in the United States until the Unified Thread Standard form was adopted in 1948. The Unified Thread Standard form was also adopted by Canada and Great Britain.

Metric-based threaded fasteners are classified under the Metric Screw Thread Series standard. The metric thread form is the world standard. For this reason, both inch-based fasteners and metric fasteners are used in the US. Inch-based and metric-based screw threads have the same basic *profile* (shape), but the two systems are not interchangeable. The common parts of a screw thread are shown in Figure 11-1.

The basic shape of a screw thread is formed by a point curve that wraps around a cylinder in a spiral. This shape is known as a ***helix***. The ***helix angle*** is the angle formed by the incline of the helix and a line perpendicular to the vertical axis of the screw thread. The ***pitch*** of a screw thread is defined as the distance from a point on one thread to a corresponding point on the next thread. For inch-based screw threads, the pitch is measured in relation to the number of threads per inch. It is expressed using the formula *1/number of threads per inch*. For metric-based screw threads, the pitch is indicated as the actual pitch measurement and is measured in millimeters.

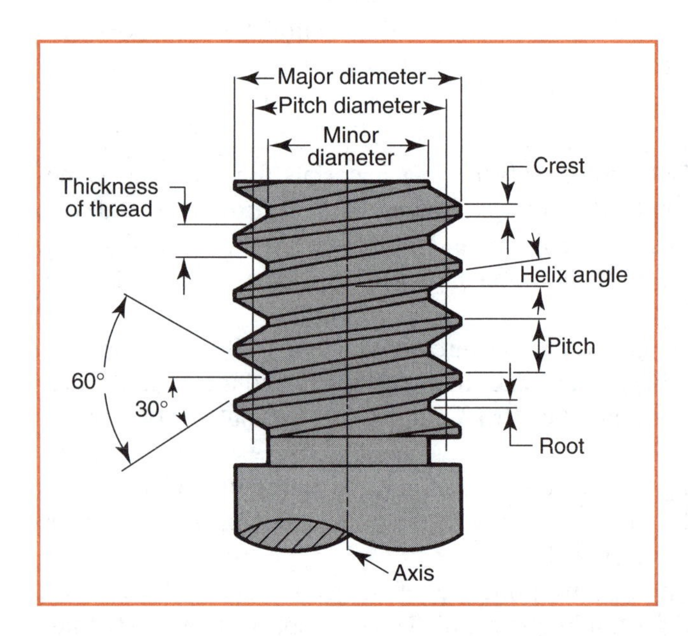

Figure 11-1.
The basic parts of a screw thread.

Thread Series

In the Unified Thread Standard system, threads are classified under ***thread series***. Thread series are used to specify diameter, pitch, tolerance, and other standard characteristics. The most common threads found in the United States are classified under the Unified graded pitch thread series. Graded pitch threads are groups of diameters that change in pitch, or the number of threads per inch. There are three common graded pitch series: ***Unified National Coarse (UNC)***, ***Unified National Fine (UNF)***, and ***Unified National Extra Fine (UNEF)***. See Figure 11-2.

Unified National Coarse threads are used for general purpose fasteners. The threads have good strength and are easy to fasten and unfasten.

Unified National Fine threads have more threads per inch than coarse threads. They are used when vibration might affect the holding power of the threads. UNF fasteners are used for cars and airplanes. Fine threads are also used when the number of threads that can be engaged between the fastener and mating part is limited.

Unified National Extra Fine threads are used for instrumentation and applications requiring fine adjustment.

Constant pitch threads are also specified in thread series under the Unified system and are used in some applications. In these series, the pitch remains constant as the diameter of the fastener changes. Common constant pitch series include the 8-, 12-, 16-, 20-, 28-, and 32-thread series. The series number designates threads per inch.

Thread Fit Classes

Threaded fasteners are also classified by ***class of fit***. The class of fit determines the allowable difference in size from the actual size (the ***tolerance***) and the tightness or looseness between two mating fasteners. There are three classes of fit: Class 1, Class 2, and Class 3.

Class 1 fits are used where the greatest amount of tolerance (looseness) is allowed. Usually, Class 1 fits are used for applications requiring frequent and rapid assembly and disassembly.

Class 2 fits are considered the standard for general purpose manufactured fasteners. Class 3 fits have very close tolerances. Class 3 fits are used for set screws, aircraft fasteners, and socket head cap screws.

STANDARD SERIES THREADS — GRADED PITCHES						
NOMINAL DIAMETER	UNC		UNF		UNEF	
	THREADS PER INCH	TAP DRILL	THREADS PER INCH	TAP DRILL	THREADS PER INCH	TAP DRILL
0 (.0600)			80	3/64		
1 (.0730)	64	No. 53	72	No. 53		
2 (.0860)	56	No. 50	64	No. 50		
3 (.0990)	48	No. 47	56	No. 45		
4 (.1120)	40	No. 43	48	No. 42		
5 (.1250)	40	No. 38	44	No. 37		
6 (.1380)	32	No. 36	40	No. 33		
8 (.1640)	32	No. 29	36	No. 29		
10 (.1900)	24	No. 25	32	No. 21		
12 (.2160)	24	No. 16	28	No. 14	32	No. 13
1/4 (.2500)	20	No. 7	28	No. 3	32	7/32
5/16 (.3125)	18	F	24	I	32	9/32
3/8 (.3750)	16	5/16	24	Q	32	11/32
7/16 (.4375)	14	U	20	25/64	28	13/32
1/2 (.5000)	13	27/64	20	29/64	28	15/32
9/16 (.5625)	12	31/64	18	33/64	24	33/64
5/8 (.6250)	11	17/32	18	37/64	24	37/64
11/16 (.6875)					24	41/64
3/4 (.7500)	10	21/32	16	11/16	20	45/64
13/16 (.8125)					20	49/64
7/8 (.8750)	9	49/64	14	13/16	20	53/64
15/16 (.9375)					20	57/64
1 (1.000)	8	7/8	12	59/64	20	61/64

Figure 11-2.
Specifications for graded pitch threads in the Unified National Coarse, Unified National Fine, and Unified National Extra Fine thread series.

In thread designations, external or internal threads are usually specified together with the class of fit. External threads are specified with the letter A and internal threads are specified with the letter B. For example, a Class 1A designation specifies Class 1 external threads.

Thread Notes

A ***thread note*** is used to specify information for threads on a print. The note includes the nominal diameter of the thread, the number of threads per inch, the thread series, the class of fit, the internal or external thread designation, and sometimes the length of the fastener. Figure 11-3 shows thread notes for both Unified and metric thread forms.

For internal threads, the tap drill size is often given with the thread note. A ***tap drill*** is a tool used to make a hole in a part for *tapping* (removing material to form threads).

The tap drill is used to assure that approximately 75% of the thread depth is formed. If 100% of the thread depth is cut, binding and perhaps breaking of the tap may occur.

Standard tap drill sizes for Unified threads are shown in Figure 11-2. If the tap drill diameter is not given with the thread note, the table can be used to determine the correct tap drill.

To read the table, first determine the diameter of the fastener. Then, determine the thread form (for example, UNC or UNF). Finally, determine the threads per inch. The column next to the Threads Per Inch column lists the correct tap drill.

Standard Thread Representations

On prints, threads are typically represented using one of three methods. Threads are drawn using detailed, schematic, or simplified representations. Because of the time required to draw detailed representations, many companies prefer to draw threads in schematic or simplified form. Methods of thread representation are shown in Figure 11-4.

Detailed Representation

A ***detailed representation*** shows the actual thread profile and an accurate representation of the number of threads per inch, Figure 11-4A. This method is very time consuming and is not often used.

Schematic Representation

A ***schematic representation*** uses schematic symbols to represent the threads. It shows the thread crest lines with thin lines and the thread root lines with thick lines, Figure 11-4B. The lines are often drawn at an approximation of the pitch rather than at a calculated pitch distance. The lines are drawn perpendicular to the major axis of the fastener.

Simplified Representation

A ***simplified representation*** is the most common method of representing threads on a print because it is easy to draw, Figure 11-4C. For external threads, the crest line is shown as an object line and the root line

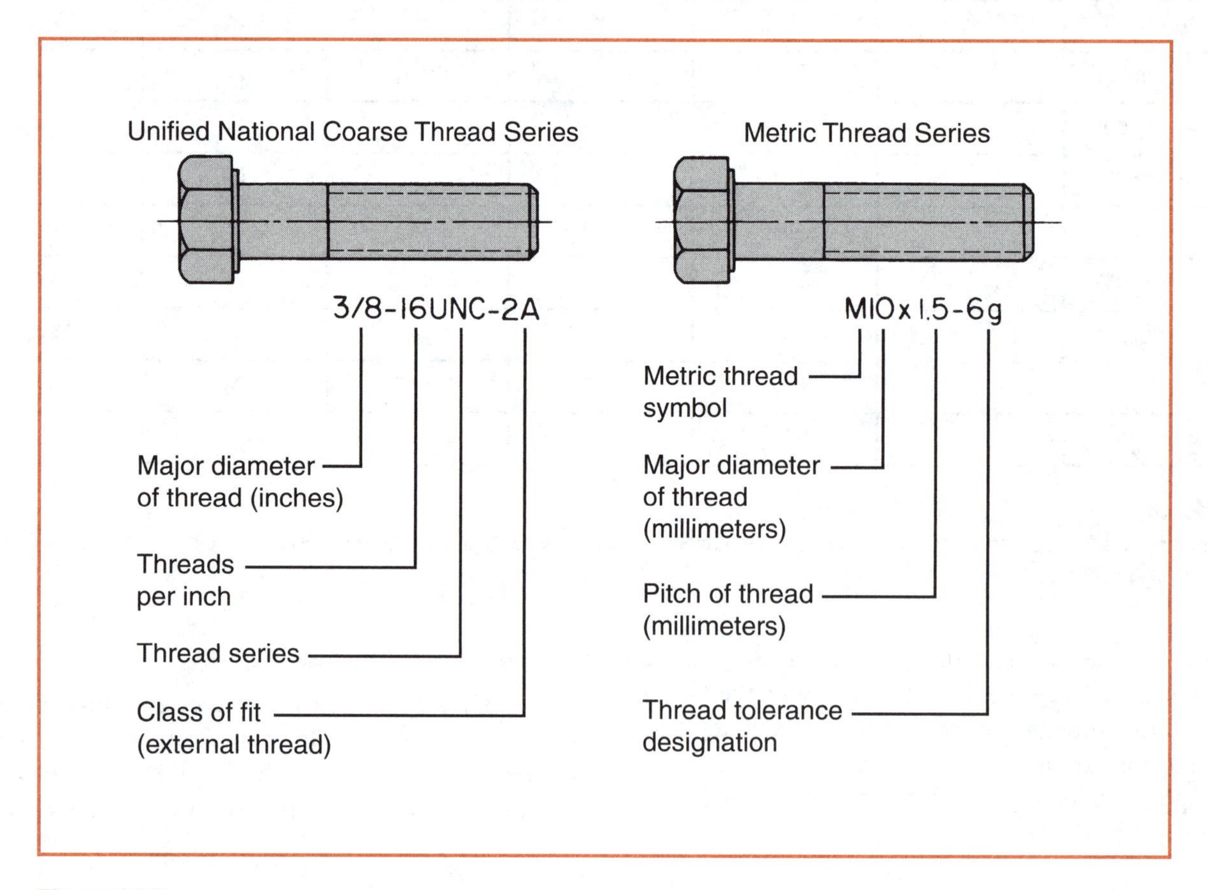

Figure 11-3.
Thread note conventions for Unified (inch-based) and metric-based thread.

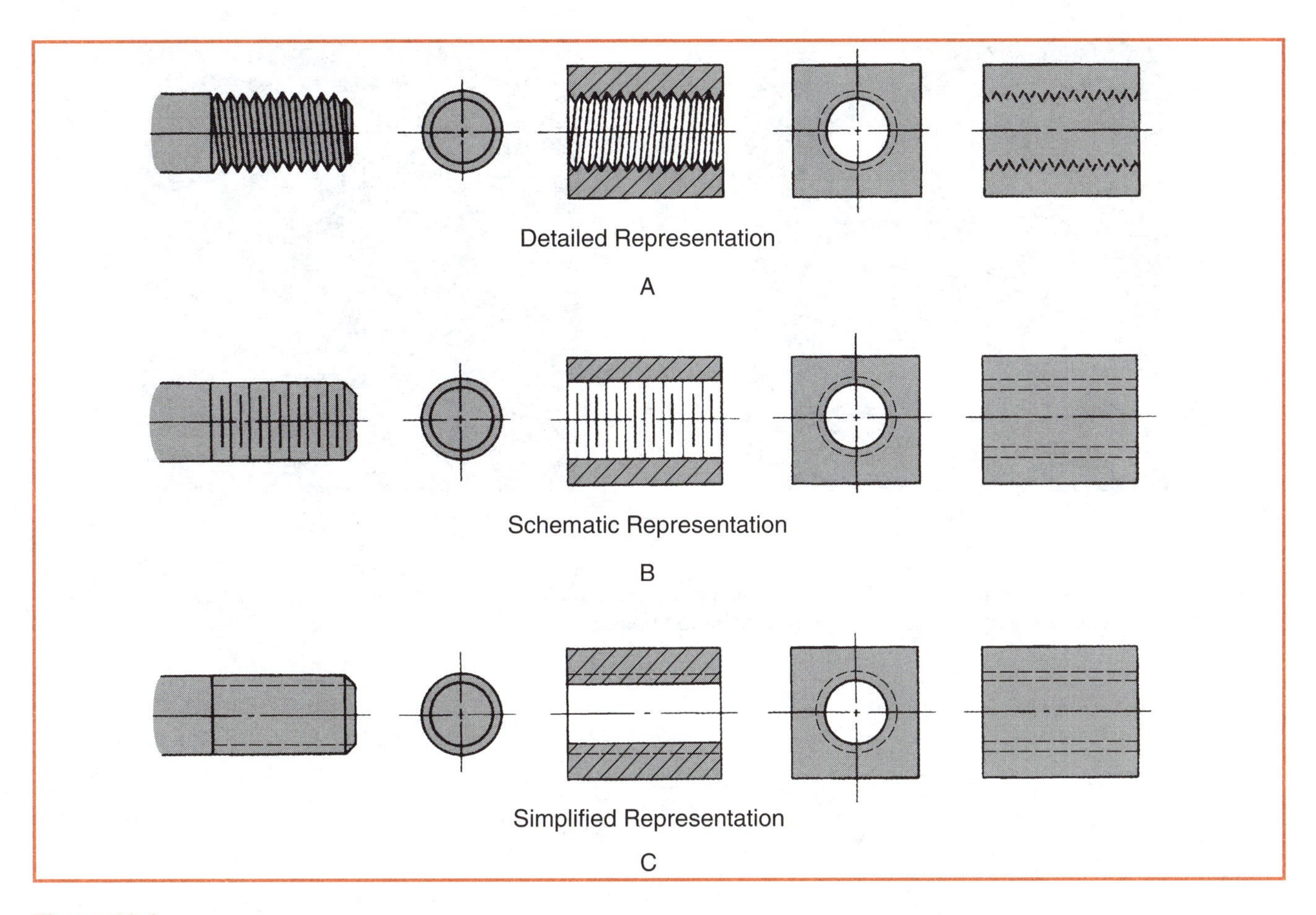

Figure 11-4.
Standard methods for representing threads on drawings.

is shown as a hidden line. For internal threads, the crest line is usually shown as a hidden line. However, other representations may be used, depending on the type of drawing.

Metric Threads

The method for showing metric threads on a print is similar to that used for Unified threads, with minor variations. All metric thread notes start with the letter M followed by the diameter. The pitch is then shown in the note following an X. For example, a 10 mm thread with a 1.5 mm pitch would be shown as M10 X 1.5.

Types of Fasteners

There are many types of threaded fasteners. The major types include bolts, nuts, and screws. Variations of each type are used for specific applications.

Bolts are used to assemble parts not requiring close tolerances. They are manufactured with hexagonal or square heads. ***Nuts*** are used to secure bolts into assemblies.

Screws are used to fasten parts and hold assemblies in place. Common types include cap screws, machine screws, and set screws. ***Cap screws*** pass through a clearance hole in one part and are screwed into a threaded hole in the mating part. ***Machine screws*** are smaller diameter fasteners and are usually used for thin parts. ***Set screws*** are used to prevent rotary motion between two parts or to hold a part in a specific location.

In some applications, fasteners are welded to parts. Bolts, nuts, or screws are sometimes joined by welding because welding makes the fasteners easy to attach. Common threaded fasteners that are welded include the *weld nut*, *stud*, and *boss*. Fasteners may be attached by arc welding or resistance welding. Joints made with weld nuts and studs are usually considered projection welds.

A ***weld nut*** is used on thin sections where there would not be enough threads engaged in the material to provide adequate holding strength. A ***stud*** is used when an external thread is required. A ***boss*** has internal threads and is used for locating a mating part in an assembly, Figure 11-5.

A

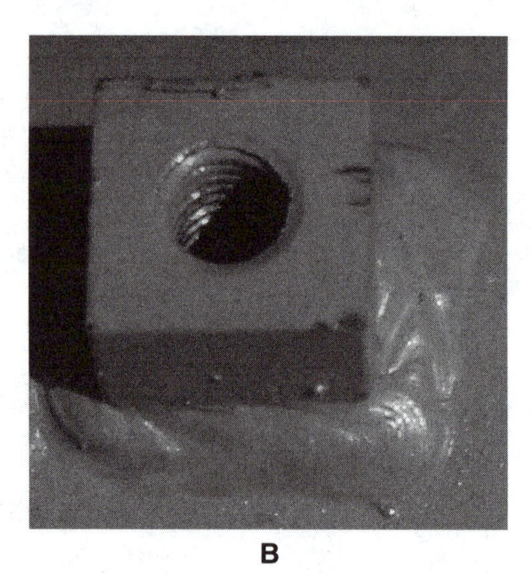

B

Figure 11-5.
Weld bosses are used when internal threads are required for an assembly. A—A round boss used to provide a mating surface. B—A square-shaped boss.

Name ______________________ Date ____________ Class ____________

Review Questions

Part I

1. Define the term *fastener*. ______________________

____

____

2. The point curve shape that wraps around a cylinder in a spiral to form the basic shape of a screw thread is known as a _____.

____

3. The thread _____ is the distance from a point on one thread to a corresponding point on the next thread.

____

4. What is the difference between coarse and fine thread? ______________________

____

____

5. In constant pitch thread series classes, the pitch remains constant as the _____ of the fastener changes.

____

6. Class _____ fits describe threads having closer tolerances than threads in other fit classes and are used for fasteners such as set screws.

____

7. In a thread note, the letter _____ is used to indicate internal threads.

____

8. The simplest method of representing a thread on a print is called a _____ representation.

____

9. The letter _____ is used at the beginning of a thread note to indicate metric thread.

____

10. A _____ is a type of fastener welded to a part to provide an external thread.

____

Part II

Use the thread note below and the table in Figure 11-2 in this unit to answer Questions 11–15.

.437-14UNC-2A

11. How many threads per inch are indicated in the thread note? ______________________

12. What thread series is specified? ______________________

____

13. Is the thread internal or external? ______________________

14. What is the diameter of the thread? ______________________

15. What tap drill size should be used for the thread?

____

Print Reading Activities

Part I

Use the support gusset in Figure 11-7 to answer the following questions.

1. List the name of the part. ____________________

 ____

2. List the overall length of the part. ____________
3. List the overall width of the part. ____________
4. List the thickness of the part. ______________
5. List the tolerance for a 2-place decimal dimension.

 ____

6. What is the maximum allowable length of the part?

 ____

7. List the material used for the part. ___________

 ____

8. List the tolerance for angles. ____________
9. Explain the note *Do not scale drawing.* _______

 ____

 ____

 ____

 ____

 ____

10. Determine the dimension for the following letters on the print:

 A. ____________________________________

 B. ____________________________________

 C. ____________________________________

Name ______________________________

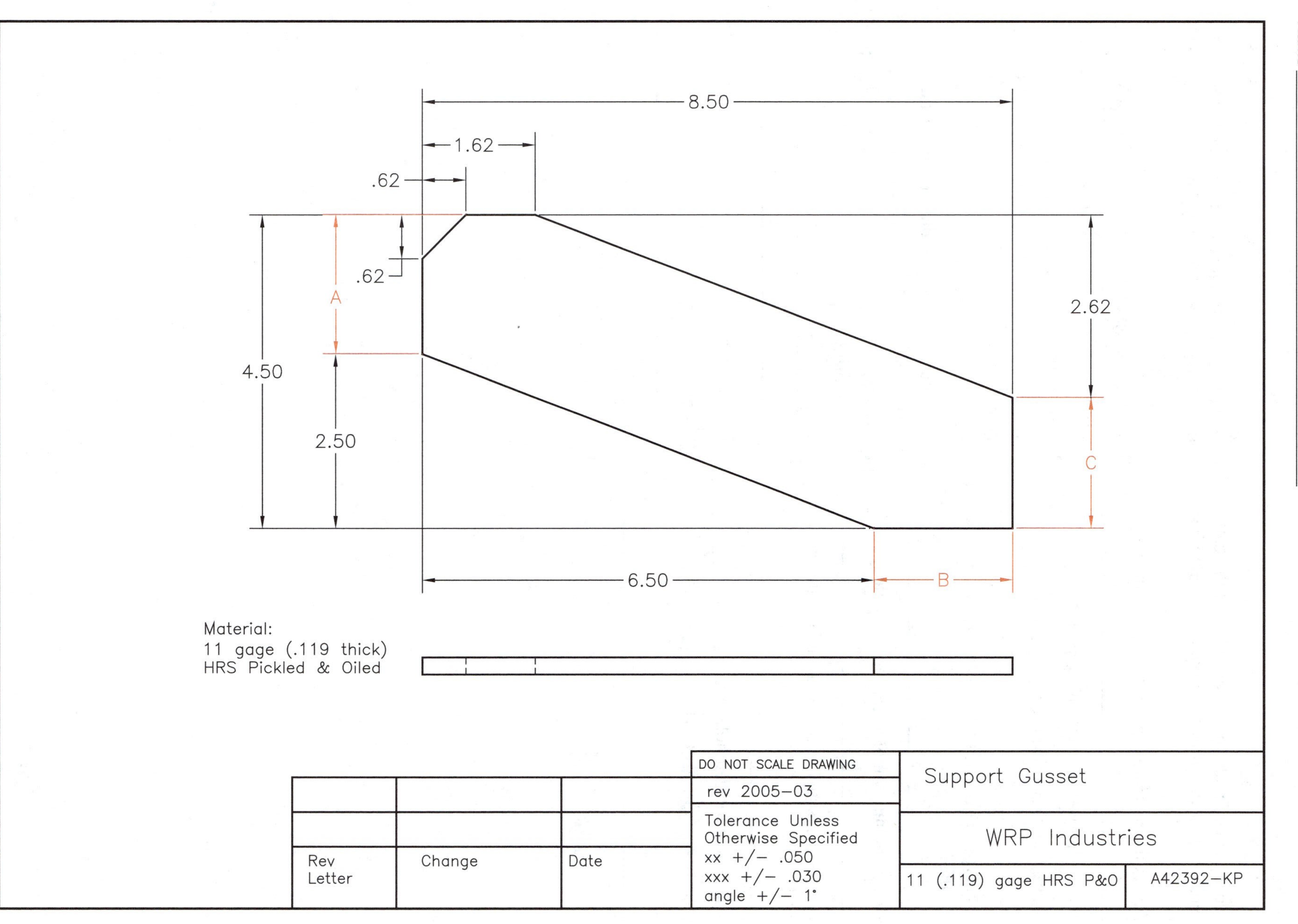

Figure 11-6.
Use this print to answer the Print Reading Activities questions in Part I.

Part II

Use the splice plate in Figure 11-8 and the table in Figure 11-2 in this unit to answer the following questions.

1. What is the maximum overall length of the part?
   ____
2. What is the overall width of the part? ________
3. What is the thread form for the .50 hole?______
   ____
4. What type of dimensioning system is used for the drawing?______________________
   ____
5. What is the maximum angle for the angle listed?
   ____
6. What is the supplementary angle for the angle listed?
   ____
7. How many threaded holes are required?_______
8. What is the tap drill size for the .25-28 UNF holes?
   ____
9. If the .50 hole required fine threads, how many threads per inch (TPI) would be cut in the hole?
   ____
10. What is the tap drill size for a .50 UNF hole?_____
11. List the scale for the drawing. ______________
12. List the material used for the part. ___________
13. What is the tap drill size for the .31-18 UNC hole?
    ____
14. How many threads would be cut in the .31-18 UNC hole?______________________
15. Determine the dimension for the following letters on the print.

    A. ______________________

    B. ______________________

    C. ______________________

    D. ______________________

    E. ______________________

Name ____

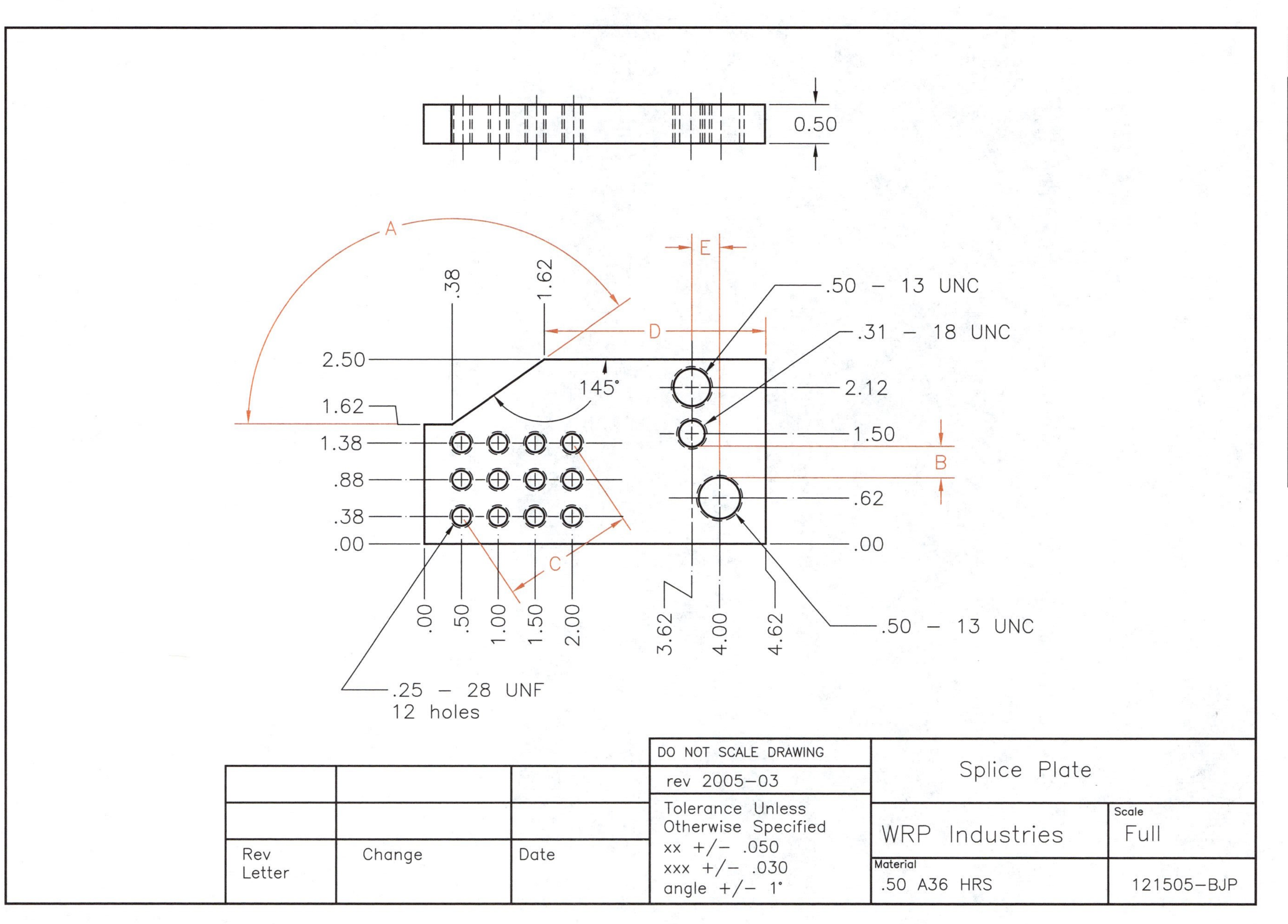

Figure 11-7.
Use this print to answer the Print Reading Activities questions in Part II.

Titanium F-14 aircraft wing center sections after they have been electron beam welded. The clam shell vacuum chamber (it can accommodate work up to 12 × 26 ft. [3.66 × 7.93m]) can ben seen in the background. (Grumman Aerospace Corp.)

Unit 12

Structural Metals

After completing Unit 12, you will be able to:

- Distinguish among the various structural shapes in which commercial metals are manufactured.
- Define common terms used to distinguish characteristics of metals and structural shapes.
- Explain the importance of knowing the characteristics of the metal(s) being welded.
- Explain the way structural shapes are specified.
- List organizations that provide specifications for metals.

Key Words

alloy
base metal
ferrous metal
metal
metal specifications
nonferrous metal
pure metal
structural shapes

Metal is available in a large range of shapes and sizes. Figure 12-1 shows the customary methods for designating or billing ***structural shapes*** on prints.

While these shapes are generally accepted, there may be differences in the way they are used by some companies. Stock structural shapes that have been cut, machined, bent, rolled, spun, stamped, or drawn and combined with castings, forgings, and extrusions can be found in all products fabricated by welding. Refer to Figure 12-2. Stock shapes are used whenever possible to keep material and machining costs at a minimum.

What Is Metal?

Metal plays an important role in welding. How would you define the term metal? ***Metal*** might be described as a tough, malleable material with high tensile strength and the ability to withstand high temperatures without melting or burning. This definition would be only partly correct. Only *some* metals have these characteristics. There are metals that do *not* have these features. For example, *mercury* is fluid at normal room temperature; body heat will melt *gallium* in the palm of your hand; and *lithium* is so soft your fingernail can scratch it.

Can you tell by looking at a piece of metal whether it is a ferrous metal or a nonferrous metal? A ***ferrous metal*** contains iron, while a ***nonferrous metal*** contains no iron. Could it be an alloy? An ***alloy*** is a mixture of two or more metals. Perhaps it is a pure metal, such as copper, tin, zinc, etc. A ***pure metal*** is a pure metallic element. When a pure metal is used as the principal alloying agent in an alloy, it is known as a ***base metal***.

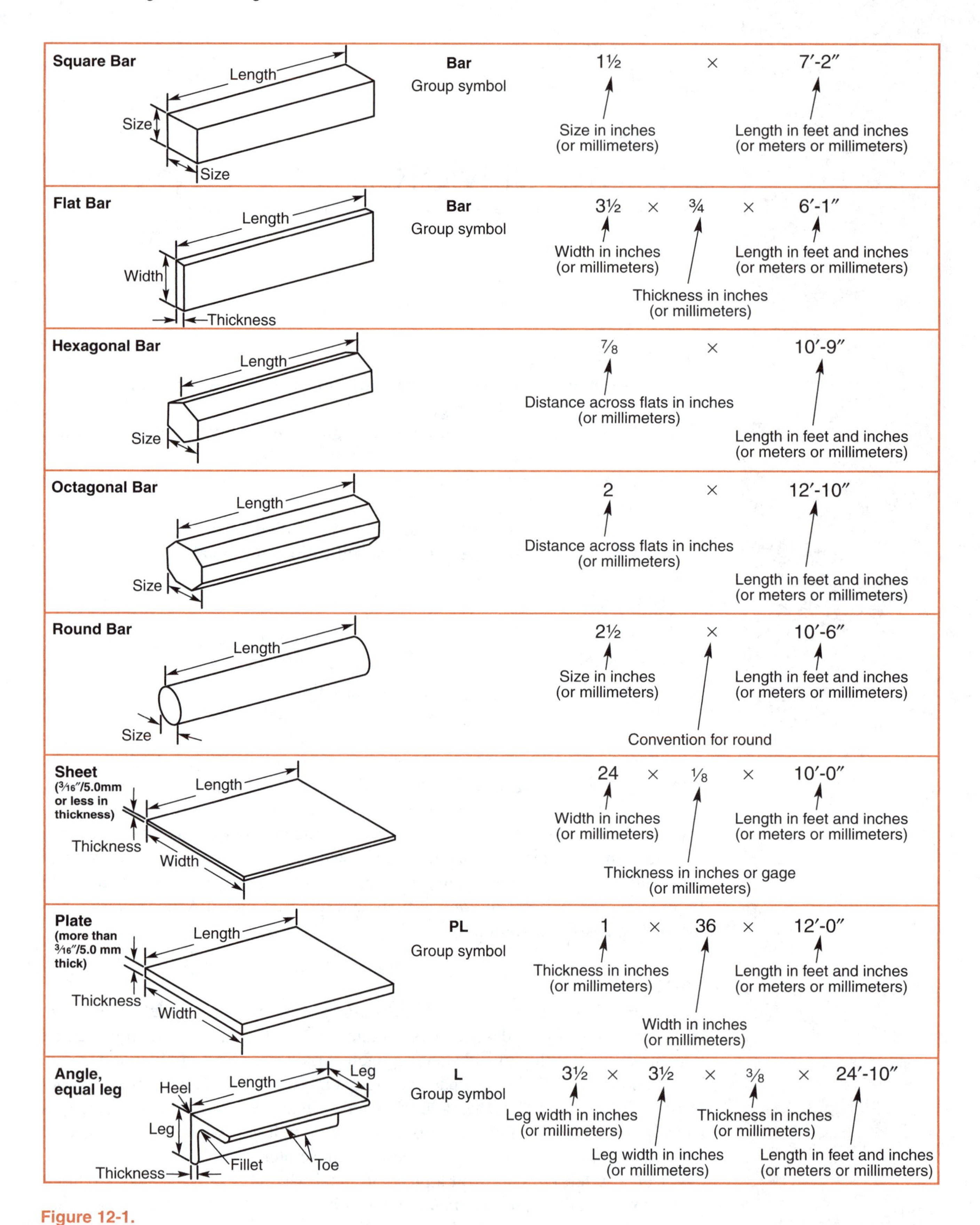

Figure 12-1.
Study the usual method of designing structural steel shapes on shop drawings. The dimensions given on the drawing are in either English or metric units.

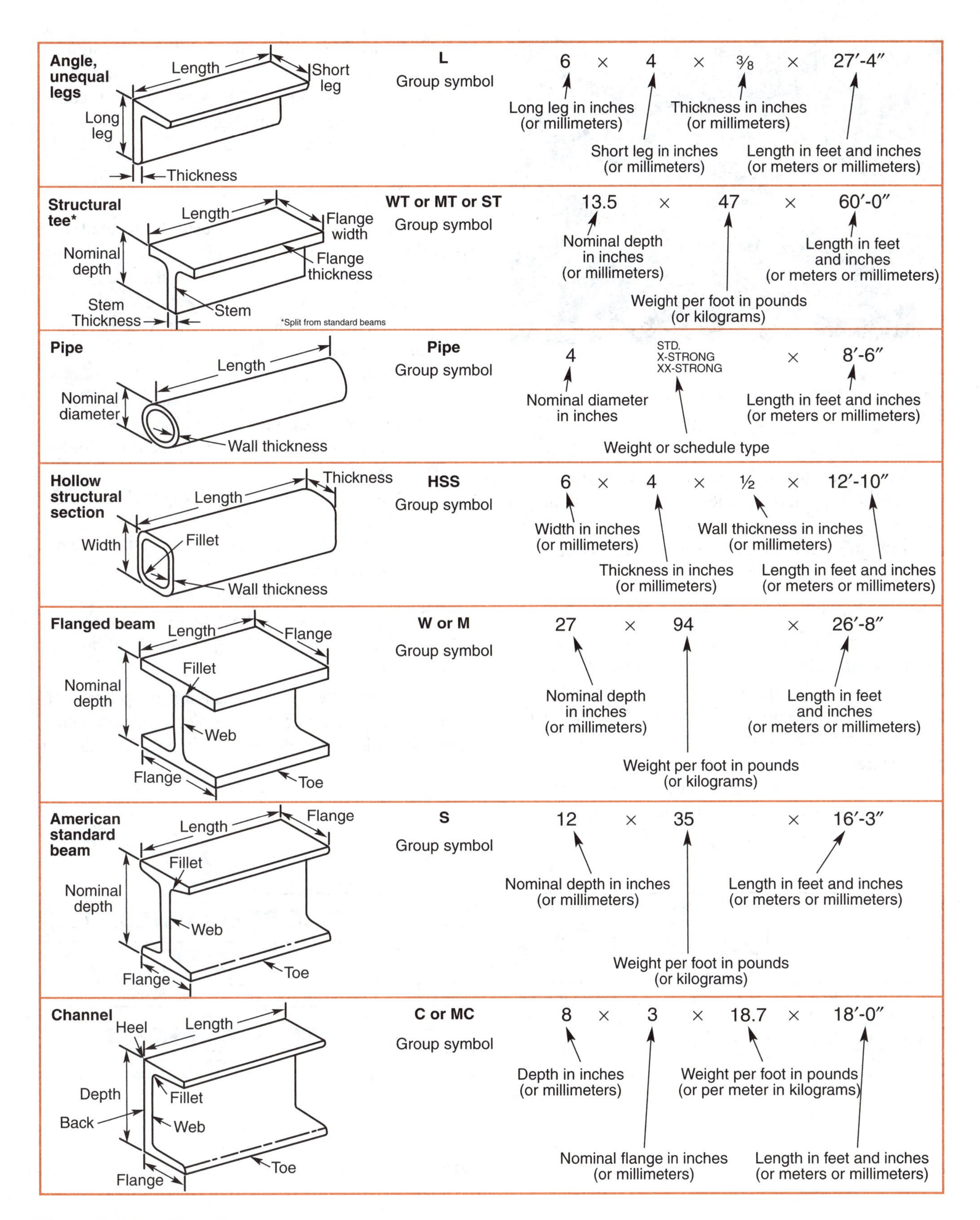

Figure 12-1. (continued)

Figure 12-2.
Many products like bridges are designed for fabrication from standard stock steel shapes.

There are no simple answers to these questions. Additional study and experience, however, can help a welder develop a working knowledge of metals to make the job safer, easier, and more successful.

Metals Supplied for Welding

The welder usually has little or no control over the metal furnished for welding. It must be assumed metal provided by the employer for a specific job meets print specifications. The metal supplier or the mill that produced the metal can certify the metallurgical characteristics of a metal.

As a welder, it is important to know the characteristics of the metal(s) you will be welding. This is important for two reasons:

- It aids in ensuring the welds meet design specifications.
- It permits you to take special safety precautions when welding metals that give off toxic fumes and residue.

Metal Specifications

Metal specifications are usually located in a special section of the title block. Figure 12-3 shows an example. There are times when the metal specs may be given elsewhere on the drawing.

Usually, the specifications are provided in one or more of the following standards:

- Military (MIL)
- American Iron and Steel Institute (AISI)
- Society of Automotive Engineers (SAE)
- American Society of Mechanical Engineers (ASME)
- American Society for Testing and Materials (ASTM)
- Some larger manufacturers have their own, like General Motors (GM)

Examples of specifications include:

- AISI-SAE 1018 Plain Low-Carbon Steel
- ASTM A36 Low-Carbon Structural Steel Plate

To secure the complete metallurgical specifications of the metal (including the chemical and mechanical properties) it is necessary to refer to the appropriate metals handbook.

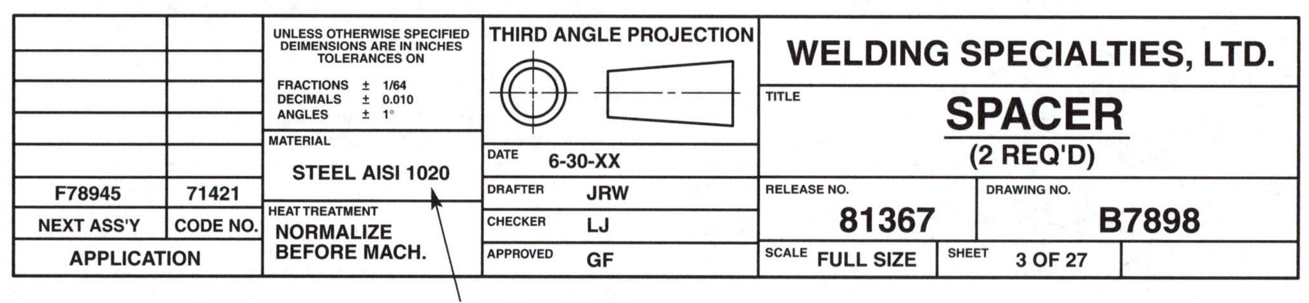

Figure 12-3.
Metal specifications are usually shown in a special section of a drawing's title block.

Name ______________________ Date ______________ Class ______________

Review Questions

1. How would you describe what a metal is?

2. Different metals are often used in a product based on the metal's special characteristics. For example, copper is used because it is easily soldered and a good conductor of heat and electricity. List some metals and identify their special characteristics.

	Metal	Characteristic
A.		
B.		
C.		
D.		
E.		

3. Define a ferrous metal and a nonferrous metal.

   A ferrous metal ______________________

   A nonferrous metal ______________________

4. Define a pure metal and an alloy.

   A pure metal ______________________

   An alloy ______________________

5. Why is it important for the welder to know the characteristics of the metal to be welded?

Supply the name of each structural shape and the group symbol used to identify it on a print. Place your answers in the spaces provided.

6. (a) Name: ______________________
   (b) Symbol: ______________________
7. (a) Name: ______________________
   (b) Symbol: ______________________
8. (a) Name: ______________________
   (b) Symbol: ______________________
9. (a) Name: ______________________
   (b) Symbol: ______________________
10. (a) Name: ______________________
    (b) Symbol: ______________________
11. (a) Name: ______________________
12. (a) Name: ______________________
    (b) Symbol: ______________________
13. (a) Name: ______________________
14. (a) Name: ______________________
15. (a) Name: ______________________
    (b) Symbol: ______________________
16. (a) Name: ______________________
17. (a) Name: ______________________
    (b) Symbol: ______________________

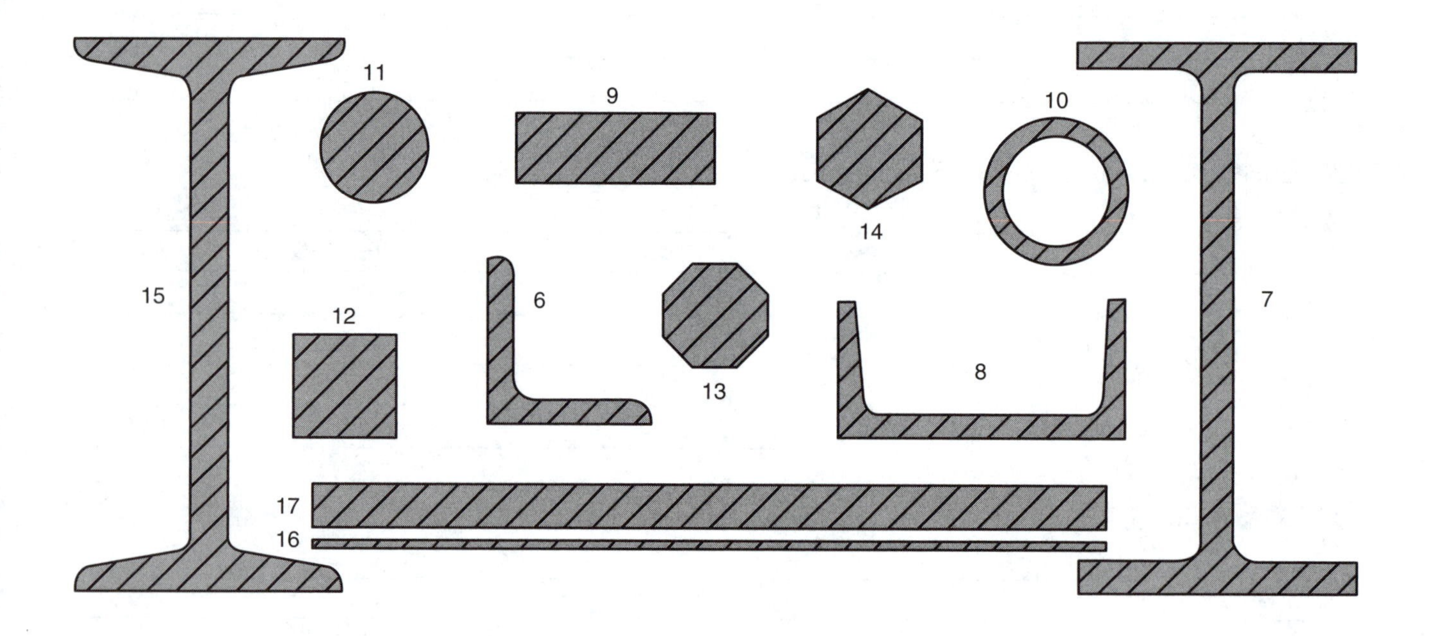

Unit 13
Common Types of Joints and Welds

After completing Unit 13, you will be able to:

- Identify basic joint designs by name and shape.
- Identify simple weld types by name and shape.
- Identify joint and groove style combinations by name and shape.
- Describe and locate common parts of fillet, groove, and butt welds by name.

Key Words

base material
bead weld
bevel angle
fillet gauge
fillet weld
groove
groove angle
groove weld
joint
parent metal
plug weld
reinforcement
resistance seam weld
resistance spot weld
slot weld
surfacing weld
weld face
weld leg
weld root
weld toe

This chapter reviews basic types of joints and welds. It is important to be able to visualize weld and joint types before studying weld symbols in the next chapter.

Joint Types

A basic ***joint*** is a way of arranging metal pieces in relation to each other so they can be welded. Common joints are shown in Figure 13-1. Combinations and variations of the basic joint designs can make other joints.

Weld Types

Each welding job will require one or more of the welds shown in Figure 13-2. Study each carefully.

A ***surfacing weld*** (known also by the nonstandard term ***bead weld***) consists of a narrow layer (or layers) of metal deposited in an unbroken puddle on the surface of the metal. Placing a series of surfacing welds tightly next to each other is also called *padding* or *lacing*.

A ***groove weld*** is a weld made in a groove on one or both surfaces to be joined.

The ***fillet weld*** is approximately triangular in shape and is used when joining two surfaces at an angle.

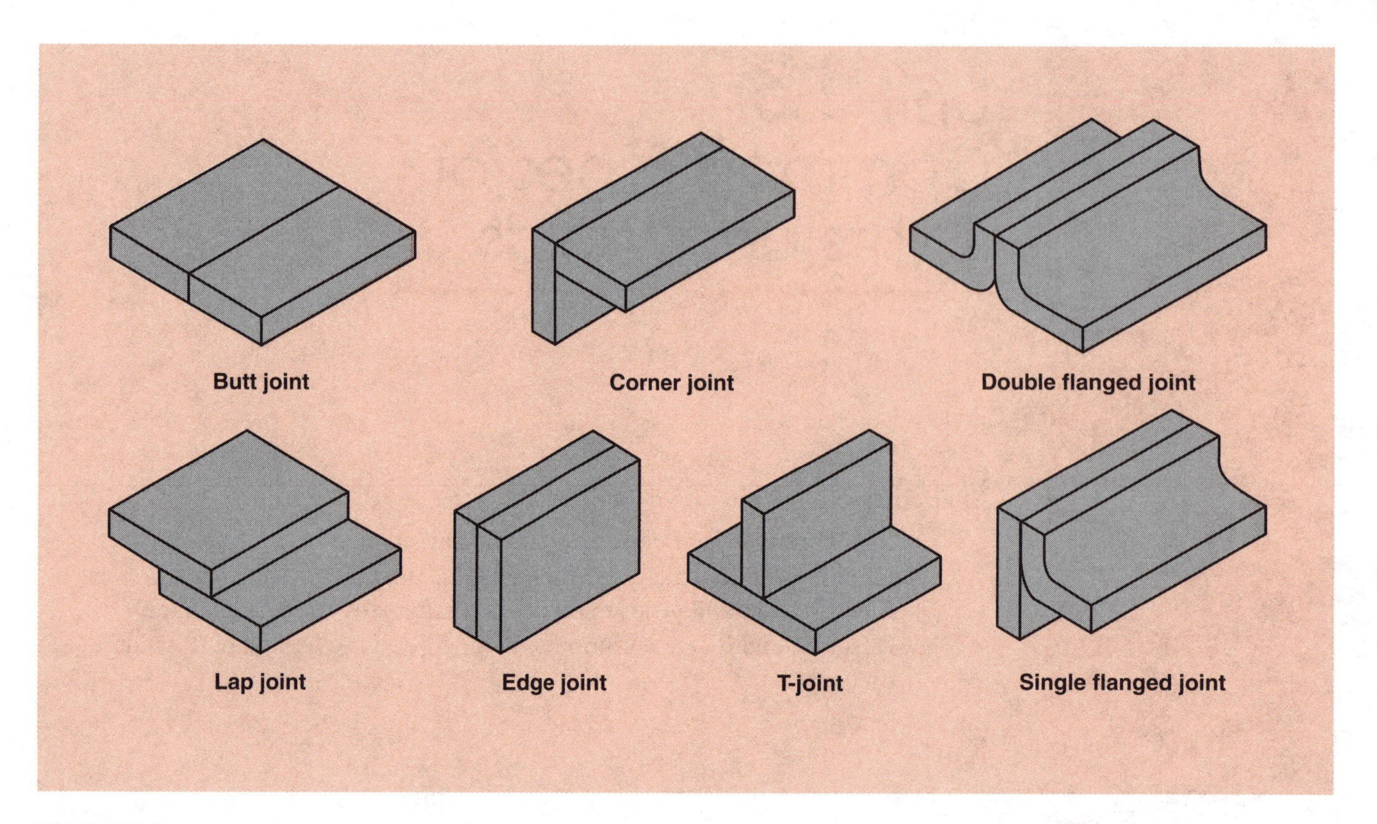

Figure 13-1.
Know the names and shapes of basic joints used in welding.

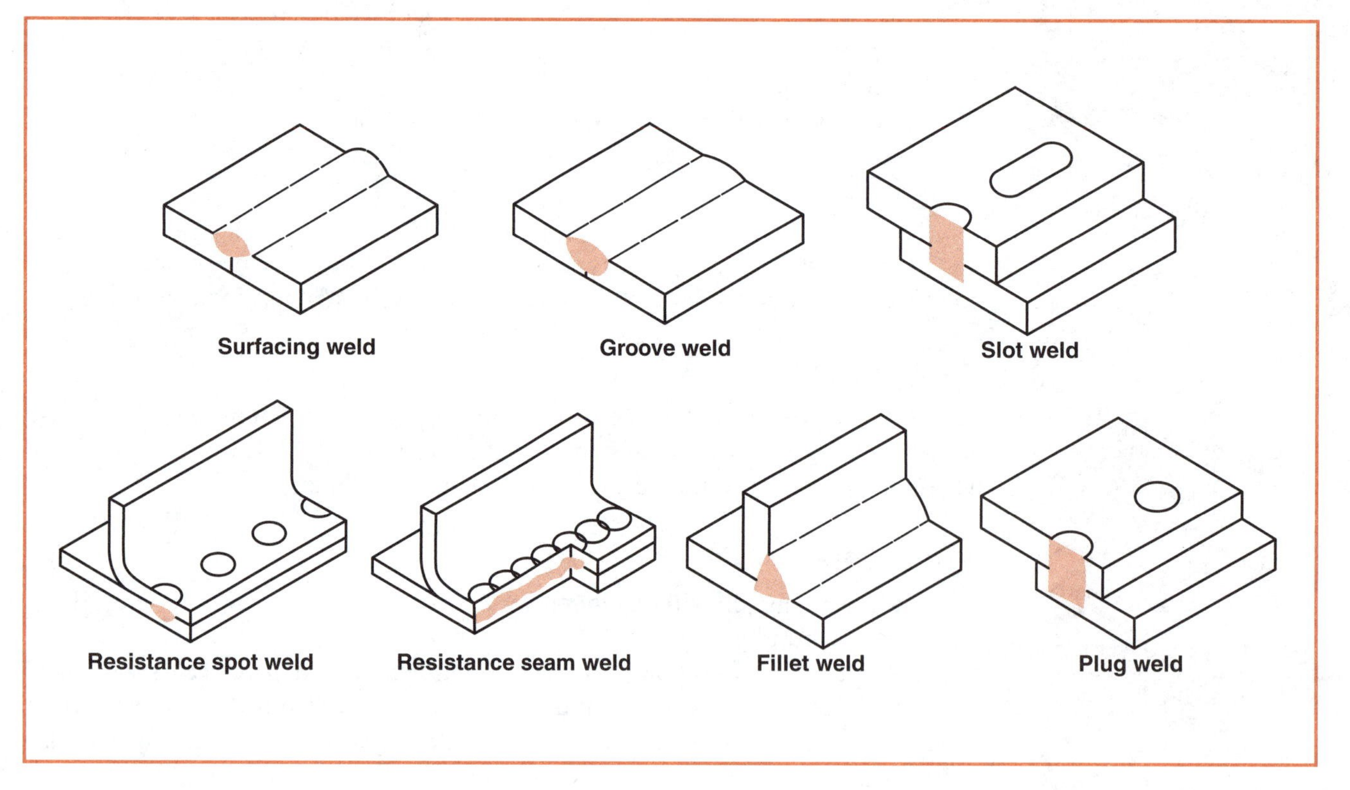

Figure 13-2.
One or more of these welds will be used on a weldment.

Plug and ***slot welds*** are welds made *through* one circular or elongated hole in a piece of metal to join it to another piece of metal. The opening may be partially or completely filled with the weld.

A ***resistance spot weld*** is an individually formed weld where the shape and size of the weld nugget is limited by the size and contour of the welding electrodes.

The ***resistance seam weld*** is a series of overlapping spot welds made progressively along the joint by rotating electrodes.

Joints and Grooves

To assure a solid weld, it is frequently necessary to combine the joint with one of the groove styles shown in Figure 13-3. The ***groove*** is the opening provided between the two metal pieces being joined by a groove weld.

Terms Used to Describe Welds

To describe a weld, names are given to common parts. Some names are specific to the type of joint or weld, but most of the names are common to all welds. For a fillet weld, five names are commonly used to describe the weld. Figure 13-4 shows the terms and their locations on the weld.

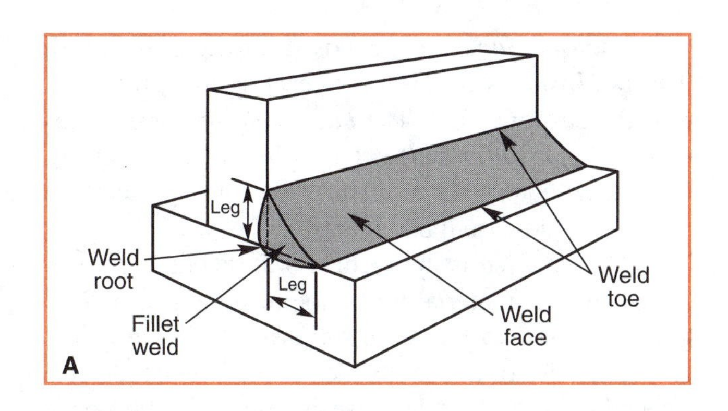

Figure 13-4.
Know the terms used to describe a weld. A—The five terms illustrated are common to all welds. B—The face of this fillet weld is exposed. (Kress Corporation)

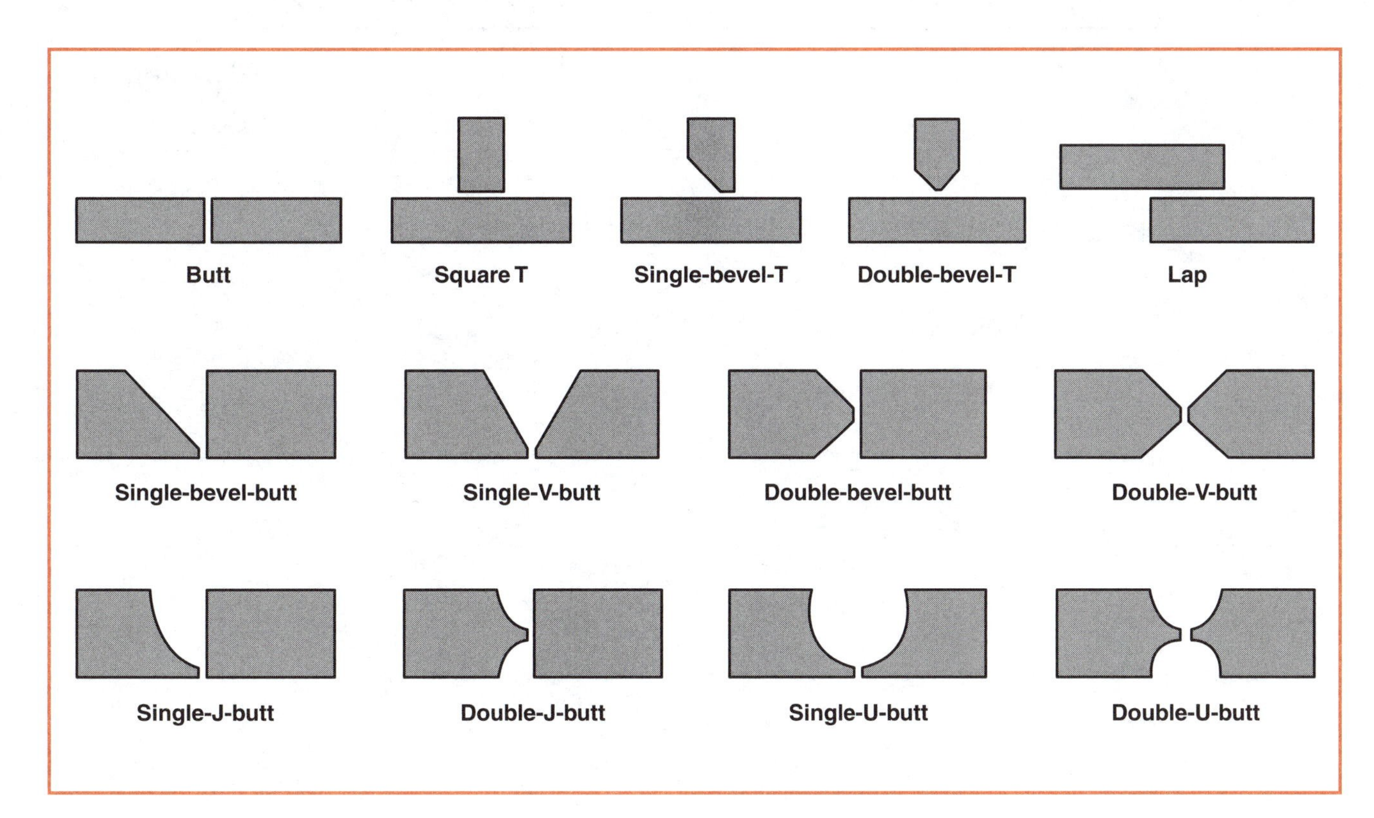

Figure 13-3.
A joint is often combined with one of these groove styles to produce a solid weld.

The ***weld face*** is the exposed surface of the weld. The ***weld root*** is the deepest penetration of the weld into the base material. The ***base material*** (sometimes called the ***parent metal***) is the material being welded, brazed, soldered, or cut. The ***weld toe*** is where the weld face and base metal meet.

The ***weld leg*** is the vertical or horizontal distance from the base material to the toe of the weld. A ***fillet gauge*** is used to measure this distance. Figure 13-5 shows the use of a fillet gauge to measure the leg size of a fillet weld. The leg size is often specified on a welding symbol.

Although most of the terms associated with a fillet weld are the same as those for a groove weld, there are a few additional terms associated with a groove weld. The weld buildup above the surface of the base material on a butt joint is called ***reinforcement.*** The angles made on the base metal before welding are called the ***bevel angles*** on a single plate or the ***groove angles*** if both plates are prepared. Terms commonly associated with a butt joint are shown in Figure 13-4.

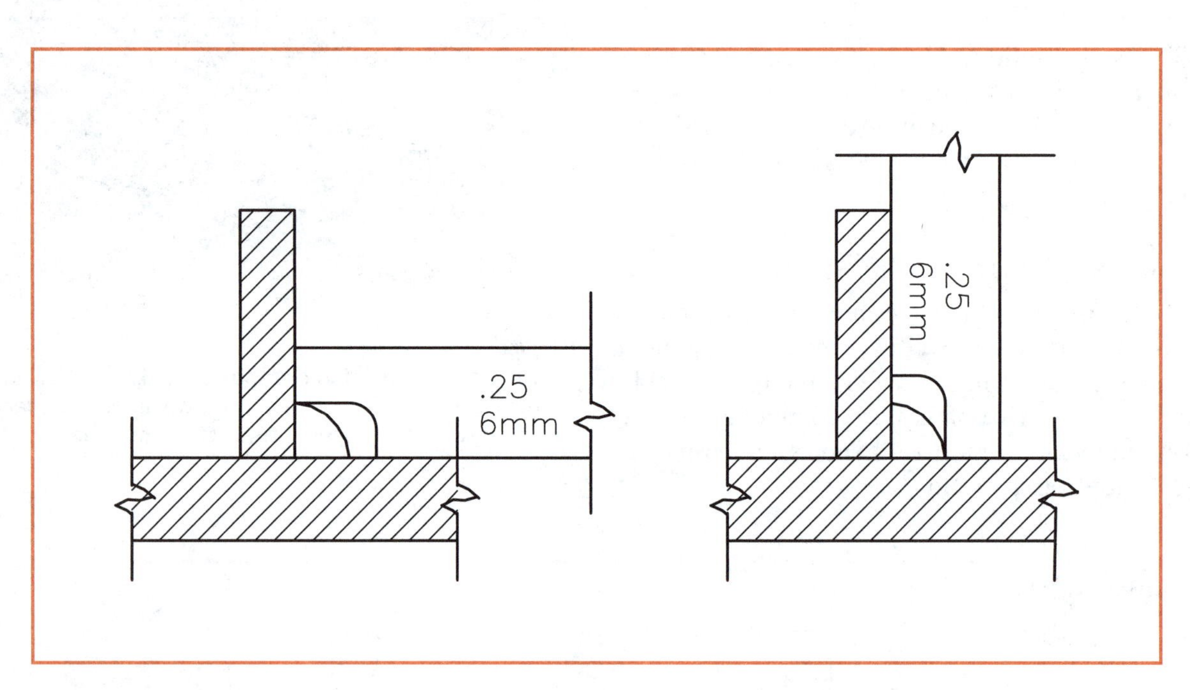

Figure 13-5.
Vertical leg and horizontal leg measurements of a fillet weld are taken using a fillet gauge.

Name ______________________ Date ______________ Class ______________

Review Questions

Identify the following types of joints and welds.

Joints

A B C D E F

G H I J K L

Welds

M N O P

Joint Combinations

Q R S T U

V W X Y Z

A. ______________________

B. ______________________

C. ______________________

D. ______________________

E. ______________________

F. ______________________

G. ______________________

H. ______________________

I. ______________________

J. ______________________

K. ______________________

L. ______________________

M. ______________________

N. ______________________

O. ______________________

P. ______________________

Q. ______________________

R. ______________________

S. ______________________

T. ______________________

U. ______________________

V. ______________________

W. ______________________

X. ______________________

Y. ______________________

Z. ______________________

Section 4 Reading Welding Prints

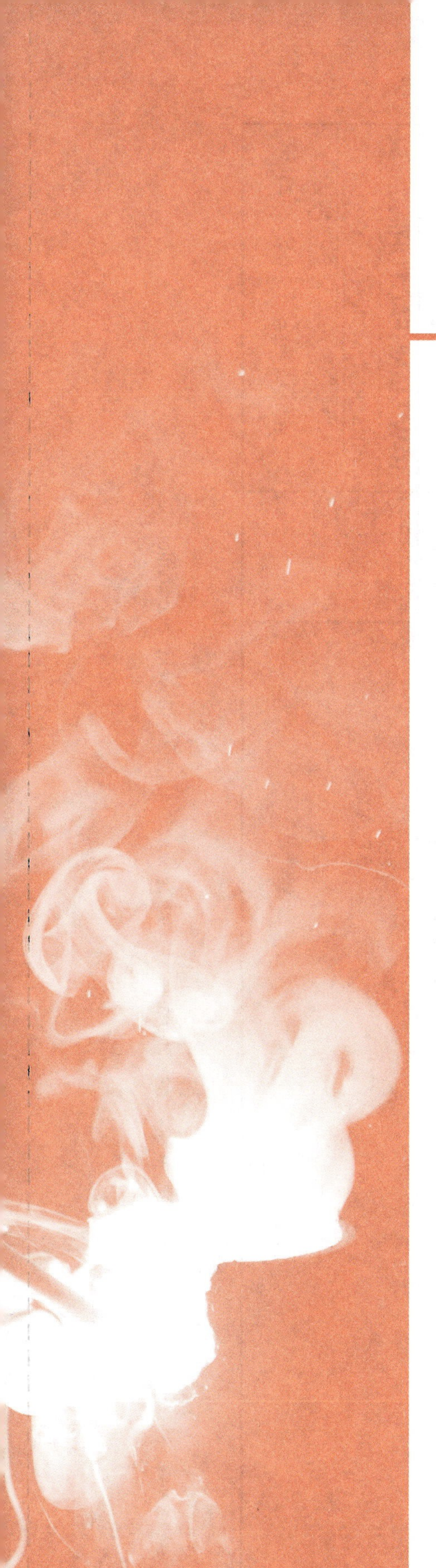

Unit 14

Welding Symbols

After completing Unit 14, you will be able to:

- ❍ List the basic elements of a welding symbol.
- ❍ Explain the meaning and use of each element of a welding symbol.
- ❍ Interpret which side of the weld joint the weld is placed.
- ❍ Identify and interpret nonpreferred weld symbols.
- ❍ Explain a welding symbol by developing a pattern to follow.

Key Words

arrow
arrow side
back weld
backing weld symbol
bent leader
broken leader
contour symbol
field weld symbol
finish symbol
joint penetration
melt-through symbol
nonpreferred weld symbols
notation
other side
reference line
supplementary symbols
tail
weld length
weld size
weld symbol
weld-all-around symbol
welding symbol

To make welds indicated on a drawing, the welder must be able to interpret welding symbols. Symbols are used to condense a large quantity of information about the weld into a small amount of space. This is shown in Figure 14-1.

A ***welding symbol*** is a graphic assembly of the elements needed to fully specify weld requirements, Figure 14-2. Welding symbols simplify communications between engineers who design the product and shop personnel who must fabricate the product. Used on drawings, symbols contain the data needed to "tell" a welder the exact type of weld wanted by the designer or engineer.

Elements of a Welding Symbol

Unless needed for clarity, all elements do *not* have to be used on every welding symbol. Only the reference line and arrow are required.

The eight possible elements of a welding symbol include:

- ❍ Reference line
- ❍ Weld symbol
- ❍ Arrow
- ❍ Tail
- ❍ Supplementary symbols
- ❍ Dimensions
- ❍ Finish symbols
- ❍ Notations pertaining to the process, filler metal, and any related standard

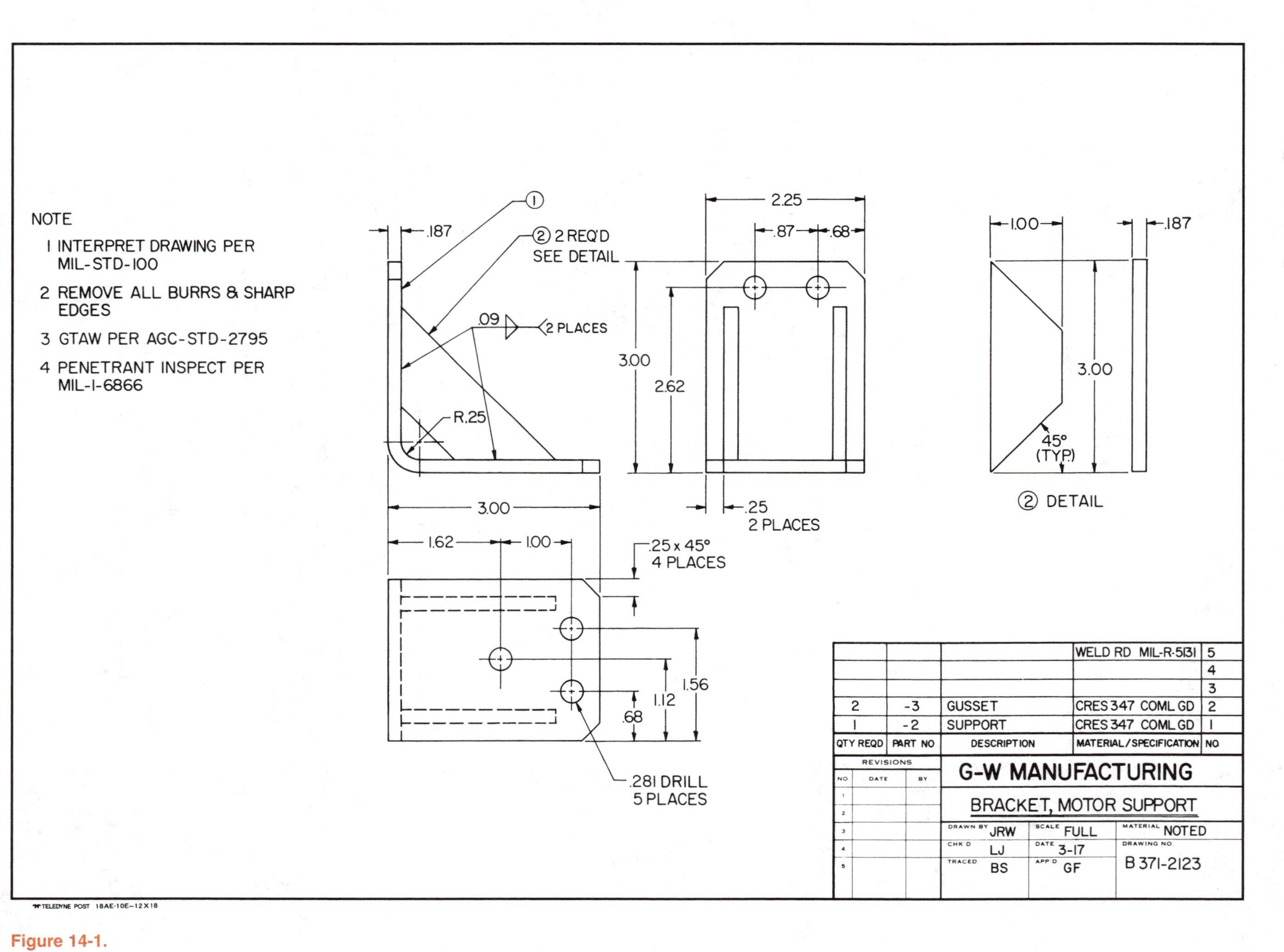

Figure 14-1.
To make the welds indicated on this drawing, the welder must be able to "read" symbols that furnish weld specifications.

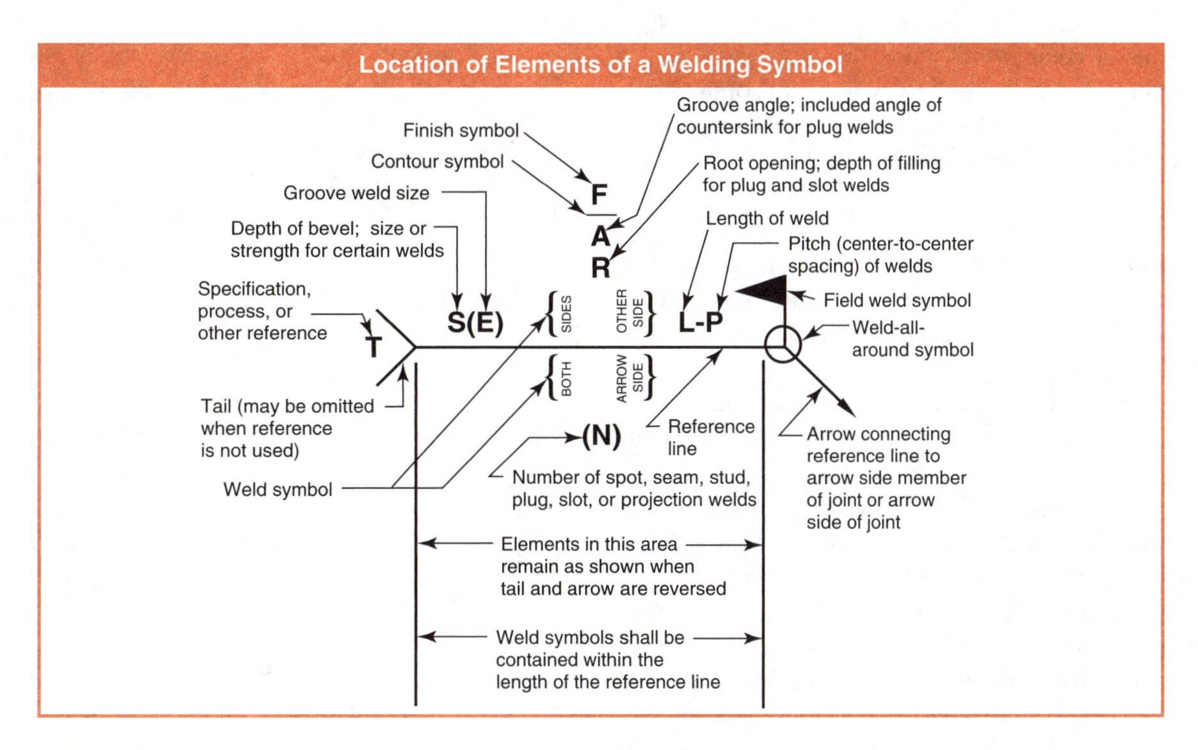

Figure 14-2.
A welding symbol is a graphic explanation of the elements needed to fully specify weld requirements.

Reference Line

The ***reference line*** is the "backbone," or required central element, of the welding symbol. It is always shown in a horizontal position. Other elements describing weld requirements are located on, above, below, and/or at either end of the reference line.

Weld Symbol

The basic weld symbol should be differentiated from the welding symbol. The basic ***weld symbol*** depicts the cross-sectional shape of the weld or joint. It is one part of the welding symbol. Study the basic weld symbols shown in Figure 14-3.

Groove							
Square	Scarf	V	Bevel	U	J	Flare-V	Flare-bevel
Fillet	**Plug or slot**	**Stud**	**Spot or projection**	**Seam**	**Back or backing**	**Surfacing**	**Edge**

Figure 14-3.
Memorize the names and shapes of the basic weld symbols. The reference line is shown as a dashed line.

Either a basic weld symbol or notations in the tail are required on a welding symbol. In some situations, however, both elements may be included on a welding symbol to furnish complete weld specifications.

Arrow

An ***arrow*** connects the reference line of a welding symbol to one side of the joint to be welded. The shape and location of the arrow are important. The use of the arrow is required.

Running the arrow from the reference line to one side of the required weld indicates the placement of *fillet, groove,* and *edge welds.*

The lower side of the reference line is called the ***arrow side.*** It indicates the same side or near side of the joint, Figure 14-4. A weld symbol *below* the reference line signifies the weld should be on the arrow side (same side) of the joint.

The opposite side of the arrow side is considered the ***other side.*** It is located on the upper side of the reference line. The other side indicates the far side of the joint, Figure 14-4. When the weld symbol is *above* the reference line, a weld is only required on the other side (opposite side) of the joint.

Weld symbols on *both* sides of the reference line indicate welds are needed on both the arrow side and other side of the joint, Figure 14-5.

The arrow for *bevel* and *J-groove welds* uses a ***bent leader*** (or ***broken leader***). The arrow head points to the particular section to be prepared, either by machining, flame cutting, air carbon arc gouging, or some other process, Figure 14-6.

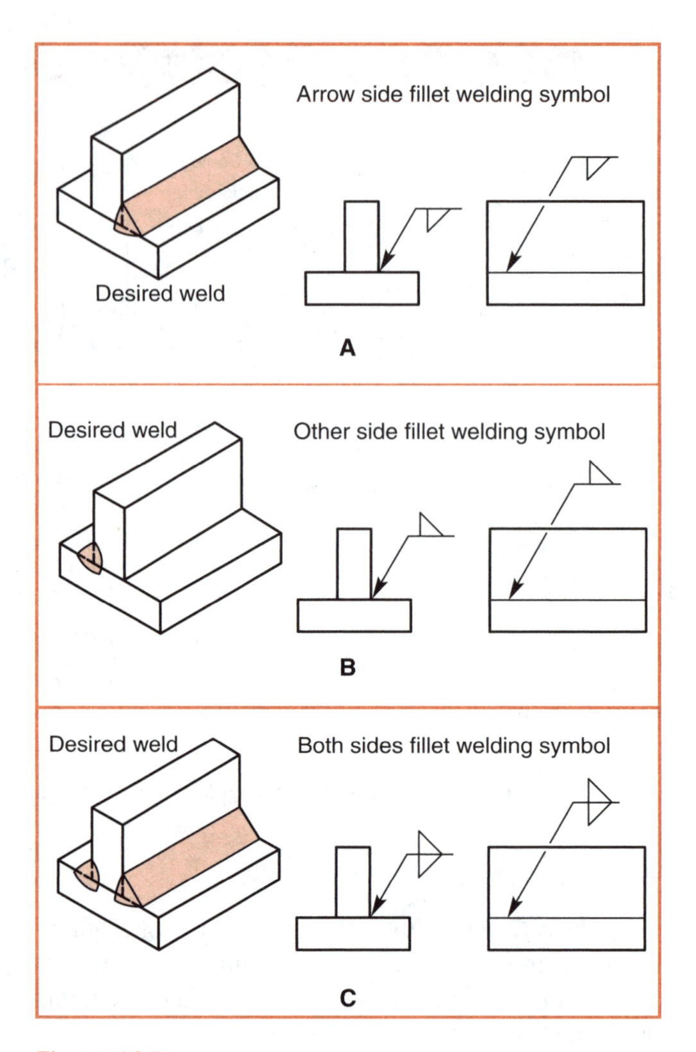

Figure 14-5.
Each fillet weld symbol shown means something different. A—The arrow side fillet welding symbol means the weld is on the same side as the arrow. B—The other side fillet welding symbol means the weld is on the opposite side of the arrow. C—This welding symbol means weld on both sides.

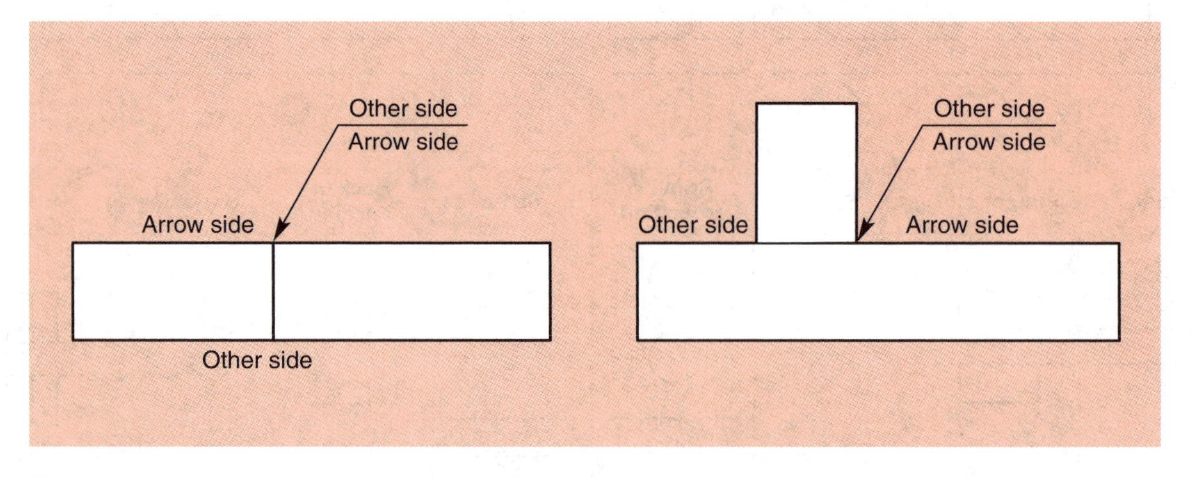

Figure 14-4.
The arrow side and the other side welds are specified in this manner.

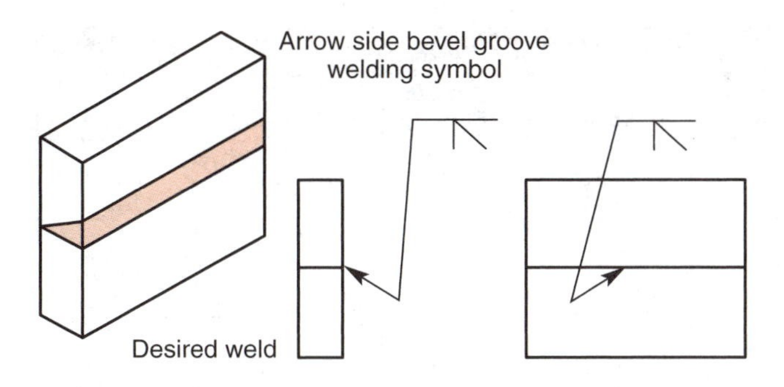

Figure 14-6.
The arrow for a bevel or J-groove weld uses a bent leader. The arrow head points to the particular section to be machined.

When *plug, slot, spot, seam,* and *projection welds* are required, the arrow points to the outer surface of one of the joint members at the center line of the desired weld. This member is considered the *arrow side member.* The other section of the joint is the *other side member,* Figure 14-7.

Some *resistance welds* have no arrow-side or other-side significance because the weld is made at the interface of the members, Figure 14-8.

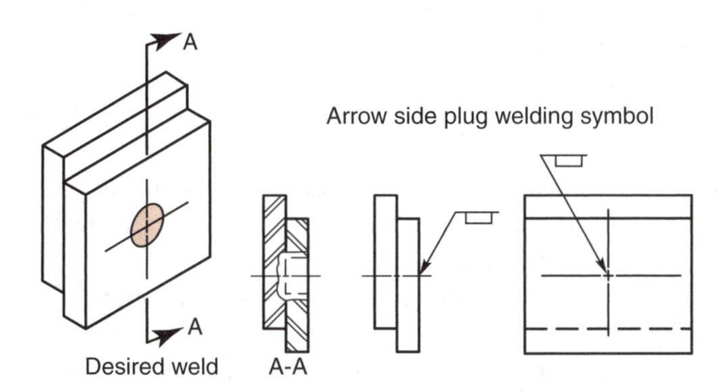

Figure 14-7.
On plug, slot, spot, seam, and projection welds, the arrow points to the outer surface of one of the joint members at the center of the weld. This member is considered the arrow side member.

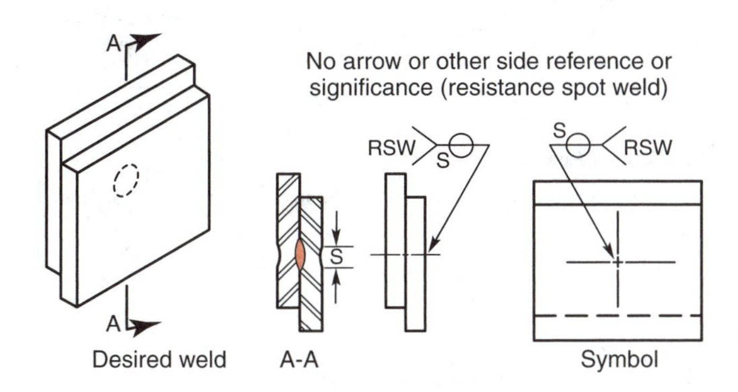

Figure 14-8.
A resistance weld is made at the interface of the members. It has no arrow side or other side significance.

Joints having more than one type of weld have a weld symbol for each weld, Figure 14-9. Multiple reference lines may also be used to indicate the sequence of welding a joint. The reference line closest to the leader shows the weld that is made first. Figure 14-10 shows the sequence of a V-groove weld and U-groove back gouge.

A chart showing basic welding symbols and their location significance is given in Figure 14-11. Learn to name each weld symbol and explain its location on the joint.

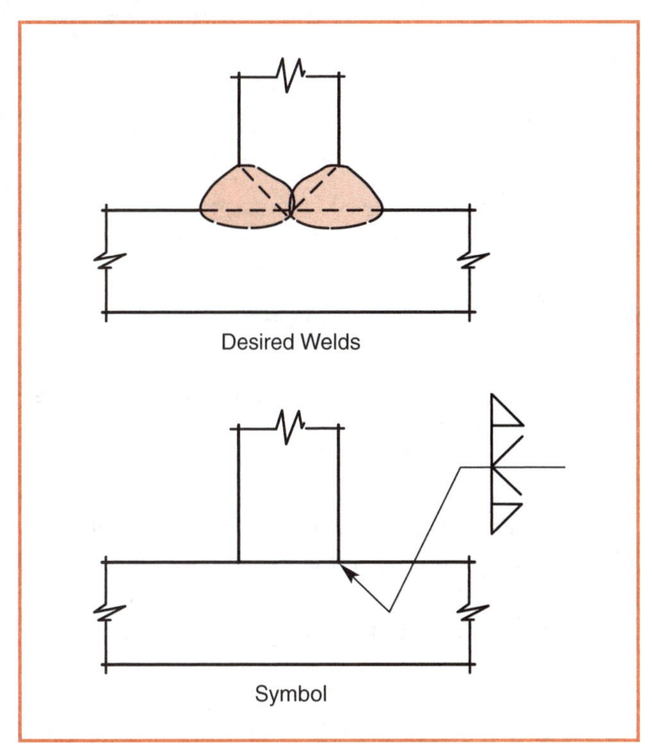

Figure 14-9.
A joint with more than one type of weld will have a weld symbol for each weld. In this example, there is a bevel weld and a fillet weld on each side of the joint.

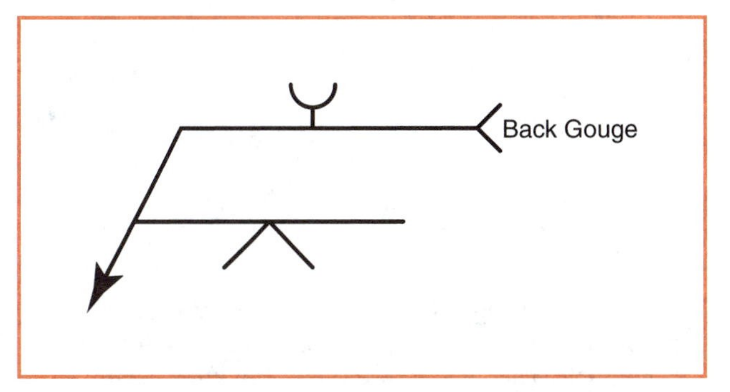

Figure 14-10.
Follow this sequence for welding a V-groove weld and U-groove back gouge.

Basic Welding Symbols and Their Location Significance								
Location Significance	Fillet	Plug or Slot	Spot or Projection	Stud	Seam	Back or Backing	Surfacing	Edge
Arrow Side								
Other Side				Not Used			Not Used	
Both Sides		Not Used	Not Used	Not Used	Not Used	Not Used	Not Used	
No Arrow Side or Other Side Significance	Not Used	Not Used		Not Used		Not Used	Not Used	Not Used
Location Significance	Groove							Scarf for Brazed Joint
	Square	V	Bevel	U	J	Flare-V	Flare-Bevel	
Arrow Side								
Other Side								
Both Sides								
No Arrow Side or Other Side Significance		Not Used	Not Used	Not Used	Not Used	Not Used	Not Used	Not Used

Figure 14-11.
It is important to know the basic welding symbols and their location significance.

Tail

The ***tail*** contains notes pertaining to the process, filler metal, and any related standards needed to establish specific weld requirements, Figure 14-12. If notations are not used, the tail element may be omitted.

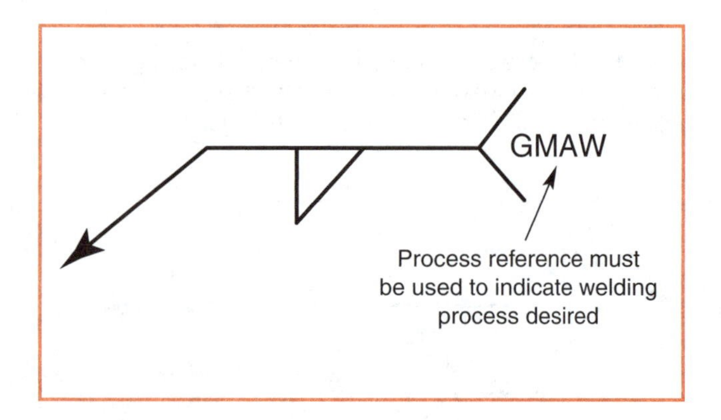

Figure 14-12.
The tail of a welding symbol may contain notes on a process, filler metal, and/or any related standards for weld requirements. This one shows gas metal arc welding.

Supplementary Symbols

Supplementary symbols are often included with basic weld symbols to provide more specific weld data not provided by other elements in the welding symbol. Supplementary weld symbols are shown in Figure 14-13.

The ***weld-all-around symbol*** signifies the weld is to be made completely around the joint without interruption, Figure 14-14.

Supplementary Symbols			
Weld-All-Around	Field Weld	Melt-Through	Consumable Insert
			(Square)
Backing or Spacer (Rectangular)	Contour		
	Flush	Convex	Concave
Backing Spacer			

Figure 14-13.
Study the supplementary weld symbols.

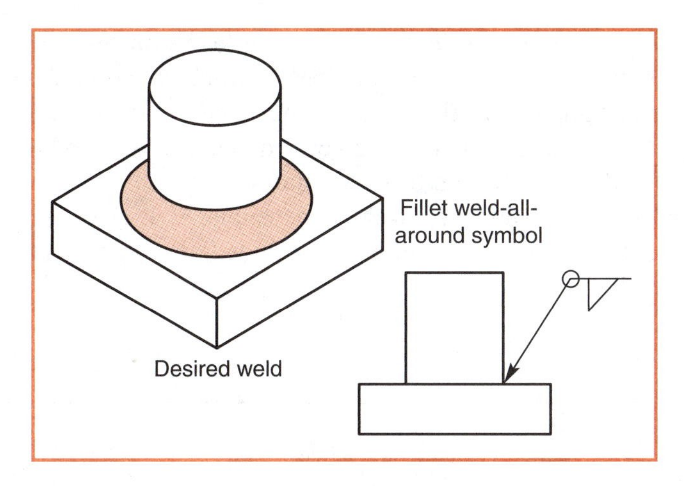

Figure 14-14.
Note the fillet weld-all-around symbol.

A ***field weld symbol*** indicates the weld is not done where the unit is initially made, Figure 14-15. Instead, it is made in the field.

The ***backing weld symbol*** indicates a bead-type backing weld on the opposite side of the regular weld. This is pictured in Figure 14-16. A ***back weld*** is made after the required weld indicated by the symbol is complete. The back weld assures full weld material through the joint (sometimes called complete joint penetration or CJP). ***Joint penetration*** refers to the distance the weld metal extends from the weld face into a joint, not including weld reinforcement. Figure 14-17 shows a back weld. Notice the groove weld is made first and the back weld is used to provide full penetration and full strength. Figure 14-18 shows the back weld symbol. Back weld is listed in the tail.

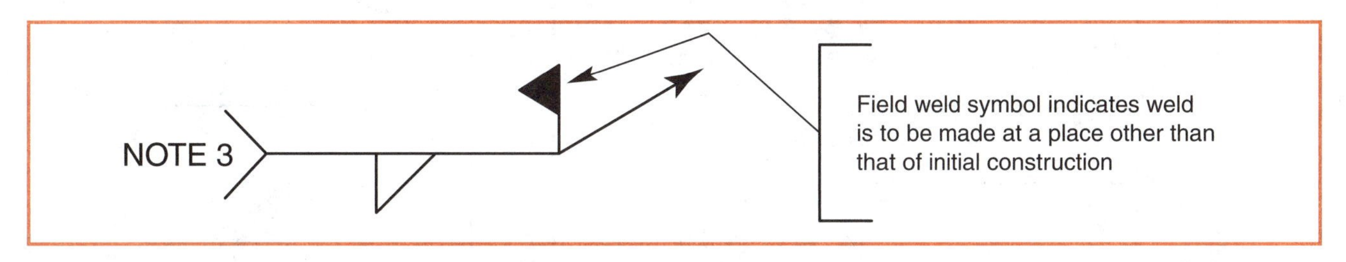

Figure 14-15.
This welding symbol indicates the weld must be made in the field.

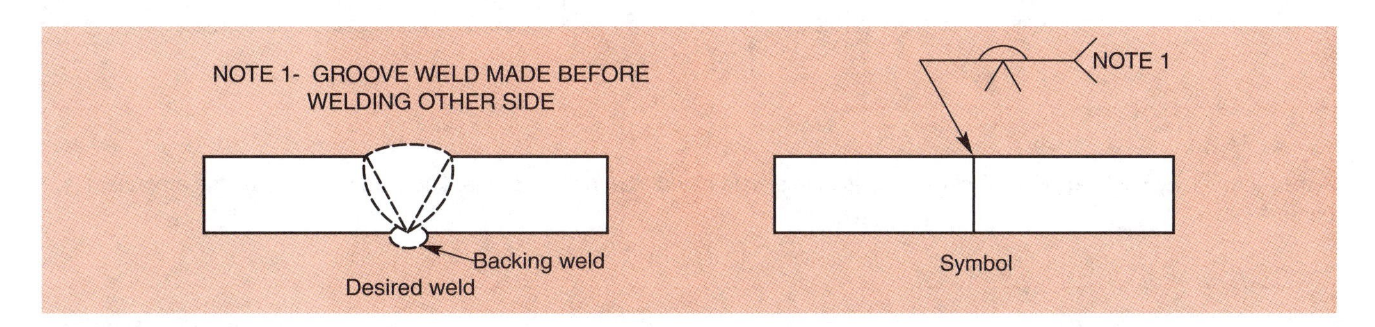

Figure 14-16.
The backing weld symbol means a bead-type backing weld is to be made on the opposite side of the regular weld.

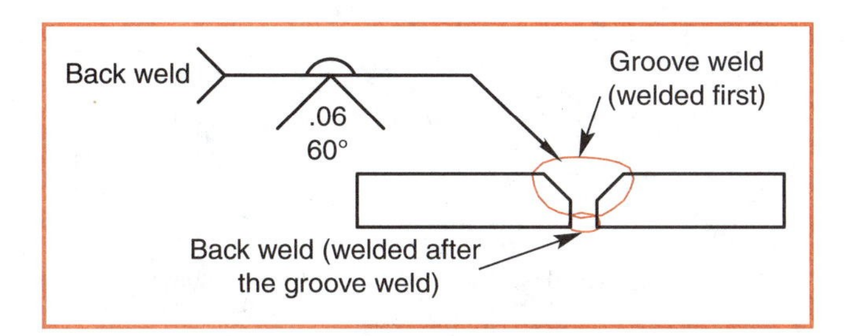

Figure 14-17.
A back weld ensures complete joint penetration.

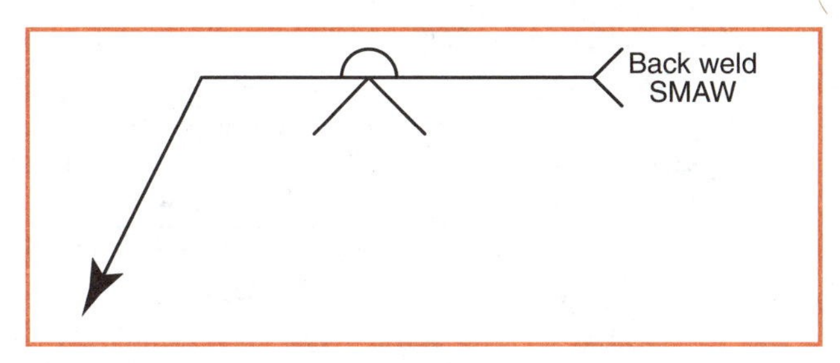

Figure 14-18.
The back weld symbol contains the weld process in the tail.

The ***melt-through symbol*** is used when complete joint penetration is required in a weld made from only one side, Figure 14-19. It also ensures 100% weld penetration.

Dimensions

Weld dimensions may be indicated in inches/fractions of an inch, or in millimeters (mm). Angles are specified in degrees.

The ***weld size*** dimension is placed on the left side of the weld symbol. ***Weld length*** is shown on the right side of the weld symbol. See Figure 14-20. No length dimension is given when the weld is to be made the full length of the joint.

Information on weld dimensions, as it refers to specific types of welds, is included in subsequent units.

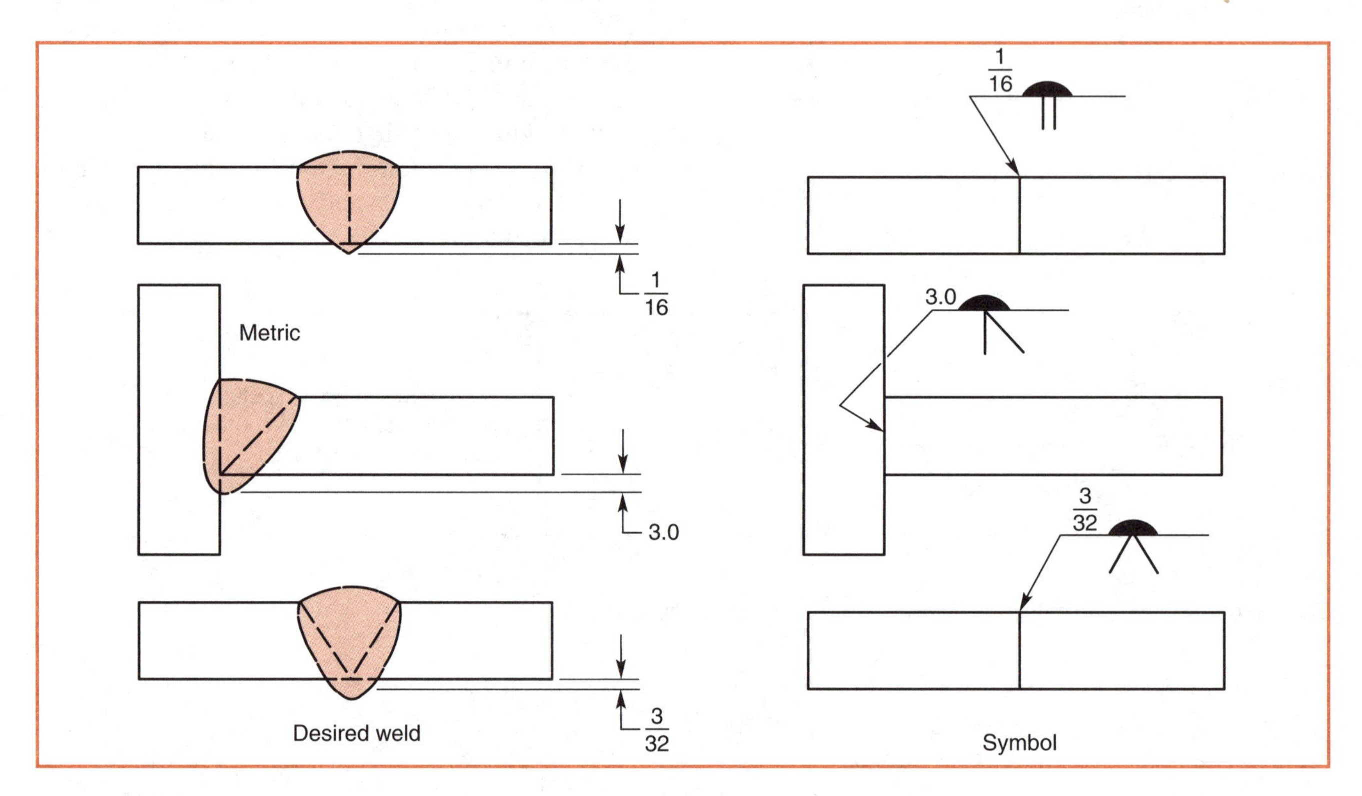

Figure 14-19.
A melt-through symbol requires full joint penetration in a weld made from only one side. The number indicates the amount of penetration.

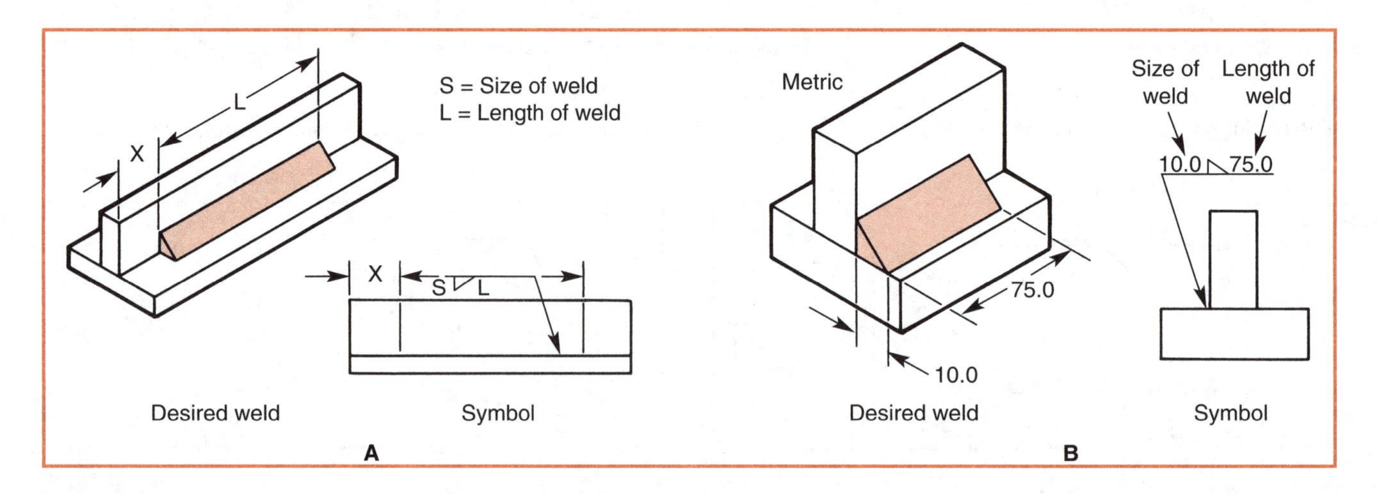

Figure 14-20.
Size and length dimensions of welds are indicated on the drawing. Weld size is placed on the left side of the weld symbol. Weld length is shown on the right side of the weld symbol.

Contour and Finish Symbols

A ***contour symbol*** is used with the weld symbol when the finished shape of the weld is important. If a weld is to be contoured or finished (other than being cleaned), a ***finish symbol*** is included with the contour symbol, Figure 14-21.

The finish symbol indicates the method, not the degree, of finish. Standard finishing methods are designated by symbols and include the following:

- C—Chipping
- G—Grinding
- H—Hammering
- M—Machining
- R—Rolling

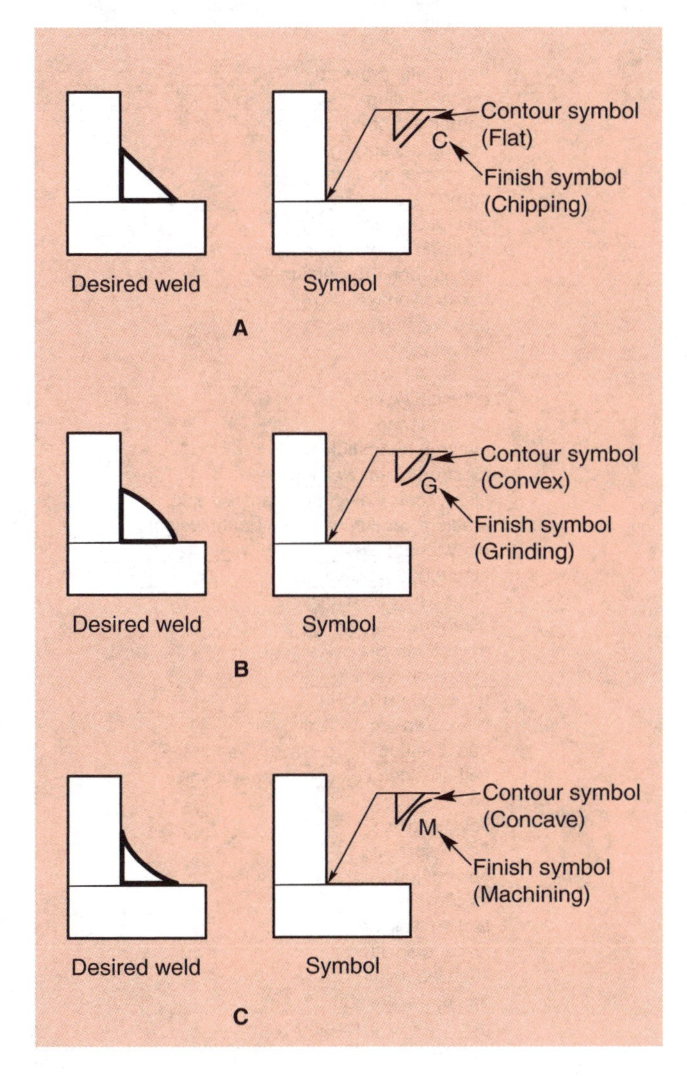

Figure 14-21.
Know symbols for contour and surface finish. Shown are symbols designating finishing methods for fillet welds.

Notations

A ***notation*** is information called out in the tail of a welding symbol. It is often in the form of abbreviations or notes. Sometimes there is insufficient room in the tail for all of the notation. In that case, a reference given in the tail states where more information on the notation can be found elsewhere on the print, Figure 14-22.

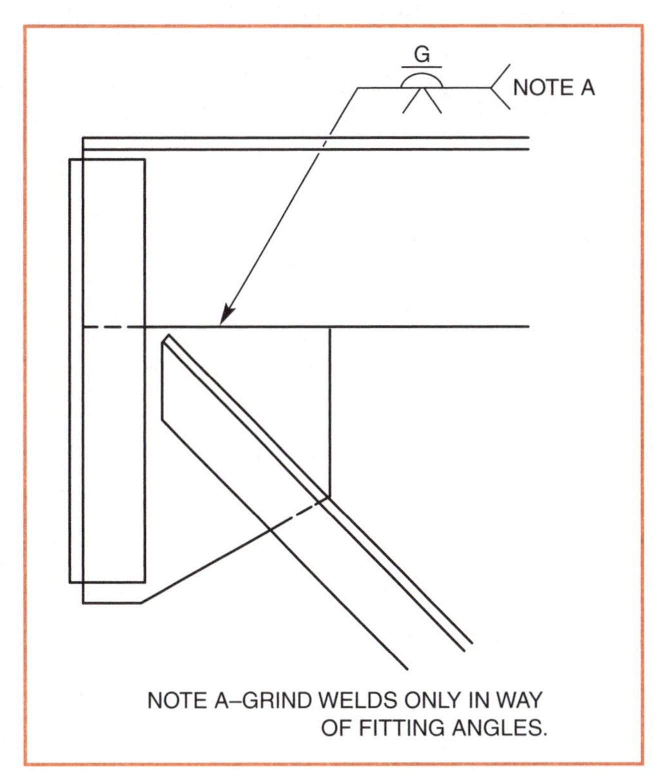

Figure 14-22.
Sometimes a note accompanies a welding symbol.

A notation may indicate a welding process by using a standard abbreviated form. Refer to Figure 14-23 for the letter designations of welding and related processes. Study this table carefully!

Nonpreferred Weld Symbols

The ***nonpreferred weld symbols*** have been replaced by new symbols. They are not used on newer prints. However, nonpreferred weld symbols may still be found on older prints and are included in Figure 14-24 for reference purposes.

Suffixes for Optional Use in Applying Welding and Allied Processes			
Adaptive control	**AD**	**Mechanized**	**ME**
Automatic	**AU**	**Robotic**	**AU**
Manual	**MA**	**Semiautomatic**	**SA**

Letter Designations of Welding and Allied Processes

Letter Designation	Processes and Variations
AAW	air acetylene welding
ABW	arc braze welding
AC	arc cutting
AHW	atomic hydrogen welding
AOC	oxygen arc cutting
ASP	arc spraying
AW	arc welding
B	brazing
BB	block brazing
BMAW	bare metal arc welding
BW	braze welding
CABW	carbon arc braze welding
CAC	carbon arc cutting
CAC-A	air carbon arc cutting
CAW	carbon arc welding
CAW-G	gas carbon arc welding
CAW-S	shielded carbon arc welding
CAW-T	twin carbon arc welding
CEW	coextrusion welding
CW	cold welding
DB	dip brazing
DFB	diffusion brazing
DFW	diffusion welding
DS	dip soldering
EBC	electron beam cutting
EBW	electron beam welding
EBW-HV	high vacuum electron beam welding
EBW-MV	medium vacuum electron beam welding
EBW-NV	nonvacuum electron beam welding
EGW	electrogas welding
ESW	electroslag welding
EXB	exothermic brazing
EXBW	exothermic braze welding
EXW	explosion welding
FB	furnace brazing
FCAW	flux cored arc welding
FCAW-G	gas shielded flux cored arc welding
FCAW-S	self-shielded flux cored arc welding
FLB	flow brazing
FLOW	flow welding
FLSP	flame spraying
FOC	flux cutting
FOW	forge welding
FRW	friction welding
FS	furnace soldering
FW	flash welding
GMAC	gas metal arc cutting
GMAW	gas metal arc welding
GMAW-P	pulsed gas metal arc welding

Figure 14-23.
Welding and allied processes are known by standard abbreviations used on prints.

Letter Designation	Processes and Variations
GMAW-S	short circuit gas metal arc welding
GTAC	gas tungsten arc cutting
GTAW	gas tungsten arc welding
GTAW-P	pulsed gas tungsten arc welding
HPW	hot pressure welding
IB	induction brazing
INS	iron soldering
IRB	infrared brazing
IRS	infrared soldering
IS	induction soldering
IW	induction welding
LBC	laser beam cutting
LBC-A	laser beam air cutting
LBC-EV	laser beam evaporative cutting
LBC-IG	laser beam inert gas cutting
LBC-O	laser beam oxygen cutting
LBW	laser beam welding
LOC	oxygen lance cutting
OAW	oxyacetylene welding
OC	oxygen cutting
OFC	oxyfuel gas cutting
OFC-A	oxyacetylene cutting
OFC-H	oxyhydrogen cutting
OFC-N	oxynatural gas cutting
OFC-P	oxypropane cutting
OFW	oxyfuel gas welding
OHW	oxyhydrogen welding
PAC	plasma arc cutting
PAW	plasma arc welding
PEW	percussion welding
PGW	pressure gas welding
POC	metal powder cutting
PSP	plasma spraying
PW	projection welding
RB	resistance brazing
ROW	roll welding
RS	resistance soldering
RSEW	resistance seam welding
RSEW-HF	high-frequency seam welding
RSEW-I	induction seam welding
RSW	resistance spot welding
RW	resistance welding
S	soldering
SAW	submerged arc welding
SAW-S	series submerged arc welding
SMAC	shielded metal arc cutting
SMAW	shielded metal arc welding
SSW	solid-state welding
SW	arc stud welding
TB	torch brazing
TC	thermal cutting
TCAB	twin carbon arc brazing
THSP	thermal spraying
TS	torch soldering
TW	thermite welding
USS	ultrasonic soldering
USW	ultrasonic welding
UW	upset welding
UW-HF	high-frequency upset welding
UW-I	induction upset welding
WS	wave soldering

Figure 14-23. (continued)

Reading and Explaining a Welding Symbol

Often you may have to explain or discuss a welding symbol with a coworker or supervisor. It is good practice to follow a consistent pattern when explaining the symbol. The symbol used in Figure 14-25 shows a fillet weld on the arrow side.

Use this pattern when explaining the symbol:

- ❍ State the weld to be made first.
- ❍ Then state the side on which the weld is to be made.
- ❍ State the size of the weld or the amount of joint preparation.
- ❍ State the angle of preparation, depth of penetration, and finish requirements.
- ❍ State other supplemental information, such as weld-around or field weld.

Refer to Figure 14-26. Using the pattern just given, it may be explained as follows: "This symbol is a V-groove weld on the arrow side, with .25 depth of penetration at an included angle of 60°, made flush after welding."

Plug or Slot	Arc Seam or Arc Spot	Projection	Resistance Spot	Resistance Seam	Flash or Upset	Field Weld

Figure 14-24.
Nonpreferred weld symbols are no longer standard but may still be present on older prints.

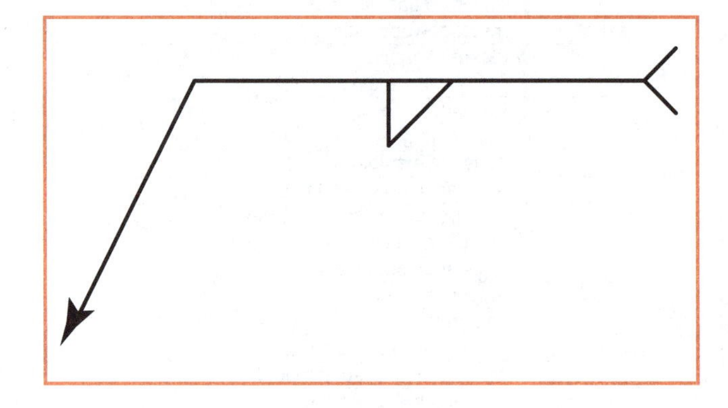

Figure 14-25.
Follow a pattern when explaining symbols such as this fillet weld on the arrow side.

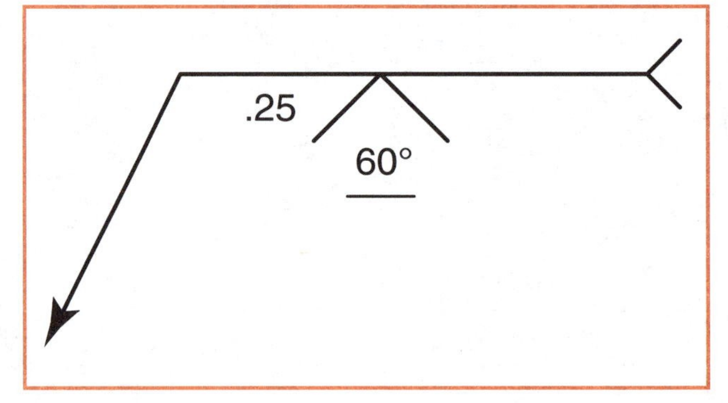

Figure 14-26.
Explain this V-groove weld.

Name ______________________ Date ______________ Class ______________

Print Reading Activities

Part I

Draw the correct symbols for the following welds.

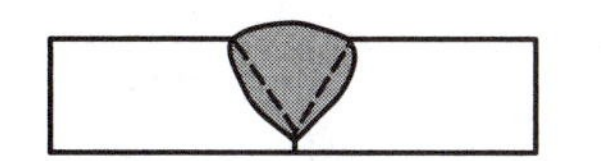

A

B

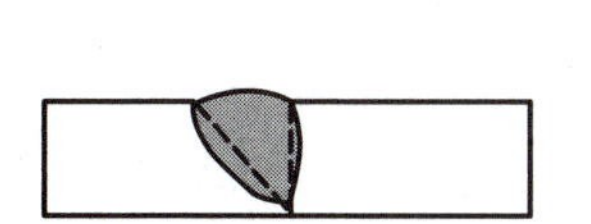

C

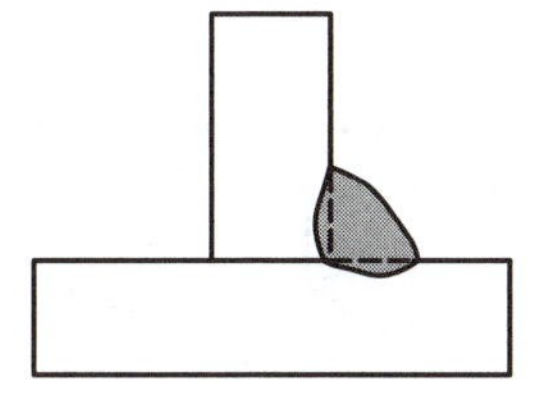

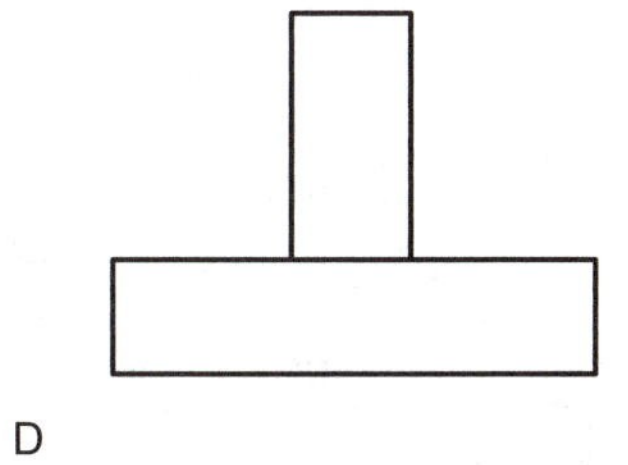

D

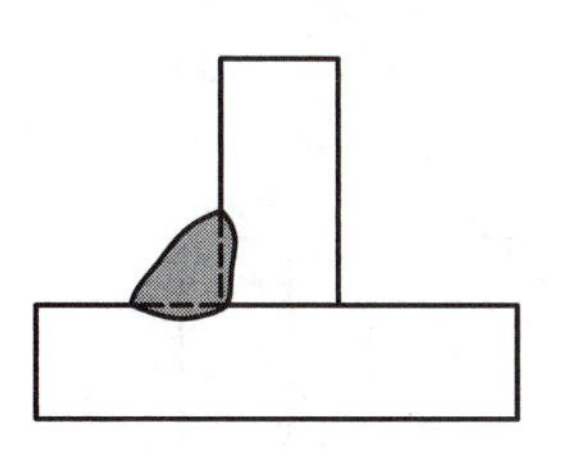

E

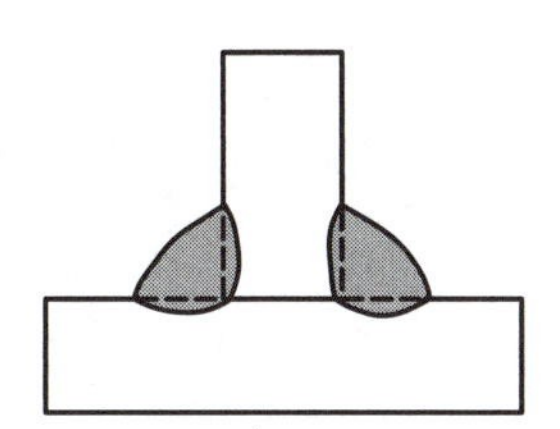

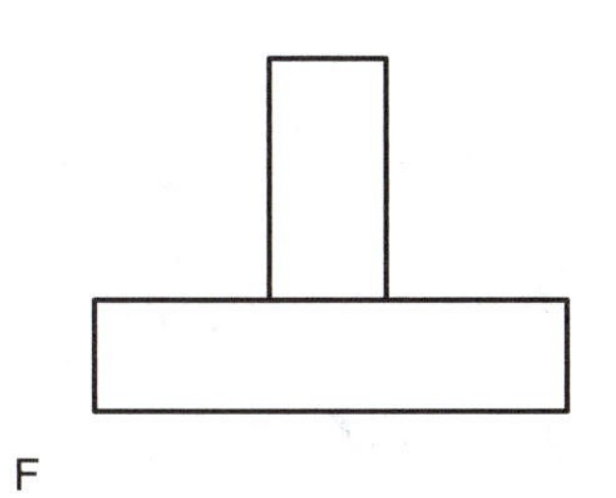

F

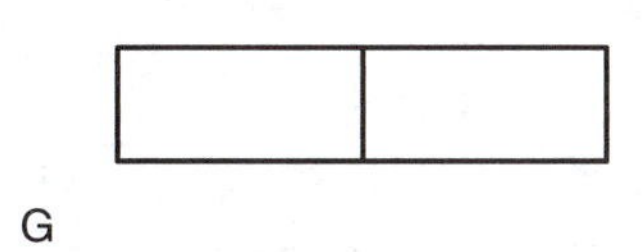

G

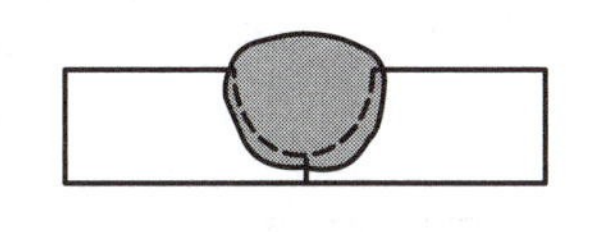

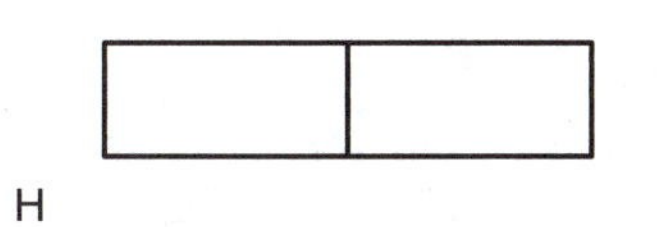

H

Part II

Draw the correct weld as indicated by the symbol.

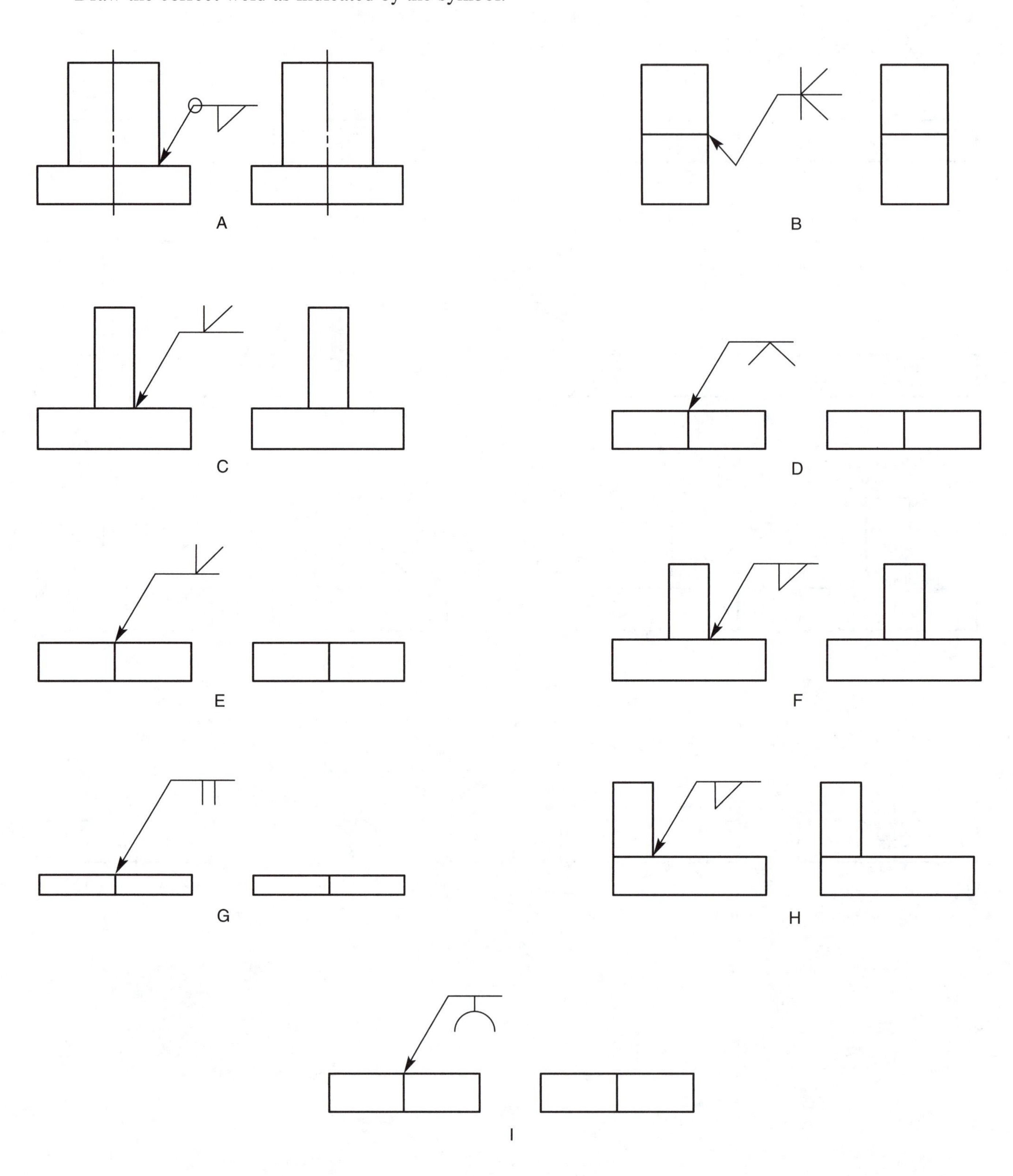

Name ______________________________

Part III

Explain the following welding symbols, using the suggestions for patterns in this unit as a guide.

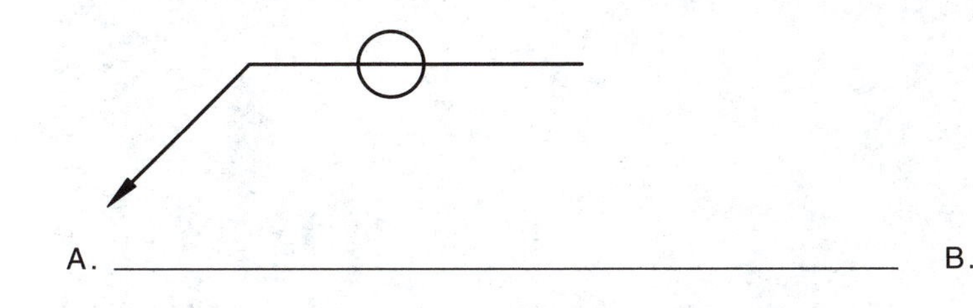

A. ______________________________

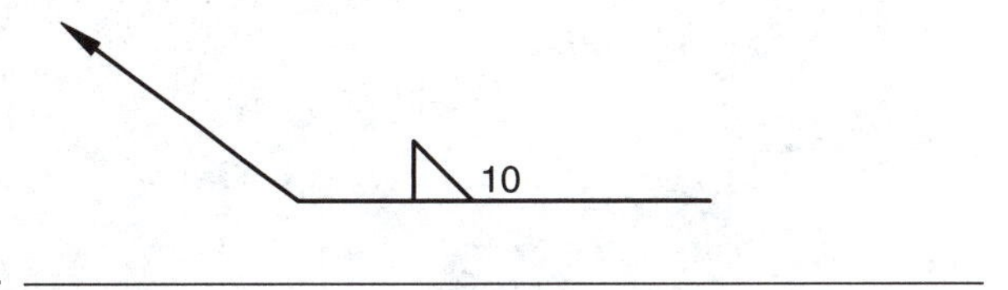

B. ______________________________

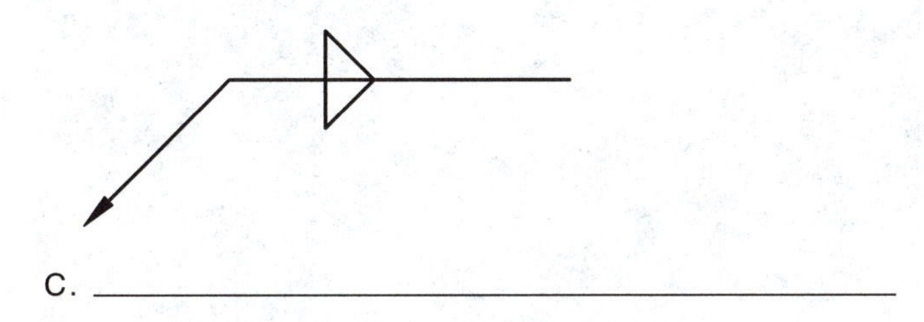

C. ______________________________

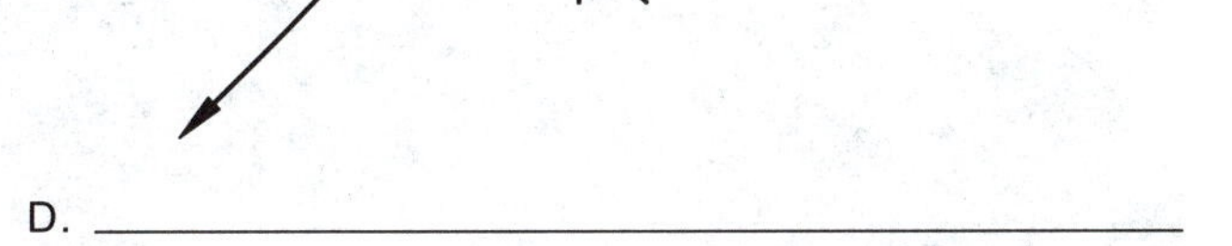

D. ______________________________

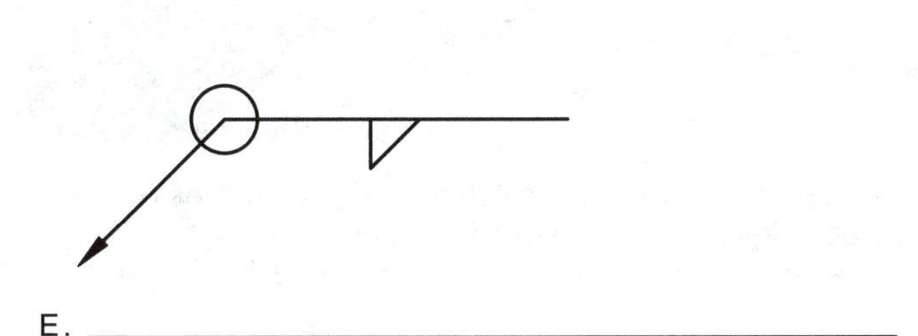

E. ______________________________

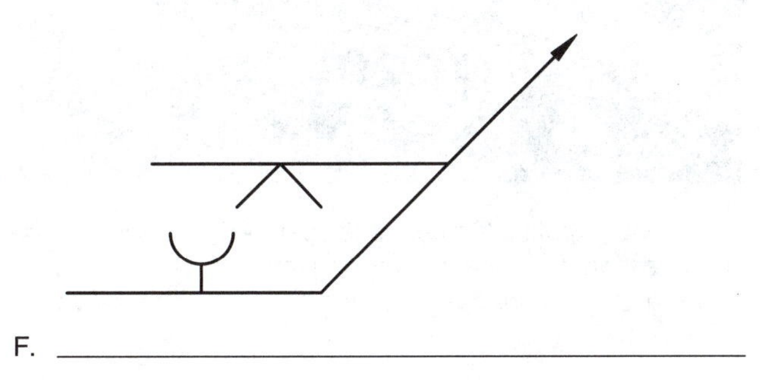

F. ______________________________

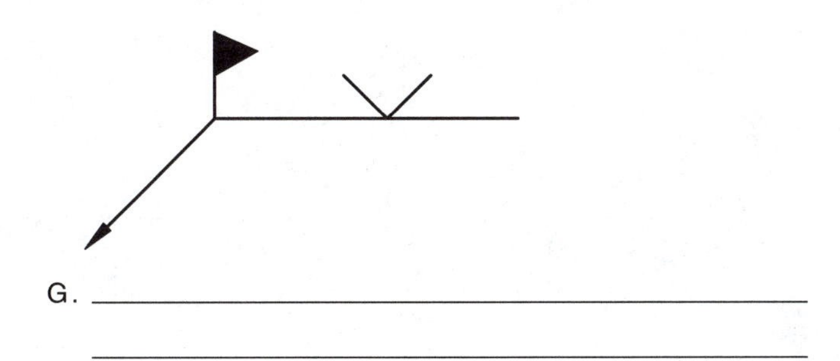

G. ______________________________

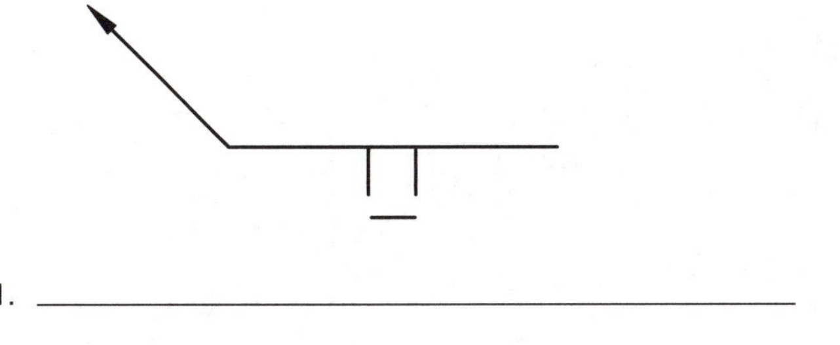

H. ______________________________

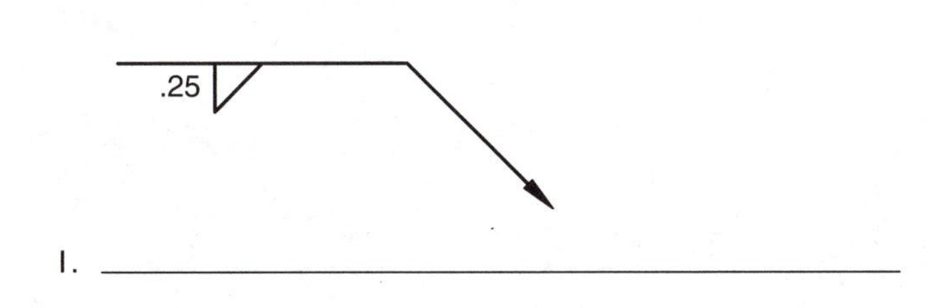

I. ______________________________

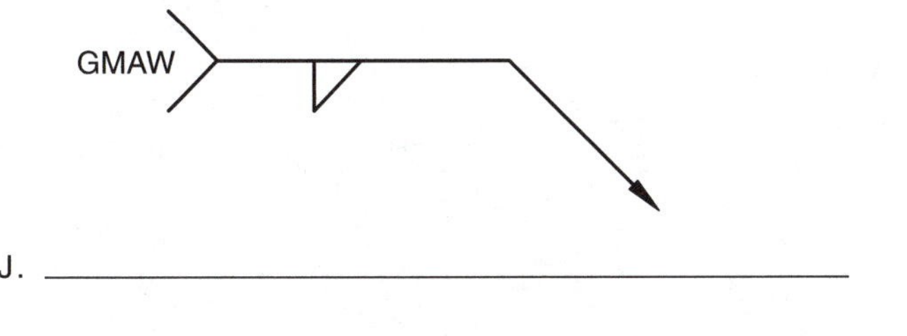

J. ______________________________

This oxyfuel multicutting head is operated from the console in front of the operator. An electronic scanner moves over a line or silhouette tracing, providing torch movement to make the desired cuts. It can also be equipped with preprogrammed shapes for which it is only necessary to key in the dimensions. (Linde Div., Union Carbide)

Unit 15

Fillet Welds

After completing Unit 15, you will be able to:

- ❍ Name the parts of a fillet weld.
- ❍ Determine weld size and weld length by interpreting the fillet welding symbol.
- ❍ Read and interpret a fillet welding symbol used for intermittent and staggered intermittent welds.
- ❍ Determine the weld face contour of a fillet weld.

Key Words

actual throat
hatching
intermittent weld
pitch
staggered intermittent weld

A fillet weld is usually used when joining two surfaces perpendicular to each other or at an angle, Figure 15-1. It is approximately triangular in shape. The parts of the weld include the weld root, leg, toe, face, (Refer to Unit 13) and throat. The ***actual throat*** is the shortest distance between the weld root and the face of the weld. The throat affects the strength of the weld.

General Use of Fillet Weld Symbol

Welding symbols for fillet welds are shown in Figure 15-2. The dimension for an *arrow side* fillet weld is placed on the arrow side of the reference line (the same side as the weld symbol). Weld size is shown on the *left side* of the weld symbol. Unless otherwise noted on the print, the minimum size of a fillet weld is the specified dimension.

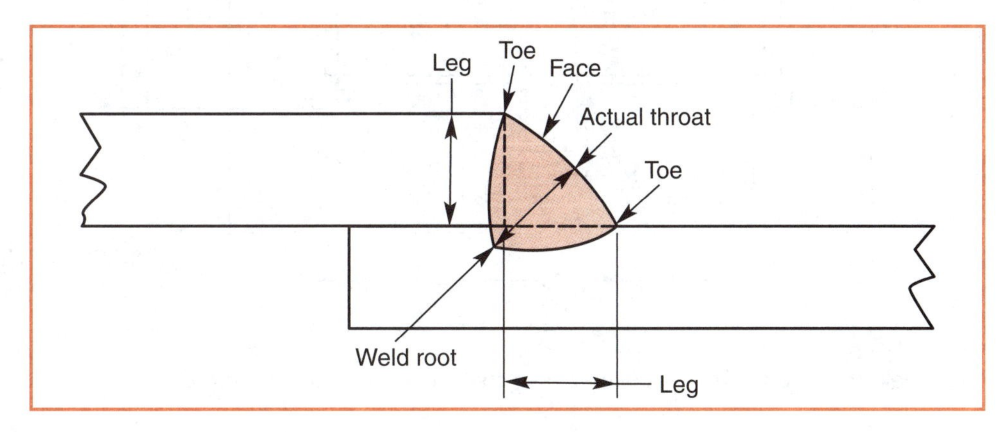

Figure 15-1.
Study the parts of a fillet weld.

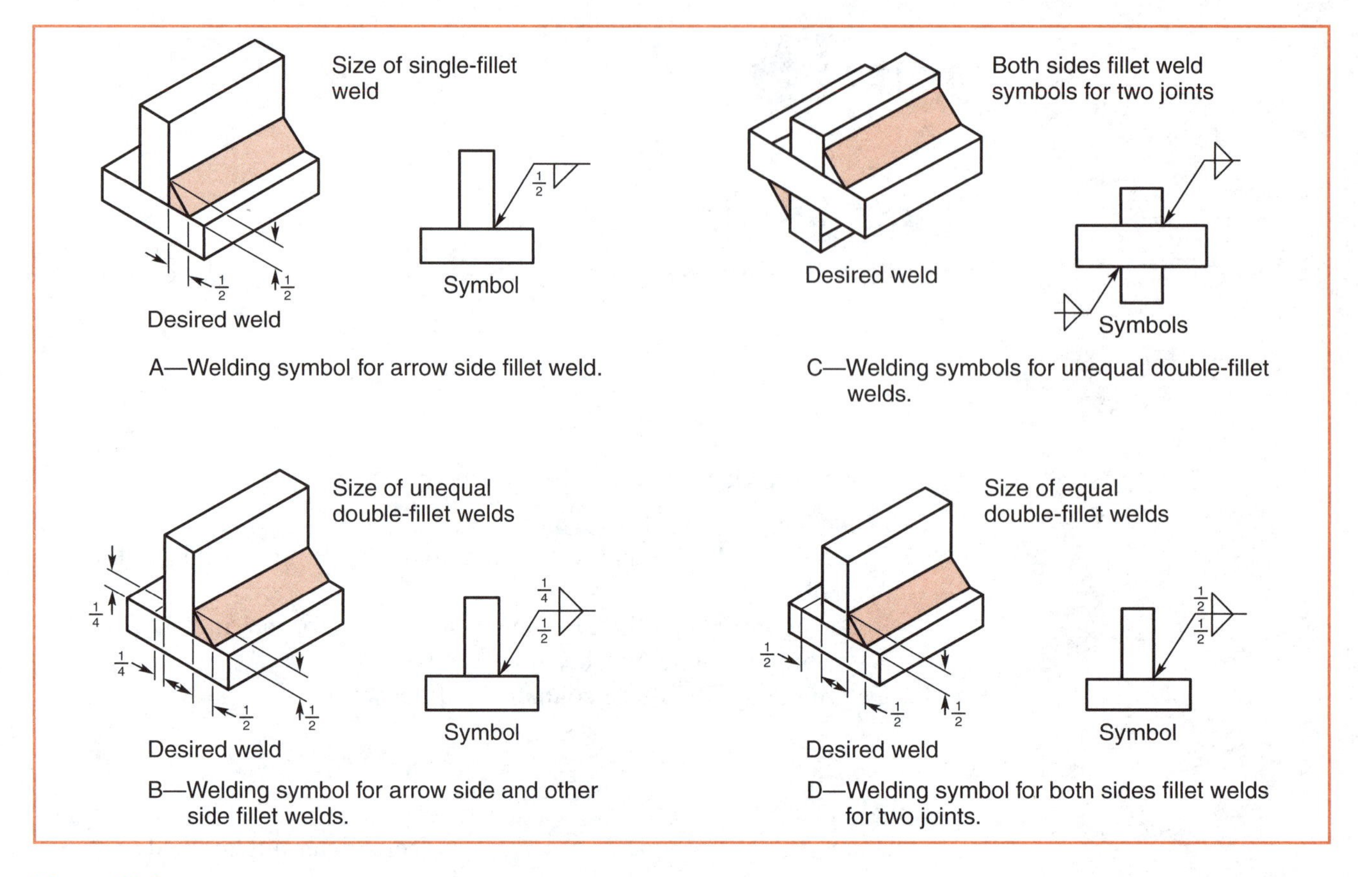

Figure 15-2.
There are variations of the welding symbol for a fillet weld.

When a fillet welding symbol does *not* include weld size, a general note governing weld size is usually found elsewhere on the print. Sometimes a company has specific specifications for fillet weld sizes without indicating size on the print. If no size is given on the print, ask your supervisor or foreman about the required size before you start to weld, Figure 15-3.

At times, fillet welds with unequal legs will be specified. Weld orientation is *not* indicated by the symbol but is shown on the print, Figure 15-4.

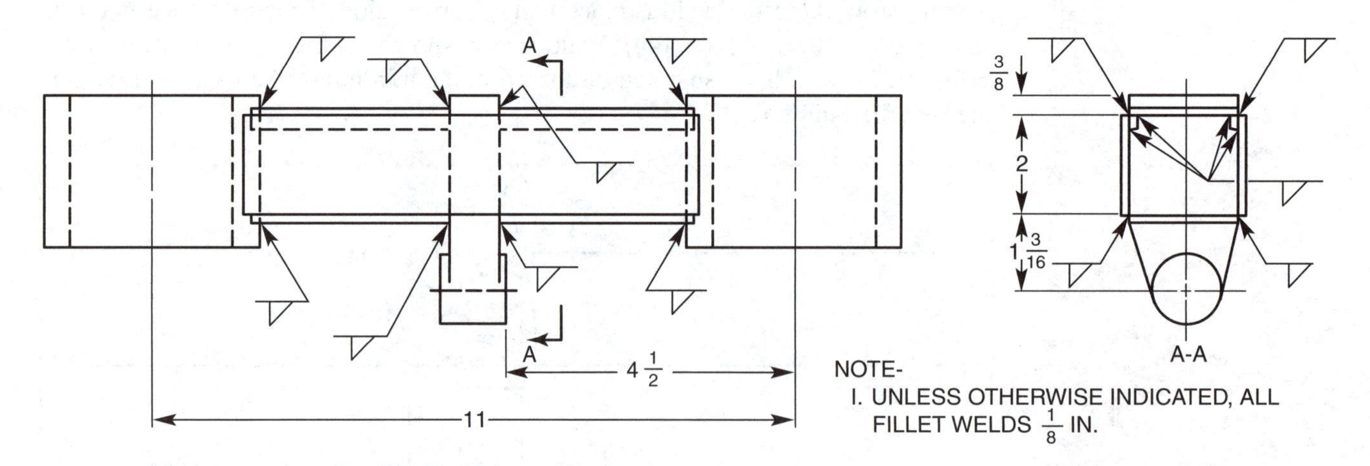

Figure 15-3.
If the weld size is not given on the welding symbol, it is found somewhere else on the print or drawing.

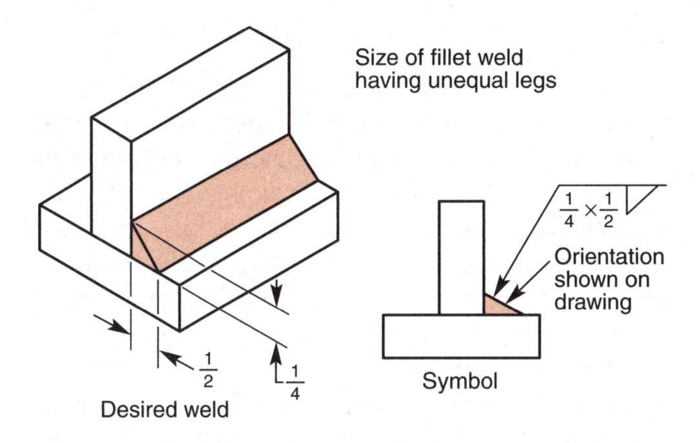

Figure 15-4.
This welding symbol is used for a fillet weld having unequal legs.

When used, the weld length is included to the right of the weld symbol. Refer to Figure 15-5. The weld is made completely across the joint when no weld length dimension is given with the welding symbol. Regardless of the geometric shape of the joint, a weld-all-around symbol means the weld is to be made continuously around the joint. Refer to Figure 15-6. For round parts, sometimes the weld length is indicated in degrees. Figure 15-7 shows an example of a fillet weld indicated for 90° of the joint. The starting and finishing location are usually indicated on the print.

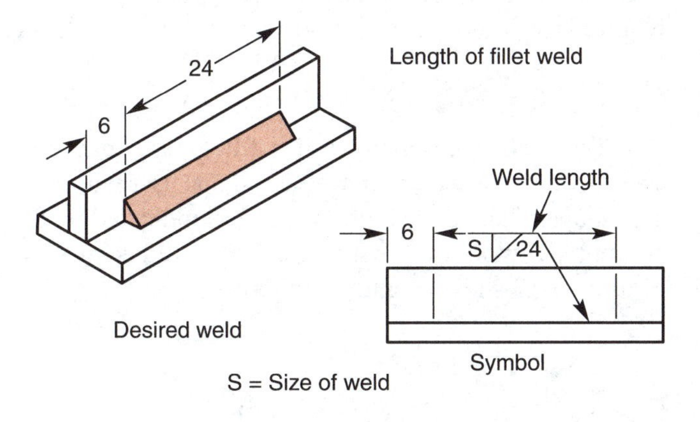

Figure 15-5.
The number to the right of the welding symbol indicates the weld length.

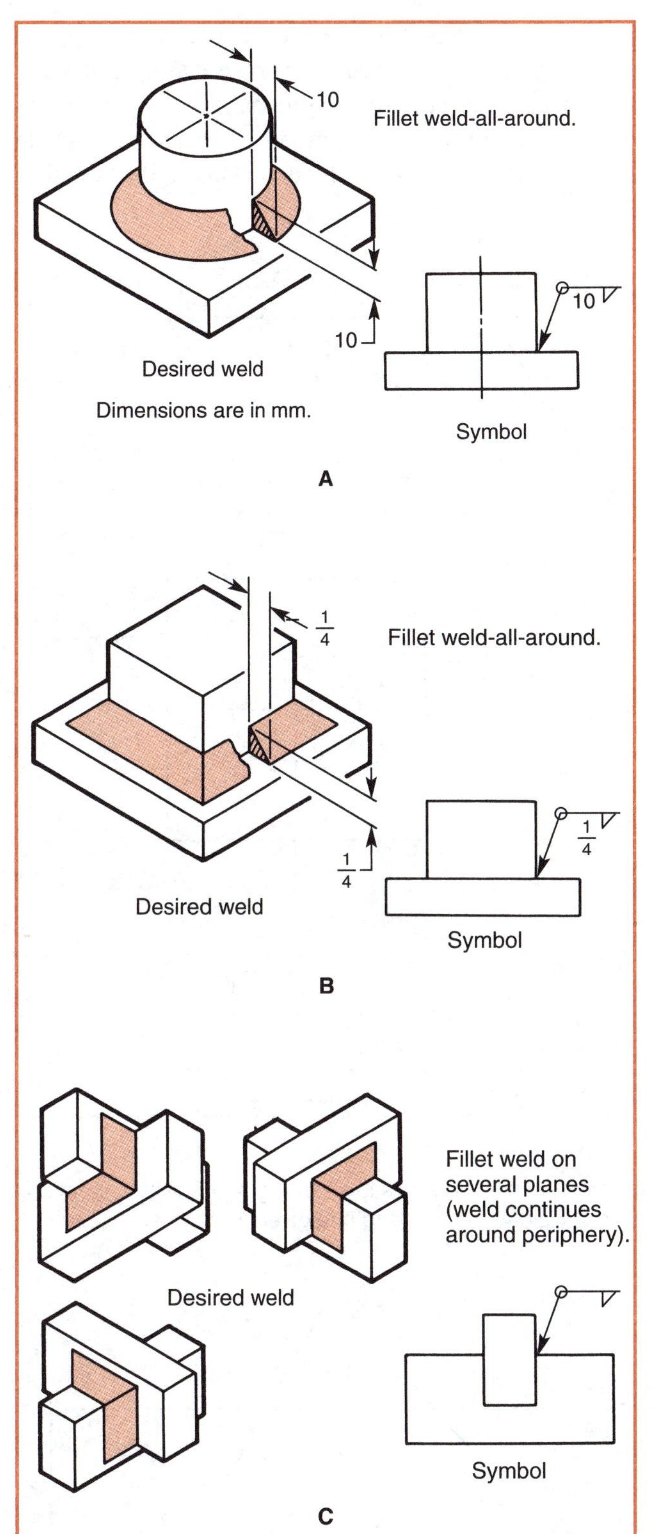

Figure 15-6.
Fillet welds applied to joints with various geometric shapes. A—Welding symbol for all-around fillet weld on round stock. B—The same welding symbol is used when stock is square or rectangular in shape. C—Note the all-around fillet weld on several planes.

The extent and location of a fillet weld may be given graphically on a print view by ***hatching.*** Hatching is shown in Figure 15-8. Additional arrows pointing to each section of the joint are used when you must make abrupt changes in weld direction. This is illustrated in Figure 15-9.

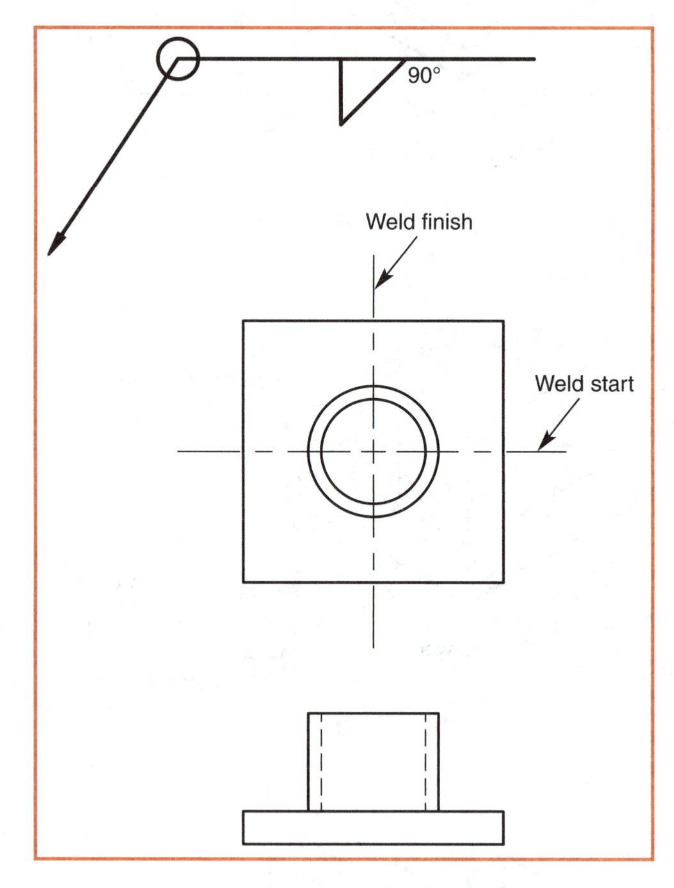

Figure 15-7.
A print usually indicates the starting and finishing location of the weld when the length is given in degrees.

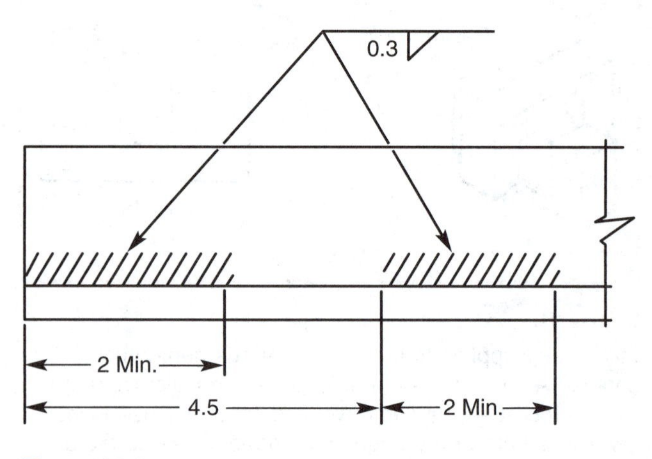

Figure 15-8.
Hatching can graphically illustrate the extent and location of a fillet weld.

Intermittent Fillet Welds

The continuity of an ***intermittent weld*** is broken by recurring unwelded spaces, Figure 15-10. The ***pitch*** (center-to-center spacing) of an intermittent weld (formerly known as a skip weld) is expressed as the distance between centers of increments on *one* side of the joint. The pitch (in inches or millimeters) is shown to the right of the length dimension. Notice the length of the weld is the first number followed by the pitch.

Figure 15-11 shows the welding symbol for staggered intermittent fillet welding. A ***staggered intermittent weld*** occurs on both sides of the joint, with weld increments on one side alternated with respect to those on the other side. For a staggered intermittent fillet weld, the side of the reference line where the weld symbol is closest to the leader indicates the side of the joint that starts with the weld. The side of the reference line where the weld symbol is furthest from the leader indicates the side that starts without the weld (the skip).

Often intermittent welds do not end evenly at the end of the joint. There may not be enough room for another intermittent weld. Be sure to ask your supervisor to explain how the space at the end of the intermittent weld should be finished.

Fillet Welds in Holes and Slots

Fillet welds are sometimes used in circular joints or slots. An all-around fillet weld specified for a hole feature is shown in Figure 15-12.

Specifying Fillet Weld Contours

The weld face contour of a fillet weld may be specified when a certain welded shape is required. The shape (flat, convex, or concave) is indicated by a contour symbol used with the fillet weld symbol. Figure 15-13 shows contour symbols.

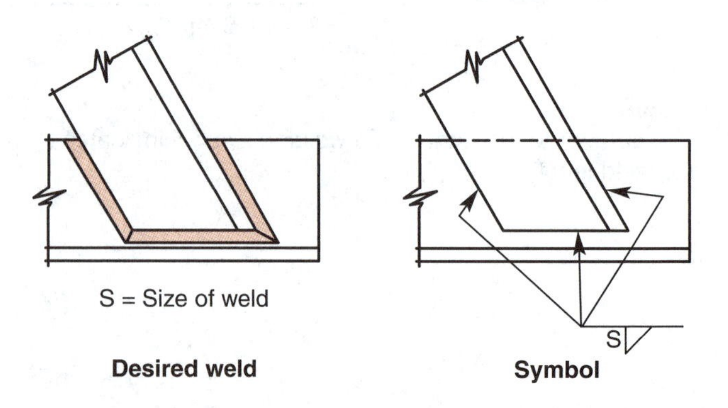

Figure 15-9.
When abrupt changes are made in the weld direction, welding symbol arrows point to each section or joint.

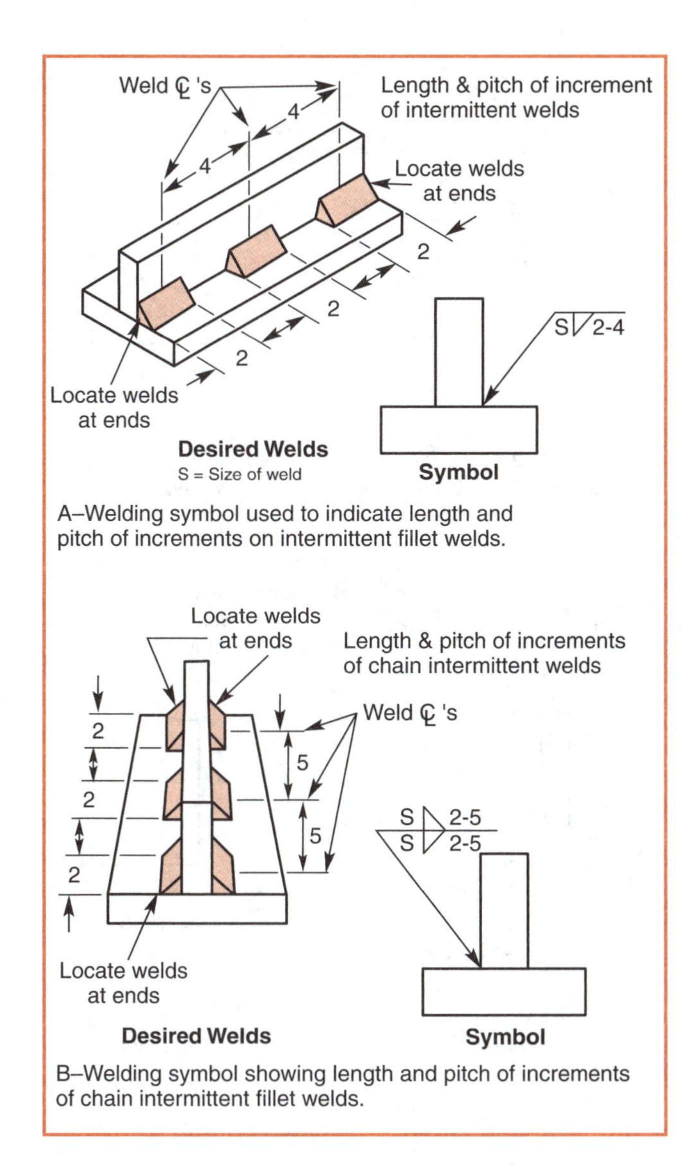

A–Welding symbol used to indicate length and pitch of increments on intermittent fillet welds.

B–Welding symbol showing length and pitch of increments of chain intermittent fillet welds.

Figure 15-10.
Note the use of the intermittent fillet welding symbol.

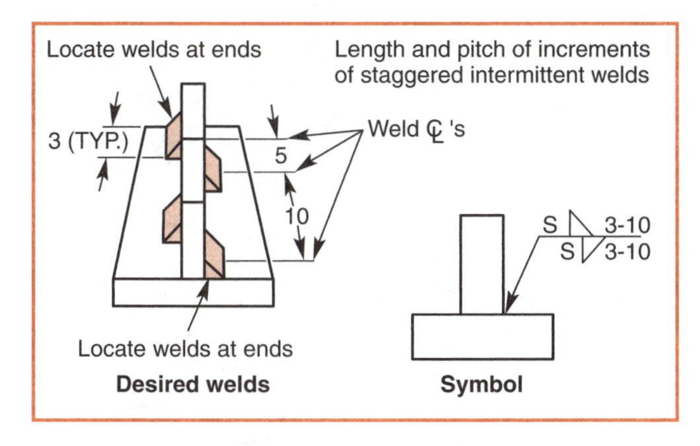

Figure 15-11.
The welding symbol for staggered intermittent fillet welds is shown.

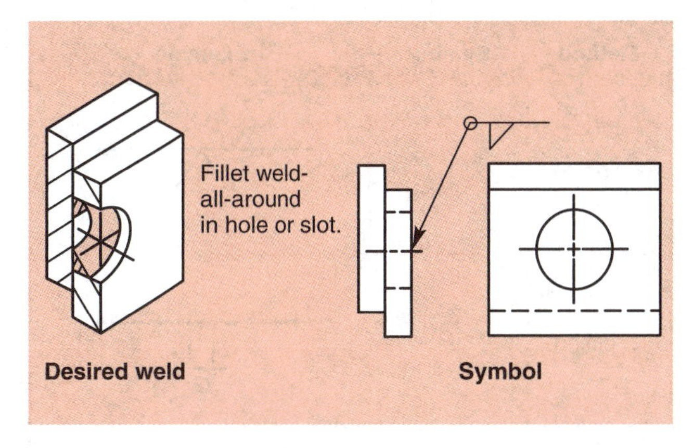

Figure 15-12.
Study the welding symbol and desired weld for an all-around fillet weld in a hole or slot.

Type	Symbol	Example	As welded
Flat	—		
Concave	‿		
Convex	⌒		

Figure 15-13.
Contour symbols used to specify the face contour of a fillet weld. The required shape is shown by the appropriate contour symbol included with the fillet weld symbol.

If the weld face contour is to be finished mechanically, a finish symbol for the method of finishing the weld is added to the contour symbol, Figure 15-14. While the degree of finish (smoothness) is *not* part of the American Welding Society standards, some prints do indicate the desired finish, Figure 15-15.

Combination Weld Symbols

When another type of weld is to be used with a fillet weld, weld symbols for both types are included on the welding symbol, Figure 15-16. Shown is a combined fillet and groove weld with specified sizes. Figure 15-17 shows specifications for a combined fillet and bevel weld.

Method	Symbol	Example
Chipping	C	C
Grinding	G	G
Hammering	H	H
Machining	M	M
Rolling	R	R
Peening	P	P

Figure 15-14.
When the face contour of a weld is to be finished mechanically, a symbol (letter) with the appropriate finish technique is included with the fillet weld and face contour weld symbols.

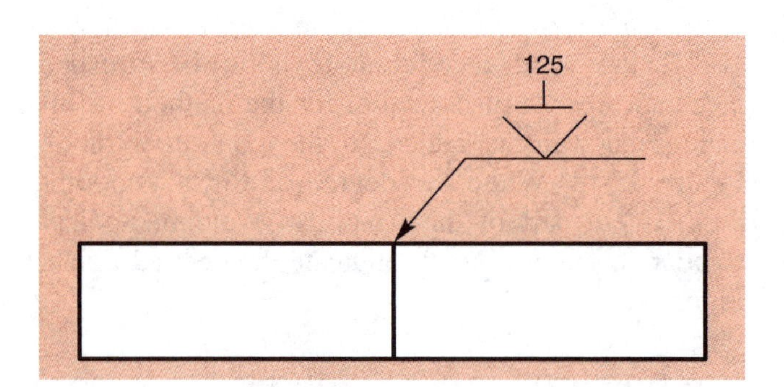

Figure 15-15.
The degree of finish or weld smoothness is indicated by a number that specifies (in microinches) the roughest surface acceptable for this particular application. The larger the number, the rougher the surface permitted. A 125 surface is smoother than a 250 finish. The symbol is seldom used.

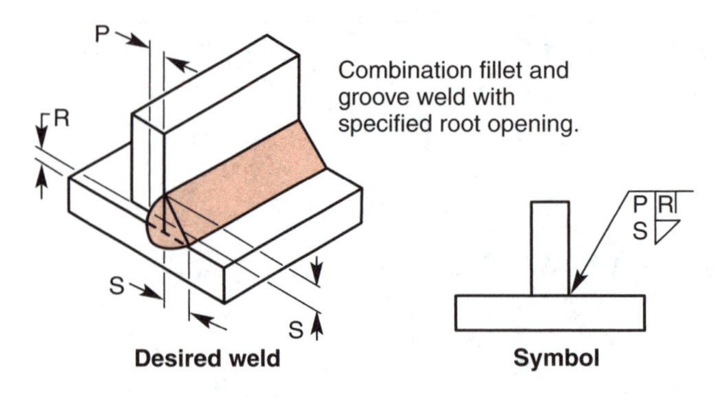

Figure 15-16.
More than one type of weld is specified for this joint. Note how both weld symbols are included on the symbol.

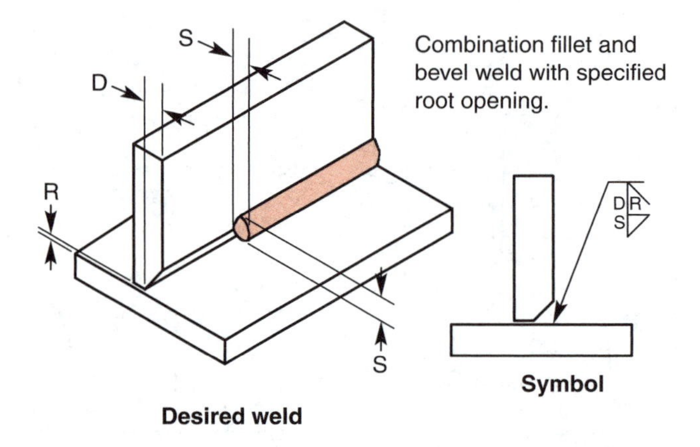

Figure 15-17.
A part requiring a combination fillet and bevel weld for fabrication.

Name ________________________________ Date ______________ Class ______________

Print Reading Activities

Part I

Refer to the drawing below to answer the following questions.

1. What is the drawing title?

   ____

2. Name the drawing number.

   ____

3. How many parts make up the assembly?

   ____

4. What welding process is to be used?

   ____

5. The gussets are located _____″ from the edges of the support.
6. The gusset is made from _____″ × _____″ × _____″ stock.
   There are _____ required in each assembly.
7. The width and thickness dimensions of the material used to make the support are _____″ and _____″.
8. What type of weld is specified? ____________
9. Name the type of welding rod that is specified.

   ____

10. What is the size of the specified weldments?

    ____

11. Sketch the specified welds on the front and side views in the drawing.
12. Each corner of the bracket is chamfered _____″ × _____°.
13. How many holes are to be drilled? ____________
14. The hole size is _____ diameter. ____________
15. The complete bracket is to be finished by _____.

    ____

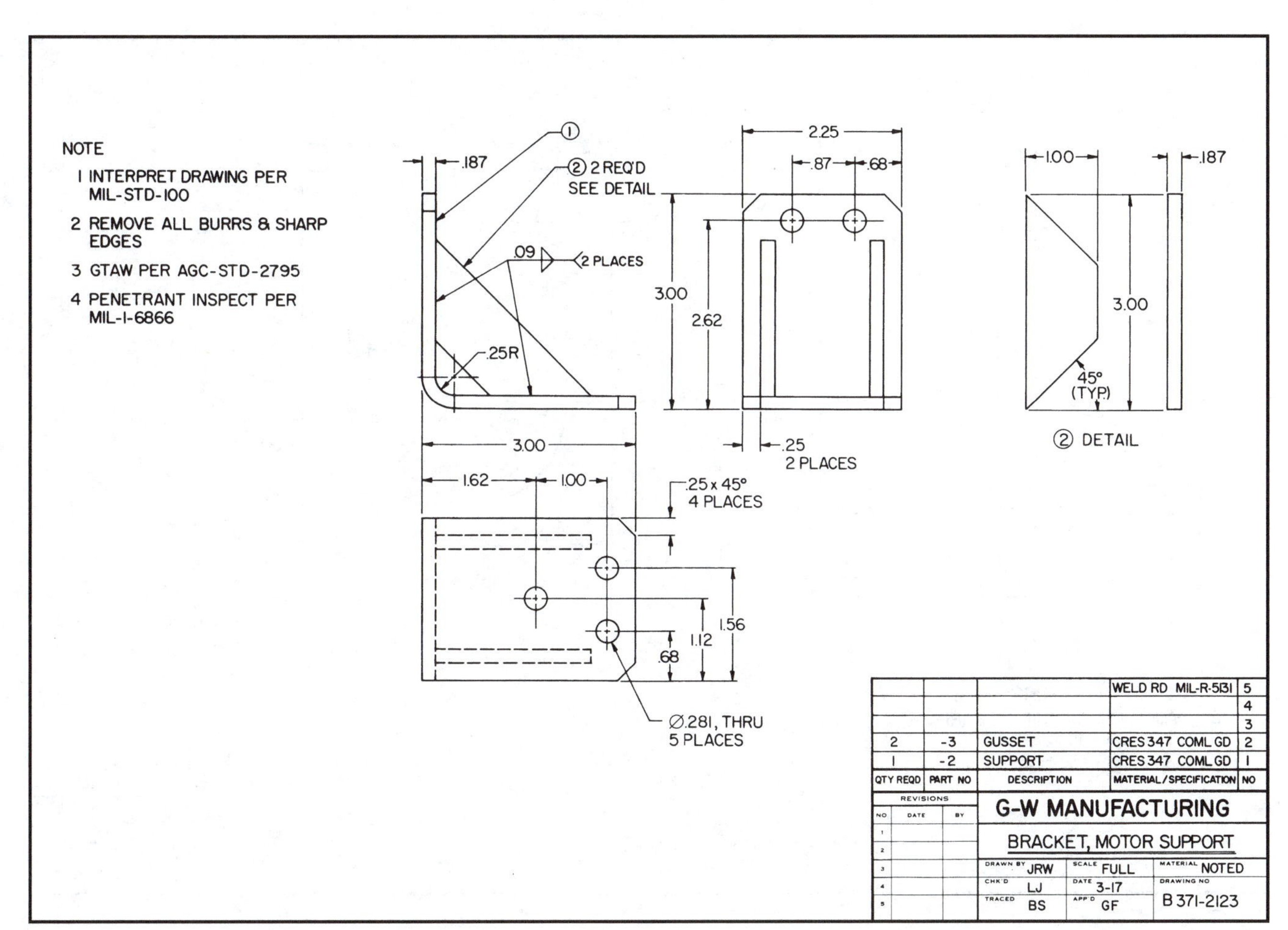

Part II

Refer to the drawing below to answer the following questions.

1. What is the name of the product to be made?

 ____

2. Name the drawing or print number.

 ____

3. List the number of parts in the assembly.

 ____

4. What material is specified to make the product?

 ____

5. What type of weld rod is specified?

 ____

6. Name the size and type of the weld specified.

 ____

7. On the drawing below, make a sketch of the required weld.

8. The size of the mounting plate is _____" × _____" × _____".

9. The width and thickness dimensions of the material required to make the curved section are _____" and _____".

10. The centerline of the curved section is located _____" from the lower end of the mounting plate on the drawing and _____" in from the side.

11. What are the number and size of holes to be made in the mounting plate?

 A. ______________________________

 B. ______________________________

12. What additional operation must be performed on the drilled holes?

 ____

13. After welding and machining, what finishing operations are specified?

 A. ______________________________

 B. ______________________________

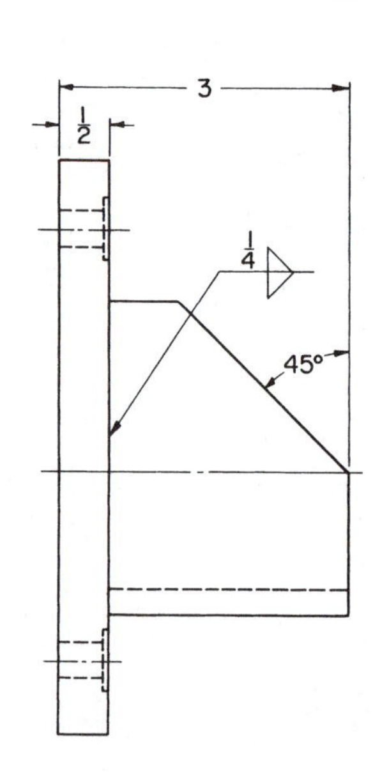

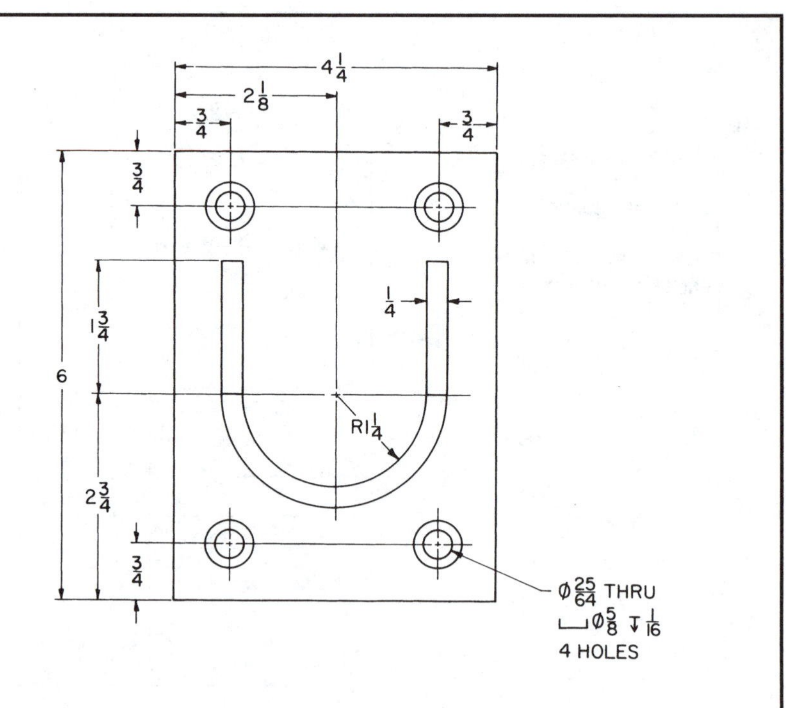

NOTE -
1. REMOVE ALL BURRS & SHARP EDGES.
2. PAINT NONMACHINED SURFACES ONE COAT SSPC-2367 (GREEN).
3. WELD-ELECTRODE E6013.

REVISIONS			GAUTHIER ENTERPRISES, INC.		
NO	DATE	BY	ROD SUPPORT		
1			DRAWN BY JRW	SCALE FULL	MATERIAL STL AISI 1020
2			CHK'D GF	DATE 2-30	DRAWING NO
3			TRACED JF	APP'D LJ	B-78675
4					
5					

Name ______________________________

Part III

Draw the correct welding symbols for these welds.

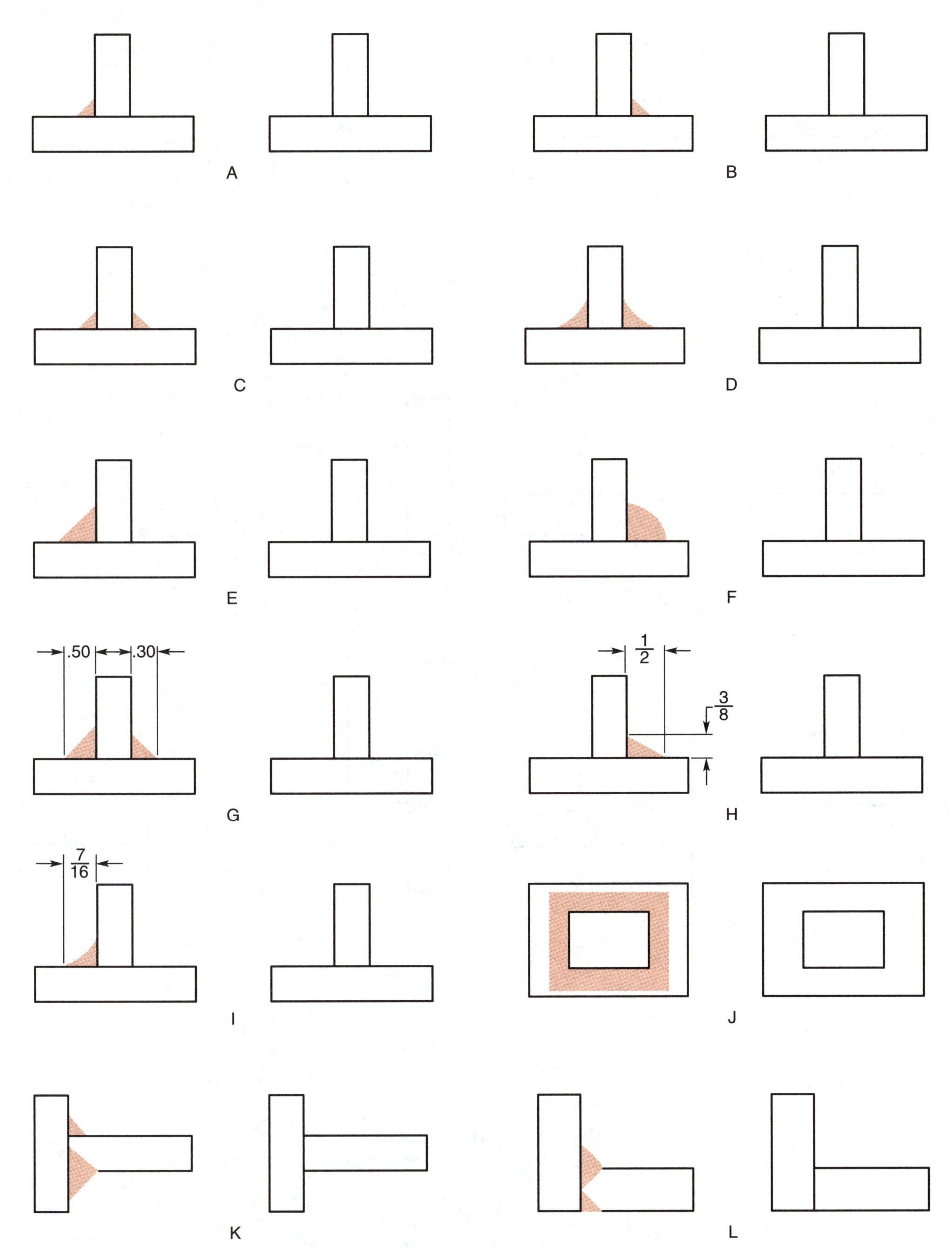

Part IV

Draw the correct weld as indicated by the welding symbol.

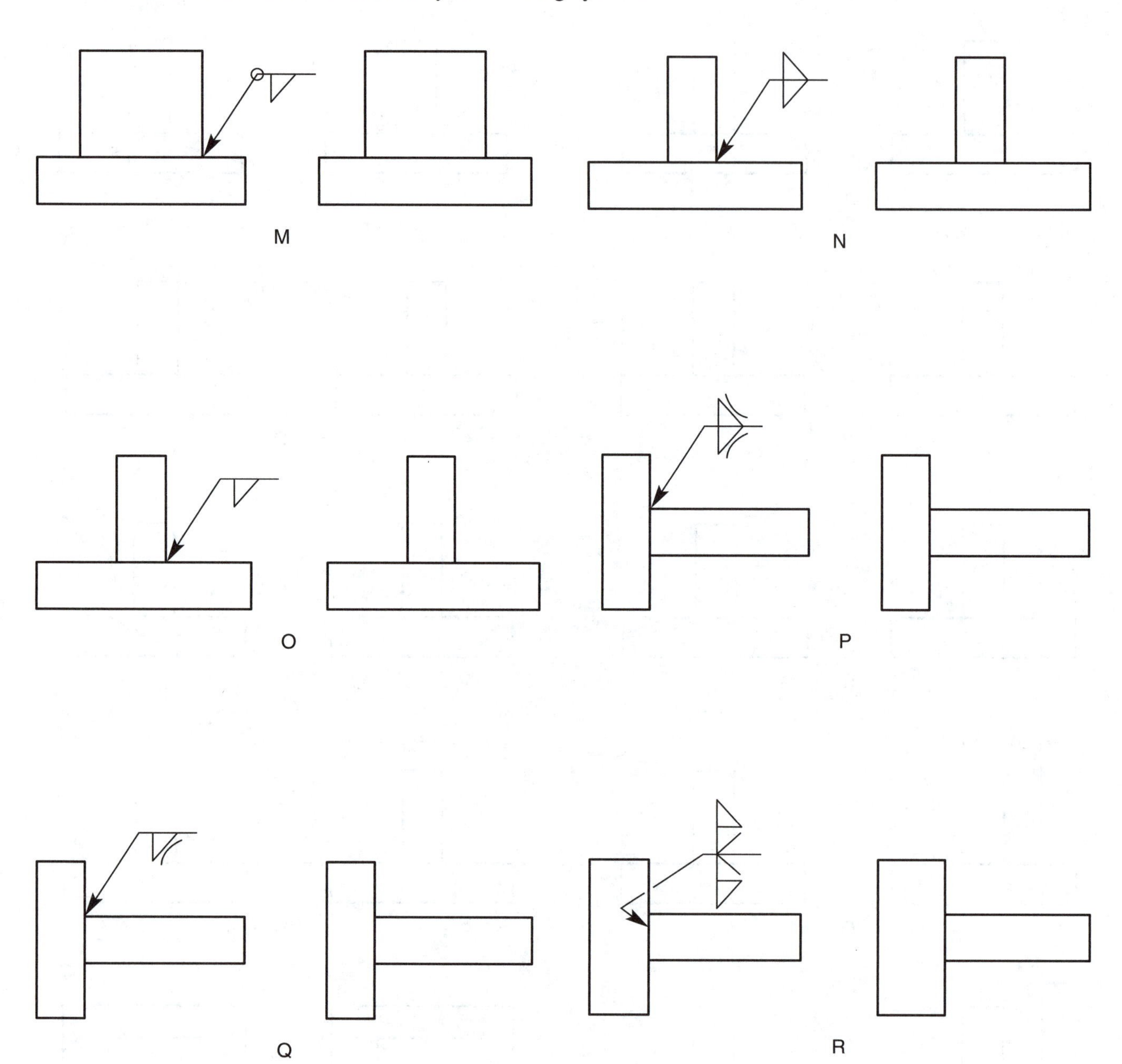

Name ____

Part V

Draw the correct welding symbol for the following welds.

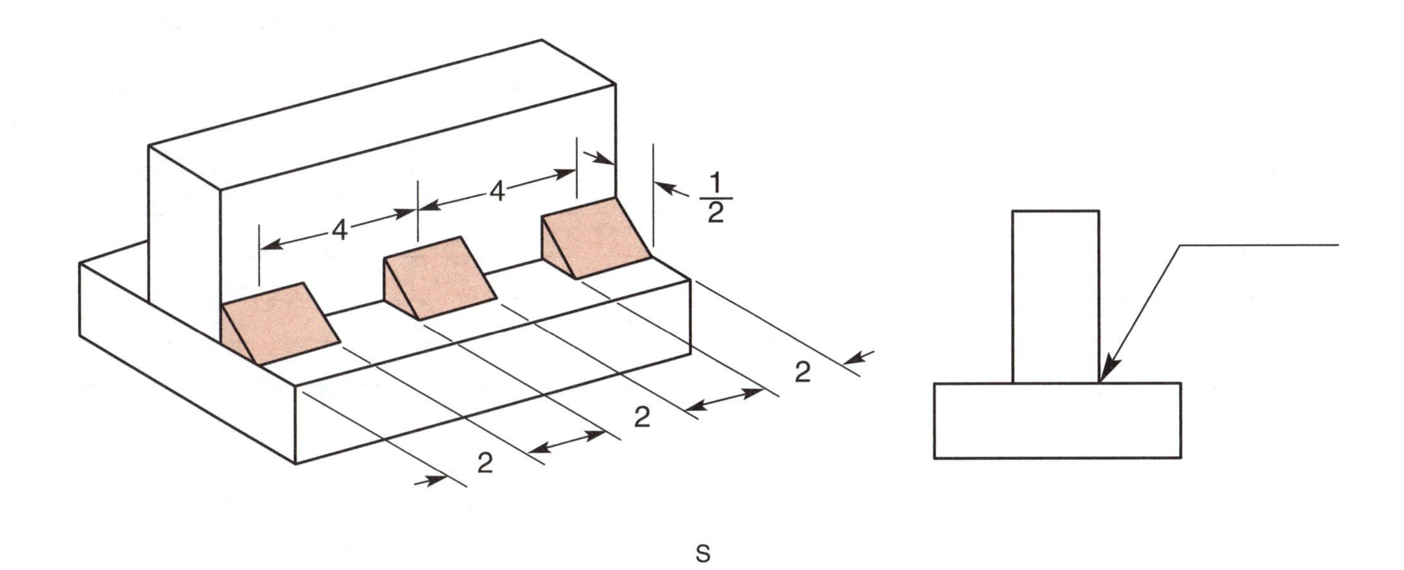

S

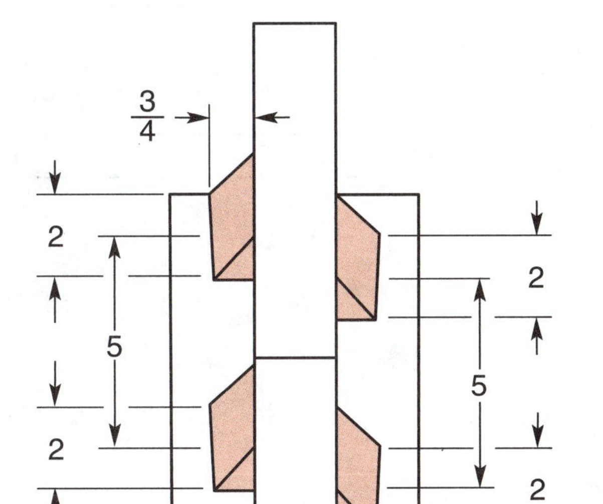

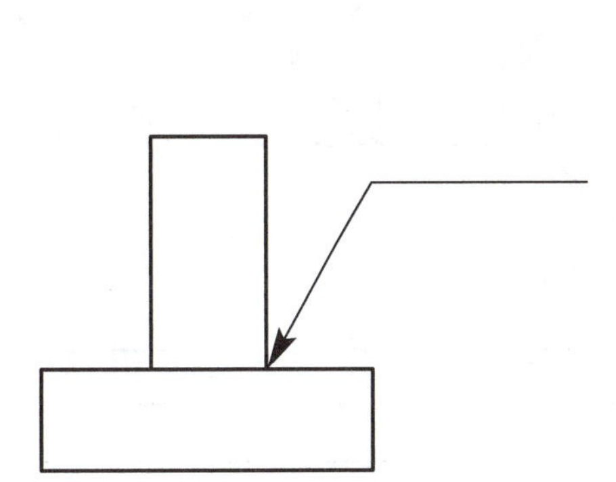

T

Part VI

Explain the following welding symbols.

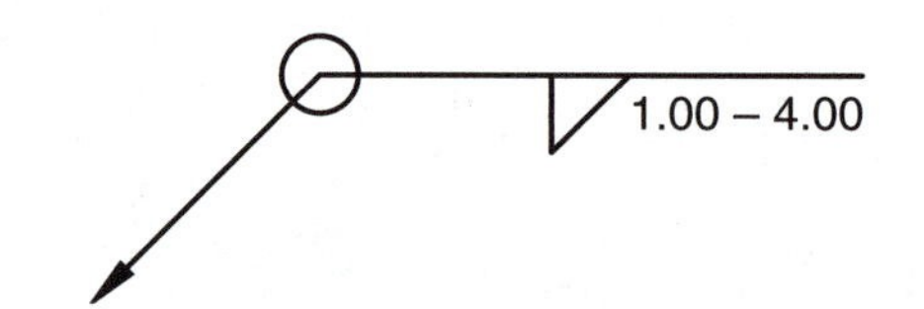

1. ______________________

.25 x .38

4. ______________________

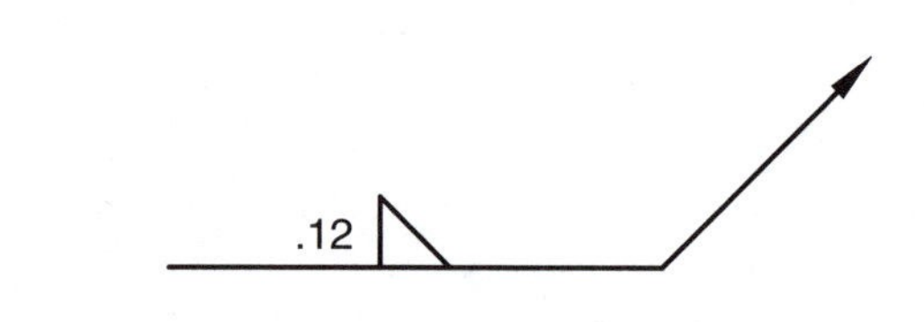

2. ______________________

.25 .50 – 1.00
.25 .38 – 2.00

5. ______________________

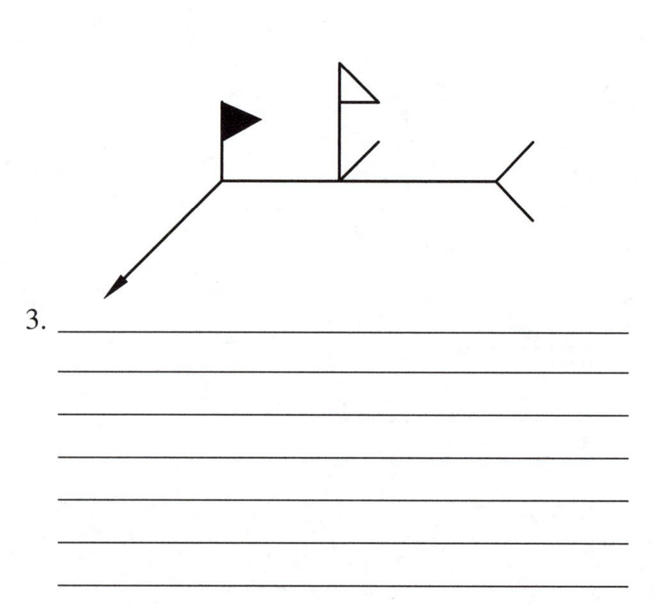

3. ______________________

Unit 16

Groove Welds

After completing Unit 16, you will be able to:

- Differentiate a groove weld from other types of welds.
- Interpret dimensions for preparing groove welds, including the depth of preparation, groove angle, bevel angle, and root opening size.
- Determine the preparation size and effective throat of groove welds.
- Apply groove weld dimensioning standards.
- Interpret surface finish and contour symbols.
- Interpret melt-through, back, and backing weld symbols.
- Explain uses for backing, joint spacers, and runoff weld tabs.

Key Words

backing
bevel angle
effective throat
flare-groove welds
groove angle
groove face
groove radii
joint root
joint spacers
root faces
root opening
runoff weld tabs

Groove welds are made in the space between two sections of metal, Figure 16-1. With the exception of the square-groove and flare-groove joints, one or more of the members being joined is prepared by removing metal to form a V-, J-, or U-shaped trough. This joint preparation provides for deeper or full penetration of

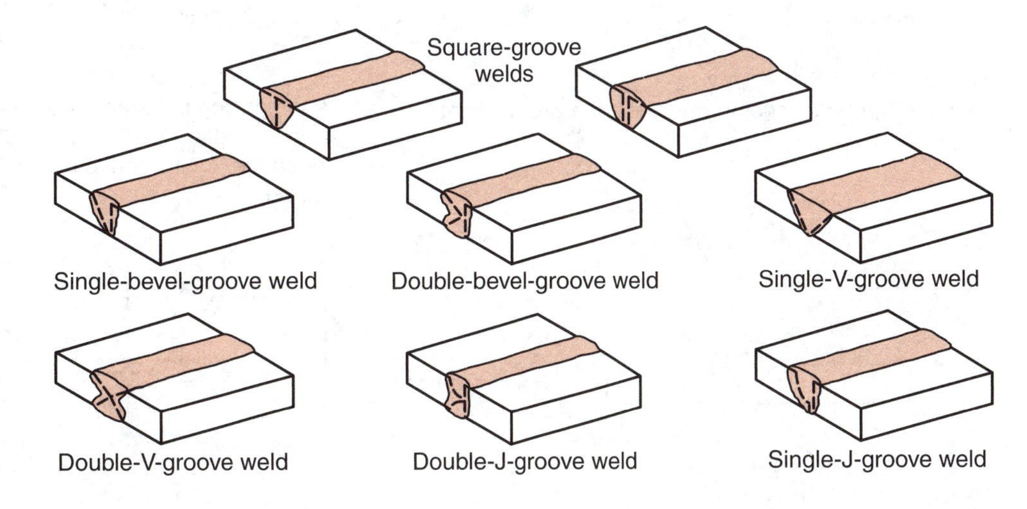

Figure 16-1.
Single-groove and double-groove weld joints are shown.

the weld into the joint and provides clearance for the electrode. Burning, grinding, arc gouging, chiseling, or machining removes the metal.

Root Opening, Groove Angle, and Bevel Angle

It is important to be familiar with the common terms associated with groove joints and the preparation of groove welds. The ***root opening*** is the gap at the joint root workpieces, Figure 16-2. The ***joint root*** refers to the part of a joint to be welded where the members align closest to each other. The root opening is used to provide access to the joint for the electrode and improved weld penetration.

When additional clearance or penetration is needed for thicker material, an angle is placed on the edge of the material. A ***groove angle*** is the total angle formed between the groove face on one workpiece and the groove face on the other workpiece. The ***groove face*** is the joint member surface included in the groove. A ***bevel angle*** is the angle formed between the bevel of one piece and a plane perpendicular to the surface of the piece. The angle may be placed on one side of the joint, as with a single-bevel-groove, or the angle may be placed on both sides, as in a V-groove.

A V-groove weld applied to the joint in Figure 16-2 is shown in Figure 16-3. Also shown is the welding symbol. Note that the distance specified for the root opening (.06) is shown inside the groove weld symbol and the groove angle information (60°) is placed outside the weld symbol. As with fillet welds, dimensions for groove welds are shown on the *same side* of the reference line as the weld symbol.

Preparation Size and Effective Throat of Groove Welds

The ***effective throat*** is the minimum distance (minus any convexity) between the weld root and the face of the weld. It describes the *weld size* (penetration), Figure 16-4. When specified for a weld, the effective throat is shown in parentheses to the left of the weld symbol. As shown in Figure 16-4, it appears to the right of the depth of bevel. The depth of bevel indicates the depth of preparation for preparing the joint.

The effective throat of a groove weld is specified when the weld extends only partially through the members being joined. Complete joint penetration is indicated when no dimension is given on the welding symbol for a single-groove or a symmetrical double-groove weld. Figure 16-5 illustrates complete weld penetration for a double-groove joint.

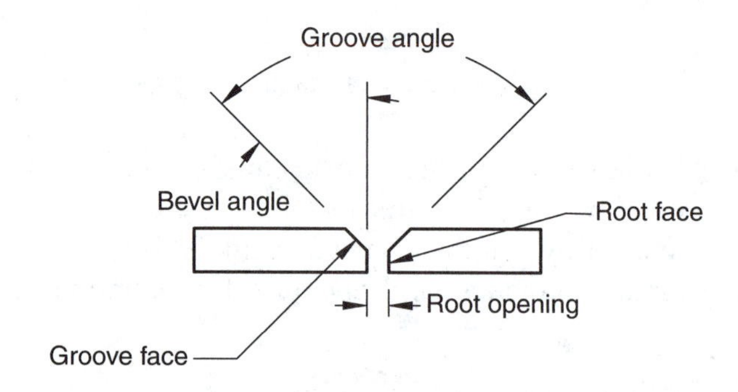

Figure 16-2.
Common terms describing the parts of a groove joint.

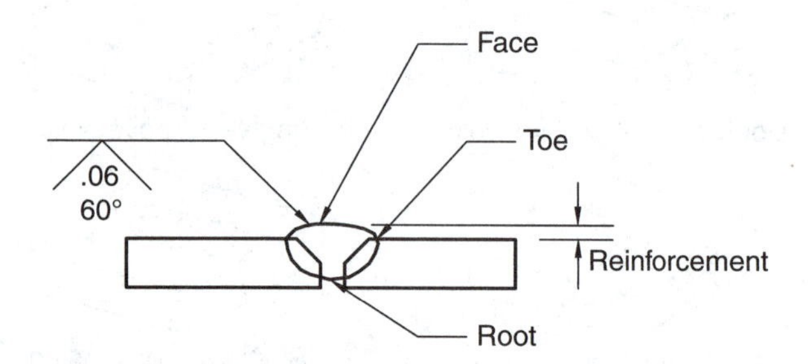

Figure 16-3.
A V-groove weld with its parts identified. The welding symbol shows the root opening size and the groove angle.

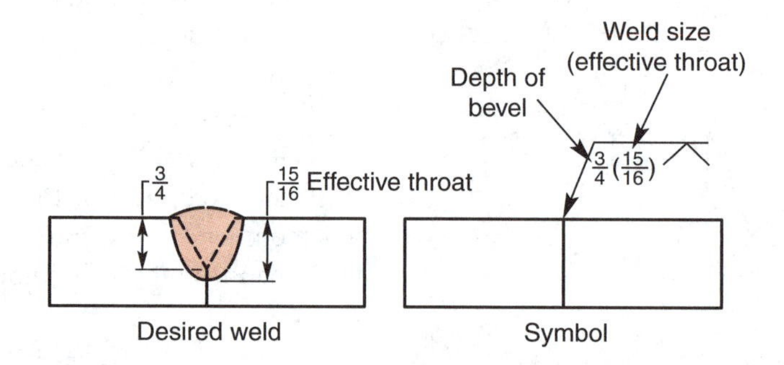

Figure 16-4.
A dimension in parentheses to the left of the weld symbol gives the effective throat (weld size) of the groove weld when the weld extends partly through the members being joined.

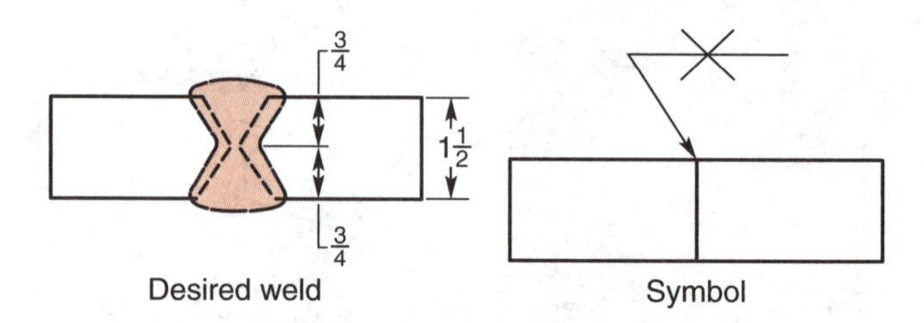

Figure 16-5.
When no dimension is given on the welding symbol, the weld should completely penetrate the joint.

A dimension *not* in parentheses on the left of a bevel-, V-, J-, or U-groove weld symbol (in cases where the effective throat is not specified, or is specified elsewhere on the print) indicates the size of the weld preparation only, Figure 16-6. *No* such dimension is needed with a square-groove weld.

Optional groove preparation with complete penetration is indicated when the letters CJP are shown in the tail of the reference line. No weld symbol is used, as in Figure 16-7.

The weld size of a flare-groove weld is considered only to the tangent point (the point where the curved surfaces meet), Figure 16-8.

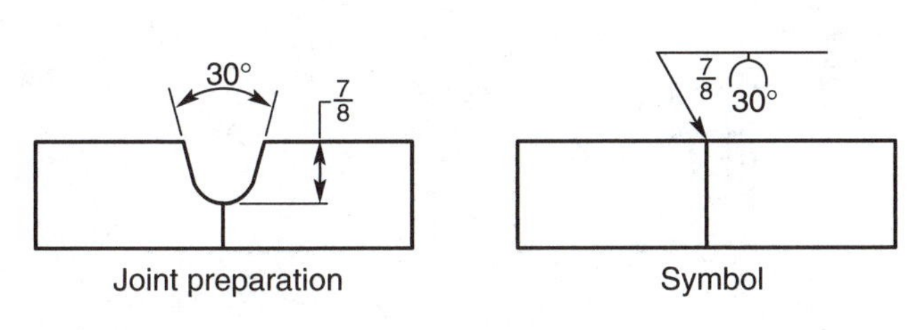

Figure 16-6.
Weld preparation only is indicated if no dimension is in parentheses to the left of the bevel-, V-, J-, or U-groove weld symbol.

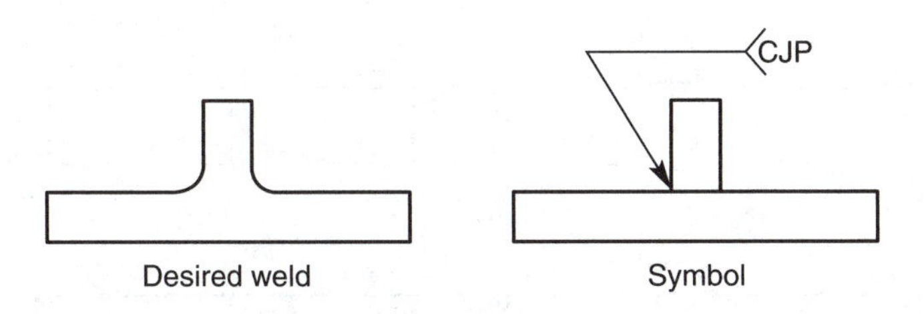

Figure 16-7.
The letters CJP in the tail of the reference line indicate optional groove preparation with complete weld penetration.

General Use of Groove Weld Symbol

Different conventions are used for groove welding symbols depending on the dimensions that are specified and the information required. As previously discussed, dimensions for the preparation of groove welds are shown on the *same side* of the reference line as the weld symbol. See Figure 16-9. This example shows a J-groove weld. The information specified includes the depth of preparation, groove angle, and root opening.

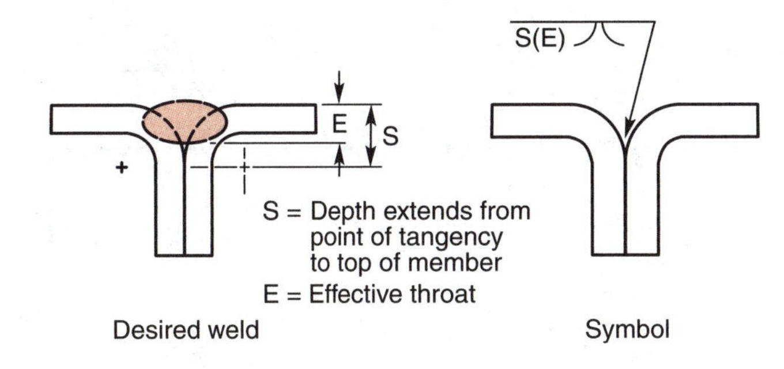

Figure 16-8.
Flare-groove weld size extends only to the tangent points of the joint members.

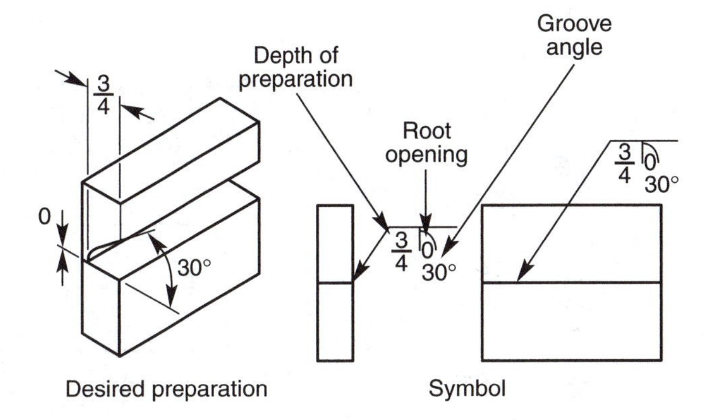

Figure 16-9.
Specifications for the preparation of an arrow side J-groove weld.

Double-groove welds are dimensioned on both sides of the reference line if no general note appears on the print, Figure 16-10. If the welds differ in size, they are dimensioned as in Figure 16-11. Groove welding symbols will *not* include dimensions when a general note determining groove weld size appears on

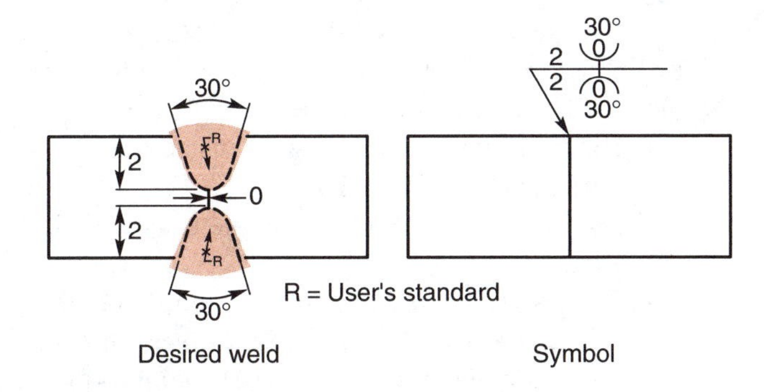

Figure 16-10.
Unless there is a general note on the print, double-groove welds are dimensioned on both sides of the reference line.

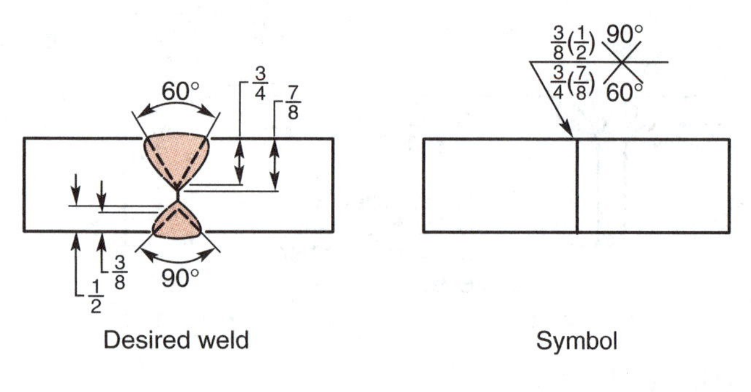

Figure 16-11.
Groove welds differing in size are dimensioned in the manner shown.

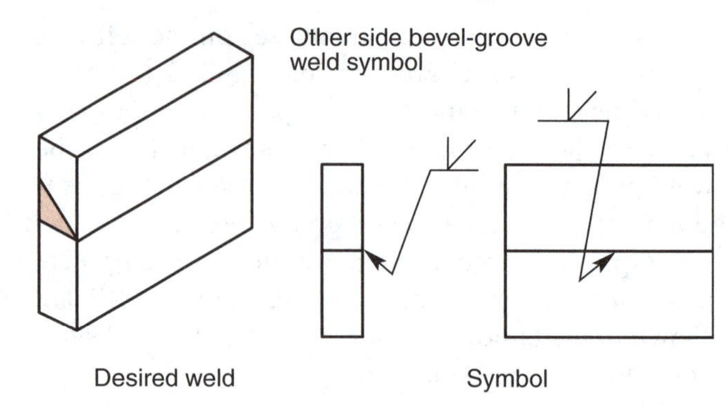

Figure 16-13.
A break in an arrow always points toward the member of the single-bevel-groove or J-groove joint to be beveled.

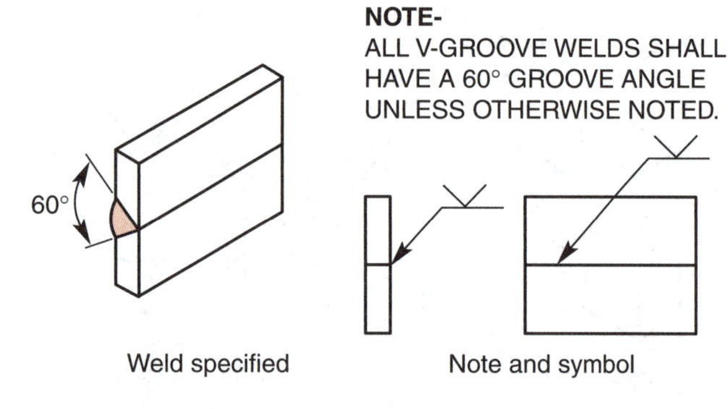

Figure 16-12.
If a general note indicating weld size is on the print, no groove weld dimensions are given with the welding symbol.

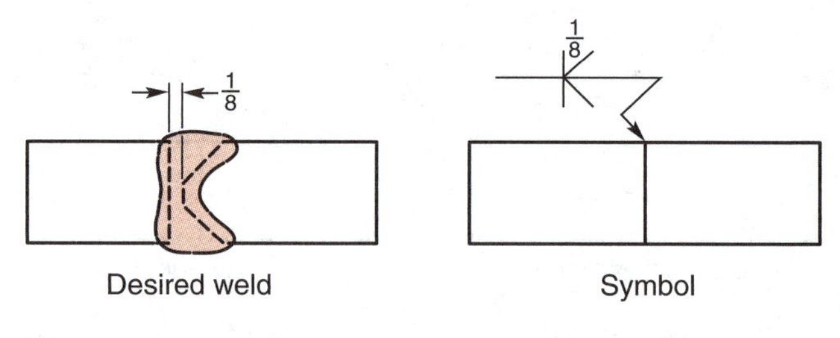

Figure 16-14.
The root opening of a groove weld is specified inside the weld symbol when standards are not otherwise indicated.

the print, Figure 16-12. When a break in the arrow is used with bevel- and J-groove welds, the arrow points *toward* the member to be beveled, Figure 16-13.

Groove Dimensions

Many companies have established their own standards for groove weld dimensions. These standards are observed unless otherwise noted on the print. When company standards for groove welds are *not* indicated, the following applies:

- The root opening is indicated inside the weld symbol, Figure 16-14.
- The groove angle or bevel angle is specified, Figure 16-15.
- The ***groove radii*** (used to form the shape of J- or U-groove welds) and ***root faces*** (the parts of the groove face within the joint root) are shown by cross section, detail, or other means with a reference on the welding symbol, Figure 16-16.

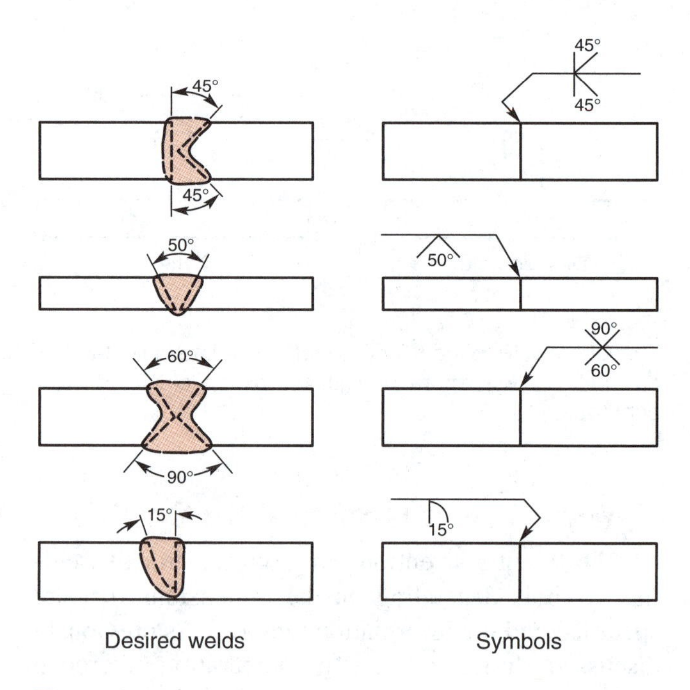

Figure 16-15.
Study how groove angles of groove welds are specified.

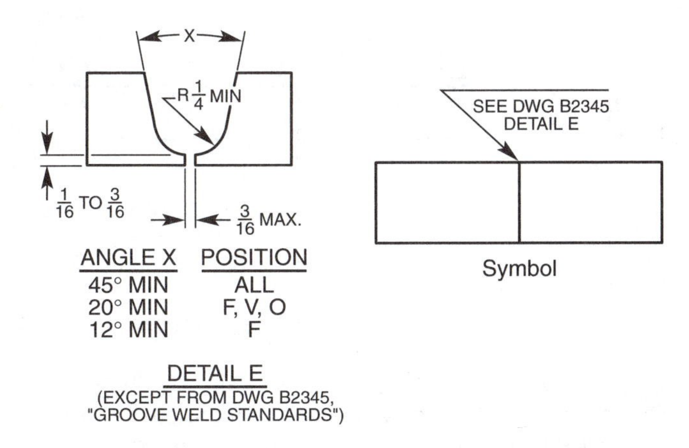

Figure 16-16.
Groove radii and root faces of U- and J-groove welds are shown by cross section, detail, or other means with a welding symbol reference.

Surface Finish and Contour of Groove Welds

The buildup of the groove weld above the surface of the base material is called *reinforcement*. Sometimes the welding symbol specifies that the reinforcement be minimized or removed.

Groove welds to be made approximately flush (but not generally finished flush mechanically) are specified by a flush contour symbol. This symbol is placed above the weld symbol, as in Figure 16-17. Groove welds to be made flush by mechanical means are specified with a flush contour symbol and the method of making the weld flush, Figure 16-18. Groove welds to be finished mechanically with a convex contour are specified by a convex contour symbol. The method of finishing the weld to a convex contour is also given, Figure 16-19.

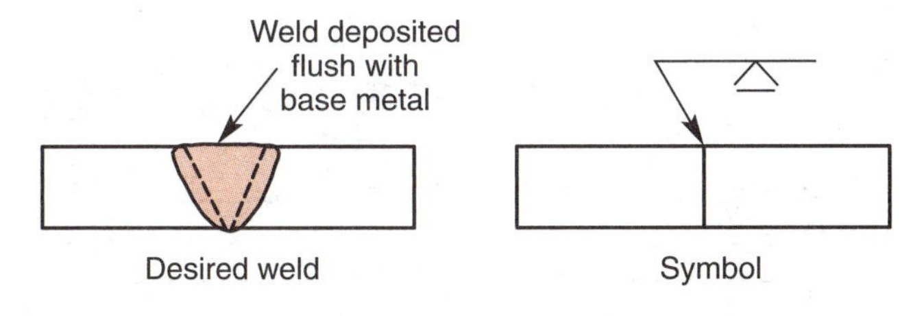

Figure 16-17.
A flush contour symbol is placed above the weld symbol when the groove weld is to be made approximately flush and without the use of grinding, chipping, hammering, or machining.

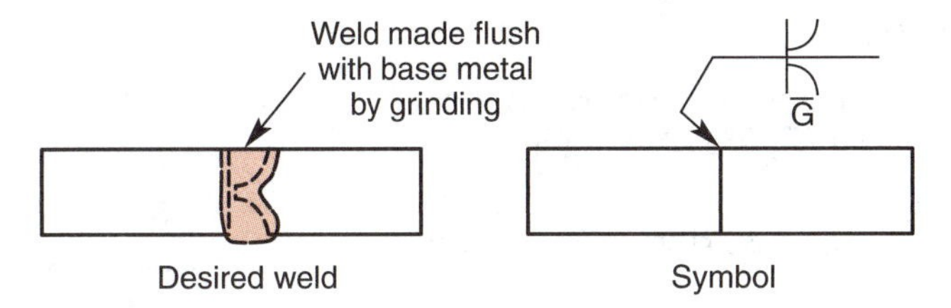

Figure 16-18.
Groove welds to be made flush mechanically are specified by a flush contour symbol and by the method to use to make the weld flush.

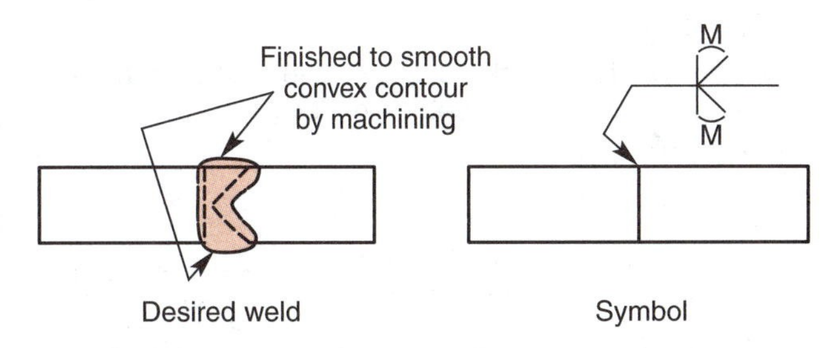

Figure 16-19.
When groove welds are to be finished mechanically to a flush contour, they are specified by a convex contour symbol and the method of finishing.

Melt-Through, Back, and Backing Welds

The melt-through, back, and backing weld symbols show that a melt-through to the other side, bead-type back, or backing weld is needed with a single-groove weld. Points to remember include:

- A back weld is made *after* the groove weld.
- A backing weld is made *before* the groove weld.
- A melt-through is a visible reinforcement produced in a groove weld from one side.

A note states whether a back or backing weld is to be made. This note is placed in the tail of the welding symbol, Figure 16-20. As shown, a back or backing weld symbol is located on the side of the reference line that is opposite the groove weld symbol.

A flush contour symbol, added to the back or backing weld symbol, indicates the weld should be approximately flush with the base metal, Figure 16-21.

If the back or backing weld is to be made flush by mechanical means, the method of making the weld flush is added to the flush contour symbol, Figure 16-22.

When a back or backing weld is to be finished to a convex contour by mechanical means, a convex contour symbol and finish symbol are added to the weld symbol.

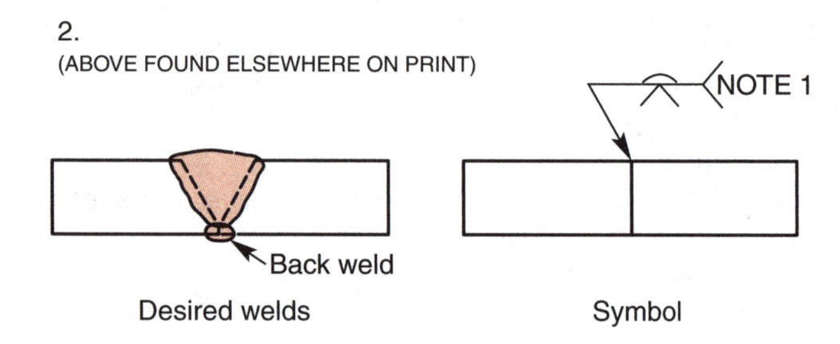

Figure 16-20.
Specifications for a back weld. A note in the tail of the reference line indicates whether to make a back or backing weld.

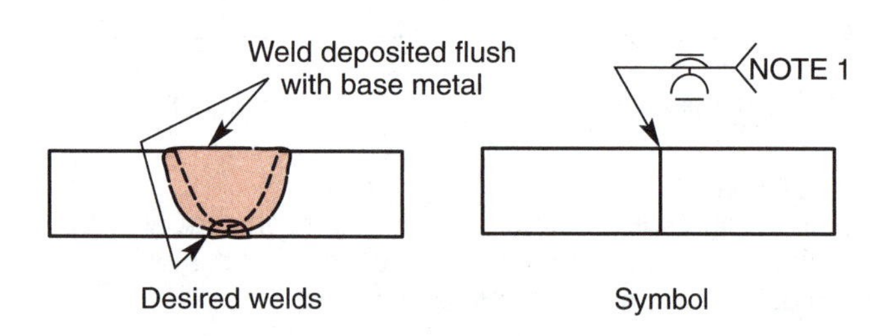

Figure 16-21.
A flush contour symbol indicates the weld is to be finished approximately flush with the base metal.

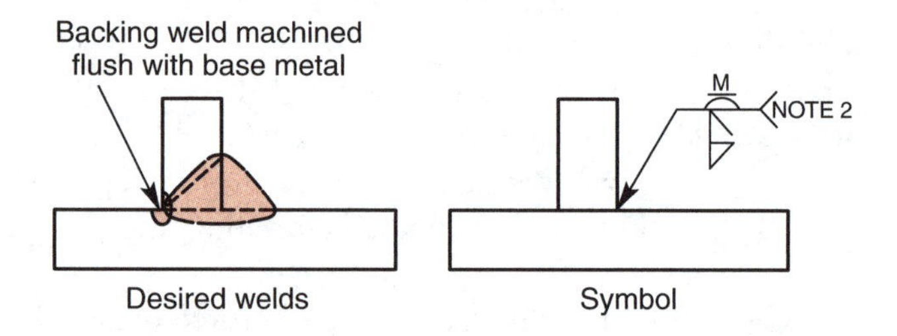

Figure 16-22.
A symbol added above the flush contour symbol identifies the mechanical method used to finish the weld flush.

With the exception of height, which is optional, no other back or backing weld dimensions are shown with the weld symbol, Figure 16-23. If other dimensions are required, they are shown on the drawing.

A melt-through weld assures full joint penetration. The melt-through weld symbol is similar to the back or backing weld symbol with the bead filled in as shown in Figure 16-24. A dimension to the left of the symbol specifies the amount of melt-through.

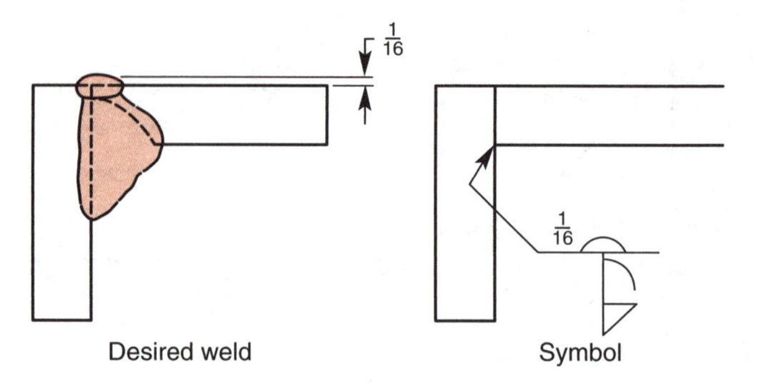

Figure 16-23.
Only the back or backing weld height dimension is shown on the welding symbol. If other dimensions are needed, they are given elsewhere on the print.

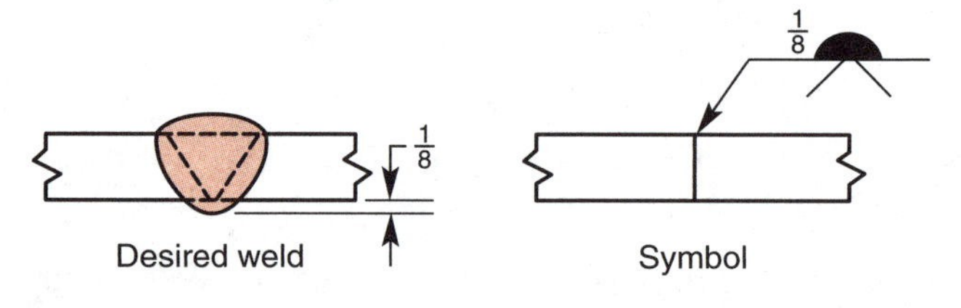

Figure 16-24.
The melt-through weld symbol resembles the back or backing weld symbol, but the bead is filled in. The height of root reinforcement may appear as a dimension specified to the left of the melt-through symbol. (American Welding Society)

Backing, Joint Spacers, and Runoff Weld Tabs

Backing is material placed against the back side of a joint to withstand molten weld metal, as shown in Figure 16-25. It is employed when full penetration groove welds are required and welding can only be done from one side. Backing is thoroughly penetrated by the weld and usually left in place.

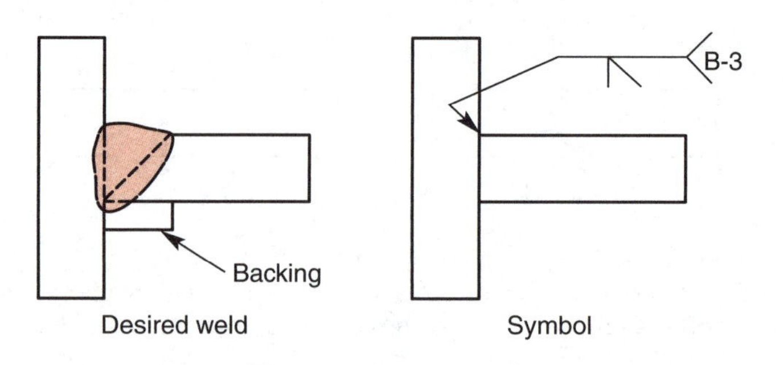

Figure 16-25.
Backing used in a joint. A specification indicating its use is shown in the tail of the reference line.

Joint spacers are metal parts inserted in the joint root as backing and to maintain the root opening during welding, Figure 16-26. Joint spacers are sometimes used, especially if the weld is in thick material and the minimum possible V-angle is specified. In such welds, the root must be gouged out completely, including the spacer bar, before the second side of the groove is welded.

Runoff weld tabs provide an extension of the groove beyond the pieces being joined when a full-length groove weld is specified, Figure 16-27. Runoff tabs provide a place to strike the arc and material at the end of the weld to eliminate the weld crater. The angle or contour of the runoff weld tab must be identical to that of the groove.

Since welding symbols give no indication of the backing, spacer, or extension bar requirements, note that unless covered by reference to AWS prequalified joints or fabricators' standards, special sketches of the weld profile are provided.

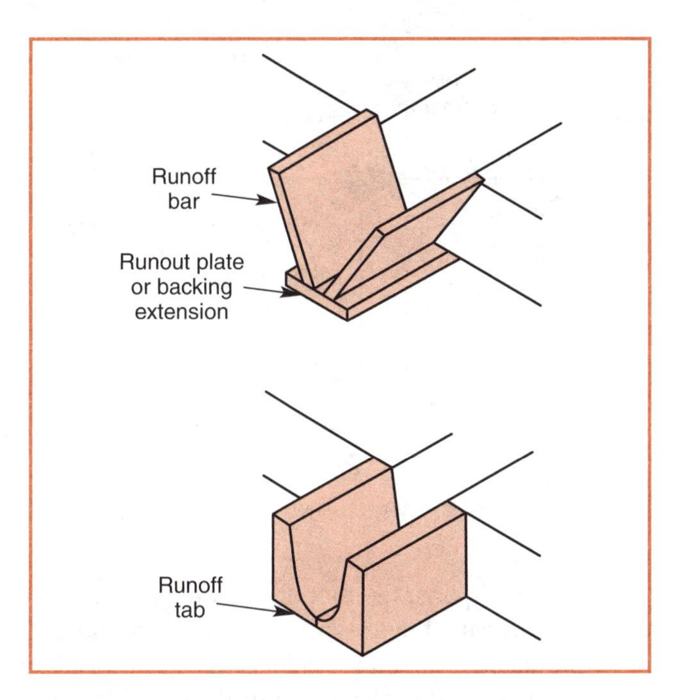

Figure 16-27.
A runoff weld tab is used when a full-length groove weld is specified. Specifications in the reference line tail or sketches on the drawing may be used to indicate a runoff weld tab.

Flare-Groove Welds

Flare-groove welds are used to join round or formed metal parts. The groove that is formed when curved surfaces are placed together does not have straight sides on one or both members. Two round steel bars laid side-by-side, such as reinforcing rod, have sides that are curved. The joint that is formed does not have straight sides like a V-groove weld.

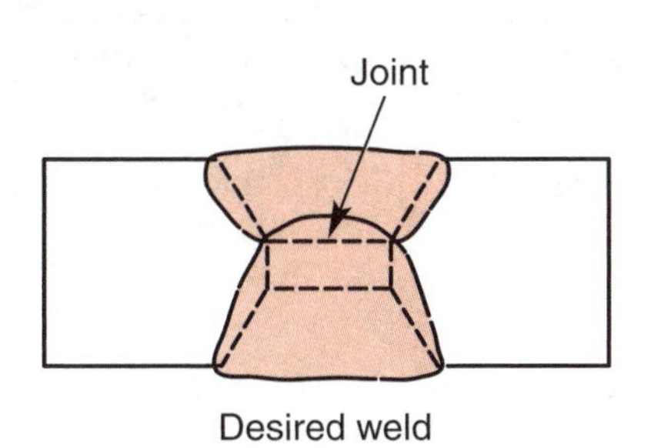

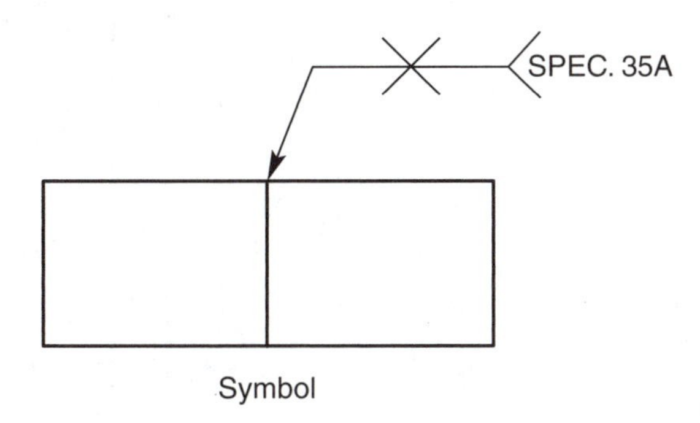

Figure 16-26.
Joint spacers may be specified when thick sections are welded. In such welds, the root and joint spacers are gouged out before the second side of the groove is welded. Specifications are shown in the reference line tail.

Figure 16-28 shows an example of a flare-V-groove weld. Either two round parts (members) or two formed parts (members) can be used to form the V-groove. The symbol for the flare-V-groove weld can be placed on a single side, or the symbol can indicate that the weld should be made on both sides.

The depth dimension for a flare-V-groove weld is given as the distance from the top of the member to the point of tangency (where it touches the other member or part). Figure 16-29 shows a dimension of .31 to the point of tangency and a weld size of .25. Notice that the weld size is placed in parentheses. The weld size is the distance from the surface of the part to the root of the weld.

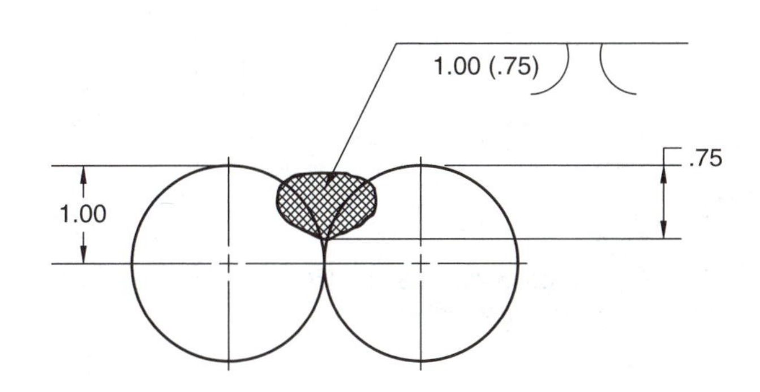

Figure 16-28.
A flare-V-groove weld applied to a joint formed by two round parts.

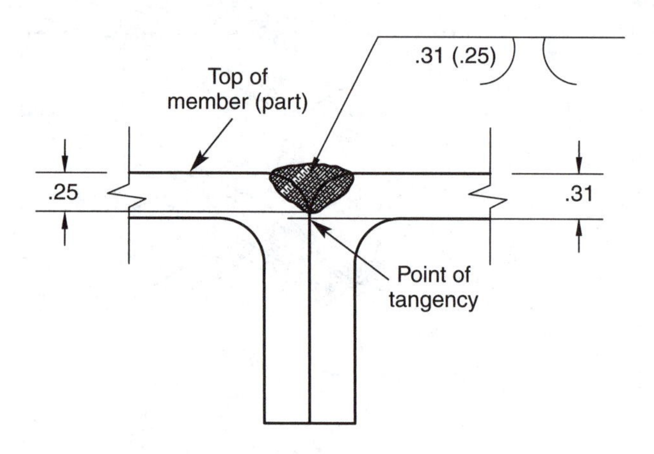

Figure 16-29.
Dimensions for a flare-V-groove weld.

For round parts, the first dimension is the radius of the round part. Figure 16-28 shows the round bar has a radius of 1.00 and a weld size of .75.

If only one part is round or formed, then the formed part and a straight part form a flare-bevel-groove weld. Figure 16-30 shows a formed part welded to a straight part. The dimensions used to describe the flare-bevel-groove weld have the same meaning as those used for the flare-V-groove weld. The first dimension indicates the distance from the top of the part to the point of tangency and the second dimension indicates the size of the weld. Notice that the first dimension for round parts indicates the radius of the round part. See Figure 16-31.

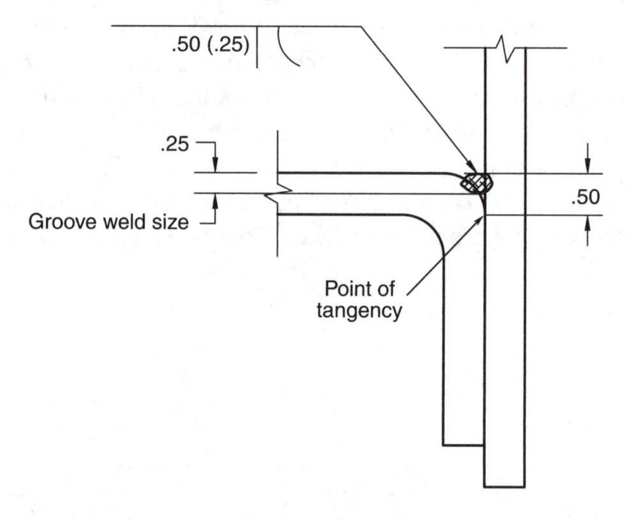

Figure 16-30.
A flare-bevel-groove weld can be formed by one round, or formed, part, and a straight part.

Combination Weld Symbols

Combination weld symbols are used when other types of welds are required in addition to groove welds. Figure 16-32 shows a flare-bevel-groove weld with two fillet welds. The weld symbols specify a fillet weld over the flare-bevel-groove weld on the arrow side, and a fillet weld on the other side. The dimensions for the combination symbol are applied in the same manner as those used for the flare-groove weld symbol and the fillet weld symbol.

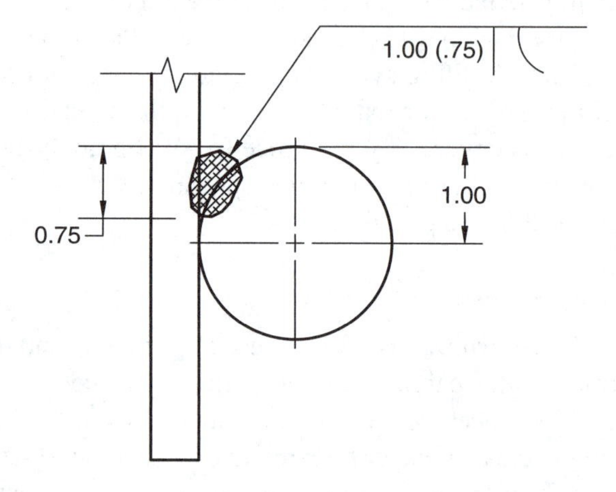

Figure 16-31.
Dimensions for a flare-bevel-groove weld joining a round part to a straight part. The first dimension in the welding symbol indicates the distance from the top of the part to the point of tangency (the radius of the round part) and the second dimension indicates the size of the weld.

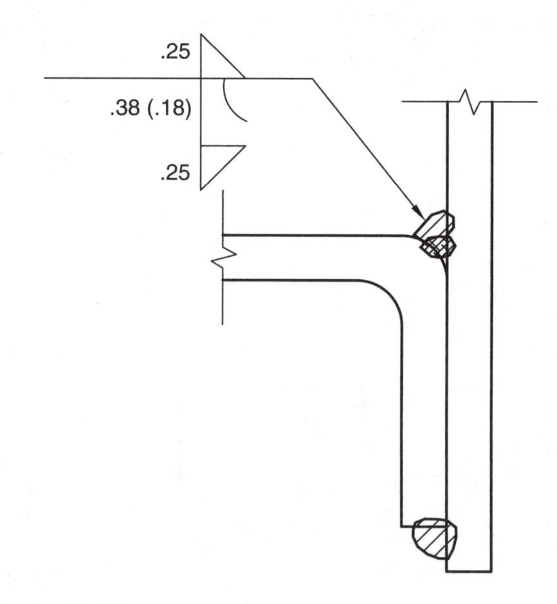

Figure 16-32.
A welding symbol with combination weld symbols for a flare-bevel-groove weld and two fillet welds. The dimensions for combination weld symbols are applied in the same manner as those for each type of weld symbol.

Name ______________________ Date ______________ Class ______________

Print Reading Activities

Part I

Identify the groove weld joints shown below. Write your answers in the spaces provided.

1. ______________________

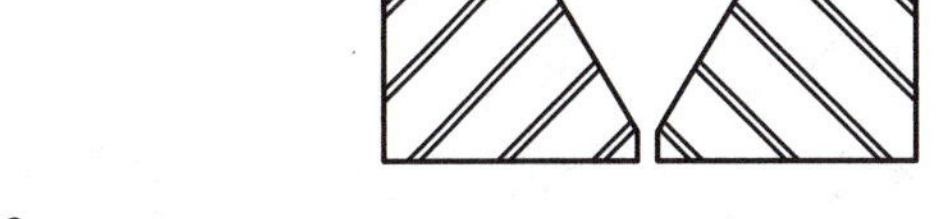

2. ______________________

3. ______________________

4. ______________________

5. ______________________

6. ______________________

7. ______________________

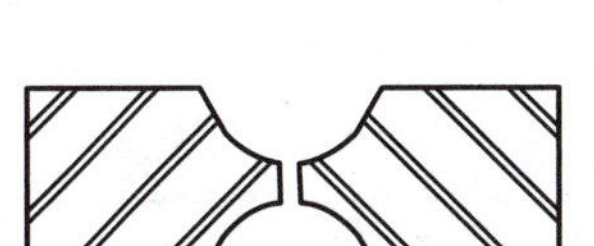

8. ______________________

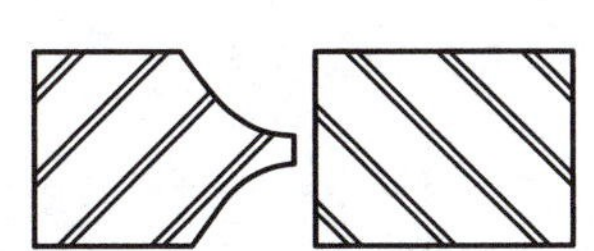

9. ______________________

Part II

Sketch in the correct welding symbol for each groove weld shown in Part I. Sketch the symbol in the correct location on the view.

Part III

Study the drawings shown and sketch in the welding symbol(s) that will describe each joint.

1.

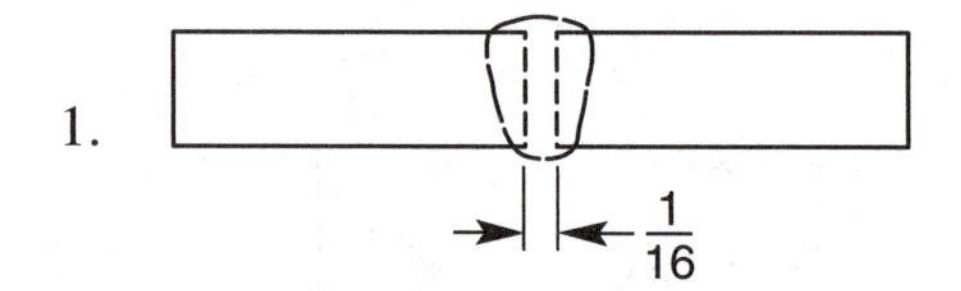

2.

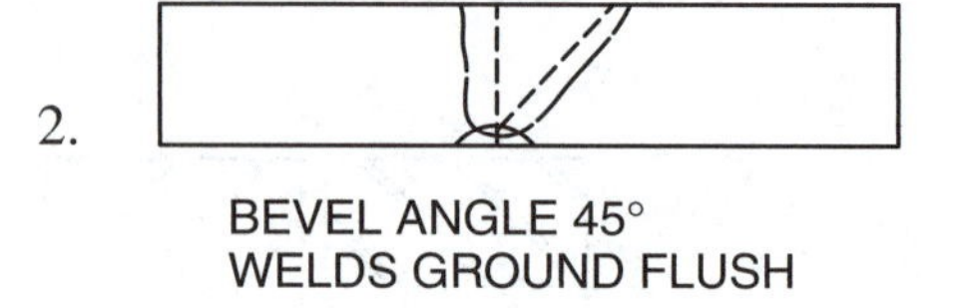

BEVEL ANGLE 45°
WELDS GROUND FLUSH

3.

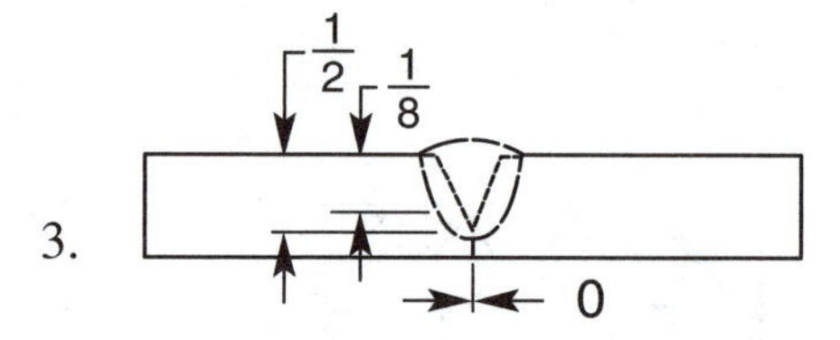

4.

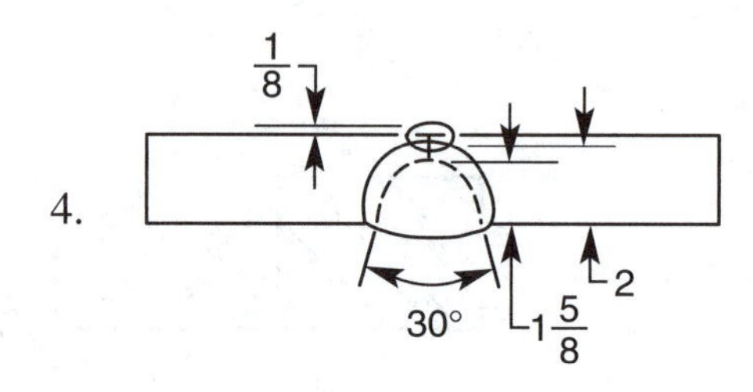

5.

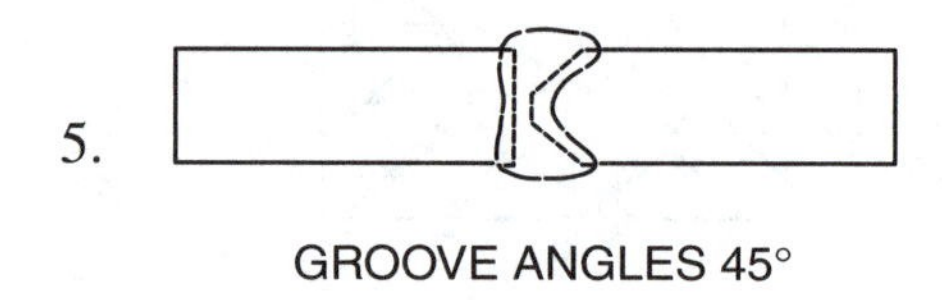

GROOVE ANGLES 45°

6.

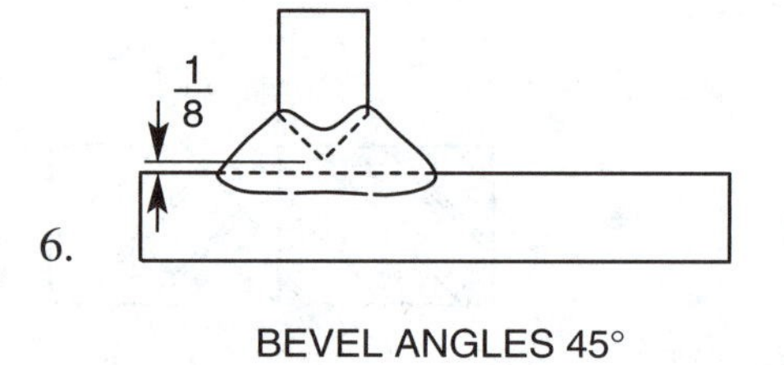

BEVEL ANGLES 45°

7.

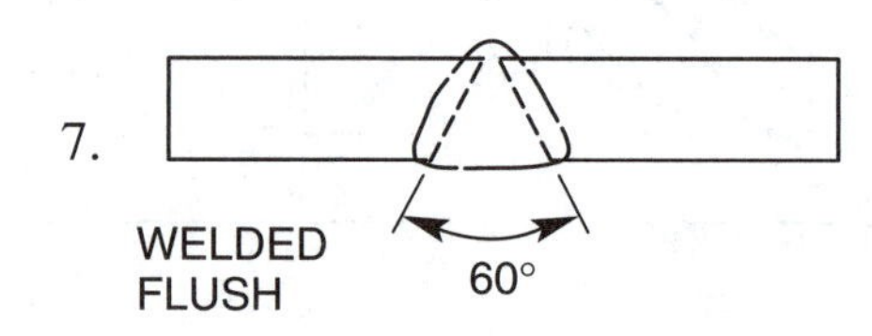

8.

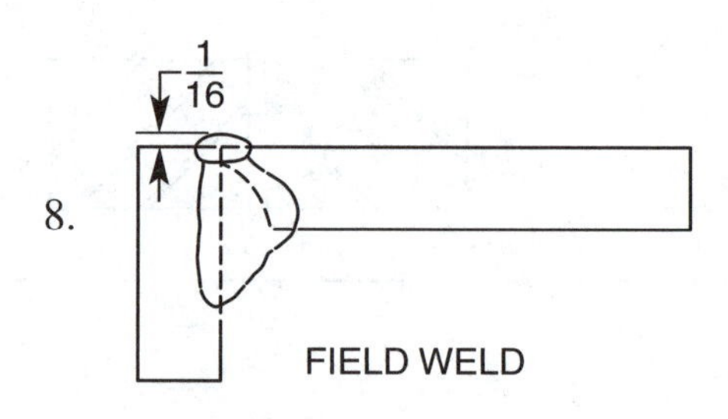

9.

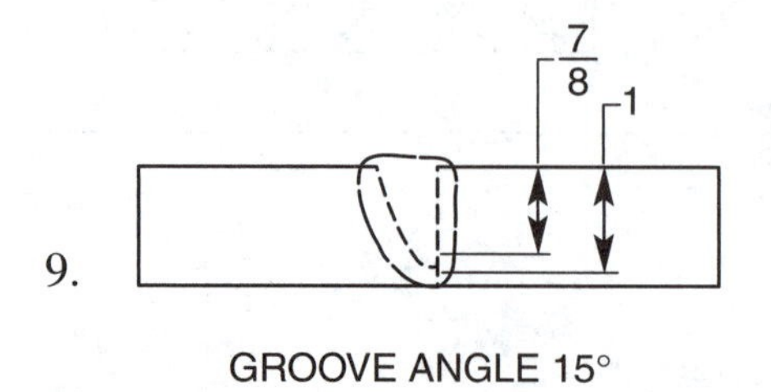

GROOVE ANGLE 15°

Name ________________________________

Part IV

Carefully study the drawing (B577891) below and answer the following questions.

1. List the name and drawing number.

   A. ________________________________

   B. ________________________________

2. What parts are to be joined by welding?

   A. ________________________________

   B. ________________________________

3. Interpret the types of welds required to make the weldment (joint of two sections.)

   A. ________________________________

   ____

   ____

   B. ________________________________

   ____

   ____

4. Joint and weld specifications can be found ______.

   ____

5. What type of welding rod is to be used? ______

   ____

6. What special requirements must be observed after the weldments are made?

   A. ________________________________

   B. ________________________________

7. Have any changes been issued against the drawing? If there have been, list the number made.

   ____

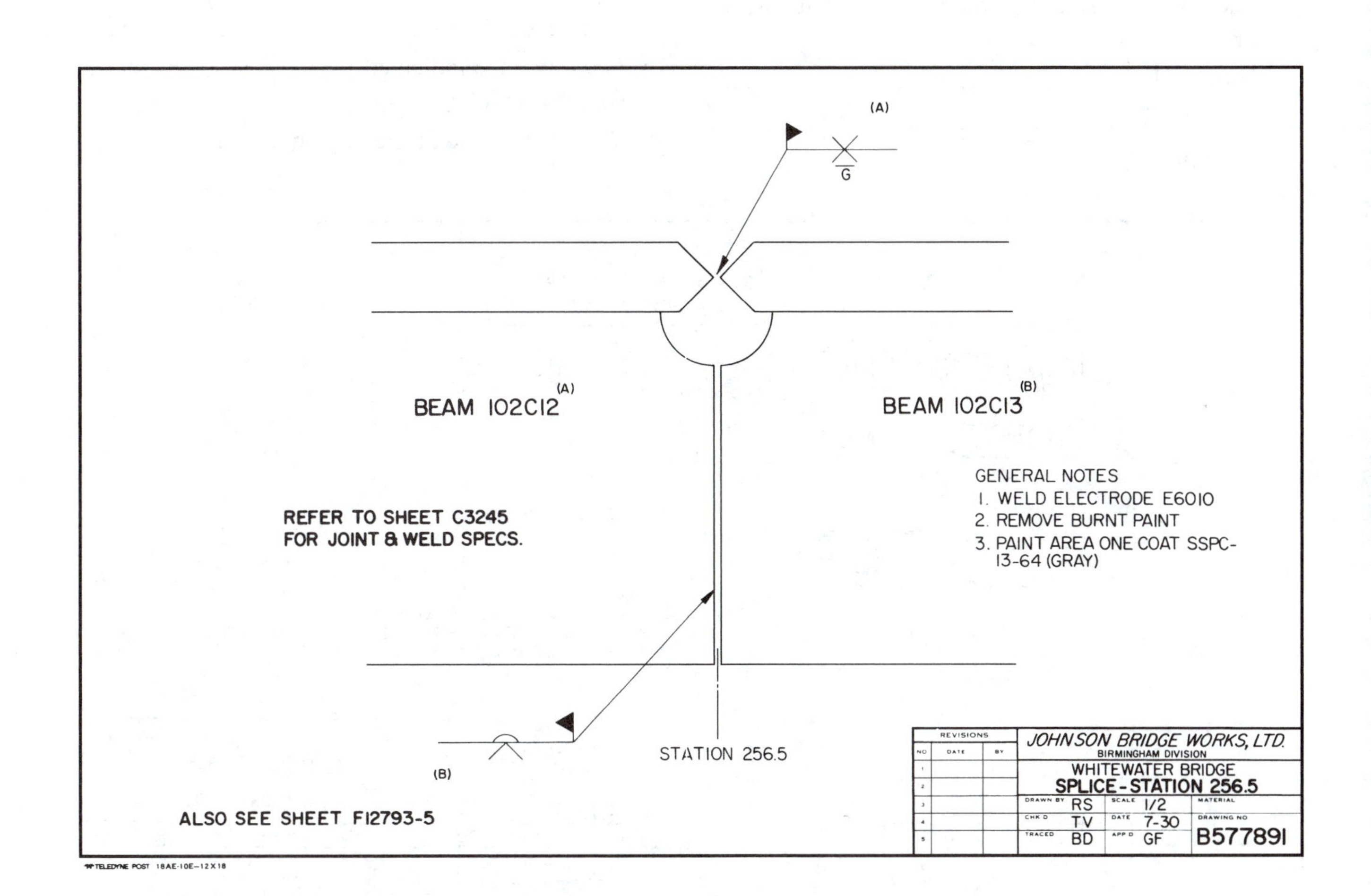

Part V

Carefully study the drawing (L-725) below and answer the following questions.

1. List the name and drawing number of the print.

   A. ______________________________

   B. ______________________________

2. How many parts make up the assembly? ______

   ____

3. What are the names of the parts that make up the assembly? ______________________________

   ____

4. Is more than one size unit indicated on the print?

   ____

5. If more than one size unit is indicated, how many are there and how is each unit identified?

   A. ______________________________

   B. ______________________________

6. List the stock size required to make each part of the assembly.

   Holder (1) ______________________________

   Holder (2) ______________________________

   Base plate ______________________________

7. Interpret the type of weld(s) required to make the weldment(s).

   ____

   ____

8. What heat treatment is required after welding?

   ____

9. How is each weld to be inspected? ______________

   ____

10. How many holes are drilled in the base? _______

    ____

11. The diameter of these holes is _____. ________

    ____

12. How many threaded holes are indicated in the holder?

    ____

13. The thread size is _____ and is tapped _____″ deep.

14. Describe how the large hole in the holder is to be made.

    ____

    ____

15. Is a tolerance indicated for the final diameter? If so, what is it? ______________________________

16. What is the size of the key on the base? _______

    ____

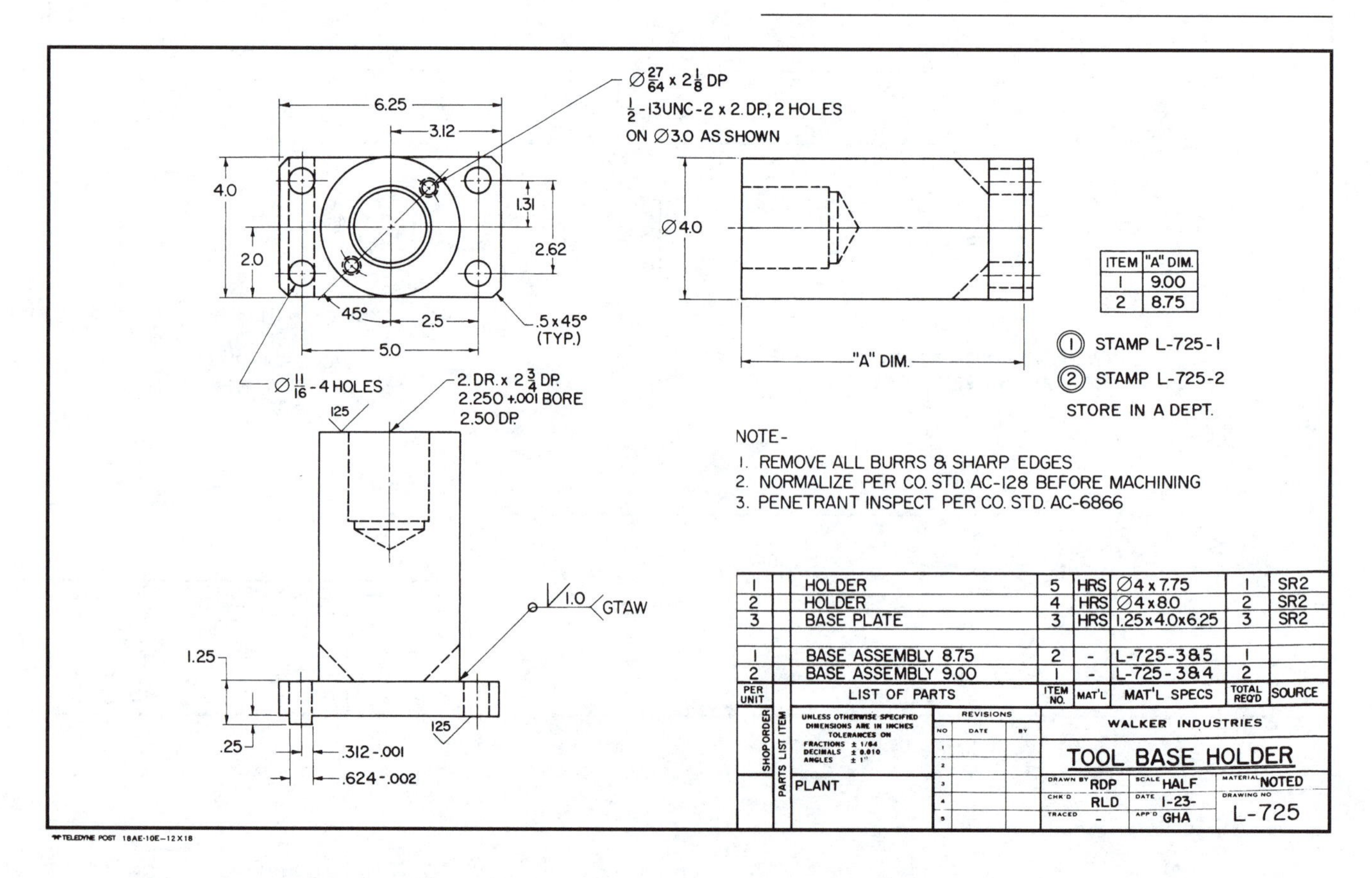

Name ______________________________

Part VI

Explain each of the following welding symbols.

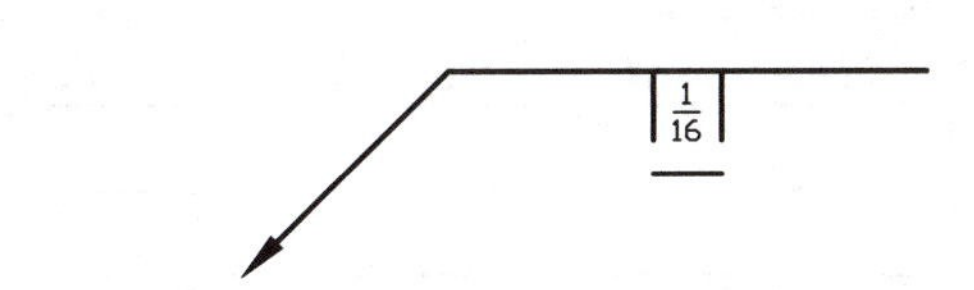

1. ______________________________

60°

2. ______________________________

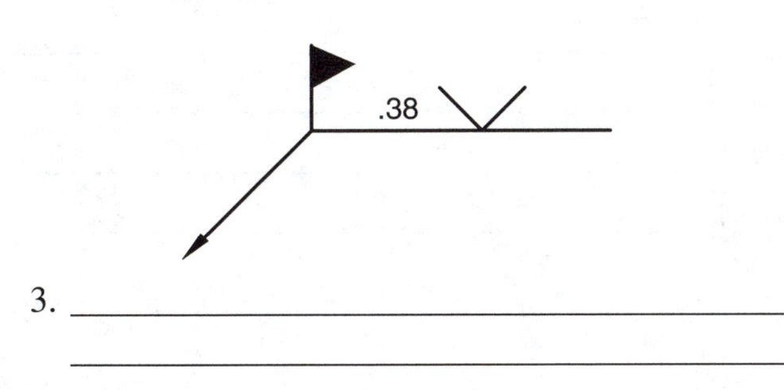

3. ______________________________

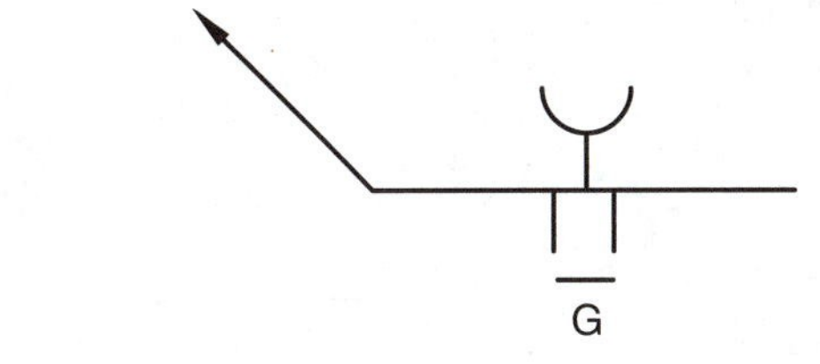

4. ______________________________

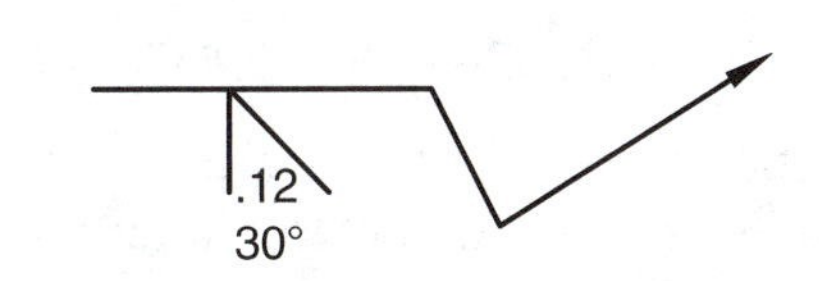

5. ______________________________

Part VII

Draw the correct weld(s) as indicated by the welding symbol.

1.

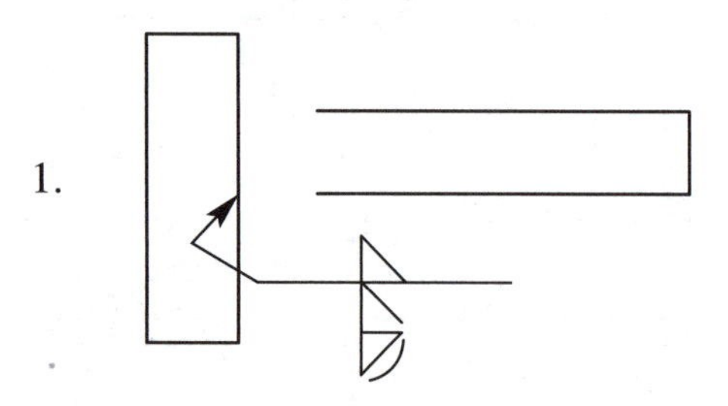

Part VIII

Use print 1104-wrp to answer the following questions.

1. List the overall length of the support bracket. ____
2. Determine the total number of holes required for the part. ____
3. Determine the overall maximum height of the part. ____
4. List the center-to-center distance of the .50 diameter holes. ____
5. List the typical plate thickness for the support bracket. ____
6. Determine the maximum angle for the 27° angle of the gusset. ____
7. What surface requires finishing? ____
8. Explain the welding symbol at K. ____
9. Explain the welding symbol at L. ____
10. Explain the welding symbol at M. ____

Part IX

Use print 1104-wrp to determine the following dimensions indicated on the print.

1. A ____
2. B ____
3. C ____
4. D ____
5. E ____
6. F ____
7. G ____
8. H ____
9. I ____
10. J ____

Name ____

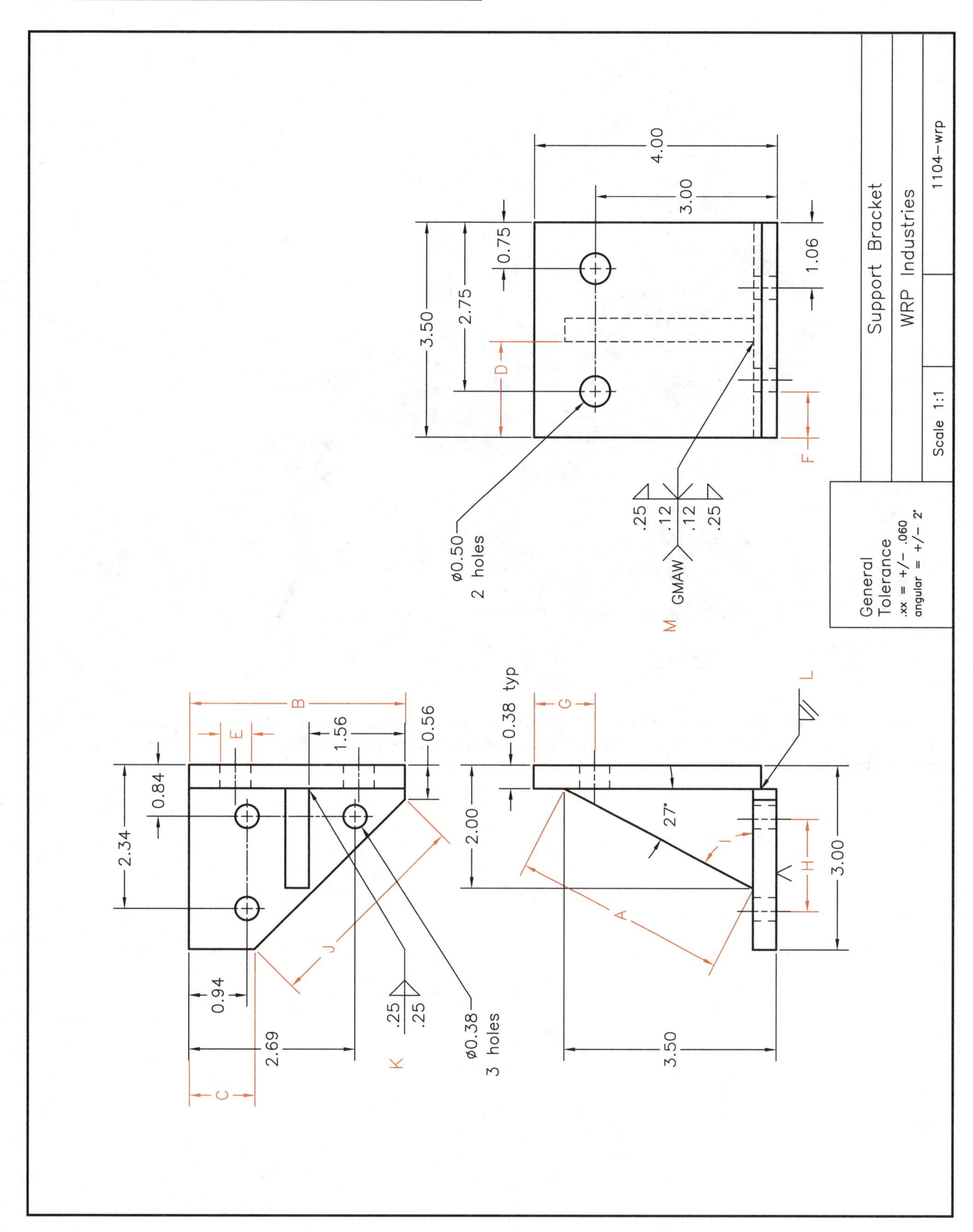
Support Bracket
WRP Industries
1104–wrp
Scale 1:1
General
Tolerance
.xx = +/– .060
angular = +/– 2°
ø0.50
2 holes
ø0.38
3 holes
GMAW
0.38 typ
27°
4.00
3.00
3.50
2.75
0.75
1.06
2.34
0.84
1.56
0.56
0.94
2.69
2.00
A
B
C
D
E
F
G
H
I
J
K
L
M

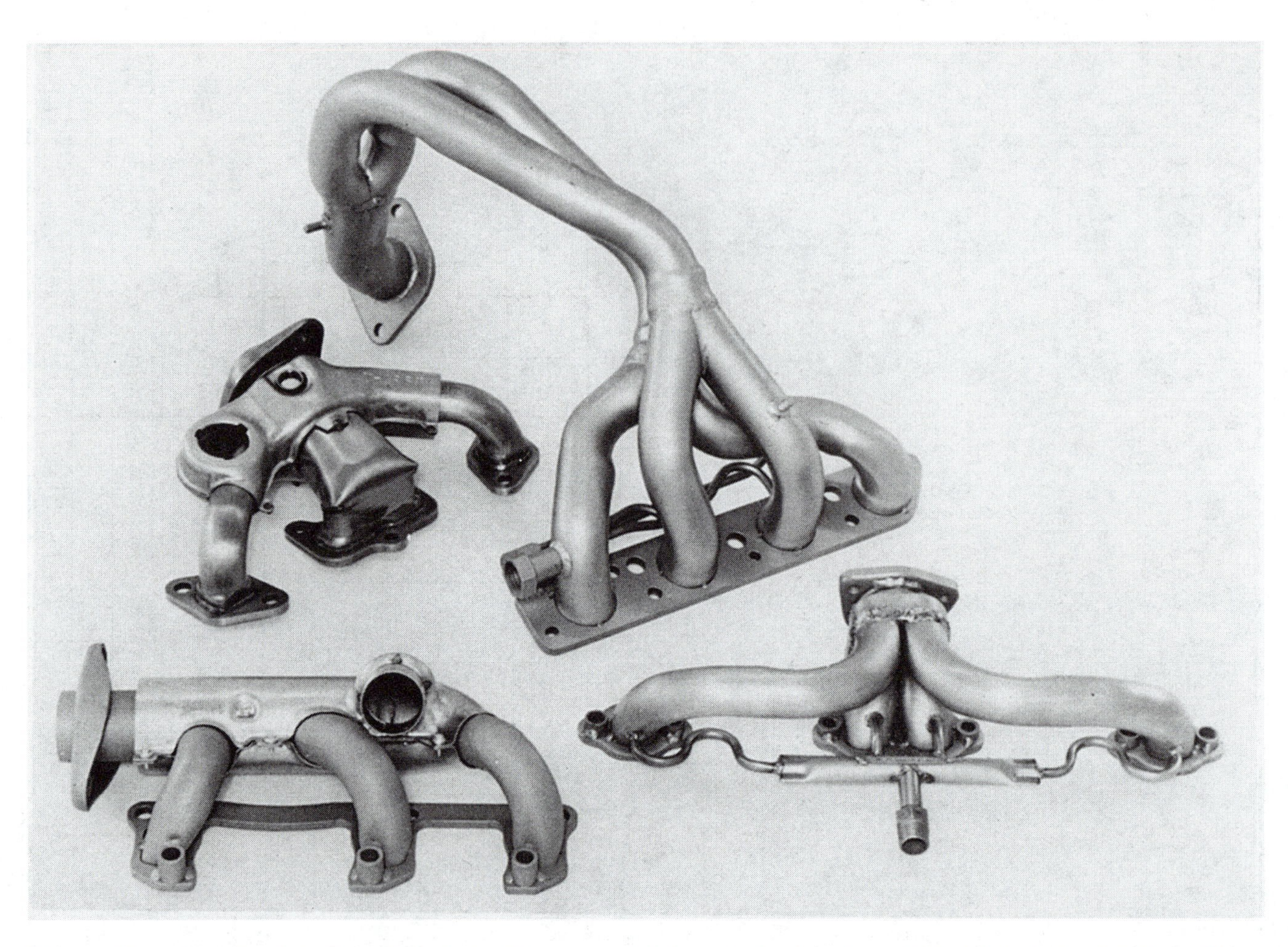

Pipe welding also extends to the auto industry. Shown are stainless steel exhaust manifolds fabricated mainly by welding. Each unit is carefully inspected because of danger of exhaust gas leakage to the vehicle's driver and passengers. (American Iron and Steel Institute)

Unit 17

Plug and Slot Welds

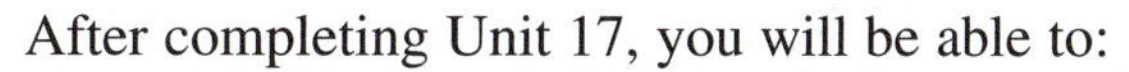

After completing Unit 17, you will be able to:

- Describe the use of plug and slot welds.
- Prepare the hole to the correct size and shape for a plug or slot weld.
- Determine the correct fill and contour requirements for plug and slot welds.
- Lay out the correct spacing for multiple plug or slot welds.

Key Words

chamfering
countersink angle
countersinking
plug weld
slot weld

A ***plug weld*** is a circular weld made through a hole in one piece of metal, joining it to another piece, Figure 17-1. A ***slot weld*** is made in an elongated cut through one member of a joint to join that part of the metal to another joint member. Figure 17-2 shows a typical example.

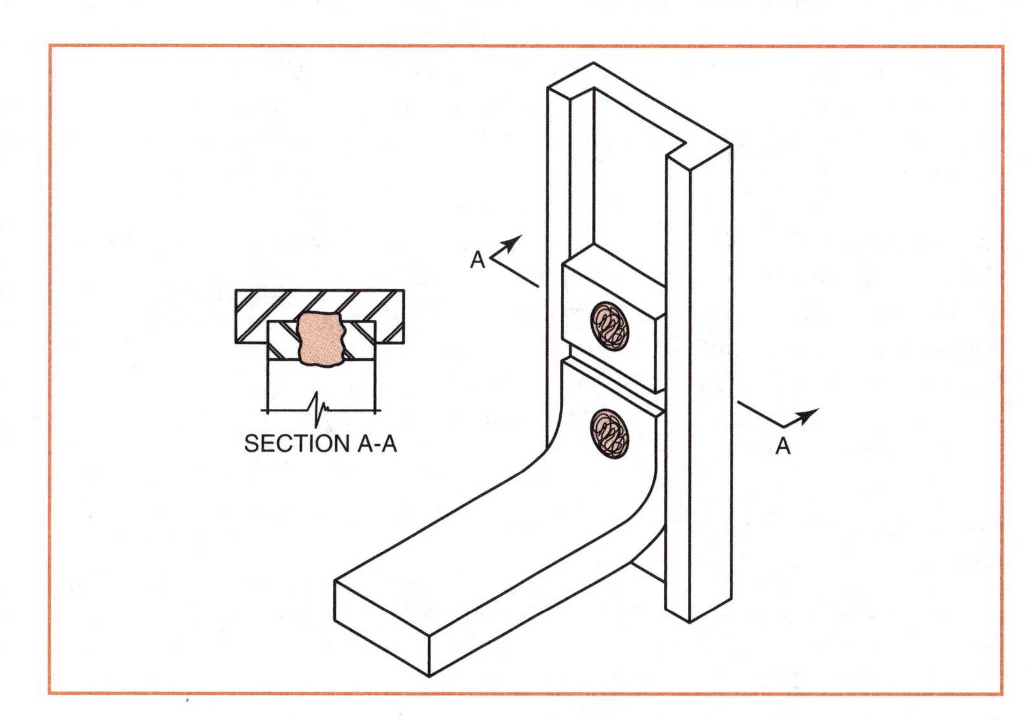

Figure 17-1.
Study this typical plug weld.

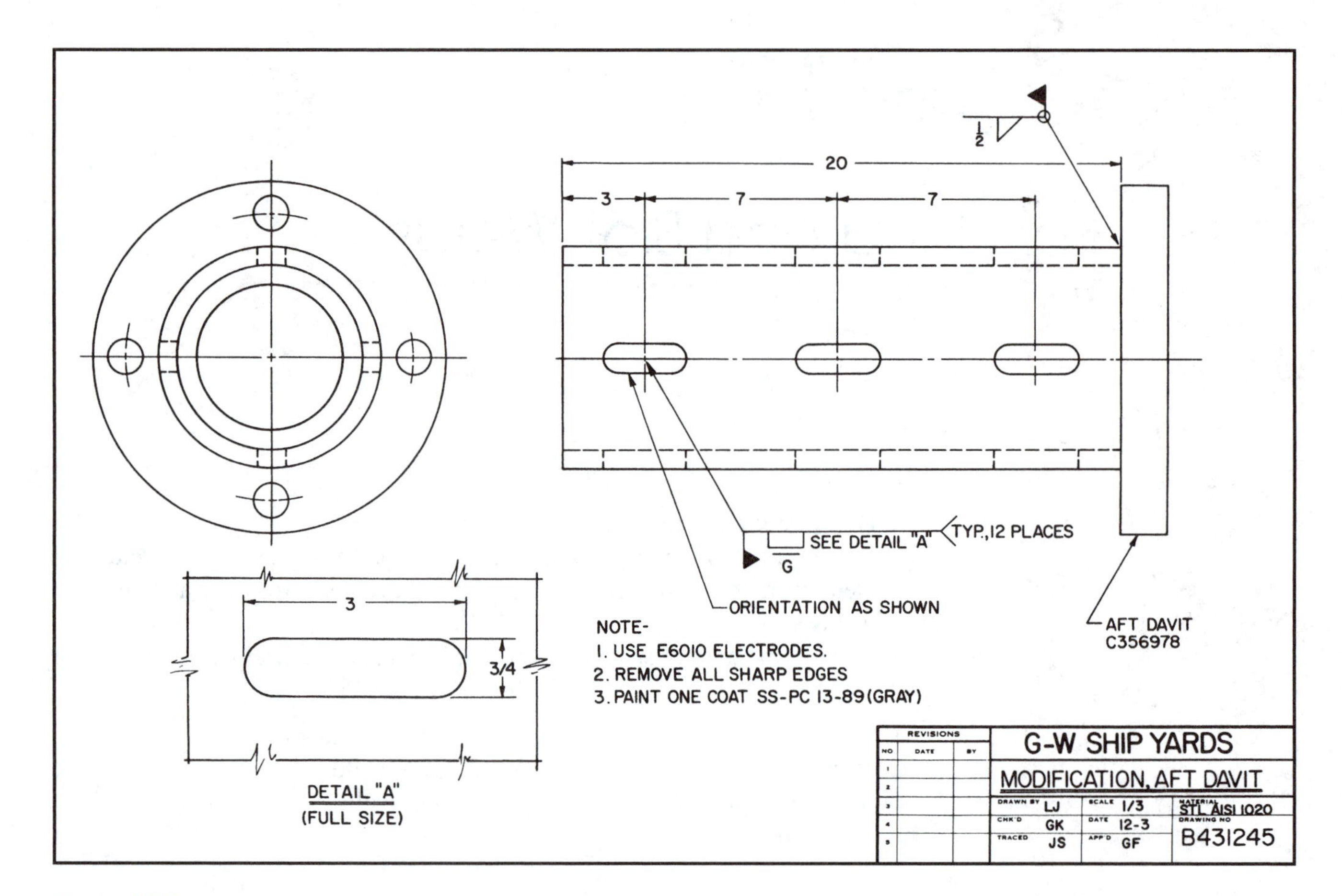

Figure 17-2.
This is a component modification drawing that requires slot welds. Where are welds to be made?

The plug and slot welds are generally used when the weld must not be seen from one side of the joint. They may also be used if joining of the components does not allow for another type of weld. The walls of the hole or slot may be parallel or angular, and the hole may be completely or only partially filled with weld metal.

NOTE! A plug weld is *not* to be confused with an all-around fillet weld made in a hole. A plug or slot weld should not be confused with a fillet weld whether the hole or slot is filled or partially filled with weld. Remember the fillet weld has a cross section that is a triangle. The plug or slot weld takes on the cross sectional shape of the hole or slot through which the weld is made.

General Use of the Plug and Slot Weld Symbol

A plug weld or slot weld in the arrow side member of a joint is specified when the weld symbol is placed on the side of the reference line *toward* the reader. This is shown in Figure 17-3.

A plug weld or slot weld in the other side member of a joint is indicated when the weld symbol is placed on the side of the reference line *away* from the reader, Figure 17-4.

Plug weld or slot weld dimensions are shown on the same side of the reference line as the weld symbol, Figure 17-5.

Size of Plug Welds

Plug weld size refers to the weld diameter at the base of the weld. It is given to the left of the weld symbol, as in Figure 17-6.

Countersink Angle of Plug Weld

Countersinking or ***chamfering*** is a process that enlarges the end of a hole conically so it is able to fit the head of a flat head screw. When the side of a plug weld hole is tapered, the groove angle of the taper is specified and shown on the welding symbol, Figure 17-7. The ***countersink angle*** is given below the plug weld symbol.

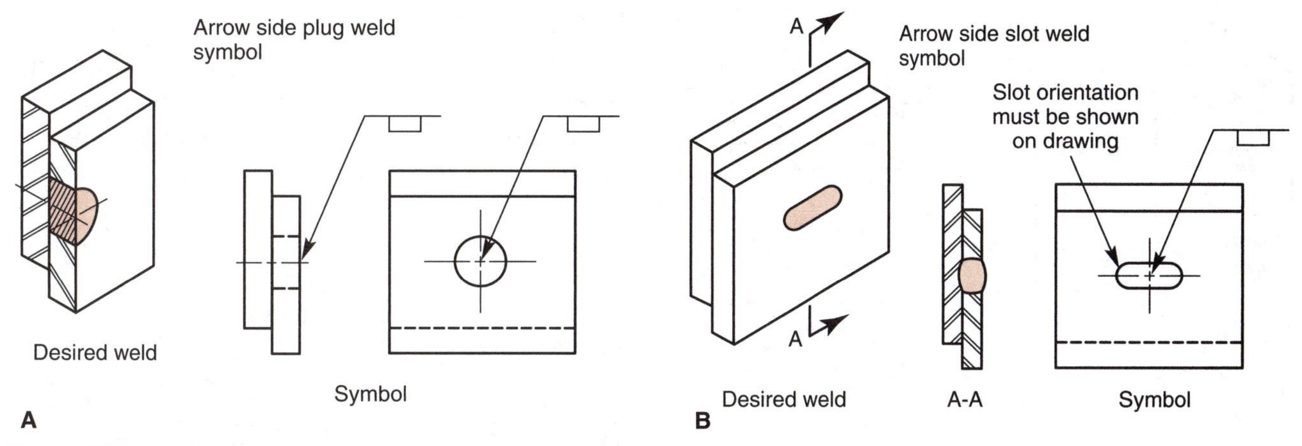

Figure 17-3.
Arrow side is specified when the weld symbol is placed on the side of the reference line toward the reader. A—Note the significance of the arrow side plug weld symbol. B—Note the significance of the arrow side slot weld symbol.

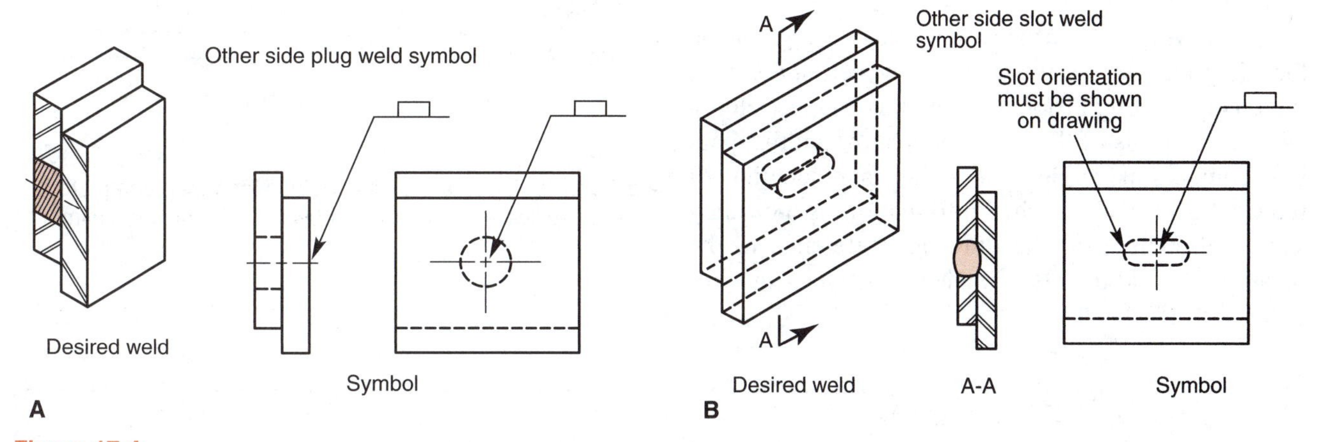

Figure 17-4.
Other side is specified when the weld symbol is placed on the side of the reference line away from the reader. A—Note the other side plug weld symbol. B—Note the other side slot weld symbol.

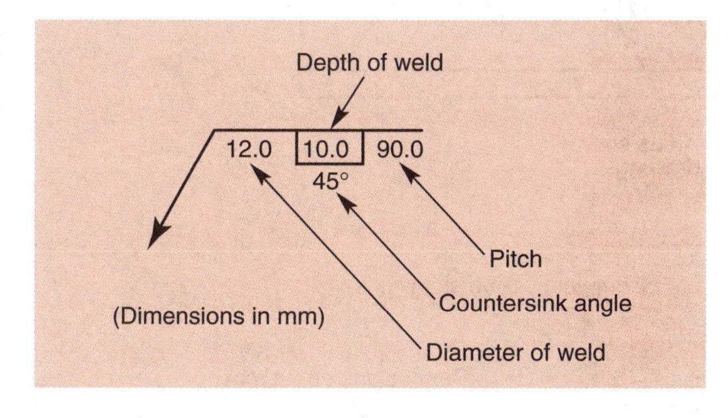

Figure 17-5.
Plug weld dimensions are on the same side of the reference line as the weld symbol. Dimensions are located as shown. Study their positions around the symbol.

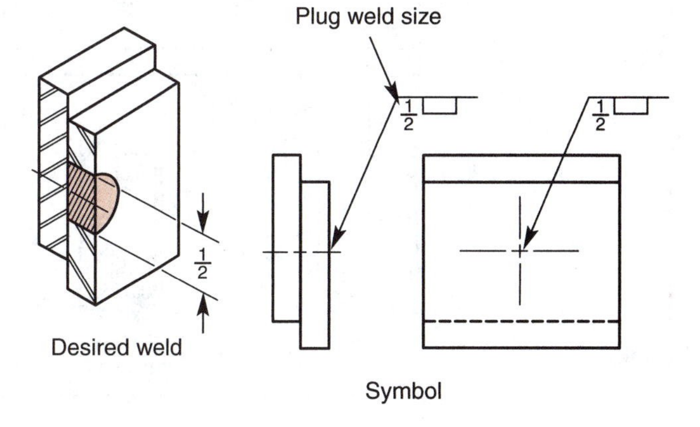

Figure 17-6.
Notice how the size of a plug weld is indicated.

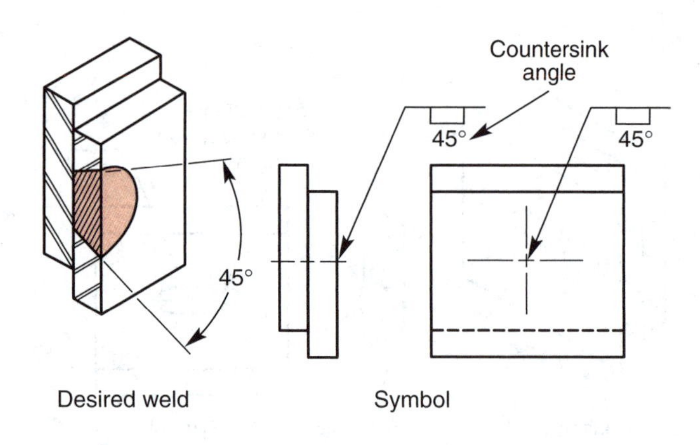

Figure 17-7.
When a plug weld is to be made in a countersunk hole, the included angle of the countersink is specified in relation to the weld symbol.

Slot Weld Details

Slot weld size information is usually given on the view that shows the weld. However, sometimes there is not enough room on the view or the welding symbol to furnish full details on length, width, spacing, countersink angle, orientation, and location of required slot welds. This information is normally shown elsewhere on the drawing. Reference to the location of weld details is made on the welding symbol, Figure 17-8.

Filling Depth of Plug or Slot Welds

Plug weld or slot weld holes are not always completely filled with weld material. When the area is *not* to be filled completely, filling depth is indicated within the weld symbol, Figure 17-9.

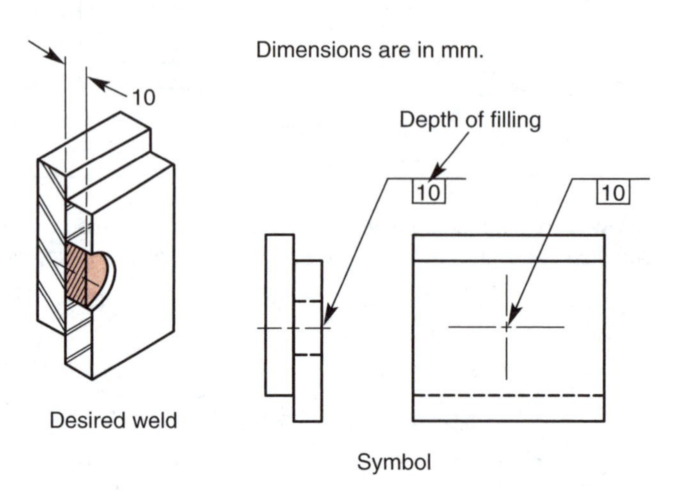

Figure 17-9.
When the depth of the plug weld is less than complete, the specified weld depth is shown inside the weld symbol.

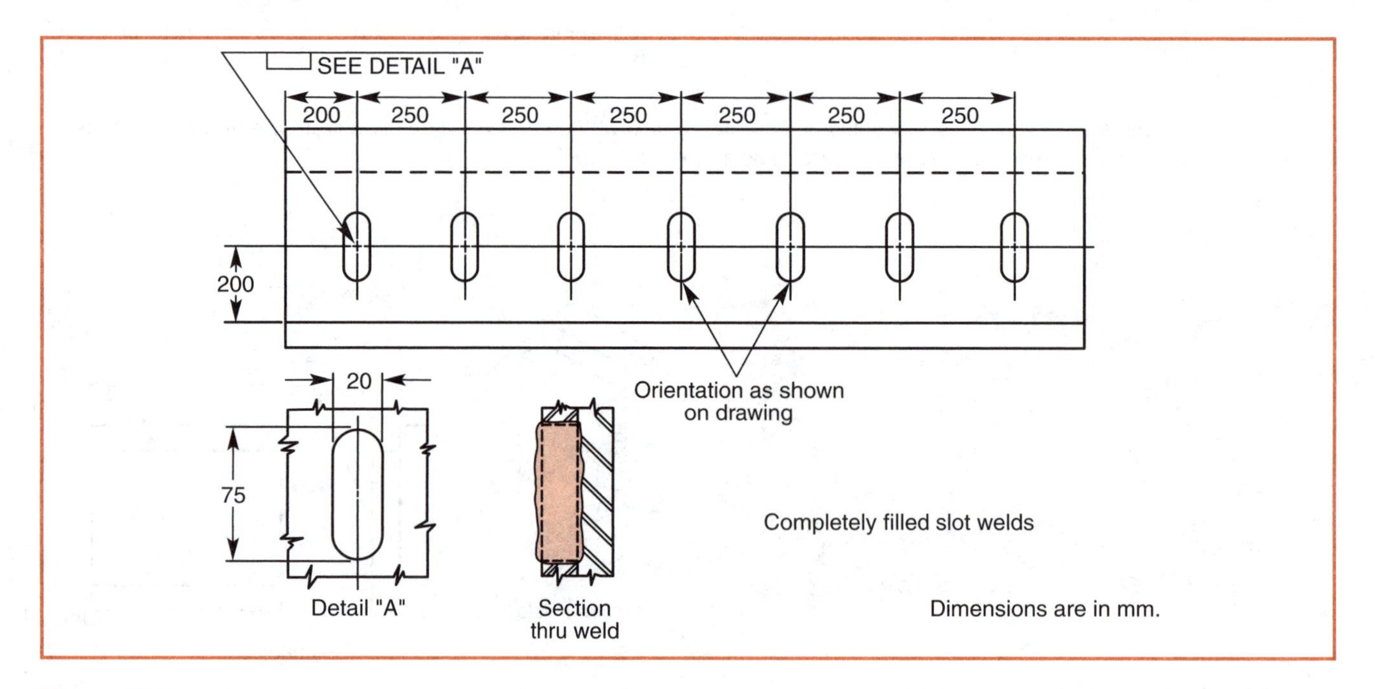

Figure 17-8.
Slot weld information is commonly shown elsewhere on the drawing. The location of information is noted on the welding symbol.

Spacing Plug or Slot Welds

The pitch (center-to-center spacing) of plug welds or slot welds is indicated by the dimension placed on the same side as, and to the right of, the weld symbol, Figure 17-10.

Figure 17-11 shows the use of multiple dimensions for a plug welding symbol. Study this illustration carefully.

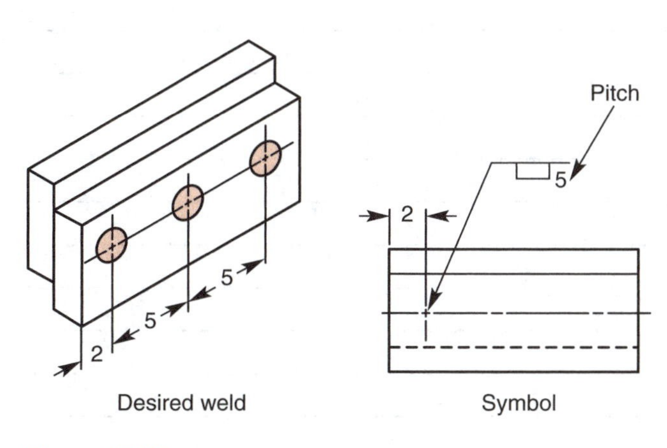

Figure 17-10.
The pitch of multiple plug welds is shown to the right of the weld symbol.

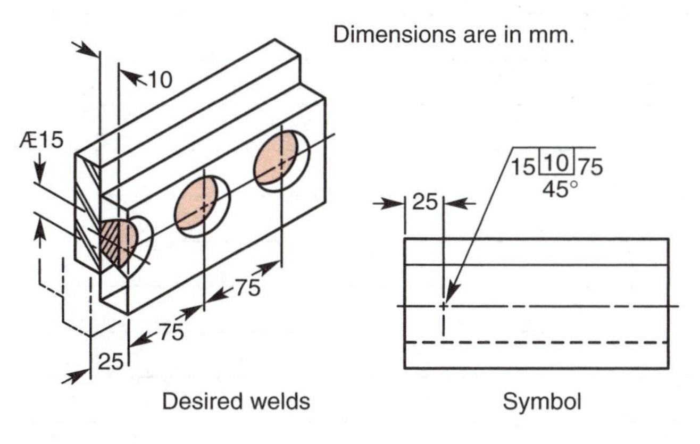

Figure 17-11.
A plug welding symbol with multiple dimensions specified. Note the position of dimensions around the weld symbol.

Surface Contour of Plug or Slot Welds

When plug or slot welds are to be welded approximately flush (with no mechanical finishing required), only the flush contour symbol is added to the weld symbol. However, if the welded surface is to be finished flush by mechanical means, the finishing technique must also be added to the weld symbol, Figure 17-12 and Figure 17-13.

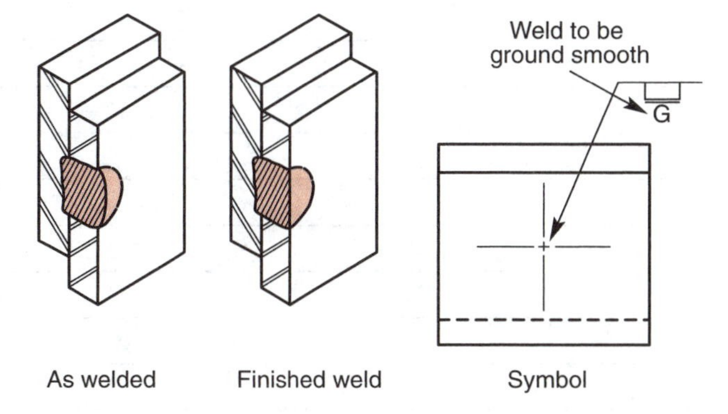

Figure 17-12.
This symbol shows how a flush surface contour plug weld is specified. When the weld surface is to be made flush with surrounding surfaces mechanically, a letter indicates the finishing technique.

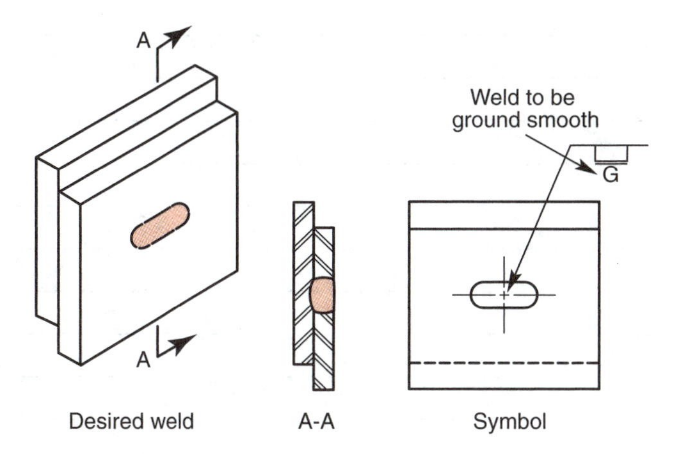

Figure 17-13.
The flush contour symbol and the method of making the weld flush are added to the slot weld symbol when made flush by mechanical means.

Notes

Name ______________________ Date ______________ Class ______________

Print Reading Activities

Part I

Refer to the drawing (167E102) on the next page and answer the following questions.

1. What is the name of the part?

   ____

2. Where is the drawing number located?

   ____

3. What is the scale of the drawing?

   ____

4. How many individual parts make up the unit? _____

5. Except as noted, what general tolerances are allowed?

   A. Decimal ______________________

   B. Fractional ______________________

   C. Angular ______________________

6. How many plug welds are required? _____

7. What specifications are given for the plug welds?

   ____

8. What other types of welds are indicated?

   ____

9. What type of welding rod is specified?

   ____

   ____

10. List the subassemblies that make up the completed unit.

	Subassembly	Part No.
A.	______________	________
	______________	________
	______________	________
B.	______________	________
	______________	________
	______________	________
C.	______________	________
	______________	________
	______________	________
D.	______________	________
	______________	________
	______________	________

11. The overall length of the unit is _____.

12. The total width of the unit is _____.

13. The length of the tongue is _____.

14. The distance from the base of the end plate to the center line of the bore is _____.

15. The bore diameter is _____.

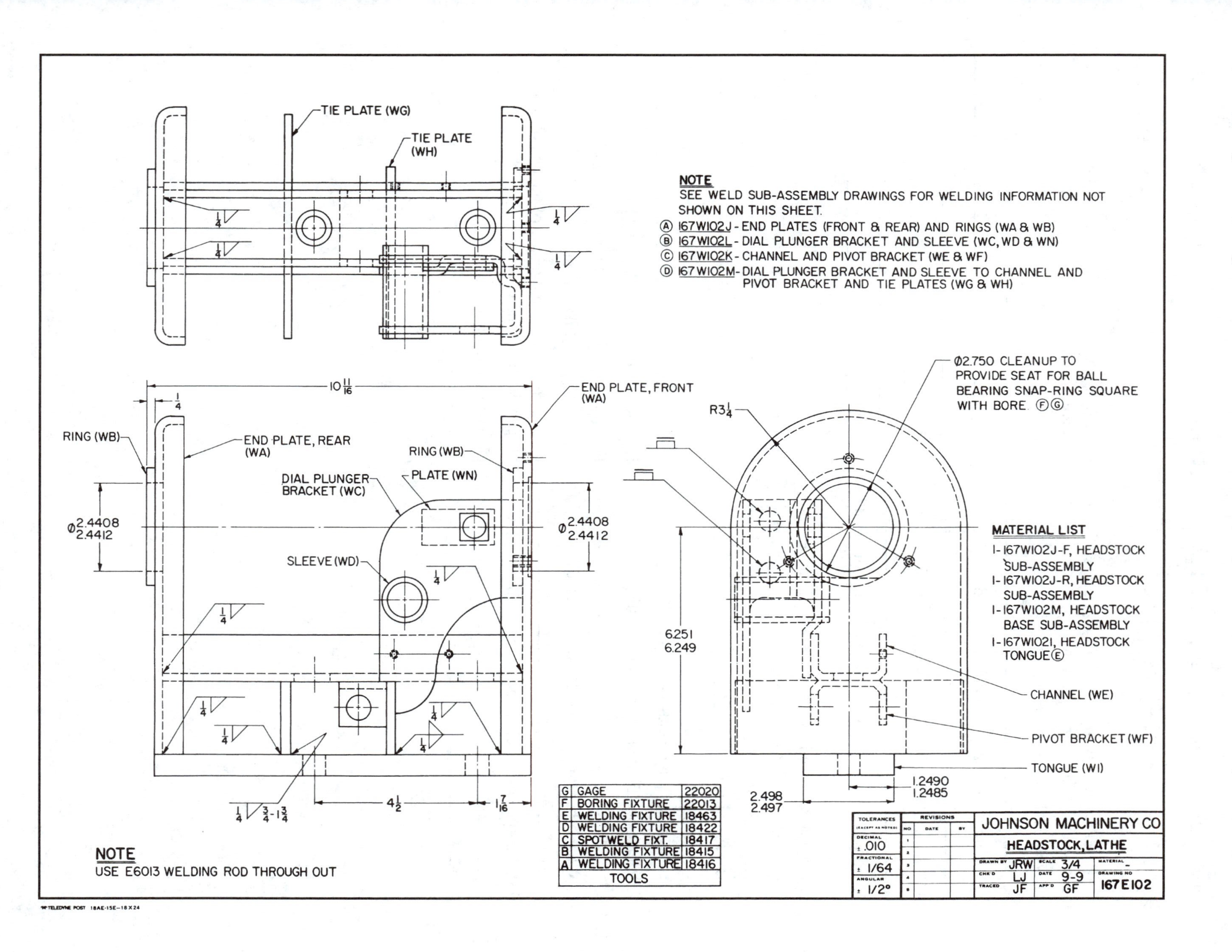
TIE PLATE (WG)
TIE PLATE (WH)
NOTE
SEE WELD SUB-ASSEMBLY DRAWINGS FOR WELDING INFORMATION NOT SHOWN ON THIS SHEET.
Ⓐ 167W102J - END PLATES (FRONT & REAR) AND RINGS (WA & WB)
Ⓑ 167W102L - DIAL PLUNGER BRACKET AND SLEEVE (WC, WD & WN)
Ⓒ 167W102K - CHANNEL AND PIVOT BRACKET (WE & WF)
Ⓓ 167W102M - DIAL PLUNGER BRACKET AND SLEEVE TO CHANNEL AND PIVOT BRACKET AND TIE PLATES (WG & WH)
10 11/16
1/4
END PLATE, FRONT (WA)
RING (WB)
END PLATE, REAR (WA)
RING (WB)
DIAL PLUNGER BRACKET (WC)
PLATE (WN)
Ø2.4408
2.4412
SLEEVE (WD)
Ø2.4408
2.4412
R3 1/4
Ø2.750 CLEANUP TO PROVIDE SEAT FOR BALL BEARING SNAP-RING SQUARE WITH BORE Ⓕ Ⓖ
MATERIAL LIST
1-167W102J-F, HEADSTOCK SUB-ASSEMBLY
1-167W102J-R, HEADSTOCK SUB-ASSEMBLY
1-167W102M, HEADSTOCK BASE SUB-ASSEMBLY
1-167W102I, HEADSTOCK TONGUE Ⓔ
6.251
6.249
CHANNEL (WE)
PIVOT BRACKET (WF)
TONGUE (WI)
1.2490
1.2485
2.498
2.497
1/4 3/4-1 3/4
4 1/2
1 7/16
G GAGE 22020
F BORING FIXTURE 22013
E WELDING FIXTURE 18463
D WELDING FIXTURE 18422
C SPOTWELD FIXT. 18417
B WELDING FIXTURE 18415
A WELDING FIXTURE 18416
TOOLS
NOTE
USE E6013 WELDING ROD THROUGH OUT
TOLERANCES (EXCEPT AS NOTED)
DECIMAL ±.010
FRACTIONAL ± 1/64
ANGULAR ± 1/2°
REVISIONS
NO DATE BY
JOHNSON MACHINERY CO
HEADSTOCK, LATHE
DRAWN BY JRW
SCALE 3/4
MATERIAL –
CHK'D LJ
DATE 9-9
DRAWING NO 167E102
TRACED JF
APP'D GF

Name ________________________________

Part II

Draw in the correct welding symbol for each plug weld shown.

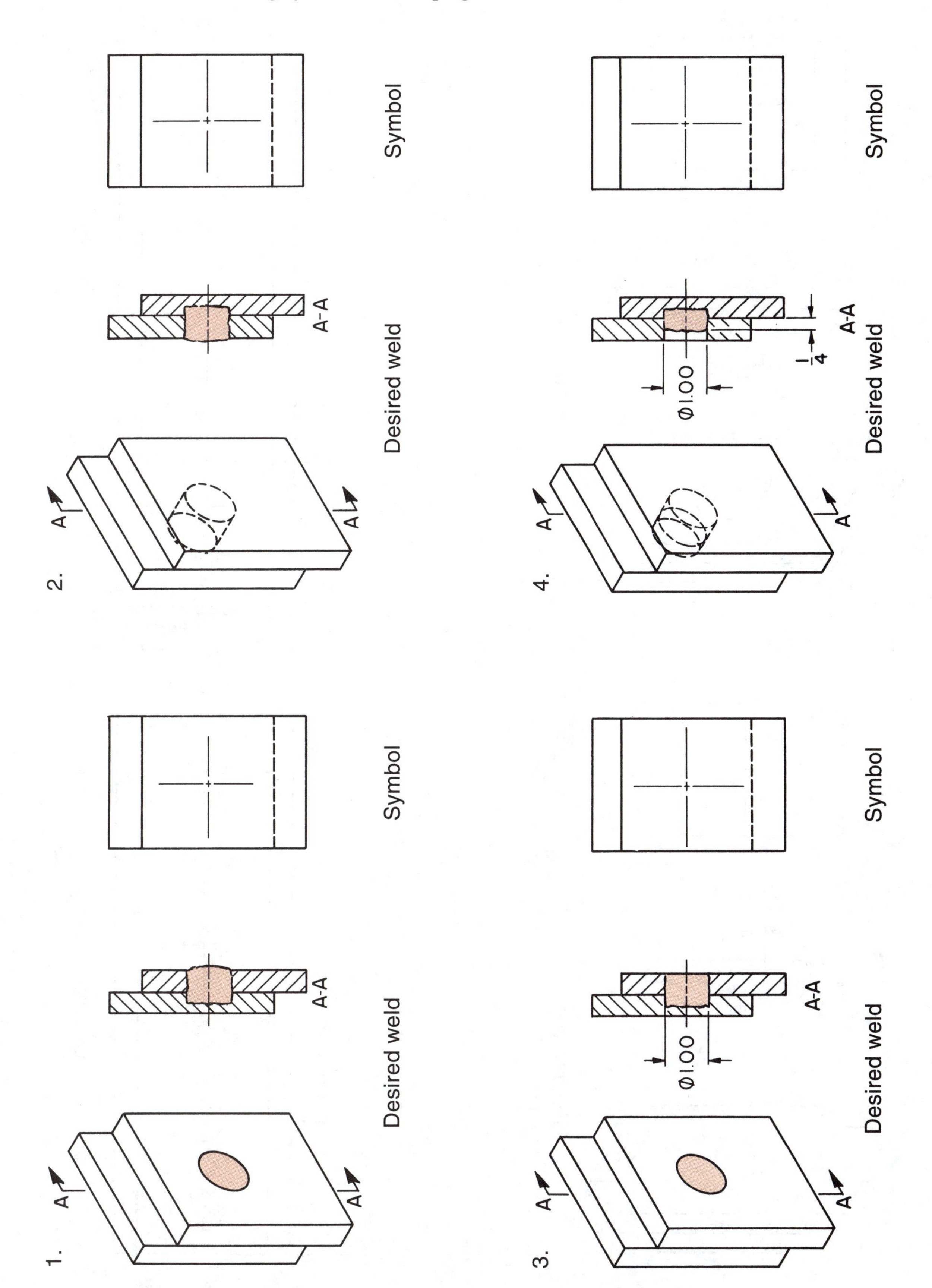

Part II (continued)

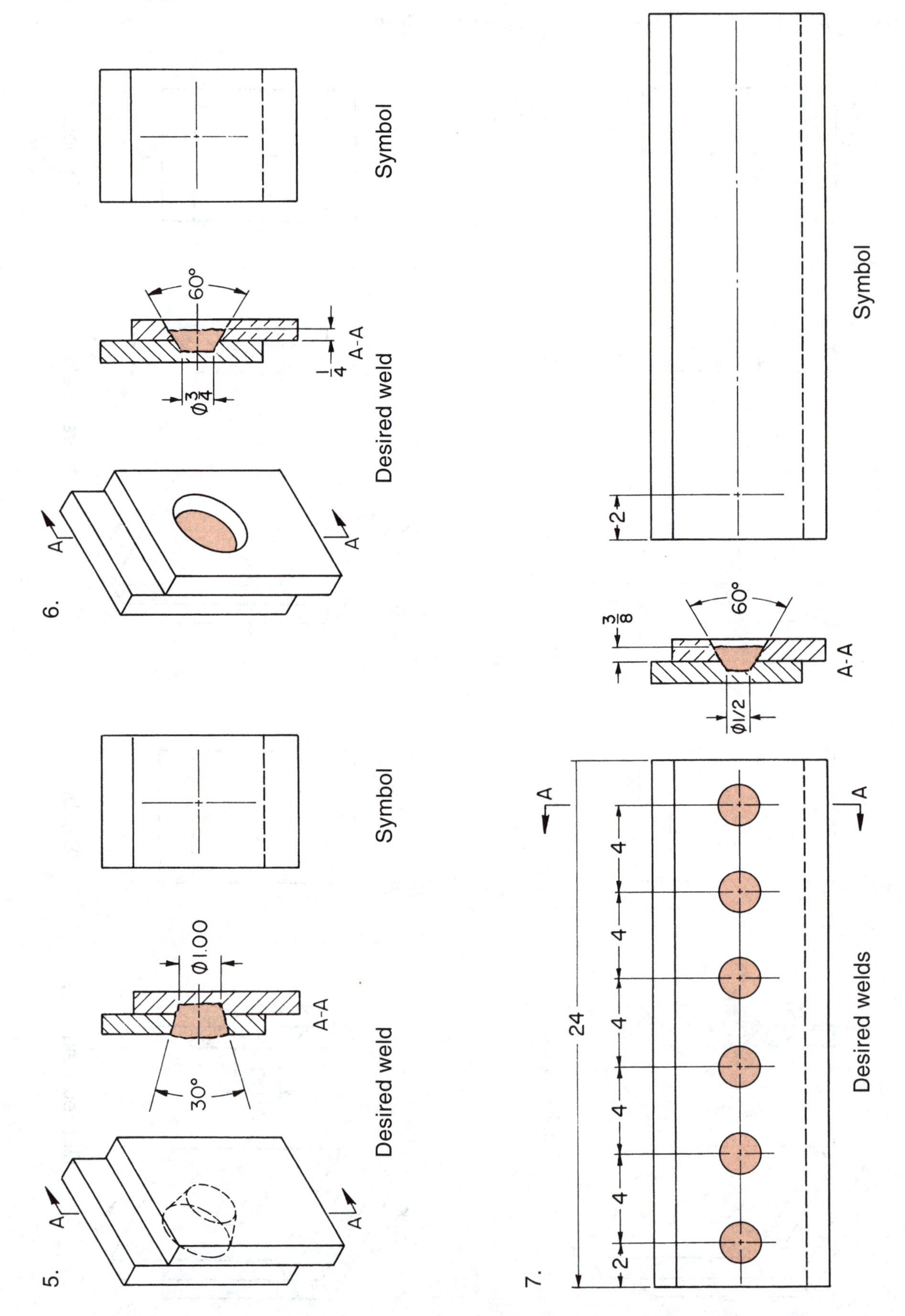

Name ____________________________________

Part III

Refer to the drawing (167E100) on the next page and answer the following questions.

1. Name the unit shown on the drawing.
____
____
2. Where is the drawing number located?
____
____
3. The scale of the drawing is _____.
4. What is the next assembly?________________
____
____
5. How many individual parts make up the completed unit? _____
6. Name the bed length of the unit shown. _______
7. What information is given on the slot welds?
____
____
8. What are the special instructions indicated in Note "A"? ____________________________
____
____
9. The distance from the base of the unit to the top of the rails is _____.
10. What special instructions are to be followed when welding the rails into place?
A. ____________________________________
____
____
B. ____________________________________
____
____
11. How many different types of welds are indicated?
____
____
____
____
12. The total width of the unit is _____.
13. The width of the mounting plates is _____.
14. What instructions are given concerning the hole positions on the mounting plate?
____
____
____
15. Welding Note "B" is concerned with _____.
____
____
____
____
____

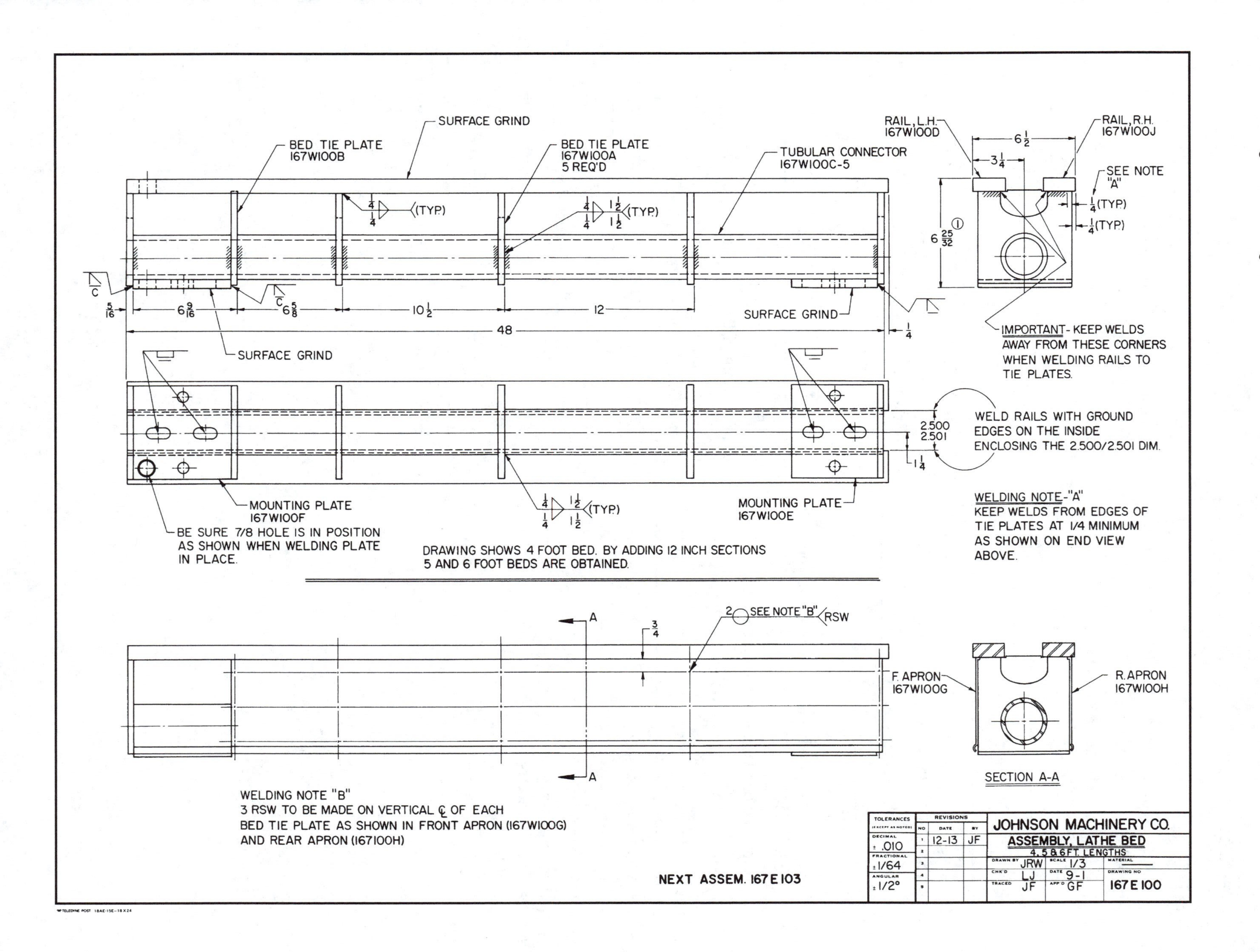
SURFACE GRIND
BED TIE PLATE 167W100B
BED TIE PLATE 167W100A 5 REQ'D
TUBULAR CONNECTOR 167W100C-5
RAIL, L.H. 167W100D
RAIL, R.H. 167W100J
6 1/2
3 1/4
SEE NOTE "A"
1/4 (TYP.)
1/4 (TYP.)
6 25/32
(TYP.)
C
5/16
6 9/16
6 5/8
10 1/2
12
48
1/4
SURFACE GRIND
IMPORTANT - KEEP WELDS AWAY FROM THESE CORNERS WHEN WELDING RAILS TO TIE PLATES.
2.500
2.501
1 1/4
WELD RAILS WITH GROUND EDGES ON THE INSIDE ENCLOSING THE 2.500/2.501 DIM.
MOUNTING PLATE 167W100F
BE SURE 7/8 HOLE IS IN POSITION AS SHOWN WHEN WELDING PLATE IN PLACE.
MOUNTING PLATE 167W100E
WELDING NOTE - "A"
KEEP WELDS FROM EDGES OF TIE PLATES AT 1/4 MINIMUM AS SHOWN ON END VIEW ABOVE.
DRAWING SHOWS 4 FOOT BED. BY ADDING 12 INCH SECTIONS 5 AND 6 FOOT BEDS ARE OBTAINED.
A
3/4
SEE NOTE "B"
RSW
F. APRON 167W100G
R. APRON 167W100H
SECTION A-A
WELDING NOTE "B"
3 RSW TO BE MADE ON VERTICAL ℄ OF EACH BED TIE PLATE AS SHOWN IN FRONT APRON (167W100G) AND REAR APRON (167100H)
NEXT ASSEM. 167 E 103
TOLERANCES (EXCEPT AS NOTED)
DECIMAL ± .010
FRACTIONAL ± 1/64
ANGULAR ± 1/2°
REVISIONS
NO DATE BY
1 12-13 JF
JOHNSON MACHINERY CO.
ASSEMBLY, LATHE BED
4, 5 & 6 FT. LENGTHS
DRAWN BY JRW
SCALE 1/3
MATERIAL
CHK'D LJ
DATE 9-1
TRACED JF
APP'D GF
DRAWING NO 167 E 100

Name ____

Part IV

Draw the correct welding symbol for the slot welds.

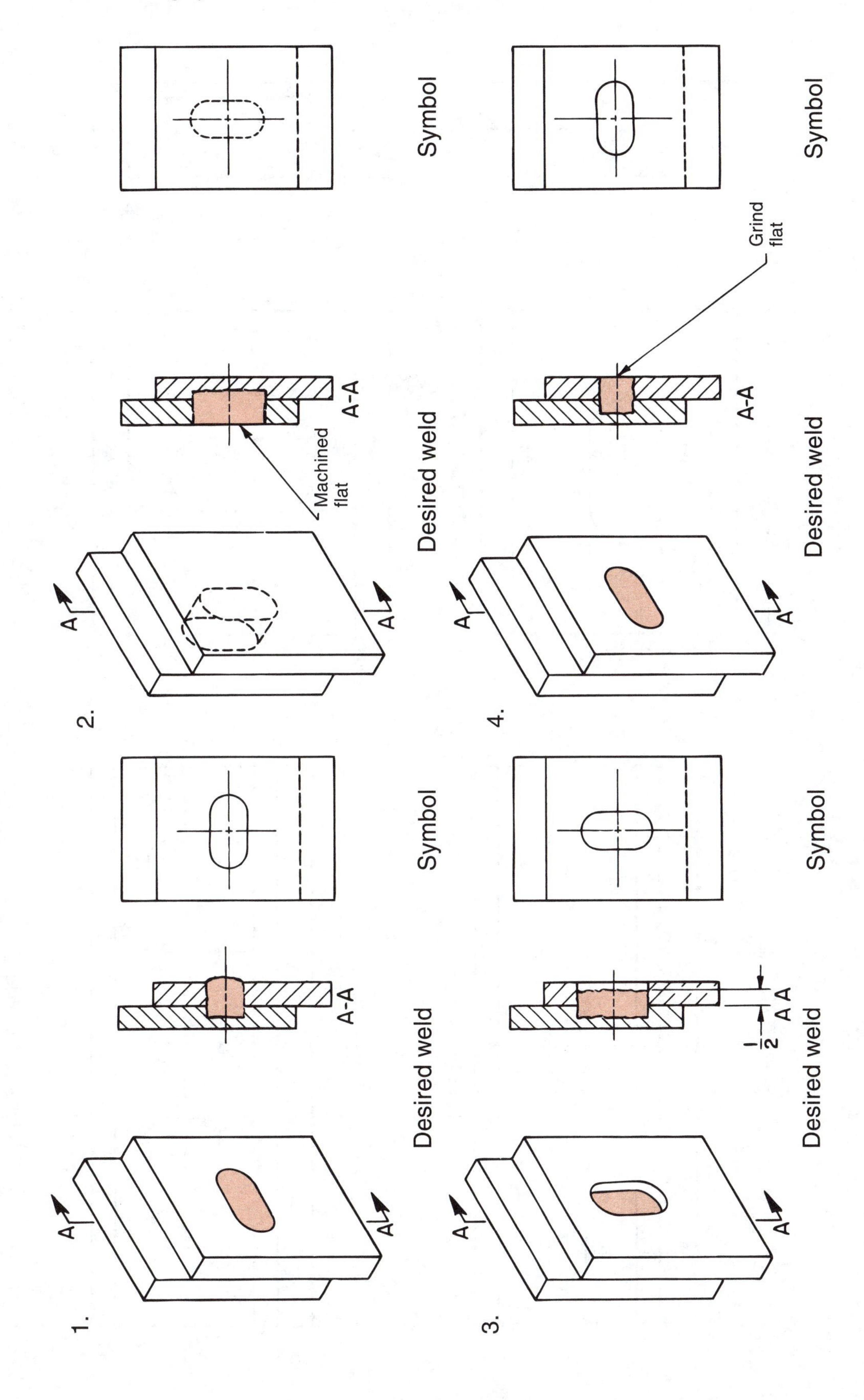

Part IV (continued)

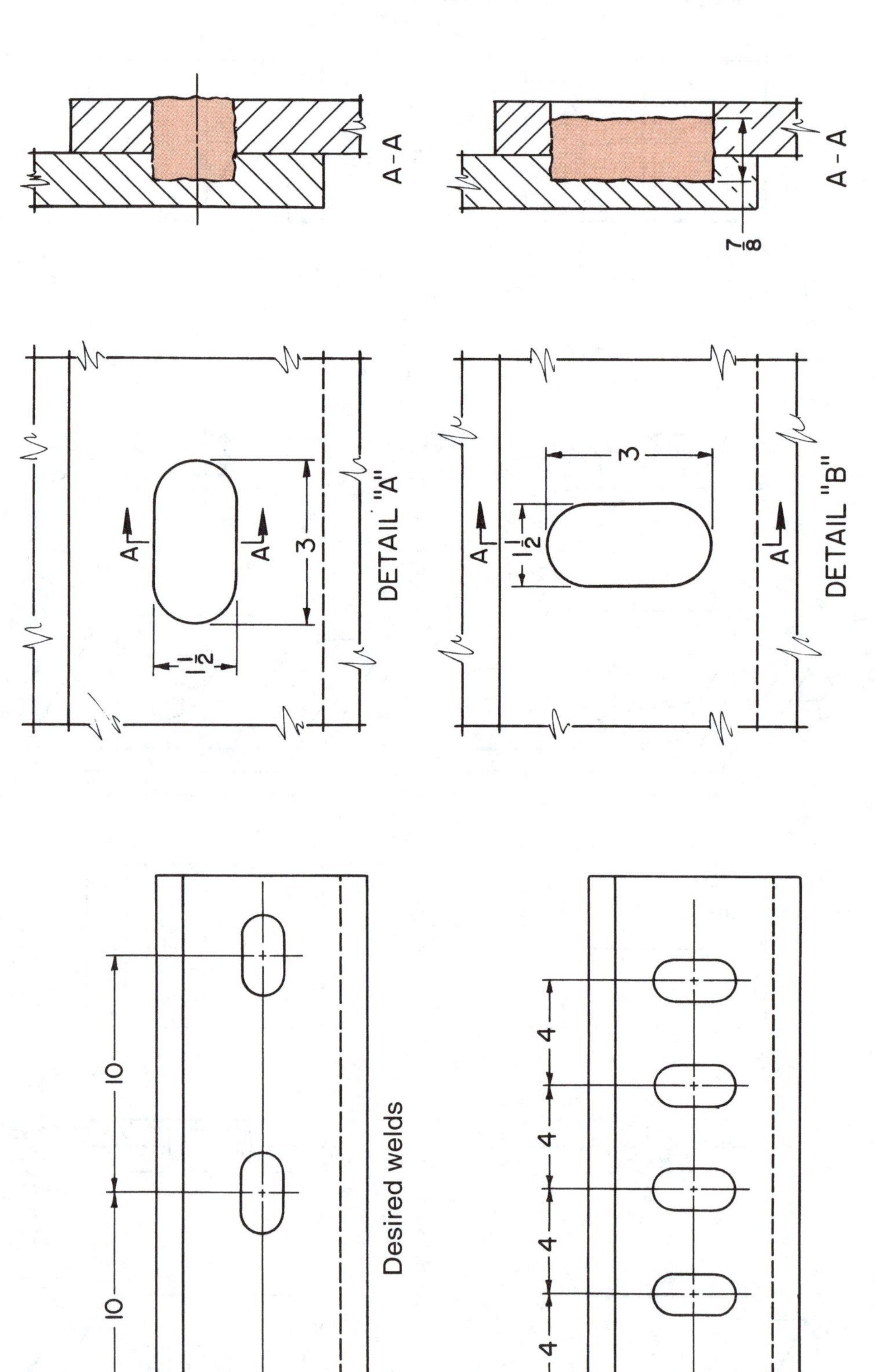

Name ______________________________

Part V

Explain each of the following welding symbols.

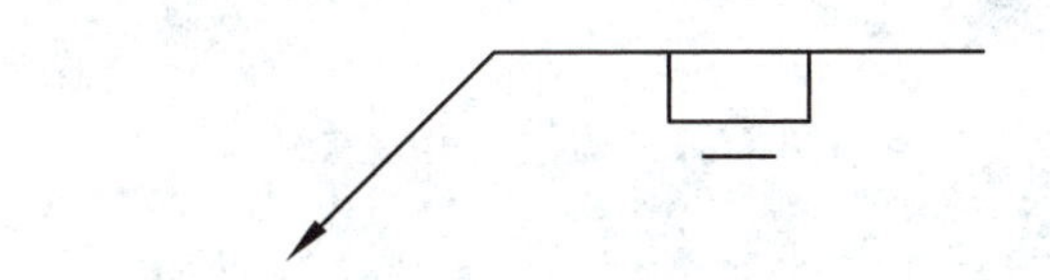

1. ______________________________

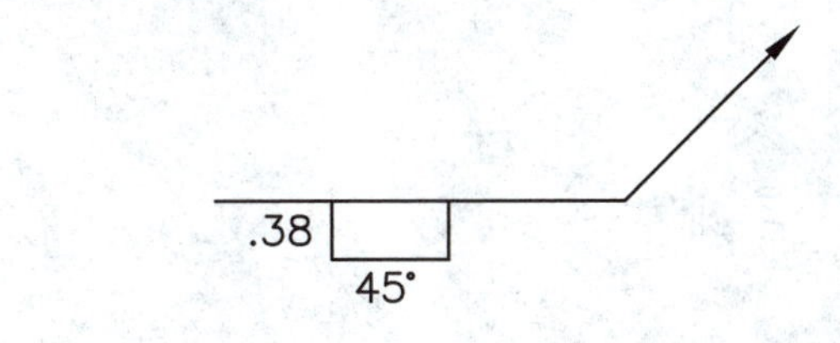

2. ______________________________

.50 6

3. ______________________________

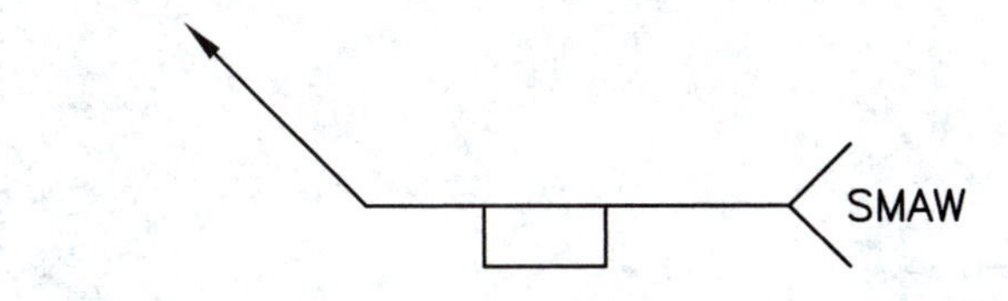

4. ______________________________

.50 8
30°

5. ______________________________

Spot welds make up a substantial number of welds performed by industrial robots during automotive body assembly. (Vasily Smirnov/Shutterstock)

Unit 18
Spot, Seam, and Projection Welds

After completing Unit 18, you will be able to:

- Explain the welding processes used to make spot, seam, and projection welds.
- Describe the similarities and differences among the welding symbols for spot, seam, and projection welds.
- Identify the location, size, strength, spacing, and number of spot, seam, and projection welds as required by the symbol.

Key Words

interface zone
projection welds
resistance welds
seam welds
spot welds
stud welding

Although several welding processes may be used to make spot, seam, and projection welds, they are generally classified as ***resistance welds***. The process of resistance welding joins metals by using heat produced from the resistance of metal to electric current passing through the parts. The parts are normally held together under pressure during the entire welding process. Figure 18-1 shows a resistance spot welding machine and Figure 18-2 shows a projection welded fastener.

Figure 18-1.
A resistance spot welding machine is shown.

Figure 18-2.
This threaded fastener was welded using the projection welding process.

Spot Welds

Spot welds are made between or upon overlapping members. Coalescence (melding of materials) may begin at the mating surfaces or may proceed from the outer surface of one member. Spot welds may be made by any of several welding processes: resistance, arc, projection, electron beam, etc. With a spot weld, the welding process is indicated in the tail of the welding symbol, Figure 18-3.

The location of the spot weld symbol in relation to the reference line of the welding symbol may or may *not* have arrow side or other side significance. Refer to Figure 18-4.

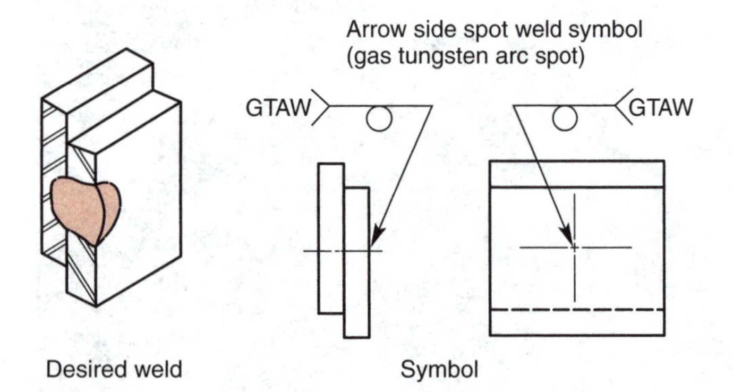

Figure 18-3.
When a spot weld is specified, the welding process is indicated in the tail of the welding symbol.

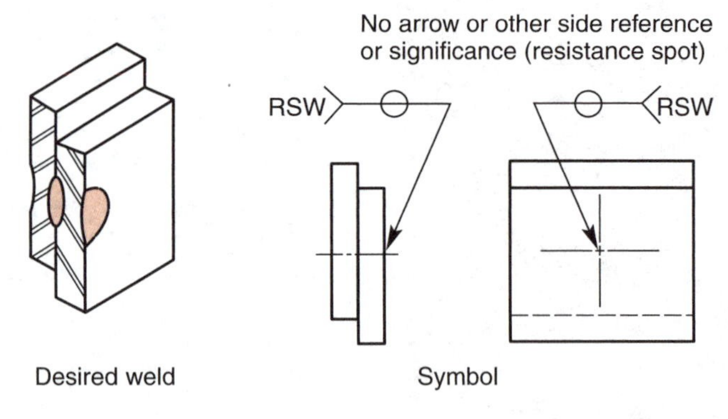

Figure 18-4.
The location of the spot weld symbol in relation to the reference line may or may not have arrow side or other side significance. (Also see Figure 18-3.)

Size and Strength of Spot Welds

A spot weld is dimensioned by either size or strength. Spot weld size is the diameter of the weld at the ***interface zone*** (point where members are joined). It is expressed in fractions, decimals, or millimeters. Spot weld size is shown to the left of the weld symbol, Figure 18-5.

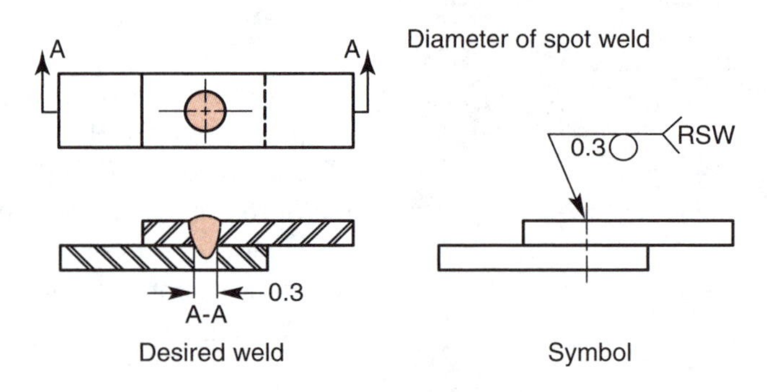

Figure 18-5.
Spot weld size is placed to the left of the weld symbol.

Spot weld strength is indicated in pounds or kilograms per spot in tension. It is also shown to the left of the weld symbol, Figure 18-6.

Spacing of Spot Welds

The pitch of spot welds is indicated to the right of the weld symbol, Figure 18-7.

Extent of Spot Welding and Number of Spot Welds

When a series of spot welds is specified for less than the full length of a joint, the extent of the welds is dimensioned as in Figure 18-8.

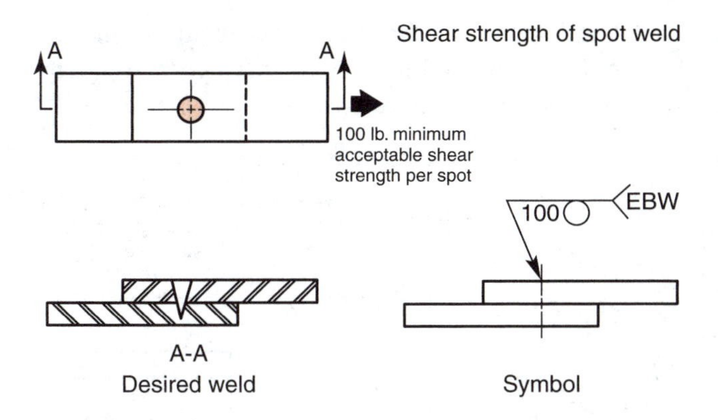

Figure 18-6.
Strength of the spot weld is placed to the left of the weld symbol.

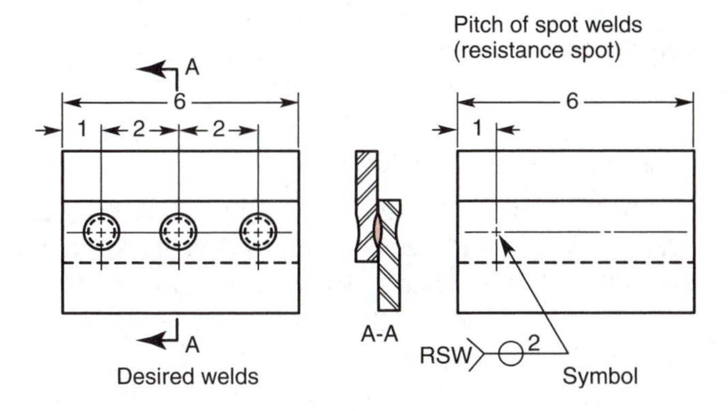

Figure 18-7.
Center-to-center spacing of spot welds is given to the right of the weld symbol.

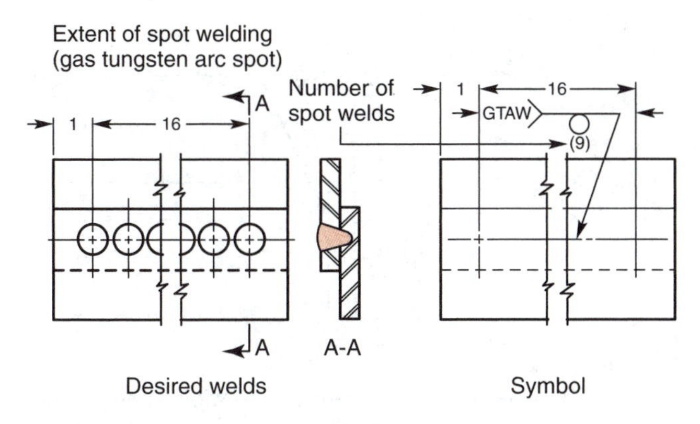

Figure 18-8.
Study the weld symbol when a series of spot welds is specified and the welds are to take up less than a full length of the joint. Also shown with the weld symbol is the number of welds required.

If a specific number of spot welds is needed for a particular joint, the number of welds is given in parentheses on the same side of the reference line as the weld symbol. Refer to Figure 18-8.

When a number of spot welds are randomly located in a group, they are indicated as shown in Figure 18-9.

The location of a group of spot welds on intersecting center lines is indicated as shown in Figure 18-10.

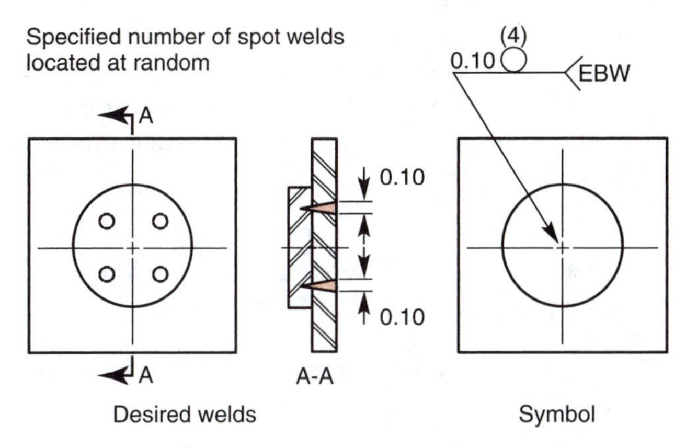

Figure 18-9.
A specific number of spot welds located randomly in a group is specified for this particular joint.

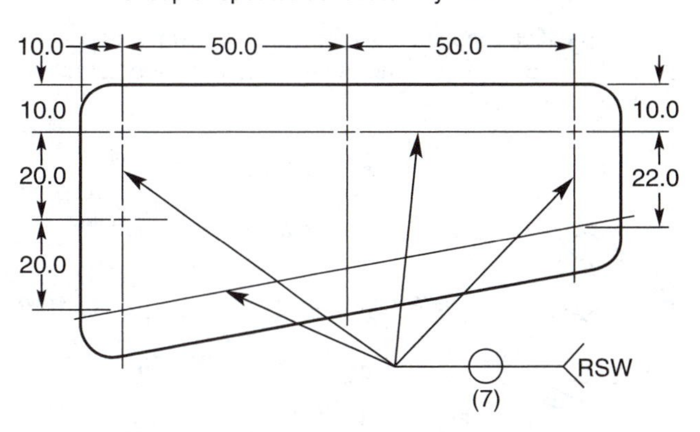

Figure 18-10.
Spot welds are specified as shown when located on center lines of a component.

Flush Spot Weld Joints

The flush contour symbol is added to the weld symbol when specifications require the exposed surface of either member of a spot welded joint to be flush. Figure 18-11 shows an example.

Multiple Member Joint Spot Welds

With multiple member joint spot welds, the same spot weld symbol is used regardless of the number of pieces inserted between the two outer members of the required joint.

Seam Welds

Seam welds are continuous welds made between or on overlapping members with coalescence starting at the mating surfaces or proceeding from the outer surface of one member. The continuous weld may be a single weld bead or a series of overlapping spot welds. A seam weld (like the spot weld) may be made by any of several welding processes. The process to be used is indicated in the tail of the welding symbol, as shown in Figure 18-12.

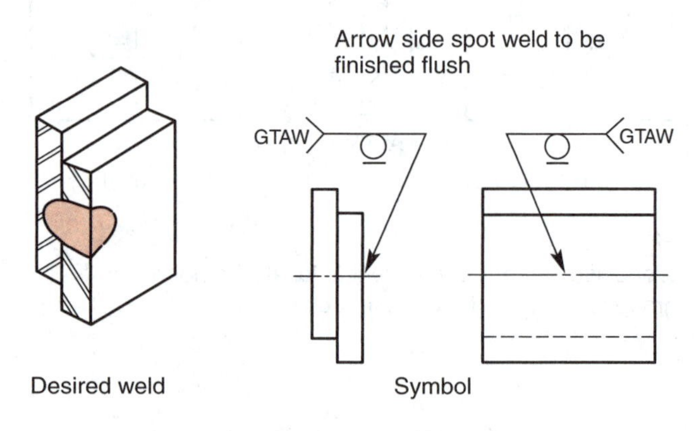

Figure 18-11.
When specifications call for the exposed surface of a spot welded joint to be flush, the flush contour symbol is added to the weld symbol.

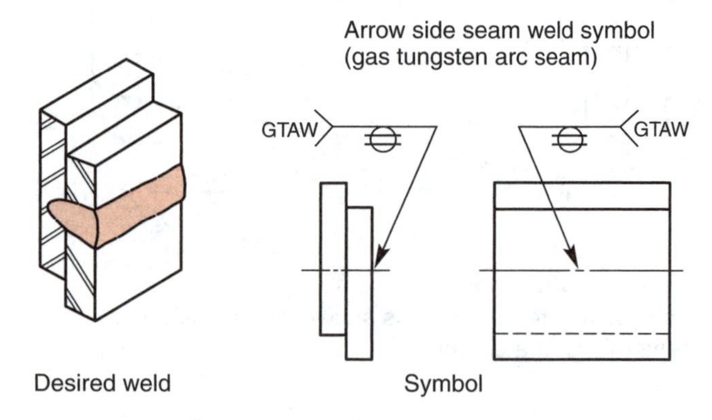

Figure 18-12.
The tail of the weld symbol contains the type of welding process for the seam weld.

The location of the seam weld symbol, in relation to the reference line, may or may *not* have arrow side or other side significance, Figure 18-13.

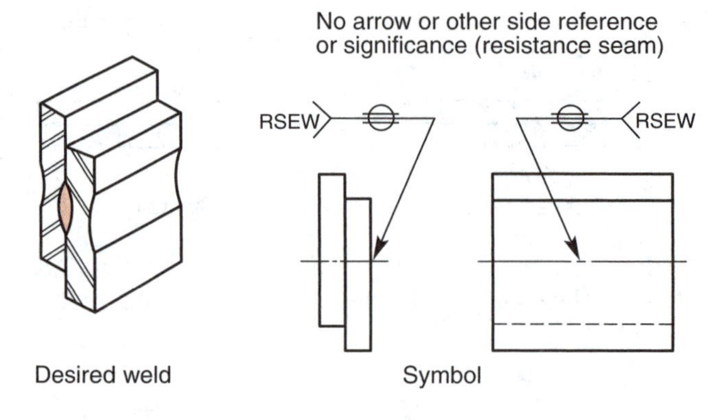

Figure 18-13.
The seam weld symbol location, in relation to the reference line, may or may not have arrow side or other side significance. (Also see Figure 18-12.)

Size and Strength of Seam Welds

Seam welds are dimensioned by either size or strength. Seam weld size is designated as the width of the weld. It is expressed in fractions of an inch, decimals, or in millimeters. Seam weld size is shown to the left of the weld symbol, Figure 18-14.

When used, seam weld strength is designated in pounds per inch of weld or in metric units. This information is also shown to the left of the weld symbol, Figure 18-15.

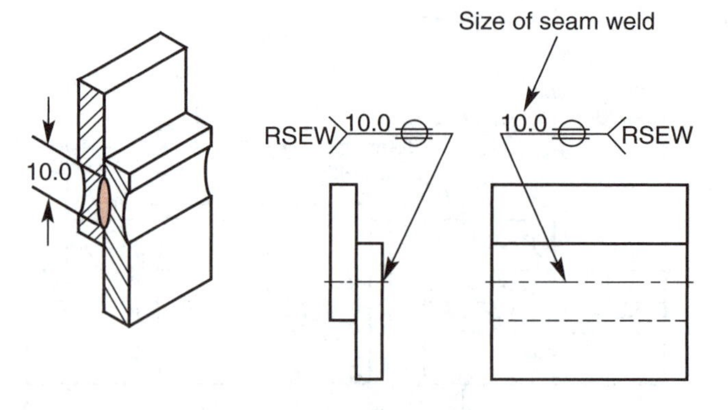

Figure 18-14.
The size of a seam weld is given as the width of the weld and is expressed in fractions of an inch, decimals, or millimeters. It is shown to the left of the weld symbol.

Length of Seam Welds

Seam weld length, when indicated on the welding symbol, is located to the right of the weld symbol, Figure 18-16.

A seam weld extending less than the full length of the joint is dimensioned as shown in Figure 18-17.

Dimensioning of Intermittent Seam Welds

The pitch of intermittent seam welds is given as the distance between the centers of the weld increments. Figure 18-18 gives an example. The pitch is shown to the right of the weld length dimension.

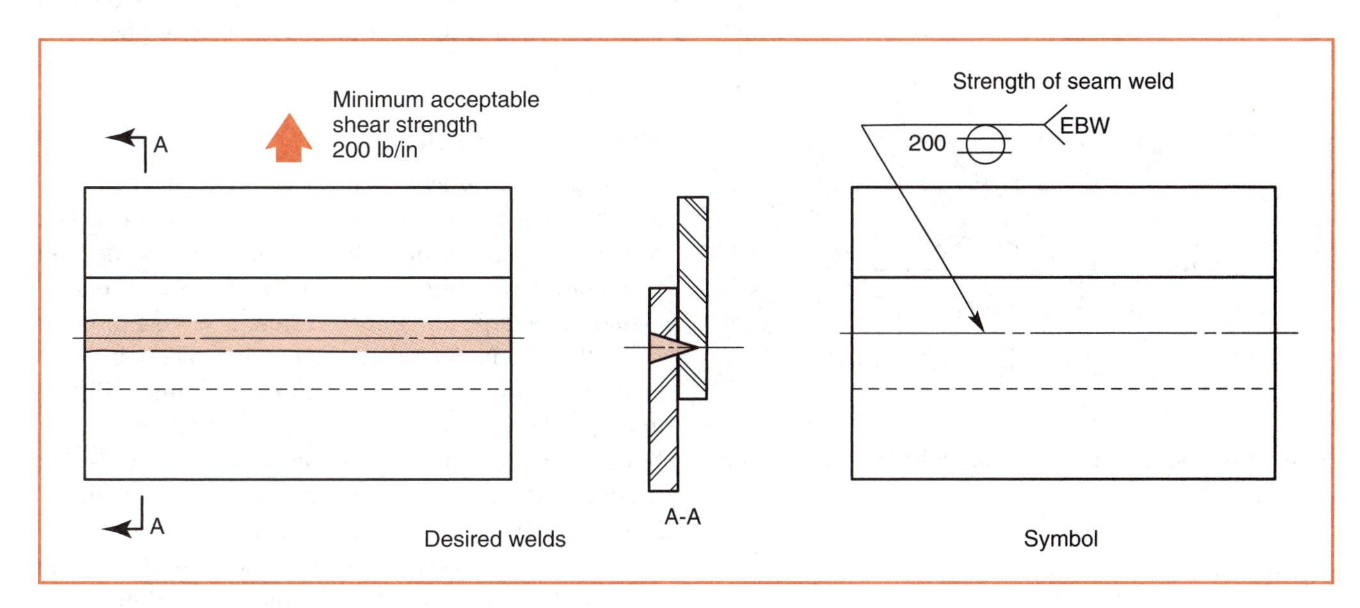

Figure 18-15.
Seam weld strength, when used, is given in pounds per linear inch or in metric units and is to the left of the weld symbol.

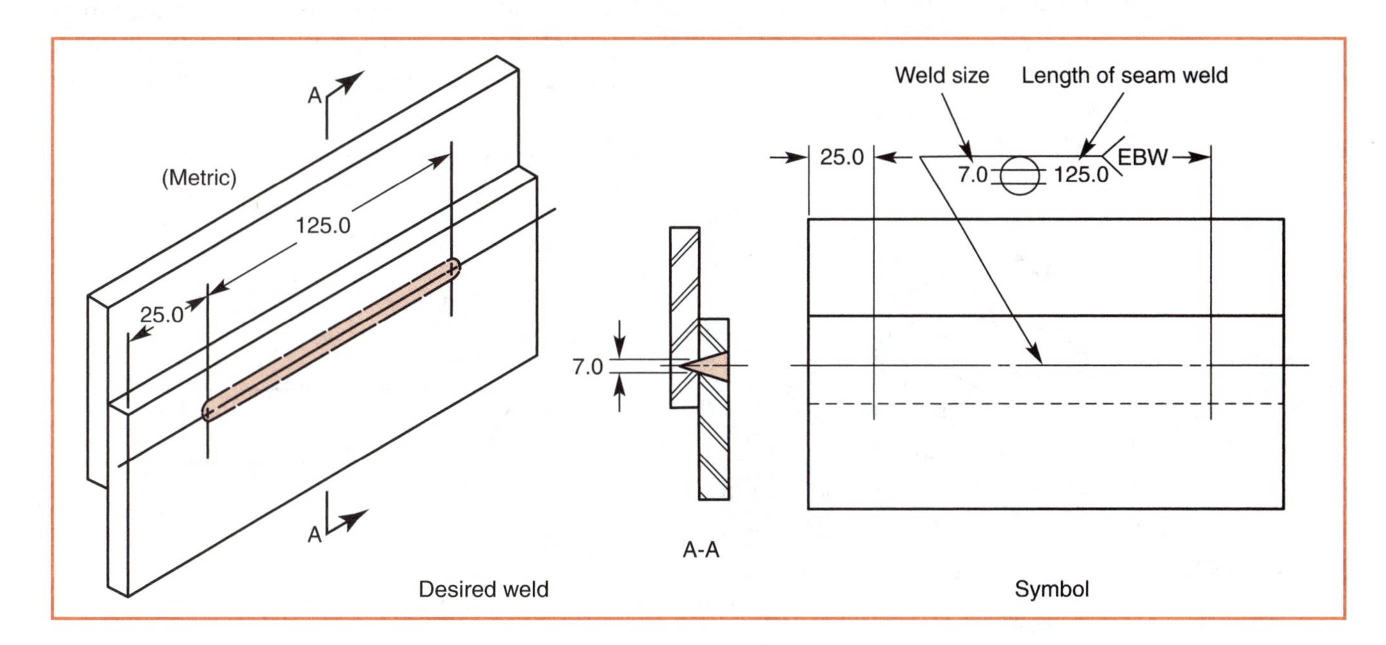

Figure 18-16.
If it is used on a welding symbol, seam weld length is on the right of the weld symbol.

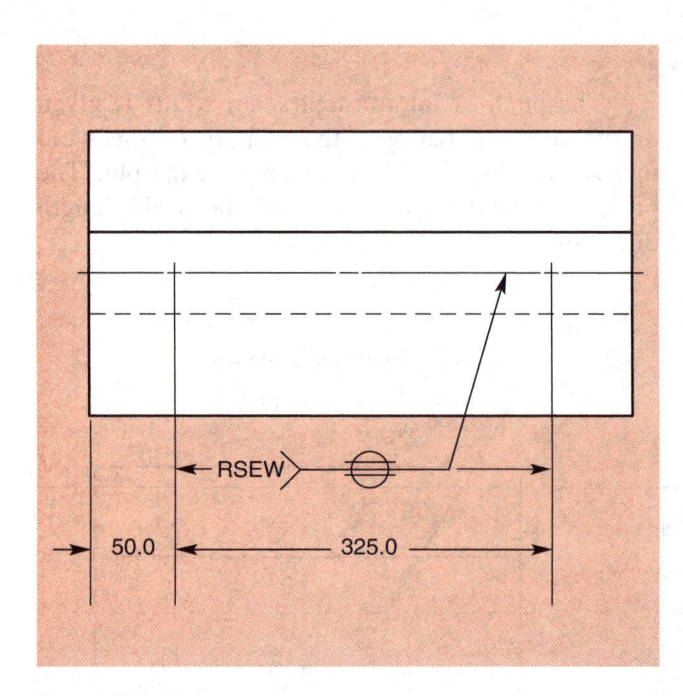

Figure 18-17.
Study dimensions for a seam weld extending less than the full length of the joint.

Orientation of Seam Welds

Unless otherwise indicated, intermittent seam welds are understood to have length and width parallel to the axis of the weld, Figure 18-19.

Flush Seam Weld Joints

The flush contour symbol is used when the exposed surface of either member (or each member) is to be finished flush, Figure 18-20.

Multiple Joint Seam Welds

No special addition to the welding symbol is required when additional pieces are inserted between the two outer members.

Projection Welds

A ***projection weld*** is produced from the heat created by the resistance to the welding current flow. The process of projection welding can be performed by resistance welding or through an arc process. Generally a large amount of current is passed through a small preformed area on the metal projection.

Projection welding is used to weld many different types of parts together. Threaded fasteners such as bolts and nuts, concrete holding studs, and cross-wire products (like fence), are all projection welded. Figure 18-21 shows a projection welded GMAW wire reel. The cross-wire welding process used to weld together the reel is a form of projection welding.

The projection weld symbol requires the welding process reference in the tail of the symbol. Figure 18-22 shows the welding process reference PW in the tail of the symbol. The symbol must also be placed either below the reference line (on the arrow side) or above

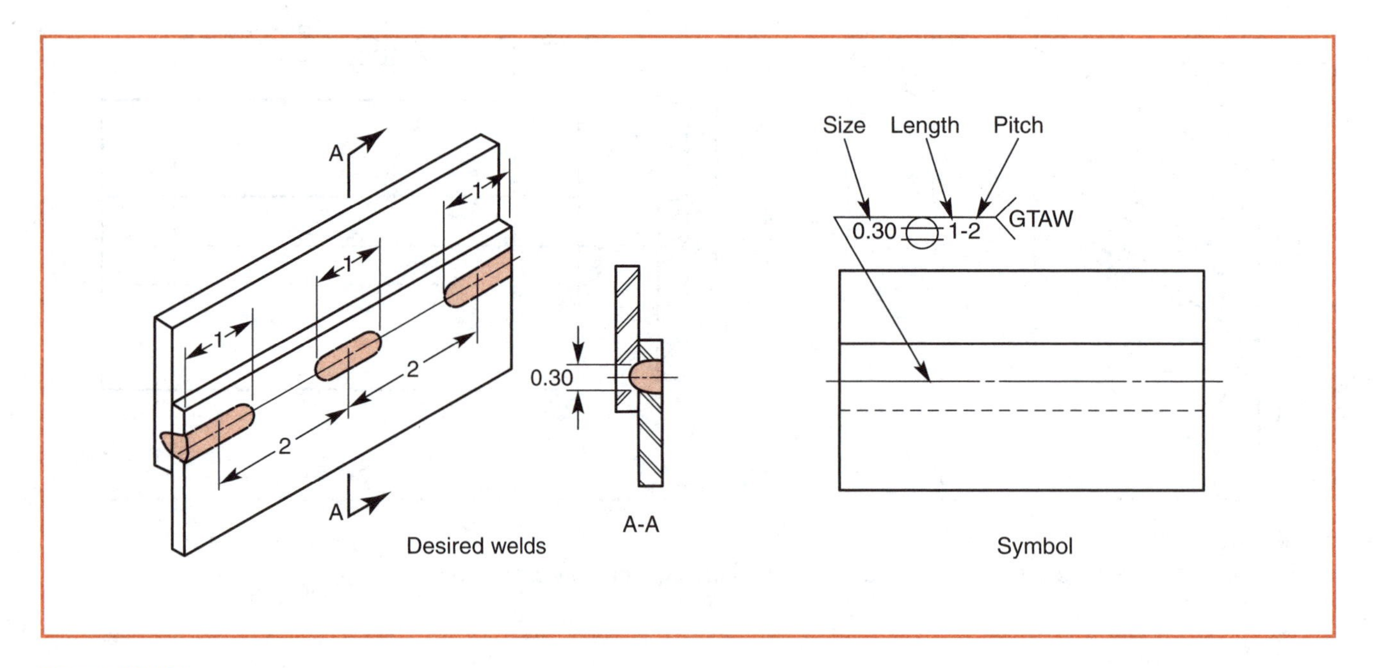

Figure 18-18.
Pitch of intermittent seam welds is shown to the right of the weld length dimension.

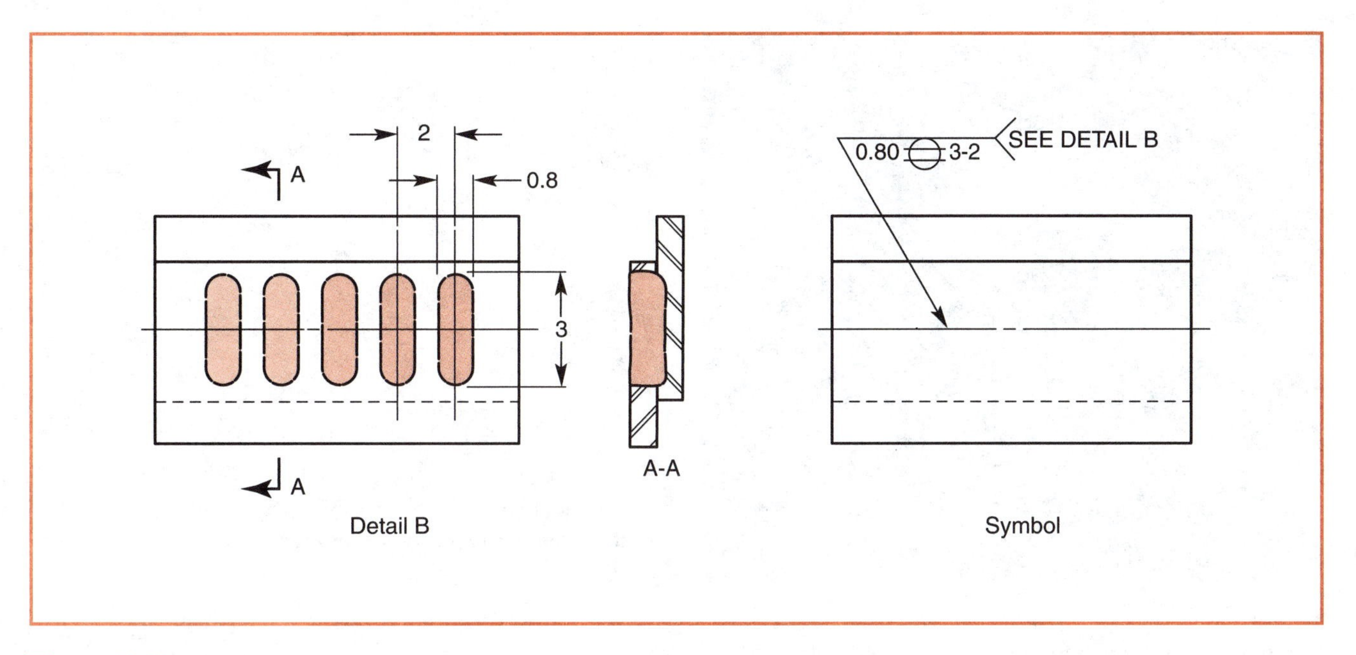

Figure 18-19.
Intermittent seam welds have length and width parallel to the axis of the weld, unless indicated otherwise.

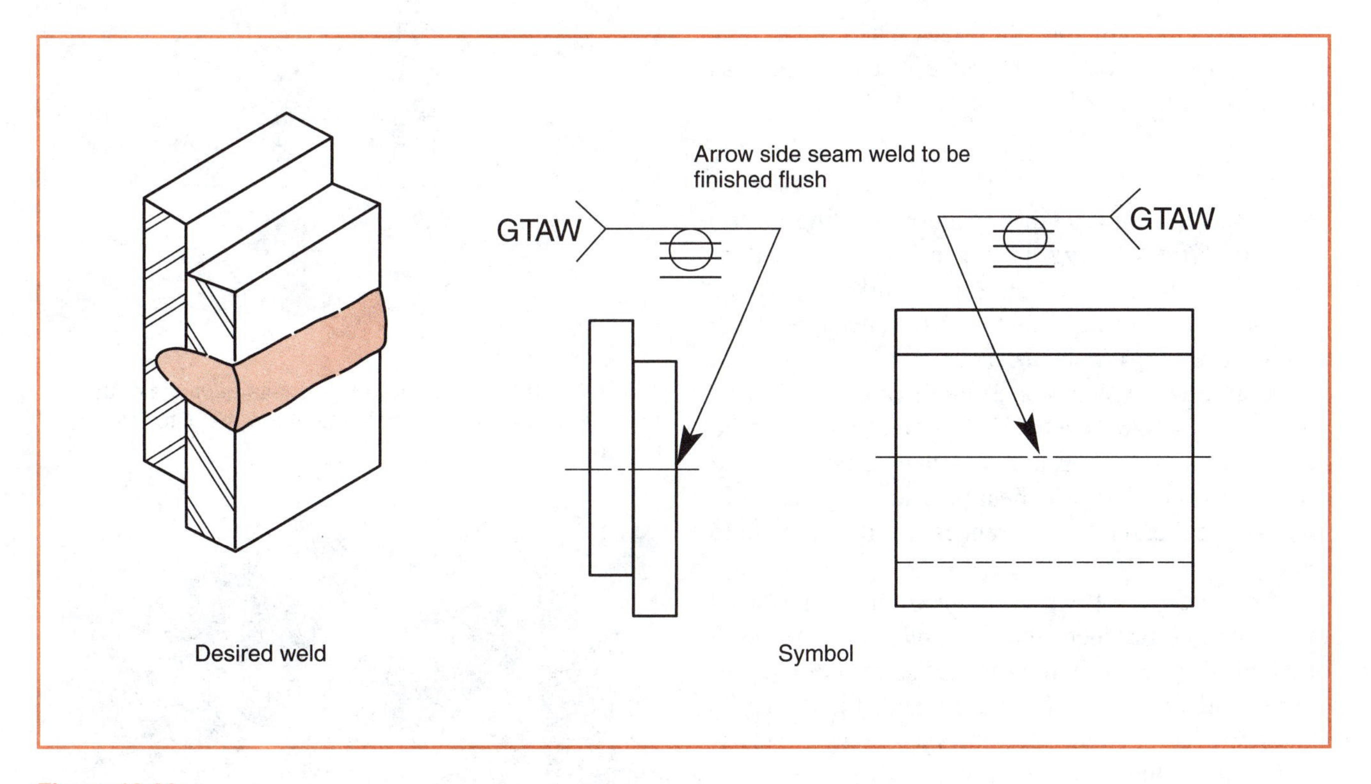

Figure 18-20.
The flush contour symbol shows the exposed surface(s) should be finished flush.

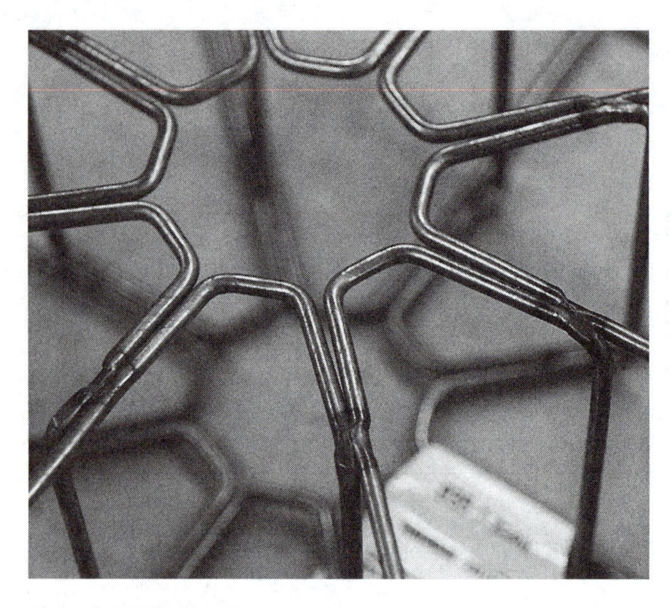

Figure 18-21.
The cross-wire welding process was used to weld together this GMAW wire reel.

the reference line (on the other side). The symbol may *not* be placed in the center of the reference line, as permitted with the spot welding symbol.

Stud Welds

A specialized type of projection welding is stud welding. ***Stud welding*** is a general term used to describe joining a metal stud to a workpiece. Studs may be in the form of threaded fasteners, concrete holding studs, or other forms.

Figure 18-23 shows an example of the preformed area on a .250-20 UNC (.250 nominal diameter, 20 threads per inch, Unified National Coarse thread form) threaded stud. The heat produced by the resistance of the metal to the current causes the preform to melt.

Pressure from the projection welding gun forces the projection (fastener, stud, etc.) into the molten pool of metal. The pool solidifies around and under the projection and fuses it to the base metal. Figure 18-24 shows projection welded fasteners joined by a capacitor discharge stud welding machine.

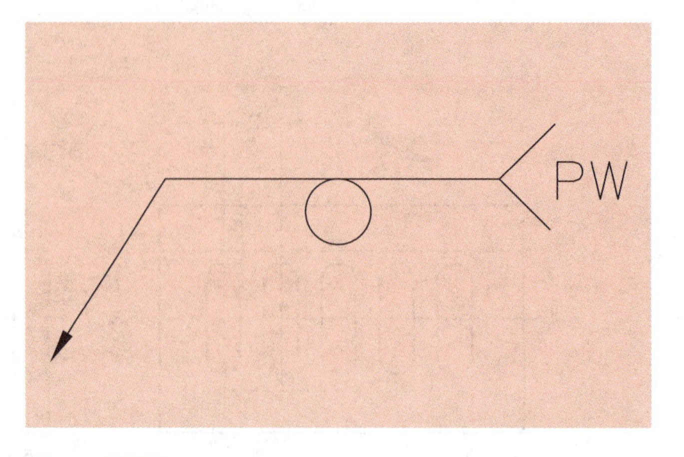

Figure 18-22.
The projection welding reference (PW) is placed in the tail of the symbol to specify the process of welding to be performed.

Figure 18-23.
This .250-20 UNC threaded stud melts because of the heat produced by the resistance of the metal to the current.

Figure 18-24.
A capacitor discharge stud welding machine was used to weld these fasteners.

Side Significance

The stud weld symbol has arrow side significance only. The weld symbol should only be placed below the reference line with the location of the stud clearly indicated. Figure 18-25 shows the stud weld symbol.

Stud Size, Spacing, and Number of Studs

The diameter of the stud is given to the left of the stud weld symbol. Figure 18-26 indicates a .50″ diameter stud. Spacing or pitch distance of the studs is indicated to the right of the stud weld symbol. Figure 18-27 shows a .38″ stud spaced with a 3″ pitch.

The number of studs is shown in parentheses and placed below the stud weld symbol. The symbol shown in Figure 18-28 requires eight studs.

Location of Stud Welds

Location information for stud welds is given on the print using standard dimensions. Generally, the center of the first stud is given from a datum or part edge and often the last stud is also dimensioned. The distance between the centers of rows of studs may also be provided. Figure 18-29 shows two rows of studs required for an assembly.

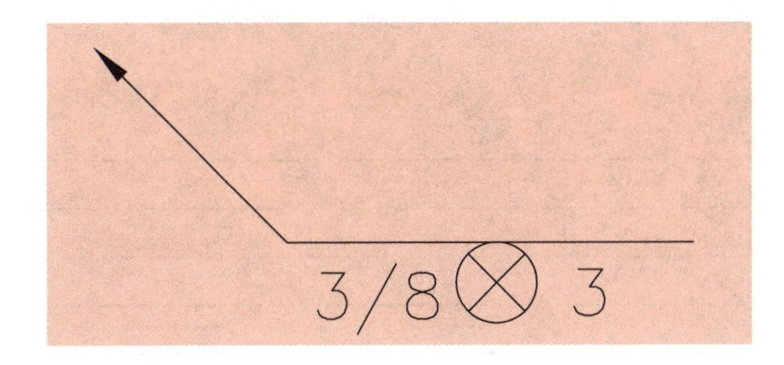

Figure 18-27.
The 3″ pitch is indicated to the right of the stud weld symbol.

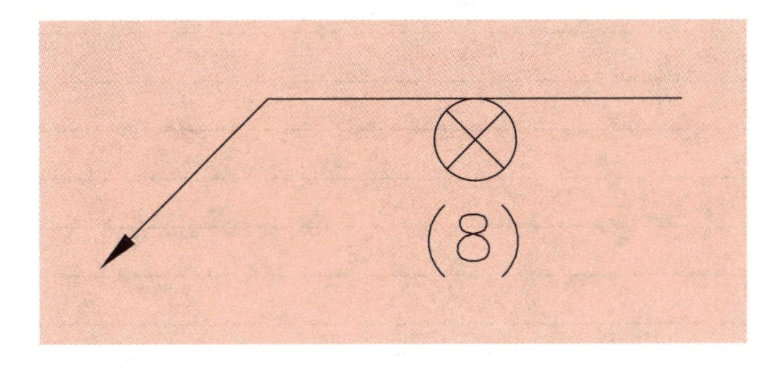

Figure 18-28.
The number of studs is placed in parentheses below the stud weld symbol.

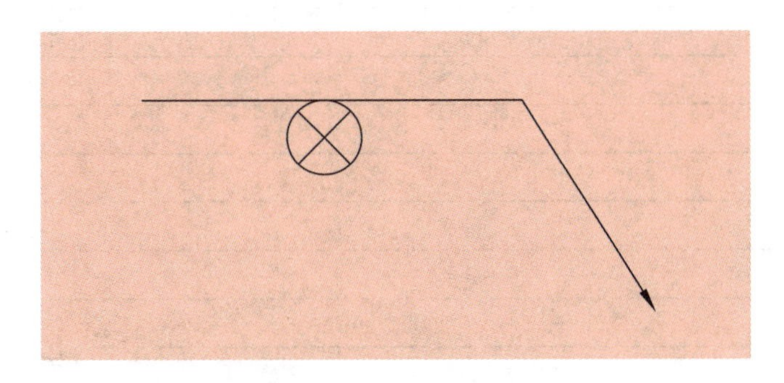

Figure 18-25.
A stud weld symbol shows arrow side significance only.

Figure 18-26.
Look for the diameter of the stud to the left of the stud weld symbol.

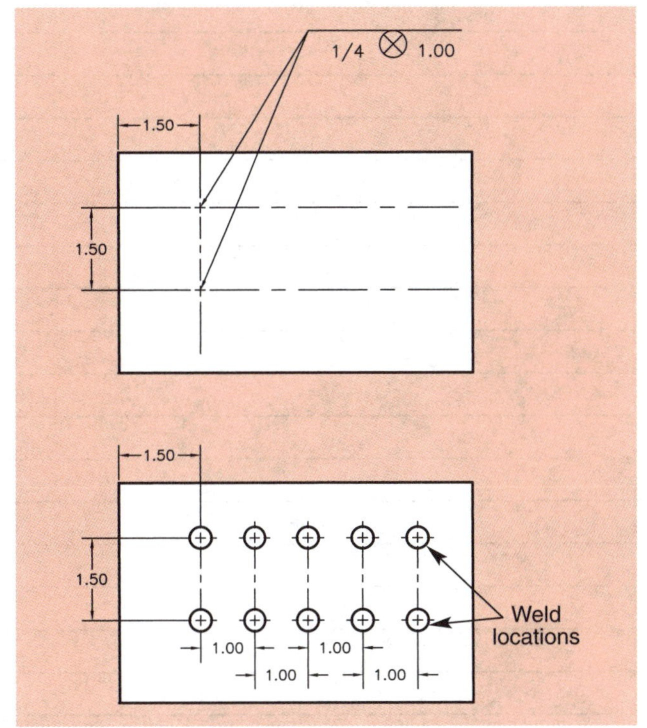

Figure 18-29.
Location dimensions for stud welds. Distance between the centers of rows of studs may be provided.

Notes

Name ________________________________

Print Reading Activities

Part I

Refer to the drawing (167E104) to answer the following questions.

1. Name the drawing title. ________________

2. What is the drawing number? ________________

3. How many total pieces make up the assembly? _____ Name them.

4. What is the welding process to be used?

5. List the specified welding sequence.

6. How many welds are specified to attach each side of the top, headstock pedestal (167W104C), to the back, headstock pedestal (167W104A), and front, headstock pedestal (167W104B)? _____ The weld pitch is _____.

7. What is the number of the next assembly? _____

8. How many welds are specified to attach each hinge (167W104E) to the front, headstock pedestal (167W104B)? _____ The weld pitch is _____.

9. What does the note concerning the placement of the hinge (167W104E) specify?

10. This welding symbol is found on the print. What does it mean?

(9)
3
RSW

RSW = ________________

(9) = ________________

3 = ________________

11. What is the full name and number of the base?

12. Section A-A is shown on the drawing. What does it show?

13. The following tolerances are allowed:

Decimal: ________________

Fraction: ________________

Angular: ________________

14. When positioning the top, headstock pedestal (167W104C), to be welded to the back, headstock pedestal (167W104A), and front, headstock pedestal (167W104B), what precaution must be observed about its placement?

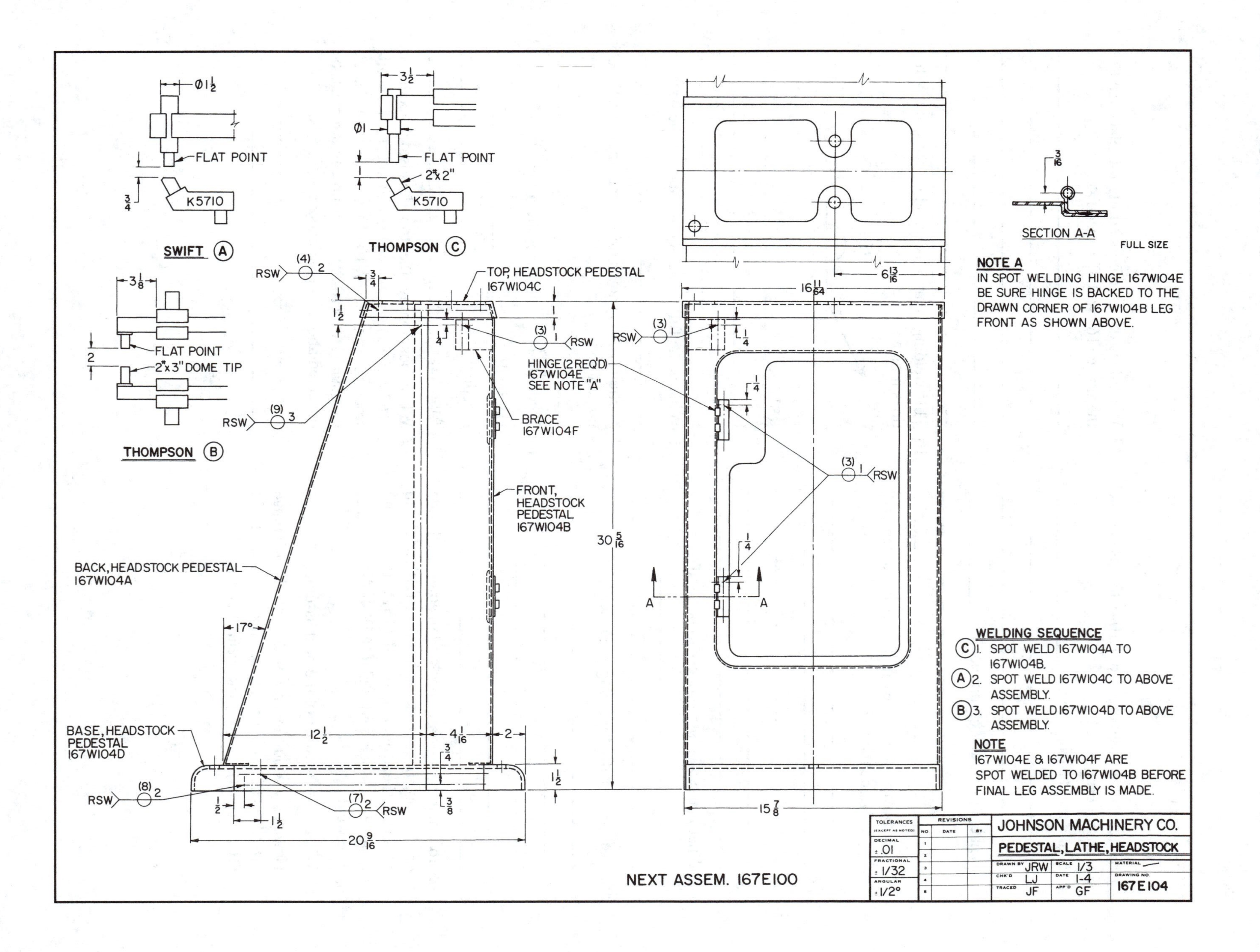
SWIFT A
FLAT POINT
K5710
THOMPSON C
FLAT POINT
2#x2"
THOMPSON B
FLAT POINT
2#x3" DOME TIP
SECTION A-A
FULL SIZE
NOTE A
IN SPOT WELDING HINGE 167W104E BE SURE HINGE IS BACKED TO THE DRAWN CORNER OF 167W104B LEG FRONT AS SHOWN ABOVE.
TOP, HEADSTOCK PEDESTAL 167W104C
HINGE (2 REQ'D) 167W104E SEE NOTE "A"
BRACE 167W104F
FRONT, HEADSTOCK PEDESTAL 167W104B
BACK, HEADSTOCK PEDESTAL 167W104A
BASE, HEADSTOCK PEDESTAL 167W104D
RSW
17°
30 5/16
16 11/64
6 13/16
15 7/8
20 9/16
12 1/2
WELDING SEQUENCE
C 1. SPOT WELD 167W104A TO 167W104B.
A 2. SPOT WELD 167W104C TO ABOVE ASSEMBLY.
B 3. SPOT WELD 167W104D TO ABOVE ASSEMBLY.
NOTE
167W104E & 167W104F ARE SPOT WELDED TO 167W104B BEFORE FINAL LEG ASSEMBLY IS MADE.
TOLERANCES (EXCEPT AS NOTED)
DECIMAL ±.01
FRACTIONAL ±1/32
ANGULAR ±1/2°
REVISIONS
NO DATE BY
JOHNSON MACHINERY CO.
PEDESTAL, LATHE, HEADSTOCK
DRAWN BY JRW
SCALE 1/3
CHK'D LJ
DATE 1-4
TRACED JF
APP'D GF
MATERIAL
DRAWING NO 167E104
NEXT ASSEM. 167E100

Name ____

Part II

1. What does the following welding symbol indicate?

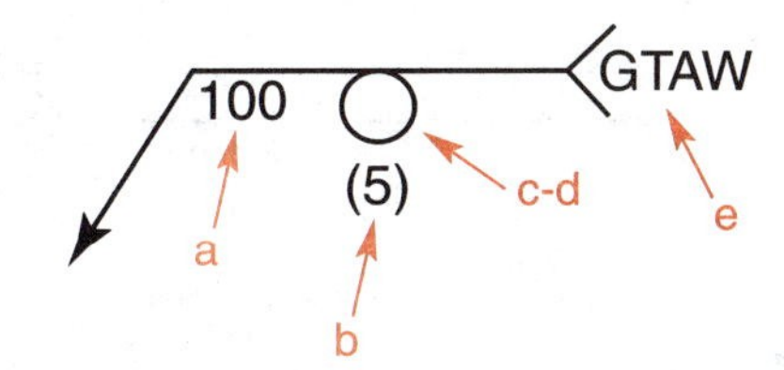

a. ____
b. ____
c. ____
d. ____
e. ____

2. Describe the weld(s) indicated by the following welding symbol.

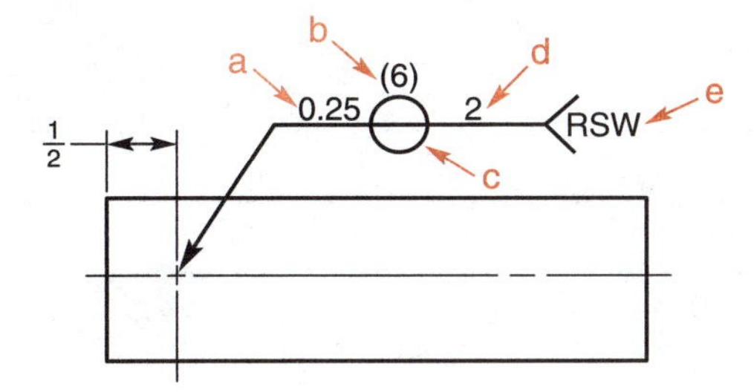

a. ____
b. ____
c. ____
d. ____
e. ____

3. Sketch in the weld(s) indicated by the welding symbol.

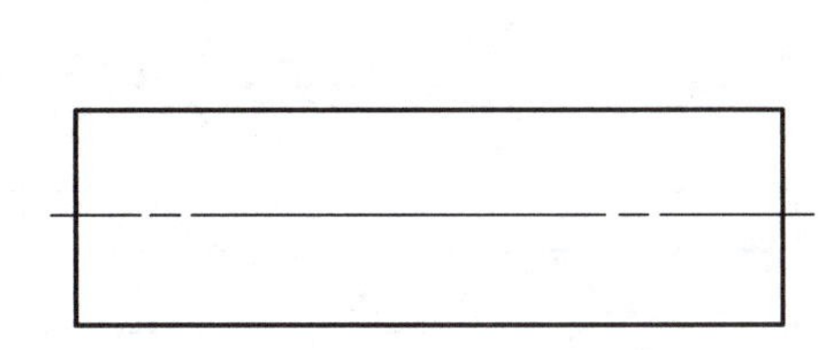

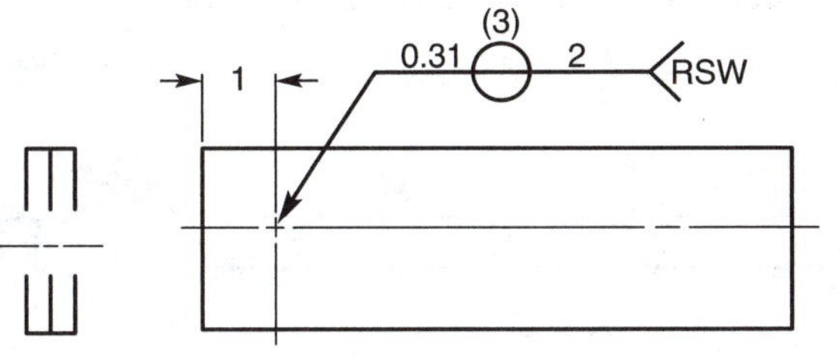

4. Describe the following weld(s).

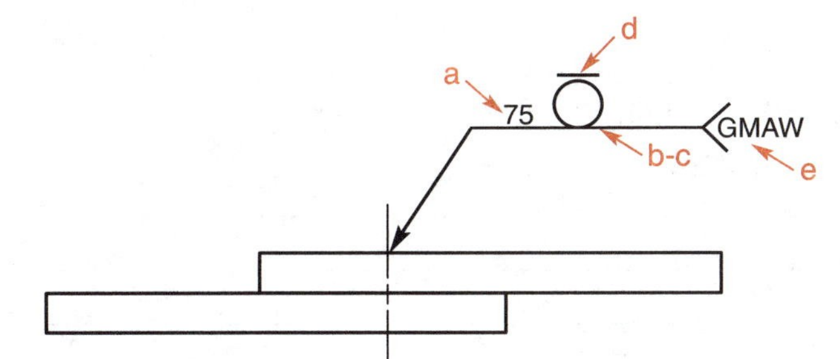

a. ____
b. ____
c. ____
d. ____
e. ____

5. Sketch the welding symbol describing the following spot weld(s).

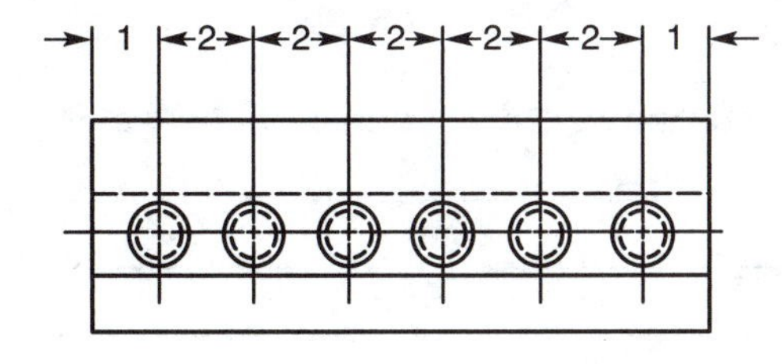

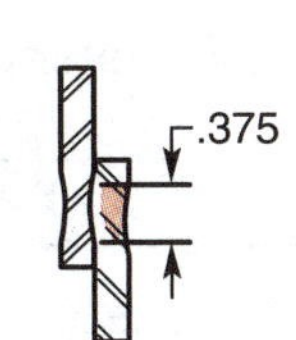

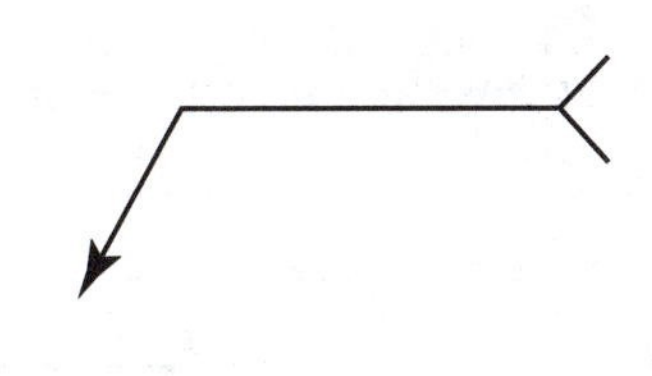

Part III

1. Sketch the welding symbol for the following seam weld on the adjacent incomplete symbol.
   A. Gas tungsten arc weld
   B. Arrow side
   C. Flush seam
   D. 0.50 seam width

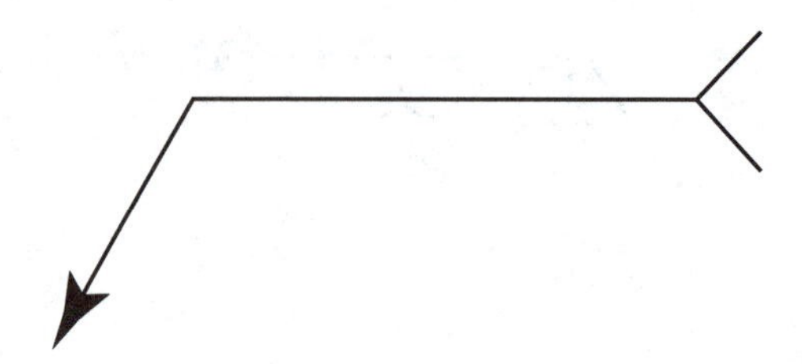

2. What does the following welding symbol indicate?

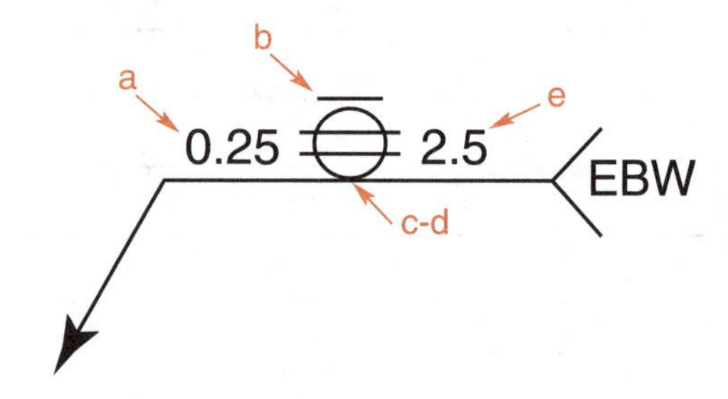

a. ___
b. ___
c. ___
d. ___
e. ___

3. Sketch in and dimension the weld(s) indicated.

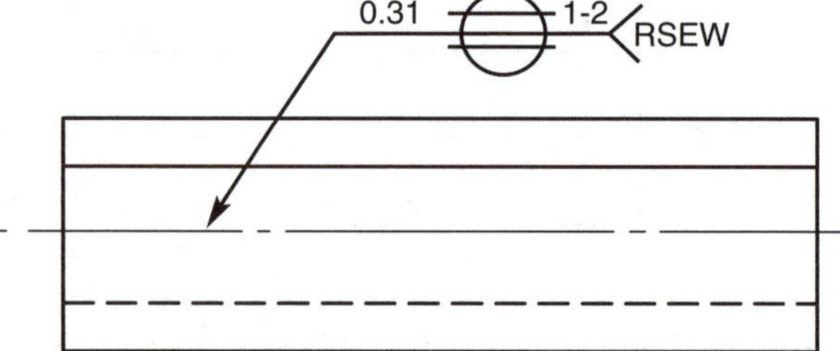

4. What does the figure (see arrow) indicate when used with a welding symbol?

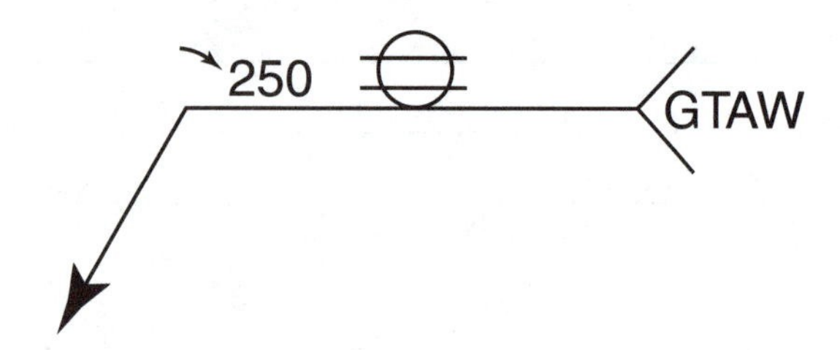

___
___
___
___
___

5. How does a welder determine orientation for intermittent seam welds (the direction the welds are to run)?

___
___
___
___

Name ________________________________

Part IV

Refer to the drawing (C24679) on the next page to answer the following questions.

1. Name the product. ________________
____
____

2. List the drawing number. ________________
____
____

3. Dimensions are in _____.
____

4. What is the scale of the drawing? __________
____
____

5. What is the scale of the stiffener details? _____
____
____

6. Name the drawing size. ________________
____
____

7. What welding process is specified? __________
____
____

8. The product is fabricated from _____ with a thickness of _____.
____

9. Name the size of the weld. _____
____

10. Dimensions of the product are _____ long, _____ wide, and _____ high.

11. The height of the stiffener is _____ and the width formed is _____.

12. What is the number of stiffeners specified? _____

13. The center line of the first stiffener is located _____ in from the edge of the shelf. The center line of the second stiffener is _____ in from the edge of the shelf.

14. The note on the drawing specifies:

    A. ________________________________
    ____

    B. ________________________________
    ____

15. Study the drawing (C24679) on the next page carefully and prepare one question to be answered by members of the class.
____
____
____

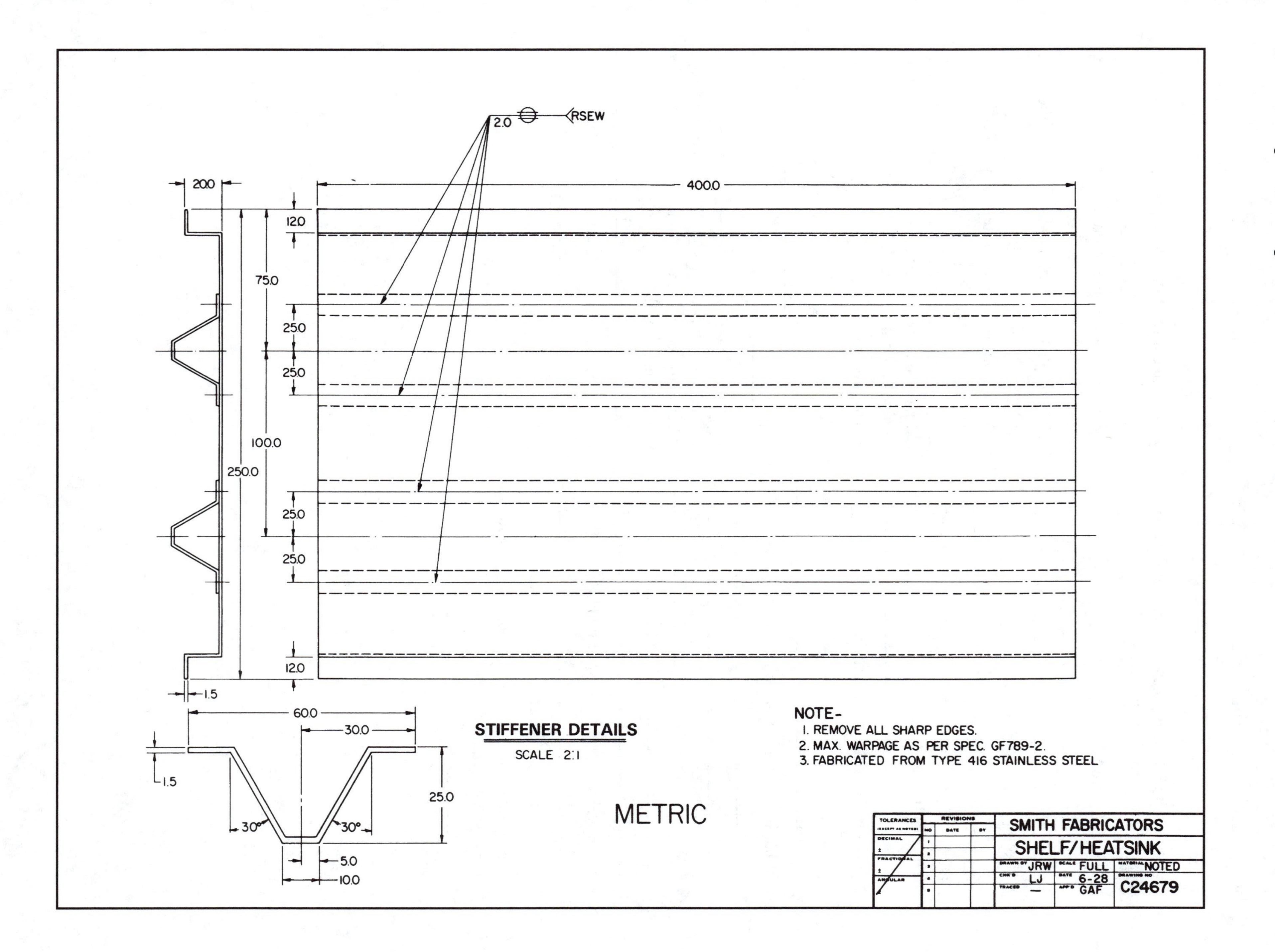

2.0
RSEW
20.0
400.0
12.0
75.0
25.0
25.0
100.0
250.0
25.0
25.0
12.0
1.5
60.0
30.0
1.5
25.0
30°
30°
5.0
10.0
STIFFENER DETAILS
SCALE 2:1
NOTE-
1. REMOVE ALL SHARP EDGES.
2. MAX. WARPAGE AS PER SPEC. GF789-2.
3. FABRICATED FROM TYPE 416 STAINLESS STEEL
METRIC
TOLERANCES
DECIMAL
FRACTIONAL
ANGULAR
REVISIONS
NO
DATE
BY
SMITH FABRICATORS
SHELF/HEATSINK
DRAWN BY JRW
SCALE FULL
MATERIAL NOTED
CHK'D LJ
DATE 6-28
DRAWING NO C24679
TRACED —
APP'D GAF

Name ______________________________

Part V

Explain the following symbols.

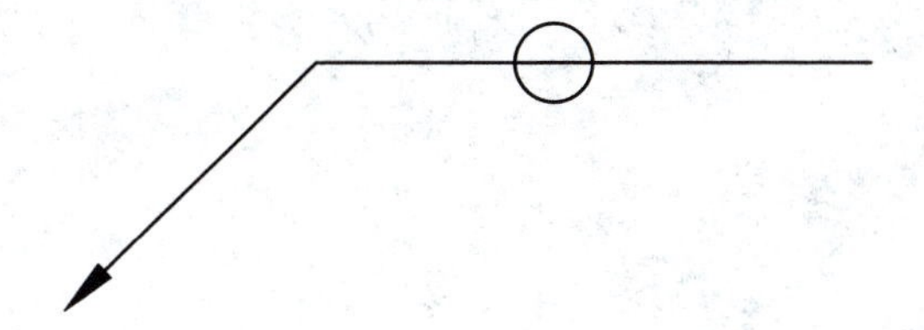

1. ____

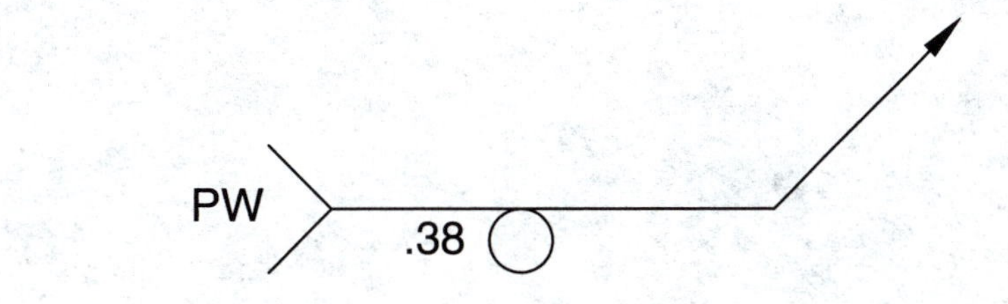

2. ____

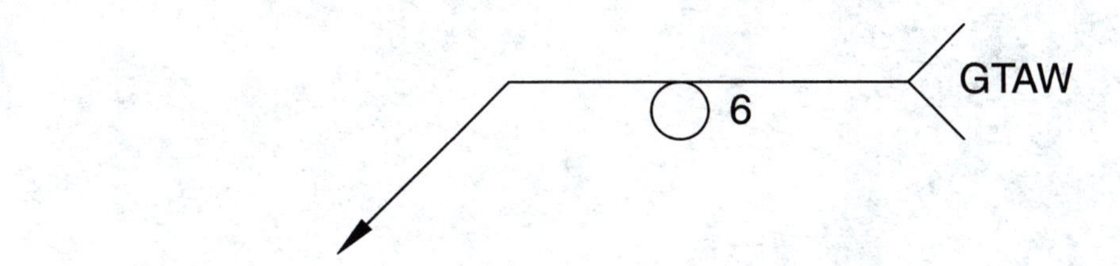

3. ____

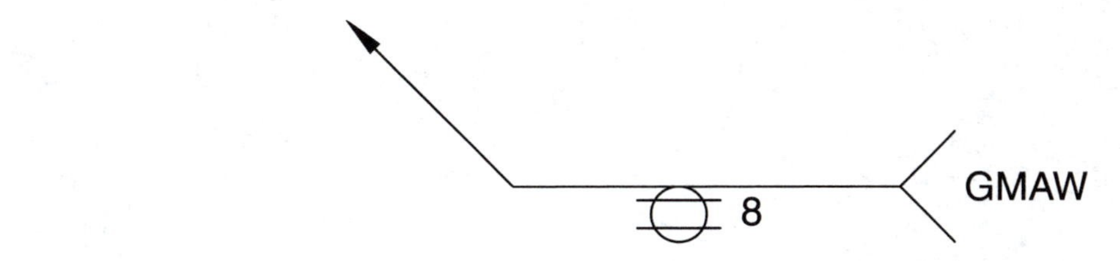

4. ____

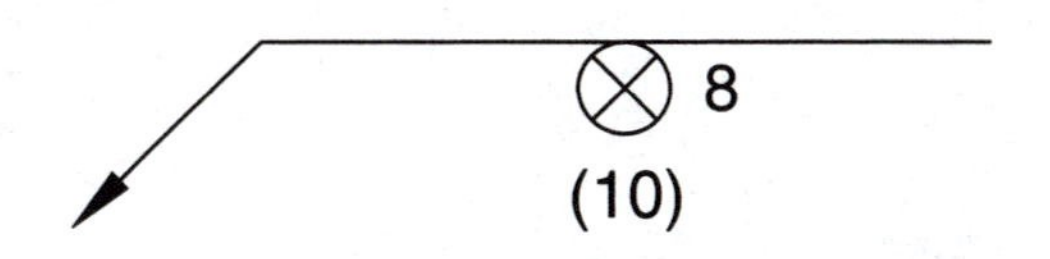

5. ____

Mining operations use surfacing welds and hardfacing to resist the constant abrasion of rocks and dirt and extend the service life of heavy machinery. (Jiri Slama/Shutterstock)

Unit 19

Surfacing Welds

After completing Unit 19, you will be able to:

- ❍ Explain the uses of surfacing welds.
- ❍ Interpret a welding symbol for a surfacing weld.
- ❍ Identify and explain surfacing weld dimensions.
- ❍ Determine the location, extent, and orientation of surfacing welds.

Key Words

buildup
buttering
hardfacing
hardness

Surfacing welds are frequently used to extend the service life of parts subjected to constant wear or abrasive conditions. They are also used to increase the dimensions of a part or to build up worn surfaces of shafts and bearing plates, Figure 19-1. When a weld is used to improve the fit of a joint or to increase the dimensions of a part, the process is sometimes called ***buttering.*** During buttering, compatible weld metals are deposited on one or more surfaces.

Hardfacing materials are generally added to improve the properties of a part (such as hardness, impact resistance, and corrosion resistance). The property of ***hardness*** allows the part to resist abrasion and wear from materials like rock, dirt, and sand.

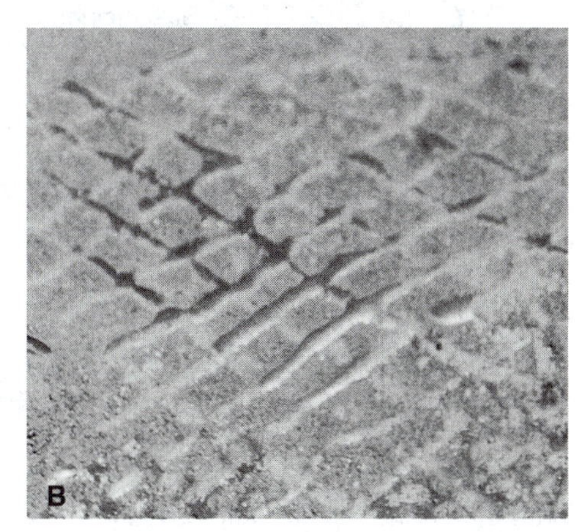

Figure 19-1.
Surfacing extends service life. A—This wheel loader is surfaced to resist the constant abrasion from dirt, rock, and sand. B—This shows a close-up of surfacing.

General Use of Surfacing Weld Symbol

The surfacing weld symbol is *added* to the welding symbol to indicate the surface(s) to be built up by welding, Figure 19-2. The same symbol applies whether the buildup is to be made by single or multiple pass welds.

Since the surfacing weld symbol does *not* indicate the welding of a joint, it has no arrow or other side significance.

The surfacing weld symbol is shown on the side of the reference line *toward* the reader. The arrow points to the surface on which the weld is to be deposited.

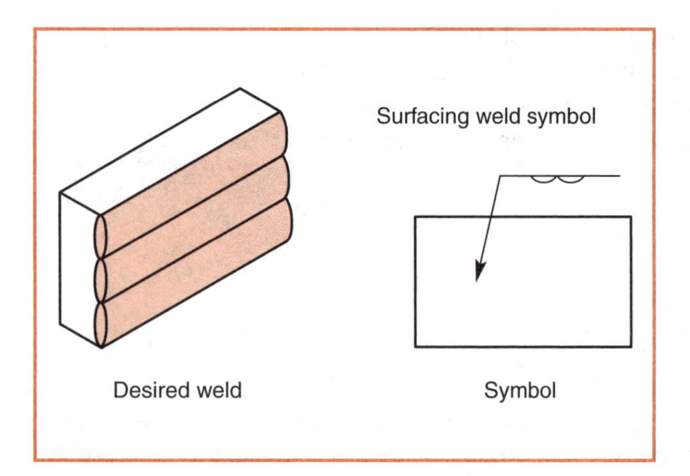

Figure 19-2.
The surfacing weld symbol shows no arrow or other side significance since no joint is welded. The arrow points toward the area to be surfaced.

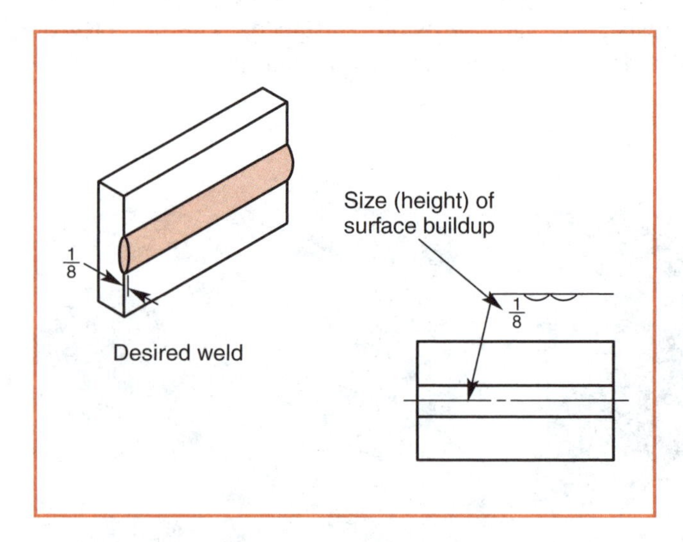

Figure 19-3.
A dimension placed to the left of the surface weld symbol shows the thickness of weld buildup.

Surfacing Weld Dimensions

Buildup is a surfacing variation in which surfacing material is deposited to obtain the desired dimensions. The buildup thickness dimension is shown on the *same side* of the reference line as the surfacing weld symbol and to the *left* of the symbol. Figure 19-3 shows an example.

The thickness dimension indicates the *minimum* height of the buildup. If no dimension is shown, no specific height of weld deposit is required. When the entire surface of a part is to be built up, the drawing is dimensioned as shown in Figure 19-4.

Location, Extent, and Orientation of Surfacing Welds

The location, extent, and orientation of the buildup area are indicated on the print when only a *portion* of the area is to be surfaced. This is shown in Figure 19-5. Multiple reference lines may be used to indicate the order of welding the surface beads. A note in the tail of the welding symbol indicates the direction of the welds as shown in Figure 19-6.

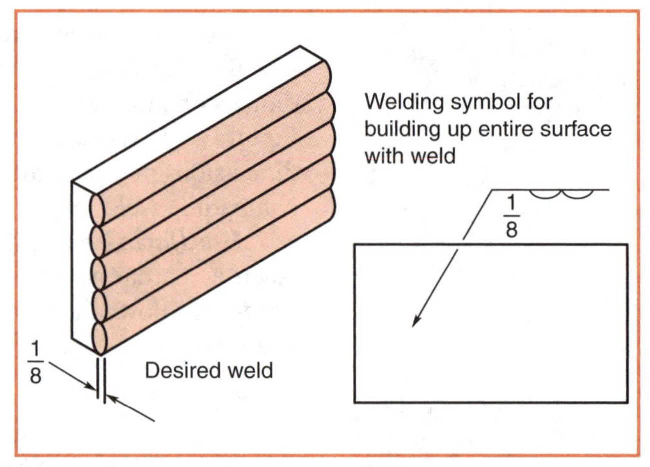

Figure 19-4.
The surfacing symbol means cover the entire surface with weld as shown.

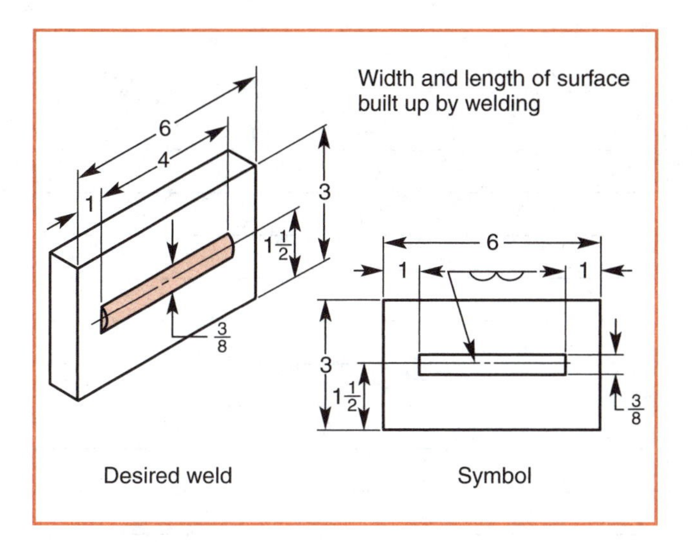

Figure 19-5.
The location, extent, and orientation of a buildup area are given on a drawing.

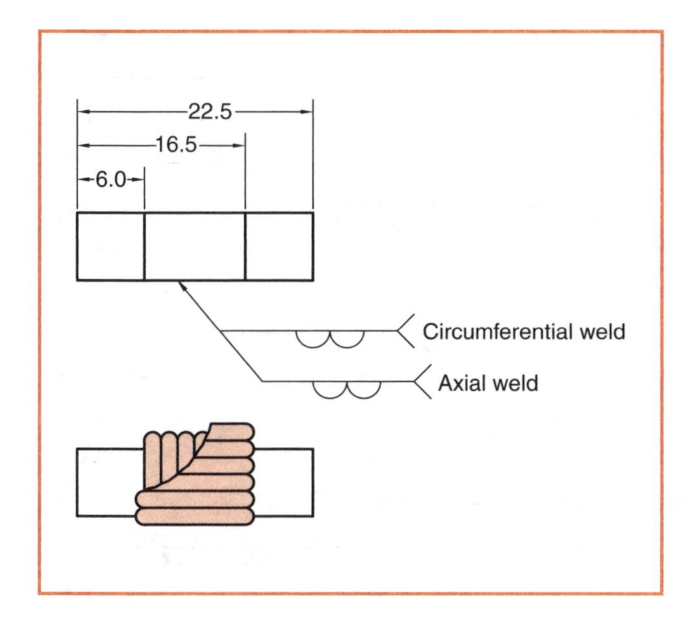

Figure 19-6.
Information regarding the direction of the welds can be found in the tail of the welding symbol.

Unless required, no dimension other than the height of the deposit is shown on the welding symbol when the entire area of a plane or curved surface is to be built up. Refer to Figure 19-4.

When only a portion of the surface is to be built up by welding, the location, extent, and orientation of the weld are shown on the drawing (not on the welding symbol). See Figure 19-7.

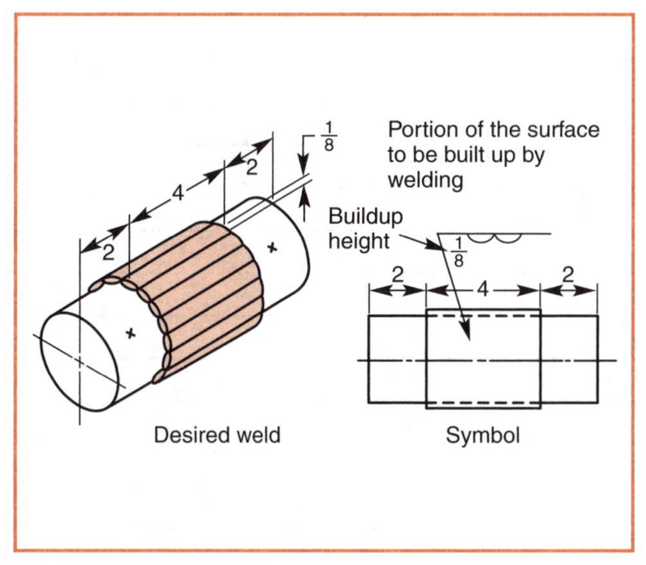

Figure 19-7.
On circular objects, the welding pattern may be linear, as shown, or circular. The weld symbol for either type of pattern is the same. Weld dimensions, other than the buildup height, are shown on the print and not on the welding symbol.

Notes

Name ______________________ Date __________ Class __________

Review Questions

Place a check next to the correct answer.

1. Surfacing welds are frequently used to:
   A. _____ extend the service life of parts subject to abrasive conditions.
   B. _____ build up worn shafts and bearing surfaces to reduce or eliminate the need for expensive new replacements.
   C. _____ add a hardfacing material to parts subject to hard wear.
   D. _____ All of the above.
   E. _____ None of the above.
2. Since the surfacing weld symbol does *not* indicate the welding of a joint, it:
   A. _____ has no arrow or other side significance.
   B. _____ has arrow or other side significance.
   C. _____ has arrow or other side significance only on special welding applications.
   D. _____ All of the above.
   E. _____ None of the above.
3. Draw the welding symbol for a surfacing weld that is 5.0 mm thick.

Complete the following statements:

4. The surfacing weld symbol is shown on the side of the reference line _____.

   ____

   ____

   The arrow points _____.

   ____

   ____
5. The thickness dimension on the surfacing welding symbol indicates _____ of the buildup.

   ____
6. If no buildup thickness dimension is shown on the surfacing welding symbol, _____.

   ____

   ____

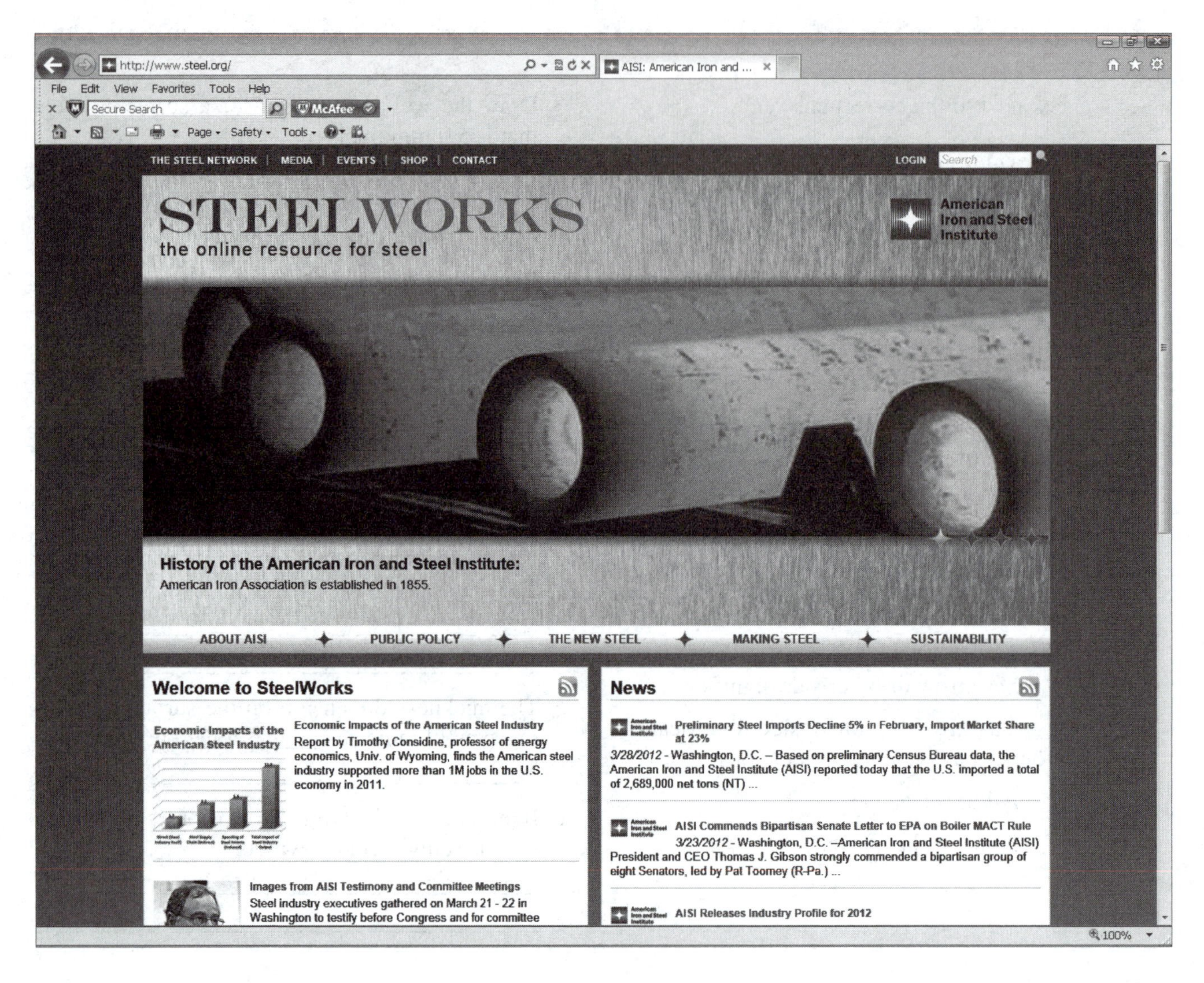

The American Iron and Steel Institute (www.steel.org) was established in 1855 to advance the iron and steel trade. Today, the Institute seeks to build a strong and sustainable steel industry in the United States and North America.

Unit 20

Edge Welds

After completing Unit 20, you will be able to:

- Describe the use of edge weld symbols.
- Identify the use of flange joints to join sheet metal.
- Describe the difference between a flanged butt and flanged corner joint.
- Determine the dimensions of edge welds.

Key Words

edge weld
flanged butt joint
flanged corner joint

Edge welds are used to join the edges of two or more members. The edges to be welded may be flared or flanged. Although edge welds may be used with plate or sheet material, they are primarily used with light gage (thin) metal in combination with a ***flanged butt joint*** or ***flanged corner joint***. A flanged butt joint is used when *both* members are flanged. Refer to A and B of Figure 20-1. A flanged corner joint is used when only *one* of the members is flanged. See C and D of Figure 20-1.

The edge weld symbol is used to specify welds on edge joints, flanged butt joints, and flanged corner joints. The size of an edge weld is found to the left of the edge weld symbol, Figure 20-1. Flange dimensions (length, radius, spacing) are considered part of the detail drawing and are not specified as part of the weld symbol.

When used to specify a flanged butt joint, the other side of the edge weld symbol is often a flare-V groove where both lines that form the symbol curve before they contact the reference line. See A in Figure 20-2. When used to specify a flanged corner joint, the other side of the edge weld symbol is often a flare-bevel groove where only one of the lines for forming the symbol curves before touching the reference line. Refer to B in Figure 20-2.

Multiple Member Edge Welds

Some assemblies require one or more pieces to be inserted between the two outer members. The *same symbol* used for the two member joints is used regardless of the number of pieces inserted. Multiple member joints requiring complete joint penetration have the melt-through symbol on the opposite side of the edge weld symbol. See C in Figure 20-2.

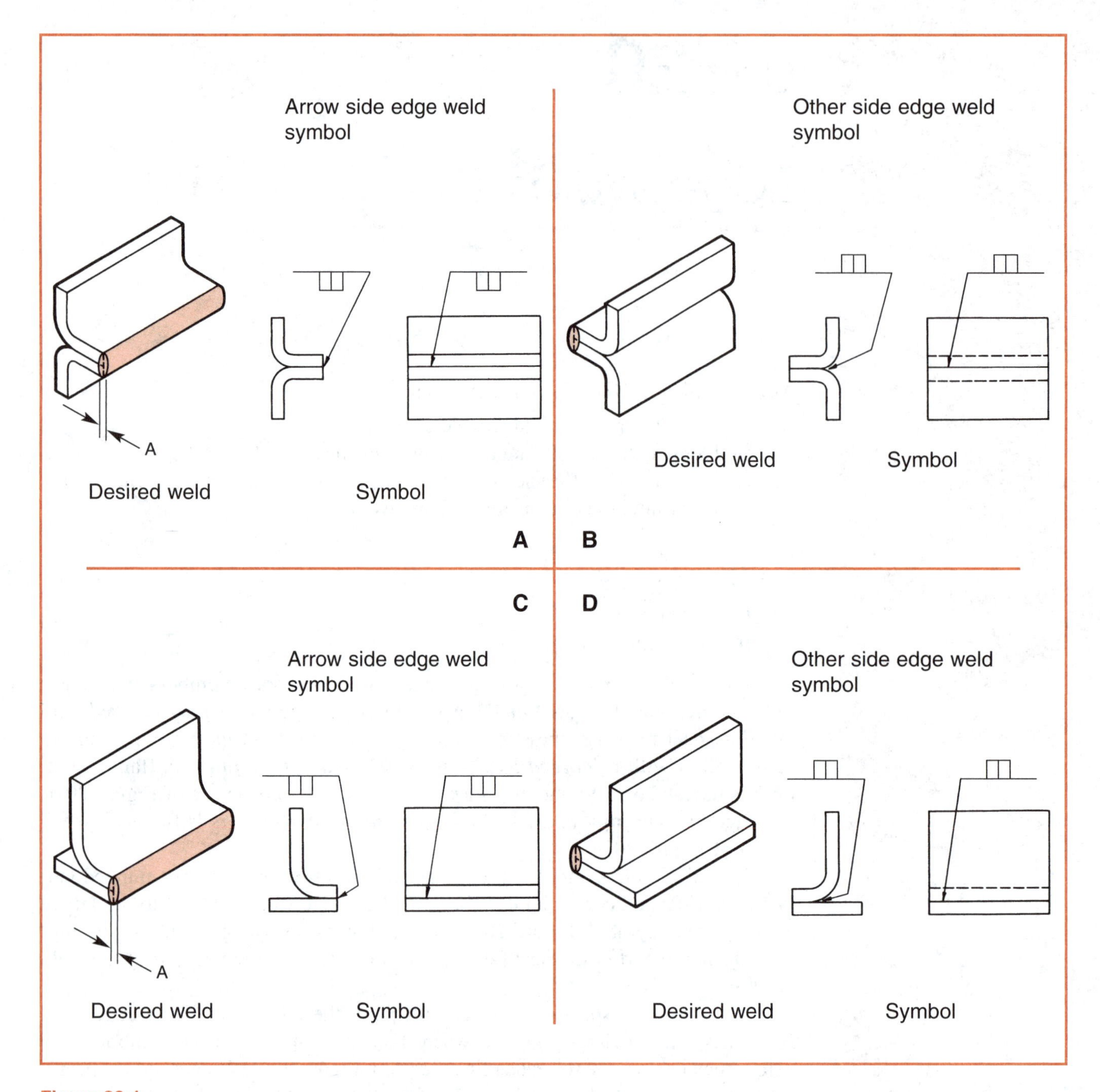

Figure 20-1.
Study how the edge weld symbol is used to show welds. The symbol does not have both sides significance. Note the difference between flanged butt and flanged corner joints.

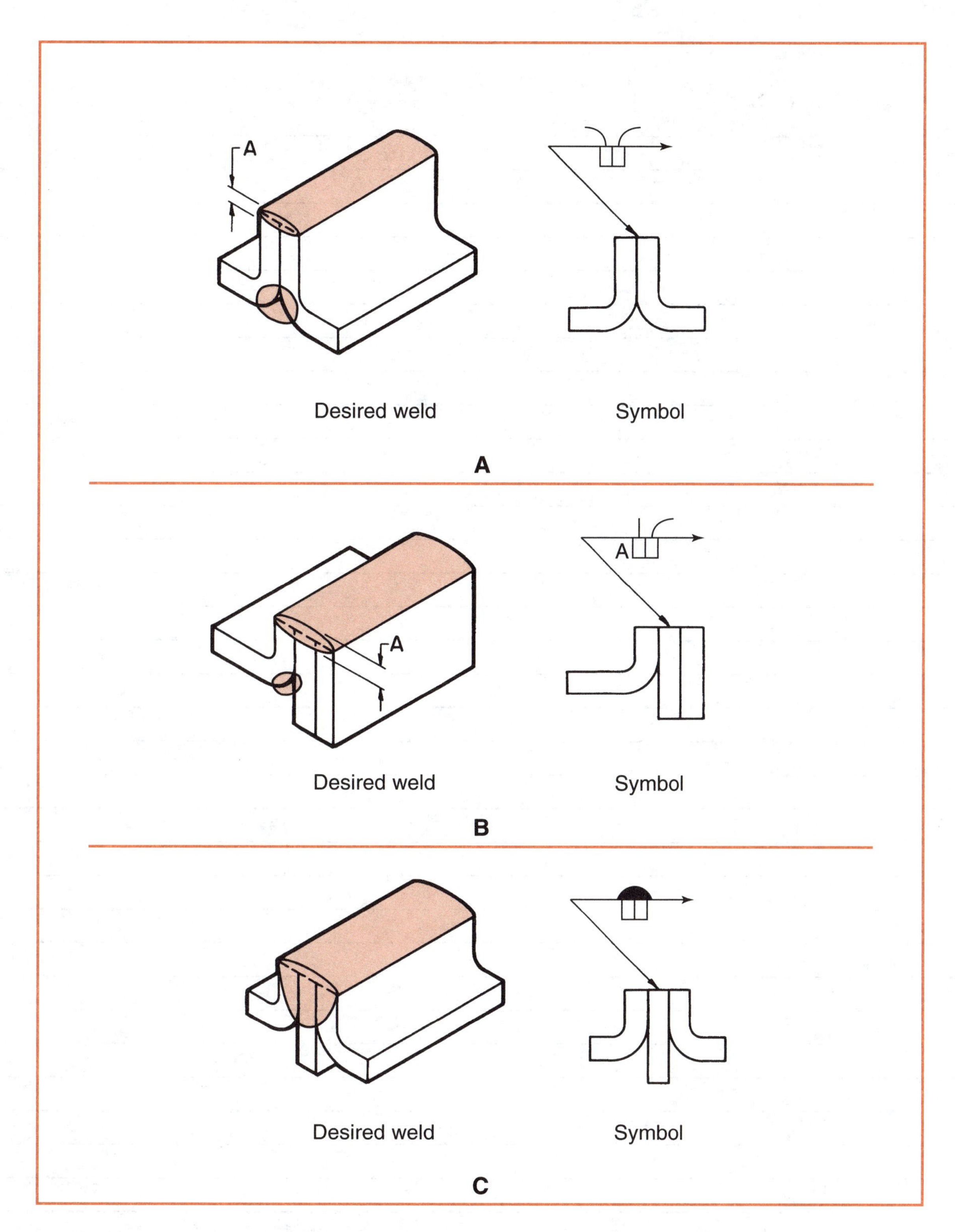

Figure 20-2.
The edge weld symbol in combination with other symbols. A—The flare-V groove symbol opposite the edge weld symbol. B—The flare-bevel groove symbol opposite the edge weld symbol. C—Three member joint requiring complete joint penetration.

Notes

Name ______________________________ Date ______________ Class ______________

Review Questions

1. Edge welds are primarily used with _____.

   ____

2. Edge weld dimensions are shown on the _____ side of the reference line as the _____.

   ____

3. Draw the edge weld symbol for this weld.

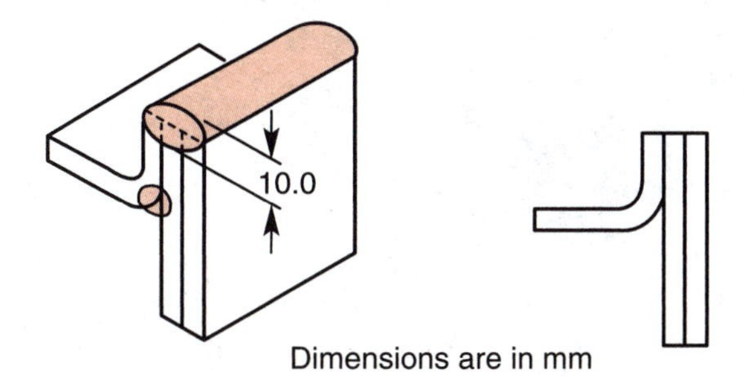

4. Draw the edge weld symbol for a joint with the same dimensions as that in Question 3. However, in this case, there are three pieces of metal between the two outer pieces.

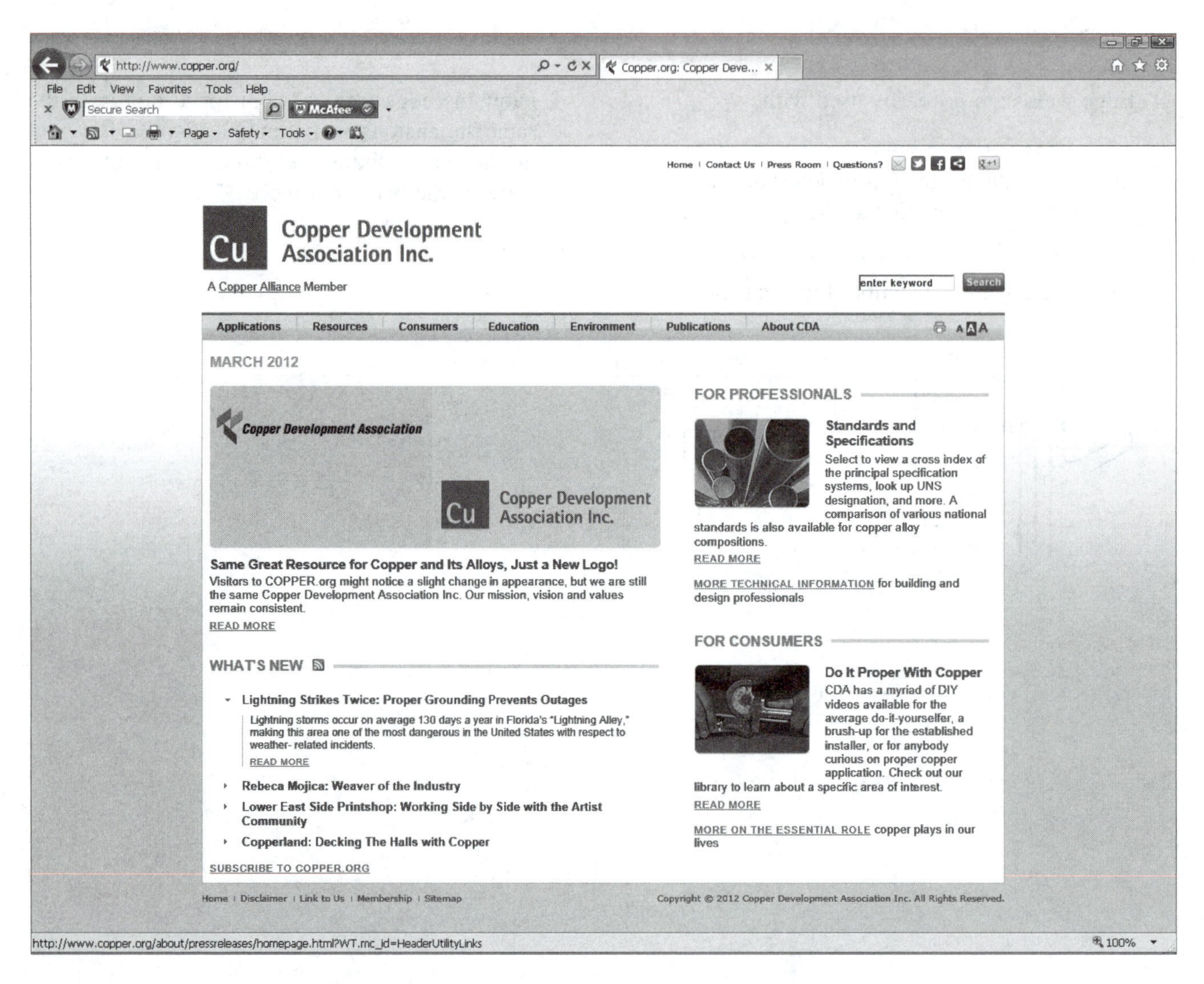

Widely used in water distribution systems, copper piping is both reliable and cost effective. The Copper Development Association (www.copper.org) outlines standards for and information on the use of copper.

Unit 21

Pipe Welding

After completing Unit 21, you will be able to:

- Identify methods of joining pipe or tube and fittings into an assembly.
- Determine pipe size and explain a pipe schedule.
- Determine how tube is measured.
- Interpret conventions of piping drawings.
- Identify conventional symbols used on a piping drawing.

Key Words

inside diameter (ID)
isometric view
oblique view
outside diameter (OD)
pipe schedule
pipe size
piping drawings
piping system

A ***piping system*** is an assembly of pipe or tube (thin wall pipe) sections with the fittings (valves, joints, etc.) necessary to direct and control fluids in either liquid or gaseous form. As shown in Figure 21-1, welding is one of the many methods used to fabricate a piping system.

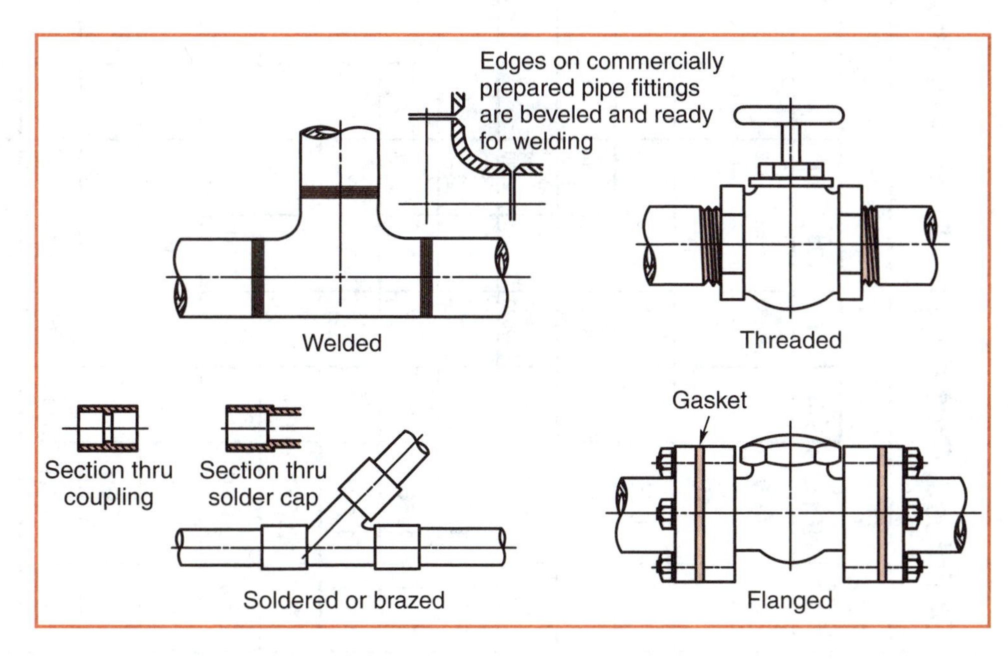

Figure 21-1.
Welding is only one of several ways of fabricating a piping system. Systems can also be joined by soldering, threaded fittings, or by flanged fittings that use gaskets and fastening bolts.

Pipe size is determined by its nominal ***inside diameter (ID)*** measurement, which differs slightly from the actual inside diameter. Pipe is also specified by a ***pipe schedule.*** A pipe schedule number indicates the wall thickness of the pipe as listed in the schedule by nominal diameter. Generally, as the schedule number

Stainless, Carbon, and Alloy Aluminum Nickel

PIPE SIZE	O.D. IN INCHES	PIPE SCHEDULES																	
		5s	5	10s	10	20	+	30	40s & STD	40	(+)	60	80s & E.H.	80	100	120	140	160	DBL E.H.
1/8	.405		.035 .1383	.049 .1863	.049 .1863				.068 .2447	.068 .2447 .0850			.095 .3145	.095 .3145 .1090			WALL THICKNESS IN INCHES		
1/4	.540		.049 .2570	.065 .3297	.065 .3297				.088 .4248	.068 .4248 .1470			.119 .5351	.119 .5351 .1850			STEEL WEIGHT PER FOOT IN POUNDS		
3/8	.675		.049 .3275	.065 .4235	.065 .4235				.091 .5676	.091 .5676 .1960			.126 .7388	.126 .7388 .2560			ALUMINUM WEIGHT PER FOOT IN POUNDS		
1/2	.840	.065 .5383	.065 .5383 .1860	.083 .6710	.083 .6710 .2320				.109 .8510	.109 .8510 .2940			.147 1.088	.147 1.088 .3760				.187 1.304 .4510	.294 1.714
3/4	1.050	.065 .6838	.065 .6838 .2370	.083 .8572	.083 .8572 .2970				.113 1.131	.113 1.131 .3910			.154 1.474	.154 1.474 .5100				.218 1.937 .6700	.308 2.441
1	1.315	.065 .8678	.065 .8678 .3000	.109 1.404	.109 1.404 .4860				.133 1.679	.133 1.679 .5810			.179 2.172	.179 2.172 .7510				.250 2.844 .9840	.358 3.659
1 1/4	1.660	.065 1.107	.065 1.107 .3830	.109 1.806	.109 1.806 .6250				.140 2.273	.140 2.273 .7860			.191 2.997	.191 2.997 1.037				.250 3.765 1.302	.382 5.214
1 1/2	1.900	.065 1.274	.065 1.274 .4410	.109 2.085	.109 2.085 .7210				.145 2.718	.145 2.718 .9400			.200 3.631	.200 3.631 1.256				.281 4.859 1.681	.400 6.408
2	2.375	.065 1.604	.065 1.604 .5550	.109 2.638	.109 2.638 .9130				.154 3.653	.154 3.653 1.264			.218 5.022	.218 5.022 1.737				.344 7.462 2.575	.436 9.029
2 1/2	2.875	.083 2.475	.083 2.475 .8560	.120 3.531	.120 3.531 1.221				.203 5.793	.203 5.793 2.004			.276 7.661	.276 7.661 2.650				.375 10.01 3.464	.552 13.70
3	3.500	.083 3.029		.120 4.332	.120 4.332 1.498				.216 7.576	.216 7.576 2.621			.300 10.25	.300 10.25 3.547				.438 14.32 4.945	.600 18.58
3 1/2	4.000	.083 3.472	.083 3.472 1.201	.120 4.937	.120 4.937 1.720				.226 9.109	.226 9.109 3.151			.318 12.51	.318 12.51 4.326					.636 22.85
4	4.500	.083 3.915	.083 3.915 1.354	.120 5.613	.120 5.613 1.942				.237 10.79	.237 10.79 3.733			.337 14.98	.337 14.98 5.183		.438 19.00 6.560		.531 22.51 7.786	.674 27.54
4 1/2	5.000								.247 12.54				.355 17.61						
5	5.563	.109 6.349	.109 6.349 2.196	.134 7.770	.134 7.770 2.688				.258 14.62	.258 14.62 5.057			.375 20.78	.375 20.78 7.188		.500 27.04 9.353		.625 32.96 11.40	.750 38.55
6	6.625	.109 7.585	.109 7.585 2.624	.134 9.289	.134 9.290 3.213				.280 18.97	.280 18.97 6.564			.432 28.57	.432 28.57 9.884		.562 36.39 12.59		.719 43.35 15.67	.864 53.16
7	7.625								.301 23.54			.500 38.04						.875 63.08	

Figure 21-2.
Studying a pipe schedule helps determine how pipe is specified.

increases (from 10–40, for example) the wall thickness for any nominal inside diameter increases. Look at 2″ diameter schedule 10 pipe in Figure 21-2. The wall thickness is 0.109. Now look at 2″ diameter schedule 40 pipe. Notice the wall thickness is 0.154.

PIPE SIZE	O.D. IN INCHES	PIPE SCHEDULES (continued)																	
		5s	5	10s	10	20	+	30	40s & STD	40	⊕	60	80s & E.H.	80	100	120	140	160	DBL E.H.
8	8.625		1.09 9.914 3.429	.148 13.40	.148 13.40 4.635	.250 22.36 7.735	.175 5.463	.277 24.70 8.543	.322 28.55	.322 28.55 9.878		.406 35.64 12.33	.500 43.39	.500 43.39 15.01	.594 50.95 17.60	.719 60.71 20.97	.812 67.76 23.44	.906 74.79 25.84	.875 72.42
9	9.625								.342 33.91				.500 48.73						
10	10.75		.134 15.19	.165 18.65	.165 18.70 6.453	.250 28.04 9.698	.279 10.79	.307 34.24 11.34	.365 40.48	.365 40.48 14.00		.500 54.74 18.93	.500 54.74	.500 64.43 22.25	.719 77.03 26.61	.844 82.29	1.000 104.1	1.125 115.6	1.000 104.1
11	11.75								.375 45.56				.500 60.08						
12	12.75	.156 21.07	.165 22.18	.180 24.16	.180 24.16 8.359	.250 33.38 11.55	.375 17.14	.330 43.77 15.14	.375 49.56	.406 53.52 18.52	.500 22.63	.562 73.15 25.31	.500 65.42	.688 88.63 30.62	.844 107.3	1.000 125.5	1.125 136.7	1.312 160.3	1.000 125.5
14	14.00	.156 23.07		.188 27.73	.250 36.71	.312 45.61		.375 54.57	.375 54.57	.438 63.44		.594 85.05	.500 72.09	.750 106.1	.938 130.9	1.094 150.8	1.250 170.2	1.406 189.1	
16	16.00	.156 27.90		.188 31.75	.250 42.05	.312 52.27		.375 62.58	.375 62.58	.500 82.77		.656 107.5	.500 82.77	.844 136.6	1.031 164.8	1.219 192.4	1.438 223.6	1.594 245.3	
18	18.00	.156 31.43		.188 35.76	.250 47.39	.312 58.94		.438 82.15	.375 70.59	.562 104.7		.750 138.2	.500 93.45	.938 170.9	1.156 208.0	1.375 244.1	1.562 274.2	1.781 308.5	
20	20.00	.188 39.78		.218 46.05	.250 52.73	.375 78.60		.500 104.1	.375 78.60	.594 123.1		.812 166.4	.500 104.1	1.031 208.9	1.281 256.1	1.500 296.4	1.750 341.1	1.969 379.2	
24	24.00	.218 55.37		.250 63.41	.250 63.41	.375 96.42		.562 140.7	.375 94.62	.688 171.3		.969 238.4	.500 125.5	1.219 296.6	1.531 367.4	1.812 429.4	2.062 438.1	2.344 542.1	
26	26.00				.312 85.60	.500 136.17			.375 102.63				.500 136.17						
28	28.00				.312 92.26	.500 146.85		.625 182.73	.375 110.64										
30	30.00	.250 79.43		.312 98.93	.312 98.93	.500 157.53		.625 196.08	.375 118.65				.500 157.53						
32	32.00				.312 105.59	.500 168.21		.625 209.43	.375 126.66	.688 230.08			.500 168.21						
34	34.00				.312 112.25	.500 178.89		.625 222.78	.375 134.67	.688 244.77									
36	36.00				.312 118.92			.625 236.13	.375 142.68	.750 282.35			.500 189.57						

Factors applicable to other products.
For nickel and alloy produced to these pipe sizes, apply these factors to the red numbers.

Nickel 200	1.1343	Monel 400	1.1272
Bucjek 201	1.1378	Inconel 600	1.0742
Incoloy* 800	1.0247	Incoloy 825	1.0389

For aluminum TUBING produced to the listed pipe sizes–apply these factors for each grade

1100 Wt. as shown	2024 Wt. times 1.02
6061 Wt. as shown	3003 Wt. times 1.01
6063 Wt. as shown	5086 Wt. times .98
2014 Wt. times 1.03	7075 Wt. times 1.03

COLUMNS + AND ⊕ ARE WALL THICKNESS PRODUCED TO PIPE TOLERANCES

*Registered Trade Mark of INCO

Figure 21-2. (continued)

Tube is measured by its ***outside diameter (OD).*** Tubes are available in almost any wall thickness. However, the wall thickness is generally less than the wall thickness of a pipe with the same outside diameter.

NOTE! The welding of critical or high-pressure pipe systems (systems where weld defects could be costly and life threatening) requires the skills of specially trained and qualified welders. Often national or local codes of construction require the welder pass a qualification test before welding on any pipe or fitting.

Piping Drawings

Piping drawings show the assembly of a pipe system. Figure 21-3 gives an example. Pipe drawings are made in either double-line or single-line drawings. Compare the two types in Figure 21-4. Double-line pipe drawings are used to make scale layouts where dimensional accuracy is important. Pipe and fittings are drawn to exact scale. Single-line drawings are diagrammatic.

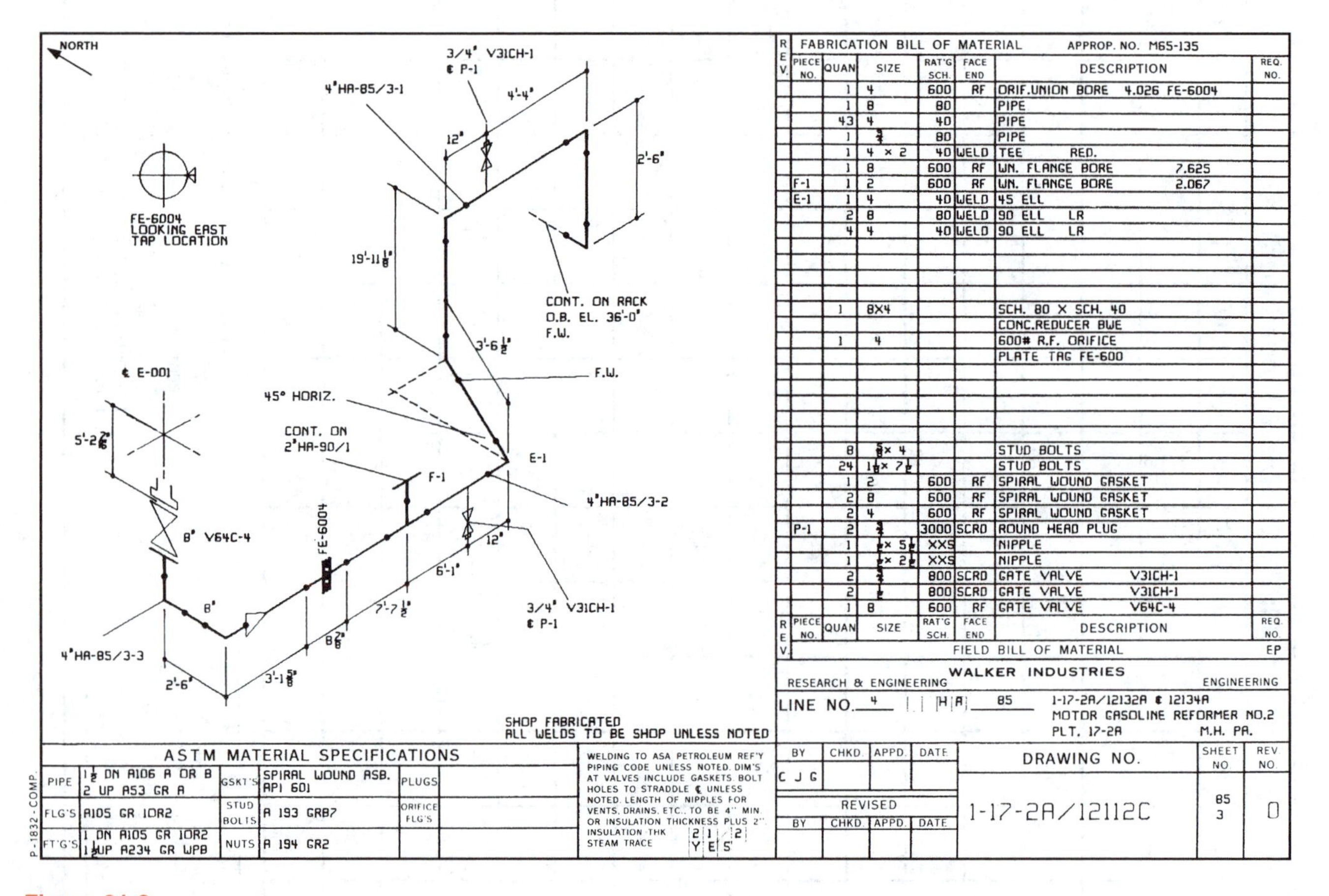

REV.	PIECE NO.	QUAN	SIZE	RAT'G SCH.	FACE END	DESCRIPTION	REQ. NO.
		1	4	600	RF	ORIF.UNION BORE 4.026 FE-6004	
		1	8	80		PIPE	
		43	4	40		PIPE	
		1	3/4	80		PIPE	
		1	4 × 2	40	WELD	TEE RED.	
		1	8	600	RF	WN. FLANGE BORE 7.625	
	F-1	1	2	600	RF	WN. FLANGE BORE 2.067	
	E-1	1	4	40	WELD	45 ELL	
		2	8	80	WELD	90 ELL LR	
		4	4	40	WELD	90 ELL LR	
		1	8X4			SCH. 80 X SCH. 40 CONC.REDUCER BWE	
		1	4			600# R.F. ORIFICE PLATE TAG FE-600	
		8	5/8 × 4			STUD BOLTS	
		24	1 1/8 × 7 1/2			STUD BOLTS	
		1	2	600	RF	SPIRAL WOUND GASKET	
		2	8	600	RF	SPIRAL WOUND GASKET	
		2	4	600	RF	SPIRAL WOUND GASKET	
	P-1	2	3/4	3000	SCRD	ROUND HEAD PLUG	
		1	3/4 × 5 1/2	XXS		NIPPLE	
		1	3/4 × 2 1/2	XXS		NIPPLE	
		2	3/4	800	SCRD	GATE VALVE V31CH-1	
		2	3/4	800	SCRD	GATE VALVE V31CH-1	
		1	8	600	RF	GATE VALVE V64C-4	

PIPE	1 1/2 DN A106 A OR B 2 UP A53 GR A	GSKT'S	SPIRAL WOUND ASB. API 601	PLUGS	
FLG'S	A105 GR 1OR2	STUD BOLTS	A 193 GRB7	ORIFICE FLG'S	
FT'G'S	1 DN A105 GR 1OR2 1 1/2 UP A234 GR WPB	NUTS	A 194 GR2		

Figure 21-3.
A piping drawing shows how to construct a pipe system. A computer-aided design (CAD) process prepared this pipe drawing. As you will learn, special symbols are used to denote components.

Most pipe layouts are single-line drawings in isometric or oblique views because orthographic projection views (three-view drawings) of piping are difficult to read. An ***isometric view*** shows an object as it is with lines showing width, length, and depth drawn full size (or in proportional scale). An ***oblique view*** shows the front view in true shape and size and the other views similar to isometric drawings.

Fittings are normally shown by standard, conventional symbols, Figure 21-5. However, some firms have developed their own series of pipe symbols that may vary slightly from the American National Standards Institute (ANSI) symbols.

Dimensions on piping drawings give the location of components. They are made to the center lines of the pipes and fittings.

A parts list or fabrication schedule may be found on the drawing or on an attached sheet. One example is shown in Figure 21-6.

Equipment installations that utilize piping are sometimes shown as a pictorial drawing (such as isometric or oblique). See Figure 21-7.

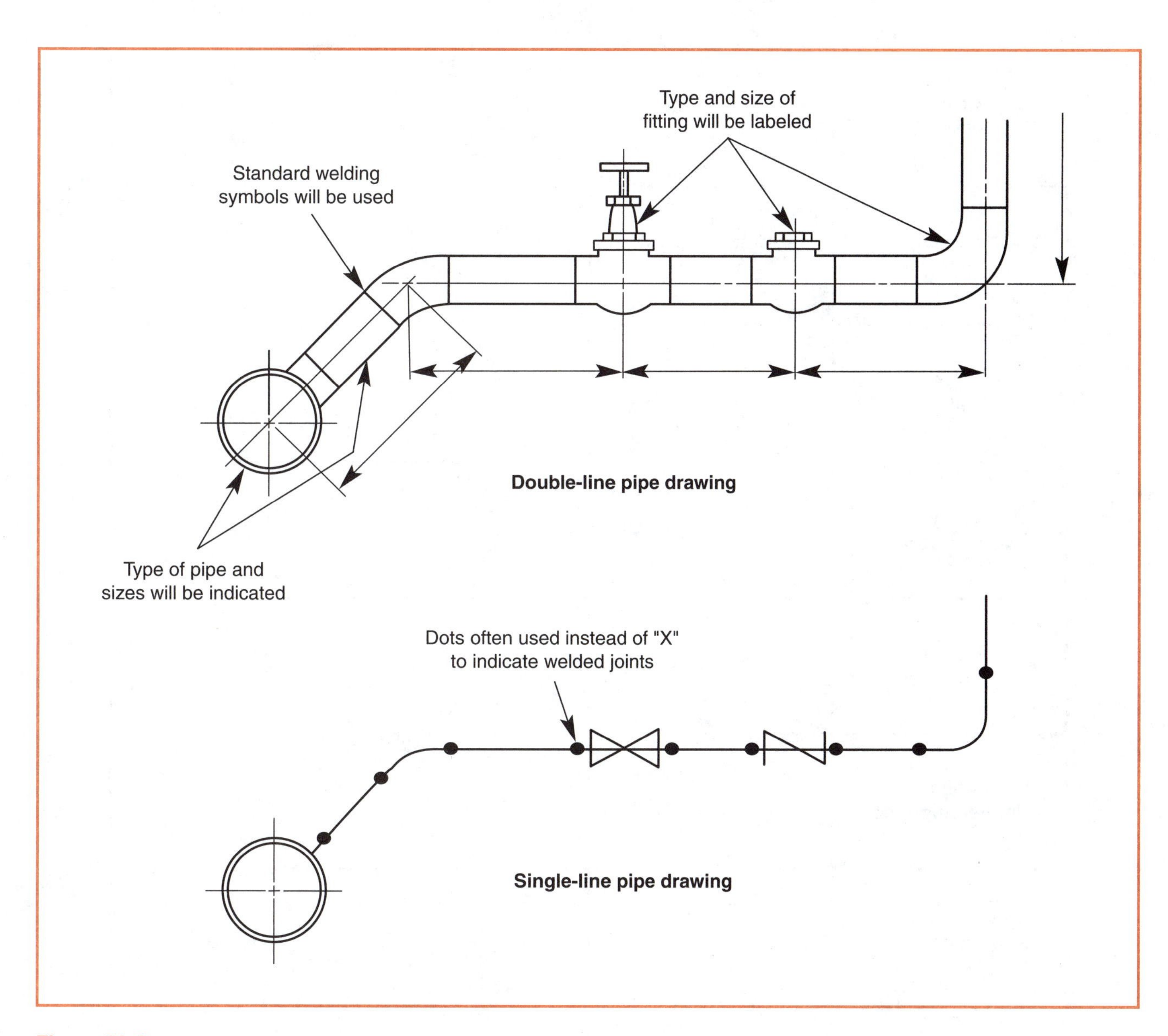

Figure 21-4.
Piping drawings may be presented as either double-line or single-line prints. Double-line drawings are used to make scale layouts where dimensional accuracy is important. Single-line drawings are diagrammatic in nature.

Fitting	Flanged	Screwed	Welded*
Bushing	None	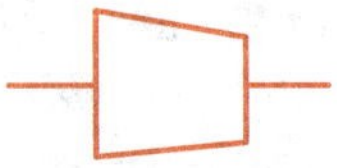	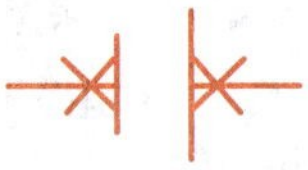
Cap	None		None
Cross, straight	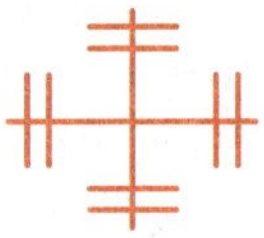	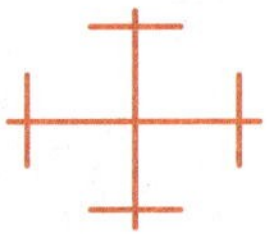	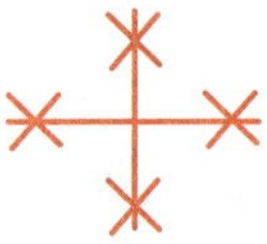
Elbow, 45°	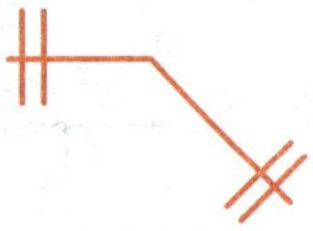	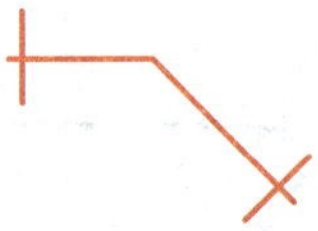	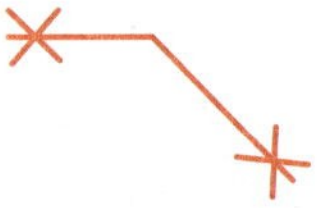
Elbow, 90°	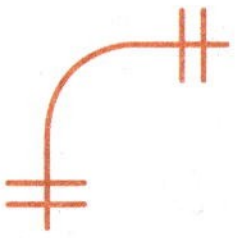	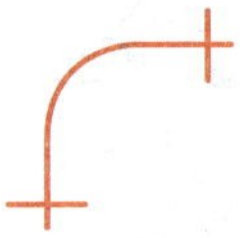	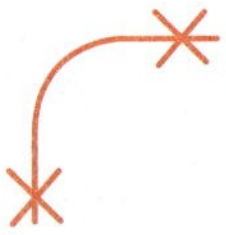
Elbow, turned down			
Elbow, turned up	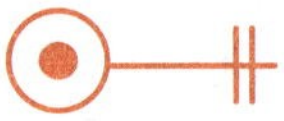	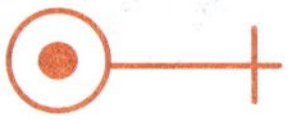	
Joint, connecting pipe			
Lateral	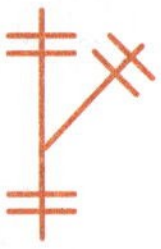		None
Pipe plug	None	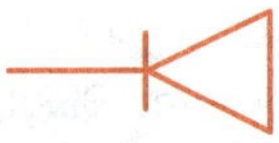	None

Figure 21-5.
Note a few of the many pipe symbols. While many companies use ANSI symbols, some large companies have devised their own symbols.

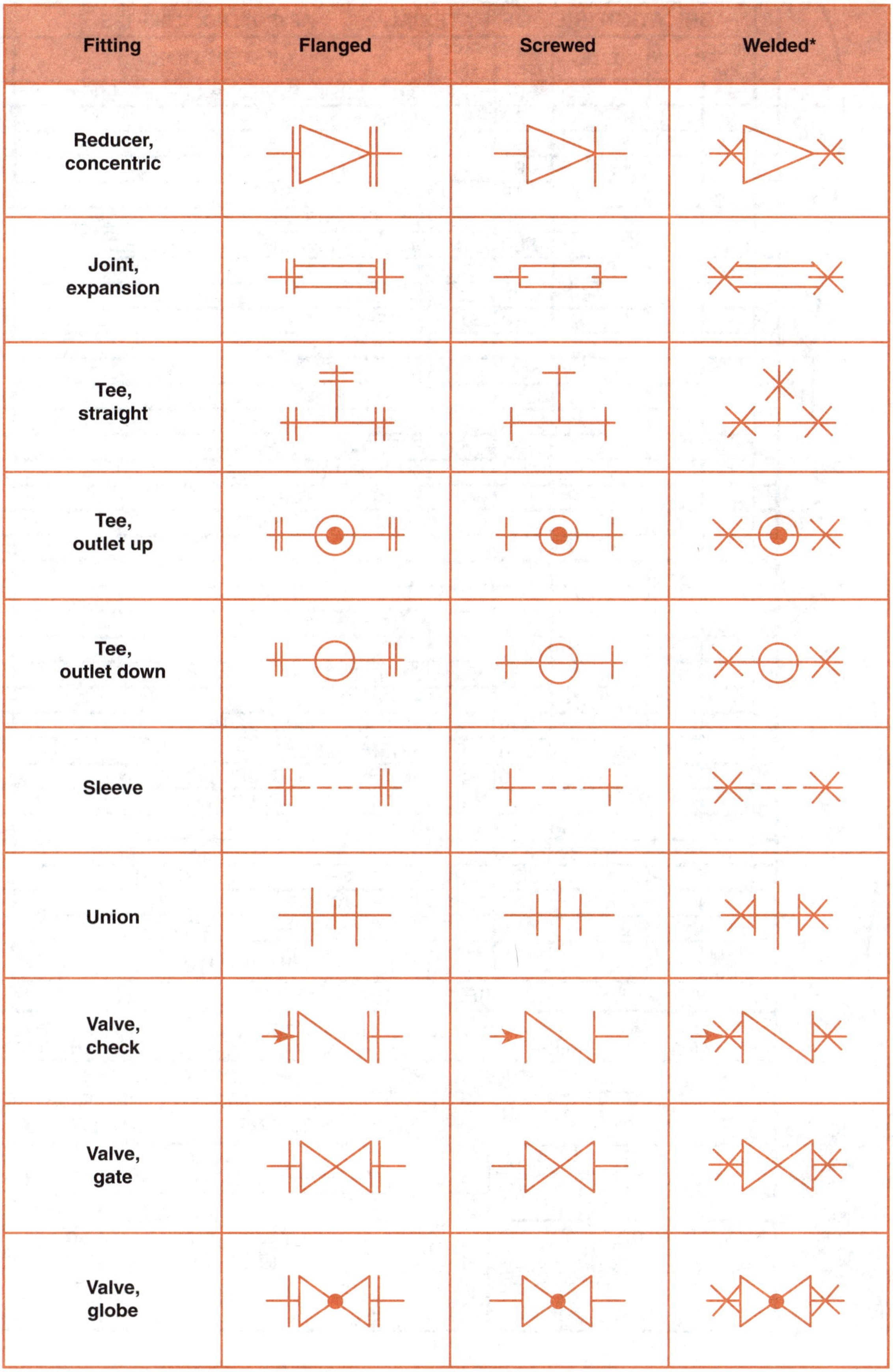

Fitting	Flanged	Screwed	Welded*
Reducer, concentric			
Joint, expansion			
Tee, straight			
Tee, outlet up			
Tee, outlet down			
Sleeve			
Union			
Valve, check			
Valve, gate			
Valve, globe			

*A ● MAY BE USED INSTEAD OF THE "×" TO REPRESENT A BUTT WELDED JOINT

Figure 21-5. (continued)

REV.	PIECE NO.	QUAN	SIZE	RAT'G SCH.	FACE END	DESCRIPTION	REQ. NO.
		FABRICATION BILL OF MATERIAL				APPROP. NO. M65-135	
		1	4	600	RF	ORIF.UNION BORE 4.026 FE-6004	
		1	8	80		PIPE	
		43	4	40		PIPE	
		1	3/4	80		PIPE	
		1	4 × 2	40	WELD	TEE RED.	
		1	8	600	RF	WN. FLANGE BORE 7.625	
	F-1	1	2	600	RF	WN. FLANGE BORE 2.067	
	E-1	1	4	40	WELD	45 ELL	
		2	8	80	WELD	90 ELL LR	
		4	4	40	WELD	90 ELL LR	
		1	8X4			SCH. 80 X SCH. 40	
						CONC.REDUCER BWE	
		1	4			600# R.F. ORIFICE	
						PLATE TAG FE-600	
		8	5/8 × 4			STUD BOLTS	
		24	1 3/8 × 7 1/2			STUD BOLTS	
		1	2	600	RF	SPIRAL WOUND GASKET	
		2	8	600	RF	SPIRAL WOUND GASKET	
		2	4	600	RF	SPIRAL WOUND GASKET	
	P-1	2	3/4	3000	SCRD	ROUND HEAD PLUG	
		1	1/2 × 5 1/2	XXS		NIPPLE	
		1	1/2 × 2 1/2	XXS		NIPPLE	
		2	3/4	800	SCRD	GATE VALVE V31CH-1	
		2	1/2	800	SCRD	GATE VALVE V31CH-1	
		1	8	600	RF	GATE VALVE V64C-4	
REV.	PIECE NO.	QUAN	SIZE	RAT'G SCH.	FACE END	DESCRIPTION	REQ. NO.
						FIELD BILL OF MATERIAL	EP

WALKER INDUSTRIES

RESEARCH & ENGINEERING — ENGINEERING

LINE NO. 4 H A 85 — 1-17-2A/12132A ℄ 12134A

MOTOR GASOLINE REFORMER NO.2

PLT. 17-2A — M.H. PA.

ESS NOTED

ROLEUM REF'Y
NOTED. DIM'S
GASKETS. BOLT
℄ UNLESS
NIPPLES FOR
. TO BE 4" MIN.
ICKNESS PLUS 2".

2 1 / 2
Y E S

BY	CHKD.	APPD.	DATE	DRAWING NO.	SHEET NO.	REV. NO.
C J G				1-17-2A/12112C	85 3	0
REVISED						
BY	CHKD.	APPD.	DATE			

Figure 21-6.
A parts list or fabrication schedule is commonly included on the pipe drawing. However, if it is too lengthy to fit on the drawing, it is presented on a second sheet. Reference to this sheet will be made on the pipe drawing.

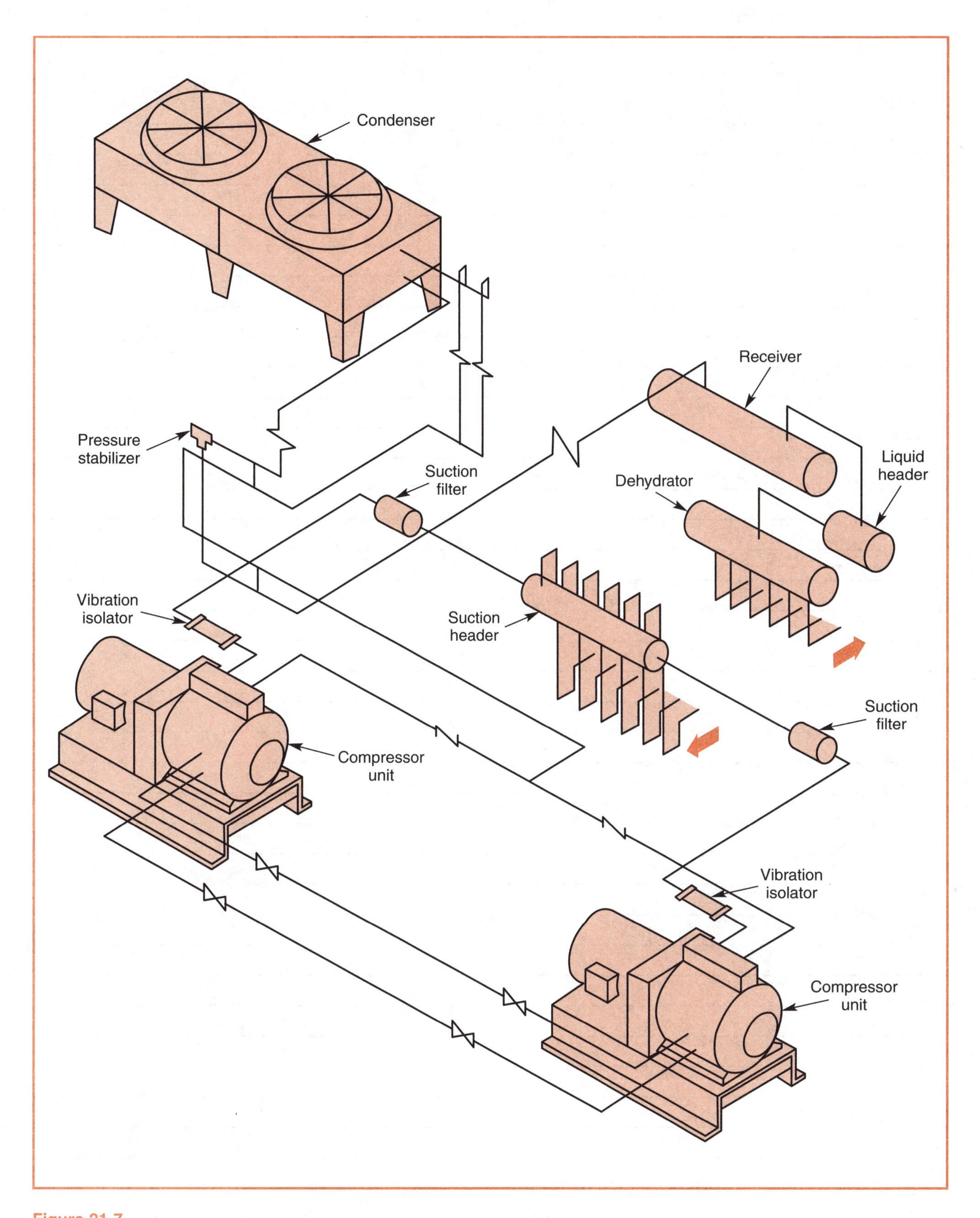

Figure 21-7.
This is a pictorial of a commercial refrigerating system with roof-mounted, air-cooled condensers. There are six refrigerant lines. Equipment installations using piping or tubing may be shown in this manner.

Notes

Name ______________________ Date ______________ Class ______________

Review Questions

Part I

Answer the following questions.

1. List at least four methods employed to fabricate pipe systems. ______________________

2. What is a piping system? ______________________

3. Pipe size is determined by _____.

4. Tube size is measured by its _____.

5. What is a pipe drawing? ______________________

6. Pipe drawings may be made as either _____ or _____ drawings.

7. Dimensions on piping drawings give the _____ of components. They are made to the _____.

8. Fittings are shown on pipe drawings by standardized symbols. Sketch the symbols for the following welded fittings. The "x" or "•" may be used to denote the weld.

A. 45° elbow

B. 90° elbow

C. Reducer

D. Expansion joint

E. Straight tee

F. Union

G. Check valve

H. Gate valve

Part II

Identify the pipe fittings on the drawing.

a. ______________________ e. ______________________

b. ______________________ f. ______________________

c. ______________________ g. ______________________

d. ______________________ h. ______________________

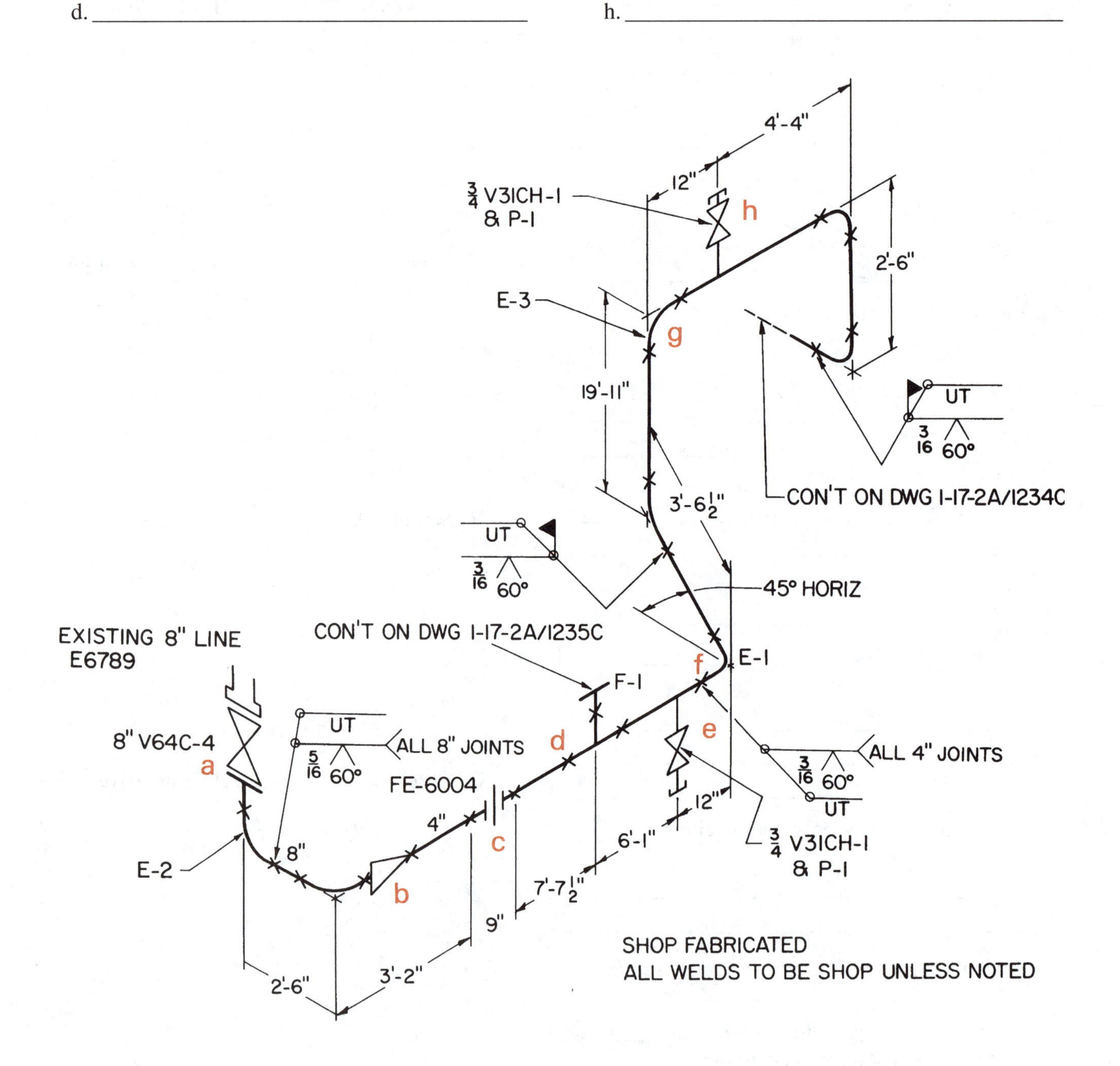

Name ____

Part III

Refer to the drawing from Part II and answer the following questions.

1. Where is the pipe system to be fabricated?

____

____

____

____

____

____

2. Welds are to be made in the _____.

____

____

____

____

____

____

3. List weld specifications for all 8″ pipe. How is the weld to be inspected?

   A. ____

   ____

   ____

   B. ____

   ____

   ____

4. Prepare a sketch of the weld specifications for all 4″ diameter pipe. The pipe has a wall thickness of ¼″.

5. Is this drawing of a complete pipe system? If not, how did you arrive at your conclusion?

____

____

____

____

____

____

6. Are all welds to be made in the shop? If not, how did you arrive at your conclusion?

____

____

____

____

____

____

Thermadyne Industries, Inc. (www.thermadyne.com) is a manufacturer of welding and cutting products, including soldering and brazing equipment.

Unit 22

Brazed Joints

After completing Unit 22, you will be able to:

- Describe the difference between brazing and braze welding.
- Identify joint designs used for brazing and braze welding.

Key Words

brazing
braze welding
capillary action
coefficient of thermal expansion
faying

Brazing is the general term given to the process in which metal pieces are joined by heating them to a suitable temperature but below their melting point. Nonferrous filler metal, having a melting point above 840°F (450°C) and below that of the base metal, is added to the weld joint. In brazing, the strength of the joint depends upon the strength of the filler metal. The filler metal (usually an alloy of copper, tin, and zinc) is distributed between the closely fitted surfaces of the joint by capillary action. ***Capillary action*** refers to the force by which the liquid filler metal is drawn between ***faying*** (tightly fitted) surfaces.

The terms brazing and ***braze welding*** are sometimes confused. As described in Unit 10, brazing is generally an automated process while braze welding is a manual process. But, in most shops, the term brazing is used to describe both the automated and the manual process.

At times, the welder encounters jobs that specify brazed joints, Figure 22-1. If such joints require no joint preparation, other than cleaning the members, only the arrow with the brazing process indicated in the tail is shown on the print.

For other brazing operations, the application of conventional weld symbols (with minor variations) to the brazed joint is used, Figure 22-2.

NOTE! Brazing can join all ferrous and nonferrous metals, including aluminum and magnesium. When dissimilar metals are brazed, and one part fits within the other, it is advisable to check the coefficient of thermal expansion, especially in the brazing temperature range. This assures proper clearance can be provided for the filler metal to enter the joint.

The ***coefficient of thermal expansion*** is generally defined as the change in length of a material as it is heated. For example, steel expands about .0000063 inches per inch per degree Fahrenheit of temperature increase through a fixed range. This is not much change, but enough to close the gap between parts if the expansion is not considered before brazing.

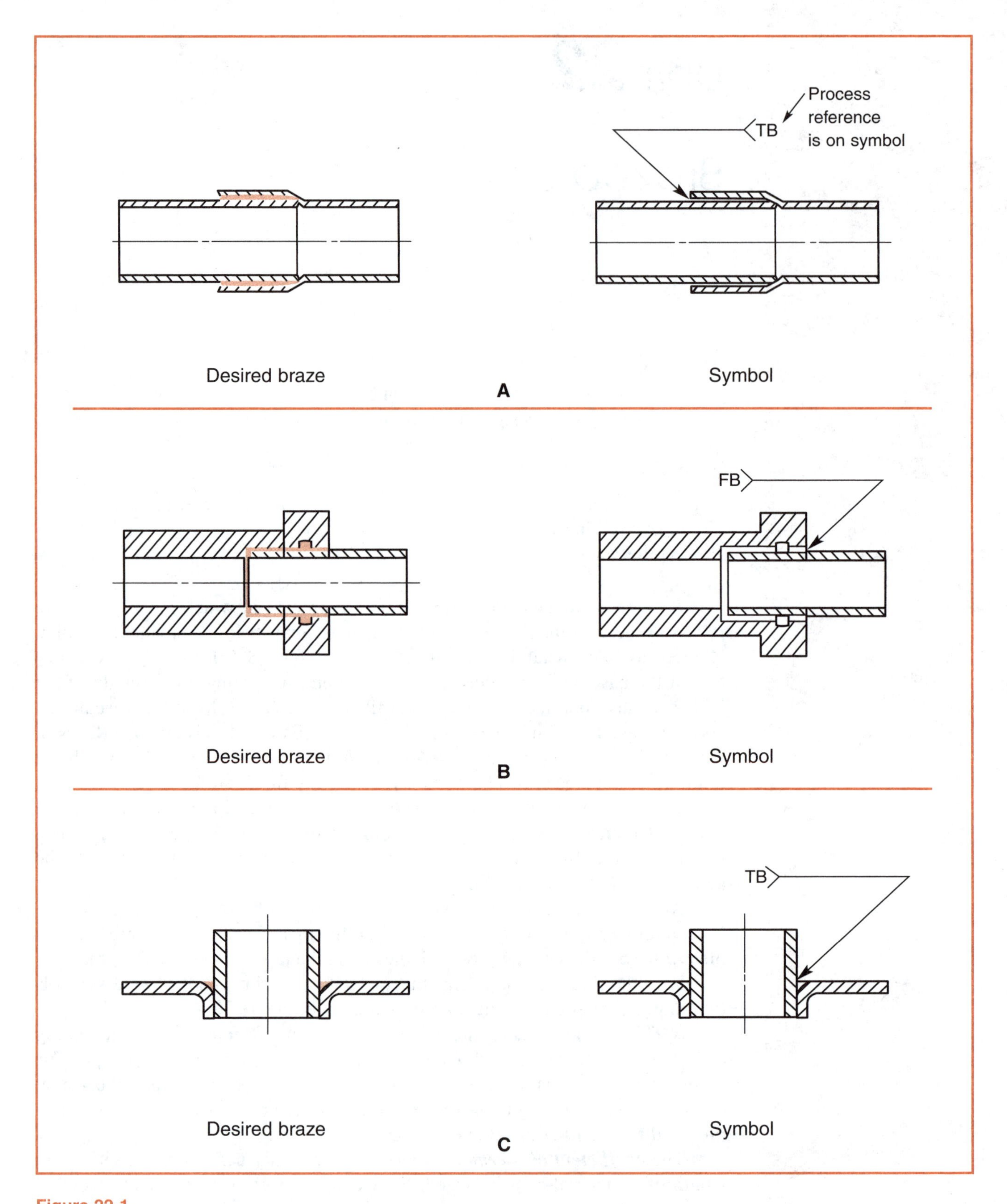

Figure 22-1.
Application of the brazing symbol is shown. When joints do not need preparation other than cleaning, only the arrow with the brazing process in the tail is on the print. A—Torch braze. B—Furnace braze. C—Torch braze.

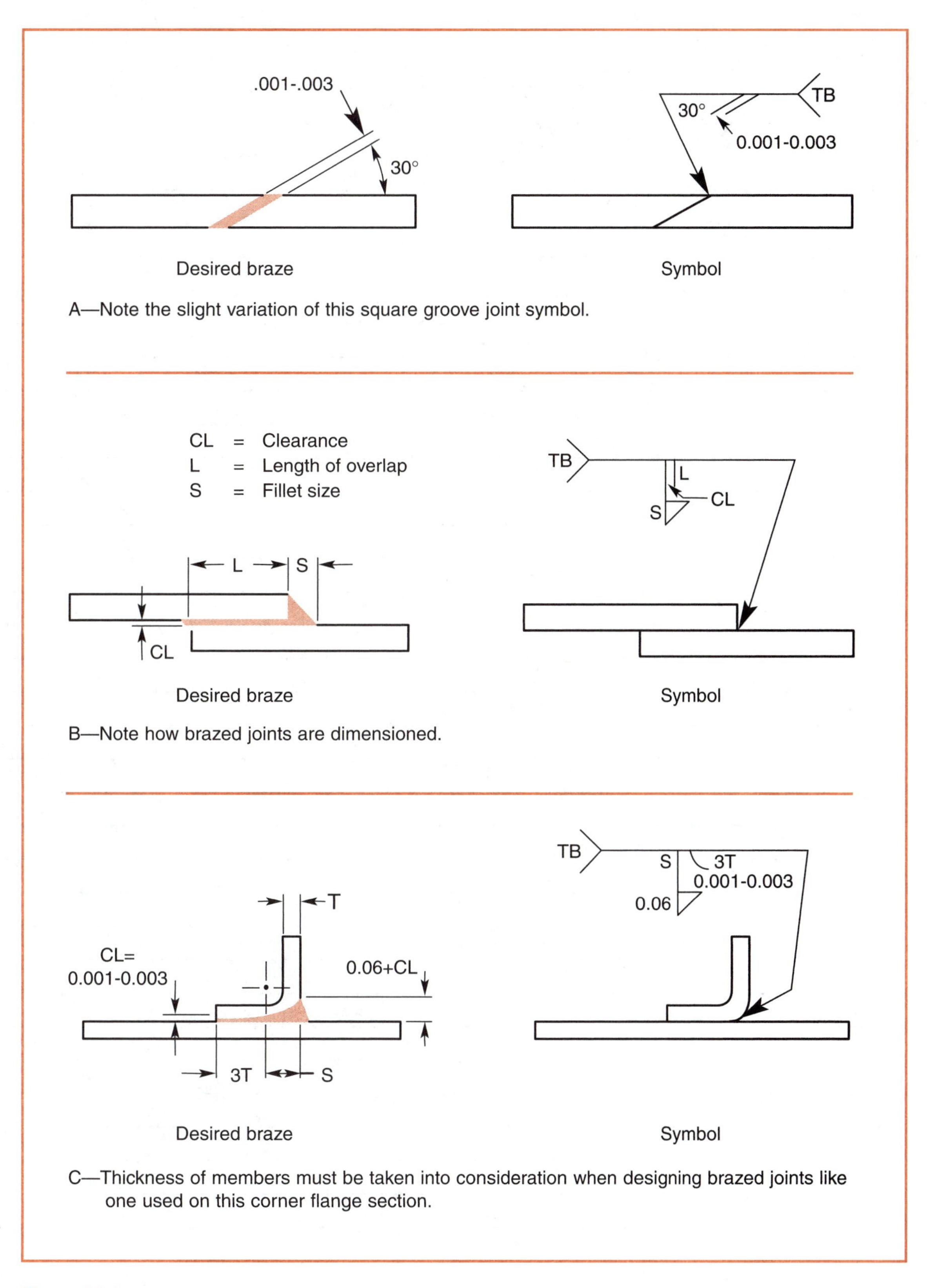

Figure 22-2.
Conventional weld symbols are used to indicate the type of brazed joint.

Notes

Print Reading Activities

When answering the following questions, refer to the provided print (B-1787-3).

1. What is the name of the part?____
____
____

2. How many components make up the assembly? _____

   Name them. _________________________________
____
____

3. What brazing process is specified?_______________
____
____

4. How many places are brazed? _____

5. What is the *maximum* acceptable length of the unit? _____

6. What is the *minimum* acceptable length of the unit? _____

7. What is the next assembly? _____

8. Must the flanges be square with the center line of the tube? _____ If so, what indicates this condition?
____
____
____

9. How far is the tube inset from the face of the flange? _____

10. Briefly explain the difference between braze welding and brazing.
____
____
____
____
____
____

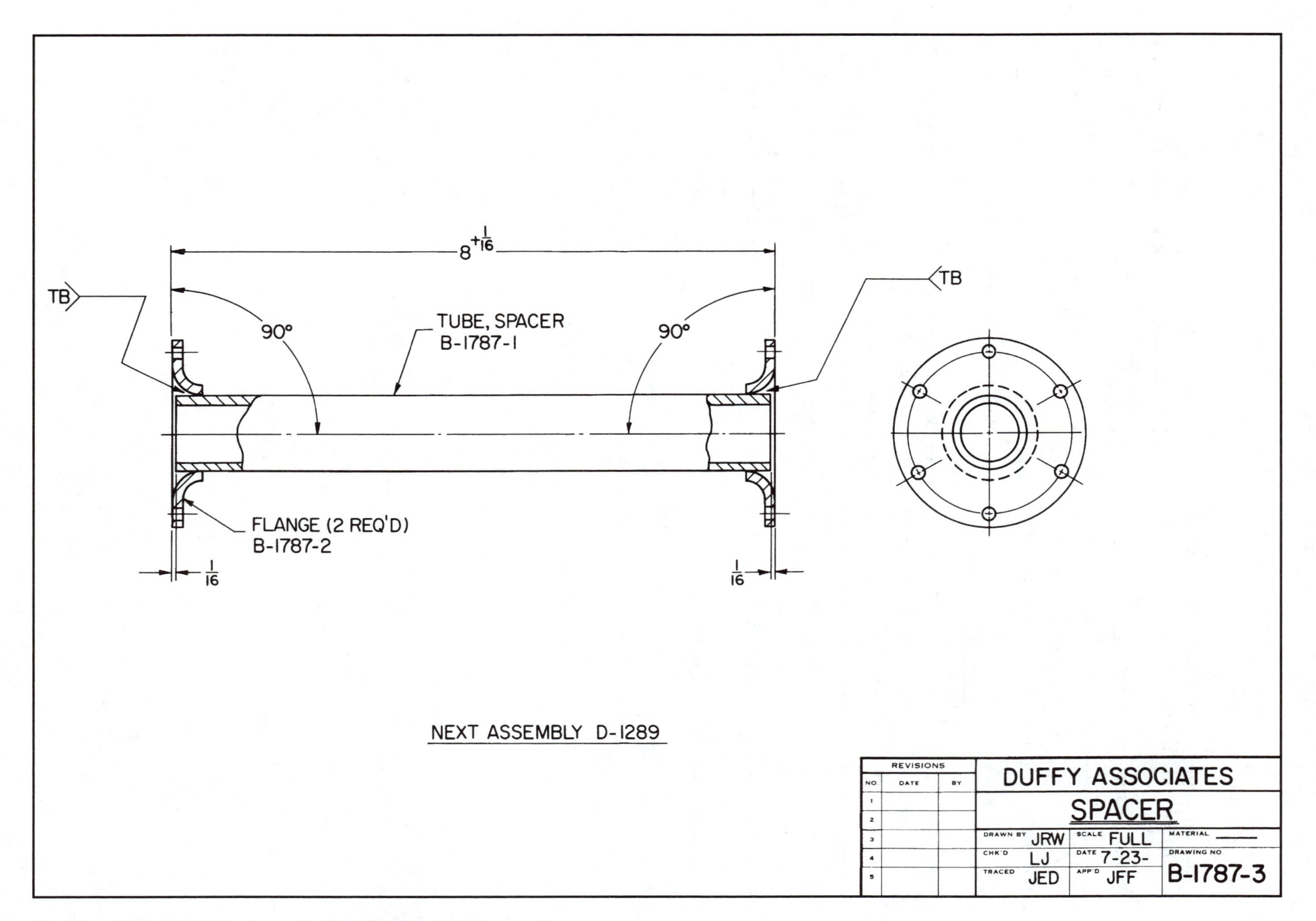

Use this print (B-1787-3) to answer the Print Reading Activities questions.

Industrial Machinery Digest (www.industrialmachinerydigest.com) is a monthly publication that features an extensive selection of industrial machinery.

Section 5 Manufacturing and Testing for Industry

Unit 23 Basic Metalworking Processes

After you complete Unit 23, you will be able to:

- Describe the major metalworking processes.
- Identify basic metalworking process machines and equipment.
- Explain the relationship between metalworking processes and welding processes.

Key Words

blanking
boring
casting
cold forming
cope
cores
counterboring
countersinking
drag
drawing (forming)
extrusion process
flask
forging
machine tool
metalworking processes
mold
reaming
refractory mold
spotfacing

Welding is but one of the techniques employed in the manufacture of the products we see and use, Figure 23-1. Most welded products are fabricated from parts produced by one or more of the basic ***metalworking processes*** (casting, forging, extruding, machining, forming, etc.).

A good welder must have an understanding of these processes. Then, he or she will be fully prepared to do complex weldments.

Castings

Casting is a metalworking process of making objects by pouring molten (melted) metal into a mold. Almost any metal able to be heated to the molten state can be cast, Figure 23-2.

The ***mold*** is a cavity made in a material suitable for holding the molten metal until it cools and solidifies. The mold cavity is almost the same shape and size of the object to be cast, Figure 23-3.

NOTE! The mold cavity must be made slightly larger than the object being cast because molten metal shrinks or contracts as it cools and becomes solid. The type of metal being cast determines how much larger the cavity must be made because each metal shrinks a different amount per foot (meter) of casting length.

Welding is a form of casting on a smaller scale. If you think of a butt joint on which a square groove weld is placed, the edges of the joint are the mold, and the weld pool is the molten metal pouring into the mold.

Figure 23-1.
Many products require welding when they are being made or repaired. (Santa Fe Railway)

Sprue (molten metal poured here)
Vent (allows hot gases to escape)
Casting
Cope
Gate
Drag
Gate (permits molten metal being poured into sprue to enter mold cavity)
Mold cavity (made with pattern)

Figure 23-3.
Study this example of a simple sand mold. The mold cavity is about the same shape and size as the object to be cast.

Permanent Mold Casting

Permanent mold castings are made in metal molds that are *not* destroyed when the casting is removed, Figure 23-4.

Figure 23-2.
The first step in making a casting is converting raw materials into molten metal. Shown is molten metal pouring from an induction furnace. It will be transported to the holding furnace shown and poured into molds. (American Foundrymen's Society, Inc.)

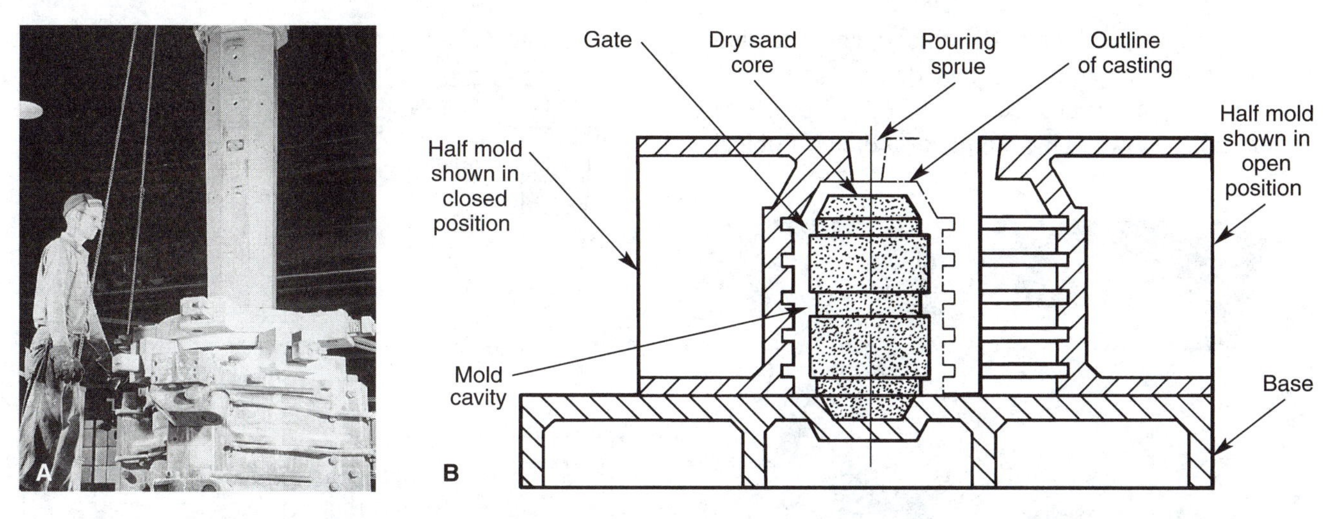

Figure 23-4. A—This worker is removing the initial core from a large permanent mold casting produced from an iron mold. Permanent molds and cores are reusable thousands of times. (Aluminum Company of America) B—Permanent molds do not have to be destroyed to remove the casting.

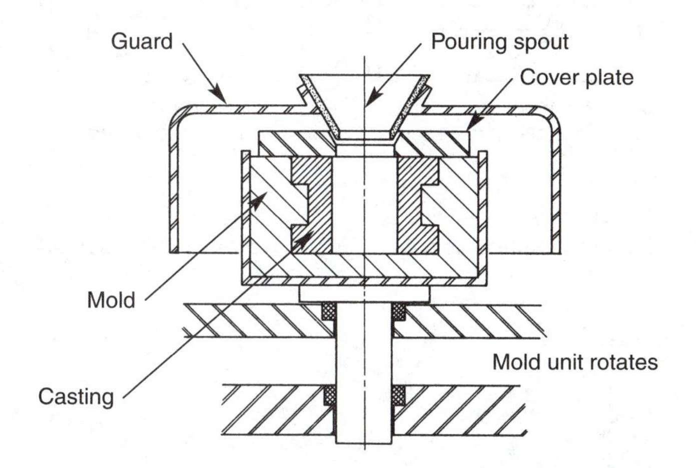

Figure 23-5.
Centrifugal force from spinning action holds molten metal against the wall of the mold until it solidifies.

Centrifugal Casting

Centrifugal casting is a form of permanent mold casting. Molten metal is poured into a spinning circular mold. Centrifugal force holds the molten metal against the wall of the mold until it solidifies, Figure 23-5.

Cylindrical objects, such as large pipes and brake drums, can be cast by this technique. Wall thickness of the object being cast is determined by the amount of molten metal poured into the mold.

Die Casting

Die casting is a variation of permanent mold casting. Molten metal is forced into a metal mold under pressure, as in Figure 23-6. After solidification, ejector pins force the completed casting from the mold or die.

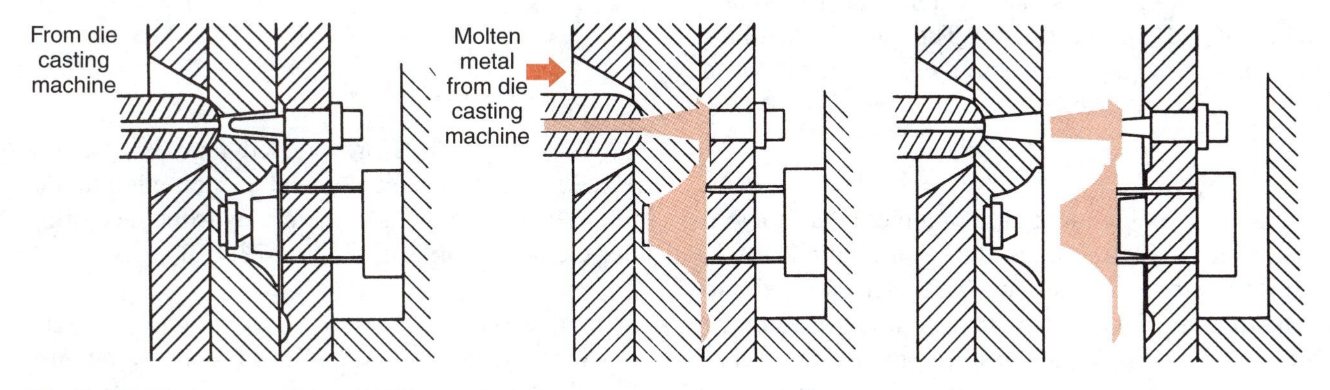

Figure 23-6.
Left: Here are two sections of a die, closed and locked to receive a "shot" of molten metal to form the casting. Center: The cavity of the mold is now completely filled. Note the metal in the overflow well at the bottom of the cavity, which provides an outlet for air entrapped in the die cavity. Right: The die is opened to permit ejection of the casting. Note the two pins that free the casting from the die.

Castings produced by this technique have smoother surface finishes, finer details, and greater accuracy than objects cast by other processes. Figure 23-7 shows one example of a die cast object.

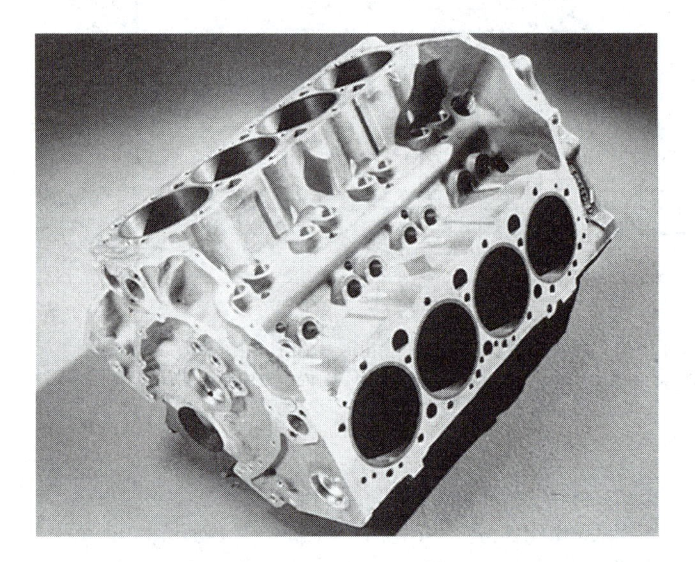

Figure 23-7.
This is a die cast aluminum engine block. The block weighs only 68 lb., including 14 lb. of cast iron cylinder liners.

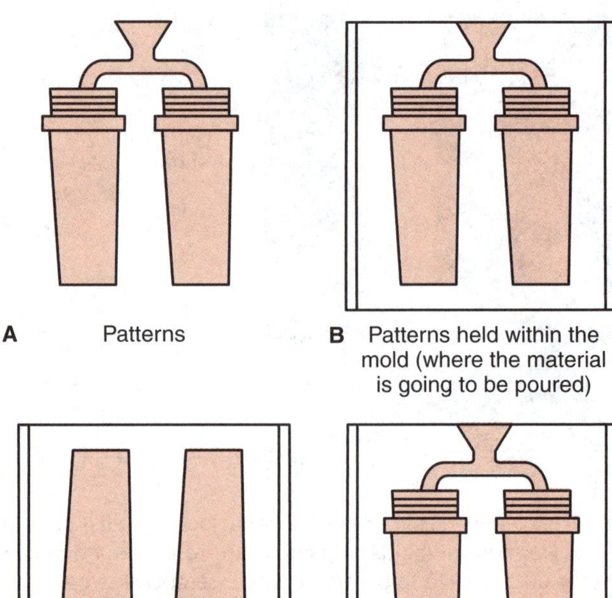

Figure 23-8.
The investment or lost wax casting process is represented.

Investment or Lost Wax Casting

Investment or lost wax casting is a foundry process used when a very accurate casting of a complex shape or intricate design must be produced. Figure 23-8 illustrates this process.

Patterns of wax or plastic are invested (placed) in a refractory mold. A ***refractory mold*** withstands high temperatures. When the mold has hardened, it is placed in an oven and heated until the wax or plastic pattern is burned out (lost). A cavity the shape of the pattern is left in the mold. Molten metal is forced into the mold and allowed to cool. The mold is then broken apart to remove the casting. Figure 23-9 shows a part made by this process.

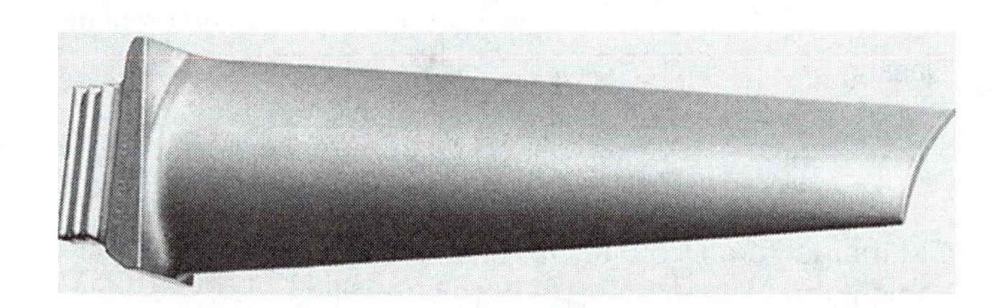

Figure 23-9.
The investment or lost wax casting process produced the complex shape of this turbine blade.

Shell Molding

Shell molding makes use of a mold that is a thin sand shell. A shell mold is shown in Figure 23-10A. A metal pattern attached to a steel plate is fitted into a molding machine. After the pattern is heated to the required temperature, a measured amount of thermosetting resin (heat causes it to take a permanent shape) and sand are deposited on it. Both halves of the mold can be made at the same time. The thin, soft shell is cured until the desired hardness is obtained. After cooling, the pattern is removed.

The shell halves are bonded together using a special adhesive. Molten metal is poured into the mold. The shell mold is broken away from the casting after the metal solidifies, Figure 23-10B.

Sand Casting

Sand casting uses a mold composed of a mixture of sand, clay, and a plastic and/or oil binder. Patterns are wood or metal (usually aluminum). Figure 23-11 shows one example.

A typical sand mold is made by packing sand around the pattern that has been positioned in a wood

Figure 23-10.
Shell molding is well suited to quantity production of quality castings. A—An example of a shell mold is pictured. B—After the metal solidifies, the shell mold is broken away from the casting. (Link-Belt)

Figure 23-11.
Patterns used for sand castings are made of wood or metal.

or metal box, called a ***flask.*** To allow the pattern to be removed, the mold is made in two parts, Figure 23-12. The top half of the mold is called the ***cope,*** the lower half the ***drag.*** The sand mold is destroyed when the casting is shaken out (removed).

Openings (holes or cavities) are often required in a casting. Inserts of sand called ***cores*** make these openings. The cores are positioned in the mold cavity before the mold is closed, Figure 23-13.

Forging

Forging is the process of using pressure to shape metal. The pressure is usually applied in a hammering action. The metal is heated almost to the melting point to improve its plasticity before pressure is applied.

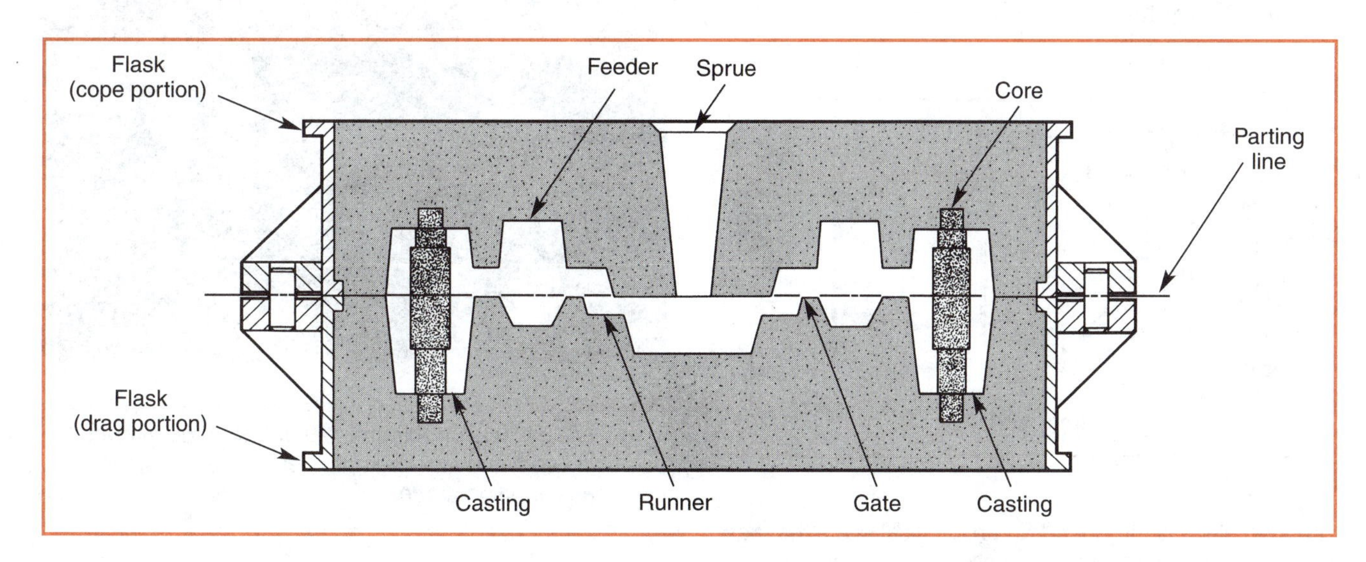

Figure 23-12.
This sectional view shows the elements of a typical two-part mold.

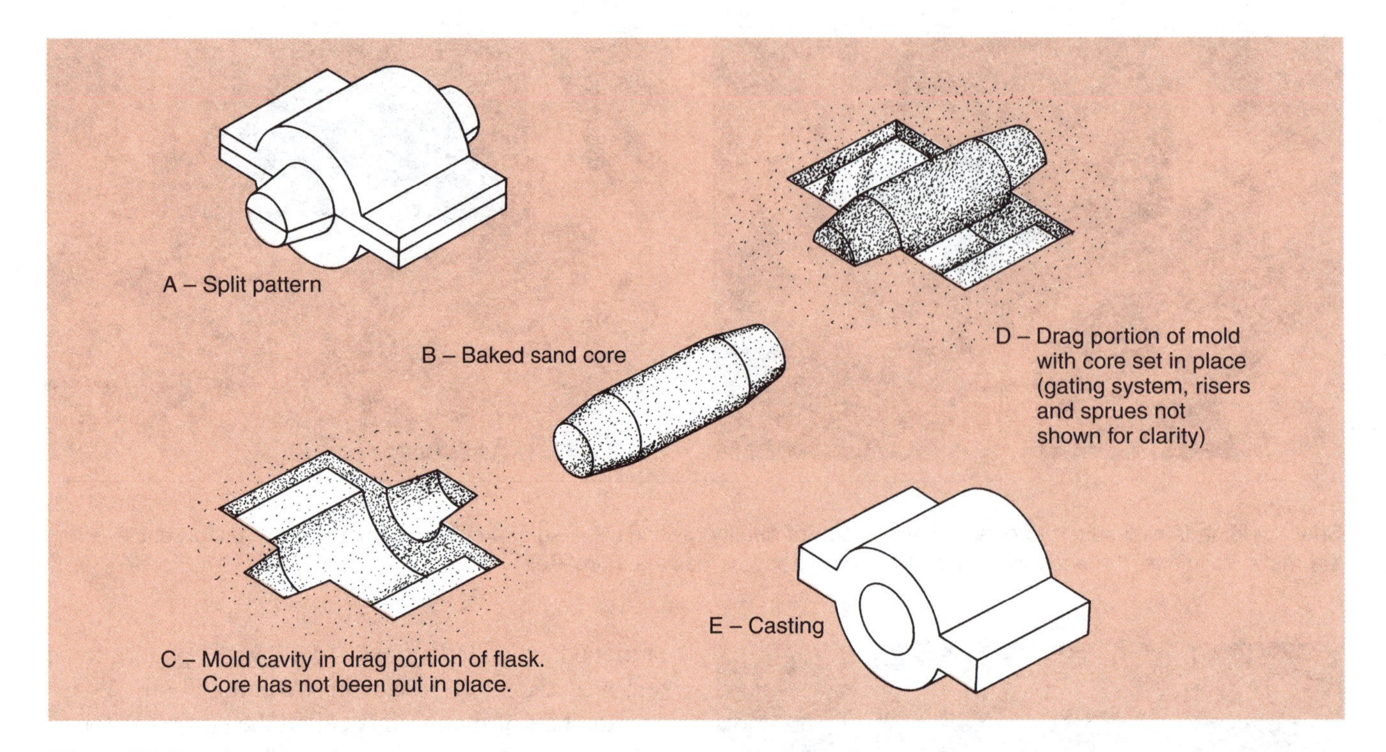

Figure 23-13.
Inserts of sand (called cores) make openings in castings. Cores are positioned in the mold cavity before the mold is closed. While a round core is shown, cores can be almost any shape.

Forging improves the physical characteristics of most metals. That is, a forged piece is stronger than an identical piece machined from a solid bar of stock or made by casting. Figure 23-14 illustrates improved grain structure in forged metal.

Dies are used with the pressure to shape the metal. Pressure may be applied hydraulically, by dropping weights (drop forging), with explosives, with expanding gases, by electrical discharge, and by hand. The various forging techniques are shown in Figures 23-15 through 23-17.

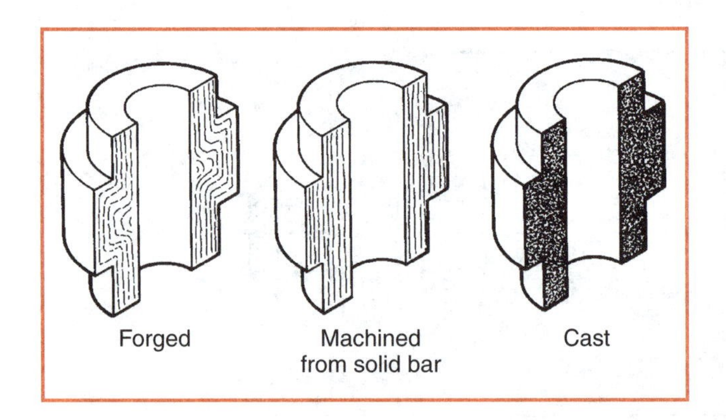

Figure 23-14.
A forged piece is stronger than an identical piece that has been cast or machined from a solid bar of metal.

Figure 23-15.
This 35,000 lb. steam hammer is used to forge aircraft parts and conventional forgings. Here, an aircraft engine crankcase is being made.

Figure 23-16.
These are two giant forging presses for making forgings. A 50,000 ton press is in the foreground and a 35,000 ton press is in the background. (Aluminum Co. of America)

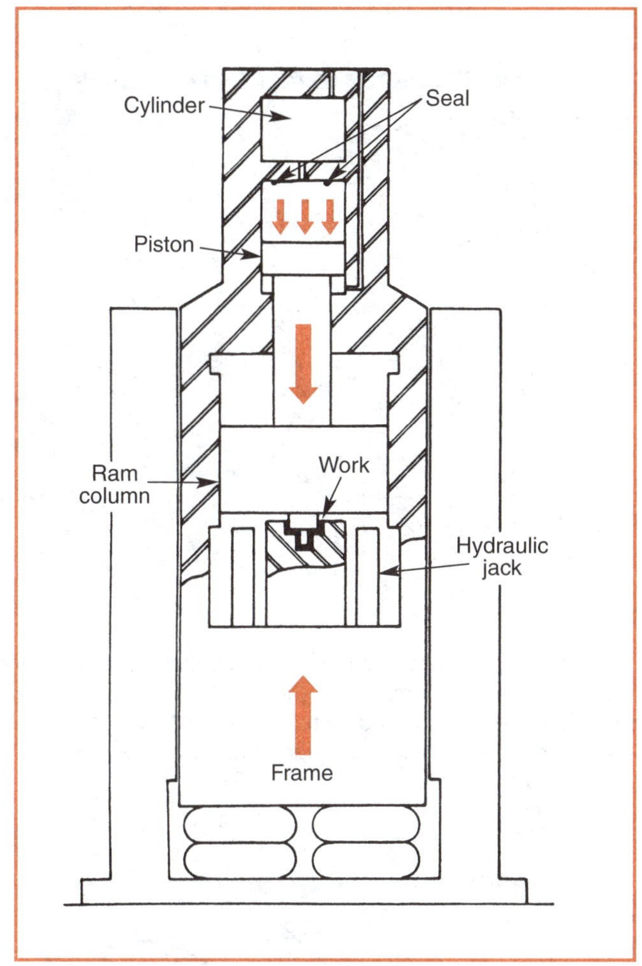

Figure 23-17.
This cross section drawing shows a pneumatic-mechanical forming press. High pressure gases are used to drive the ram.

Extrusion

The ***extrusion process*** is used to manufacture irregular shapes, both solid and hollow, that cannot be made economically by any other process. Several extruded shapes are given in Figure 23-18.

During extrusion, the metal is heated to a plastic (but not molten) state and inserted in the extrusion press. Tremendous pressures are exerted on the metal to literally squeeze it through a die, Figure 23-19. After forming, the extruded section is cut to length and straightened.

Machine Tools

A ***machine tool*** is a power driven machine used to shape metal by a cutting process. It is not hand portable. Machine tools are manufactured in a large range of styles and sizes.

Figure 23-18.
A group of aluminum sections illustrates the wide scope of extrusion applications. A—Diving board. B—Table edge trim molding. C—Rail for miniature railroad. D—Milk bottle warmer. E—Heat exchanger. F—Window jalousie section. G—Shape for pencil vending machine. H—Mine track rail. I—Price tag holder for grocery store shelves. J—Hinge section for clothes drier. K—Drip catcher for roofs of textile mills. L—Heat exchanger. M—Track rail for woolen mill. (Reynolds Metals Co.)

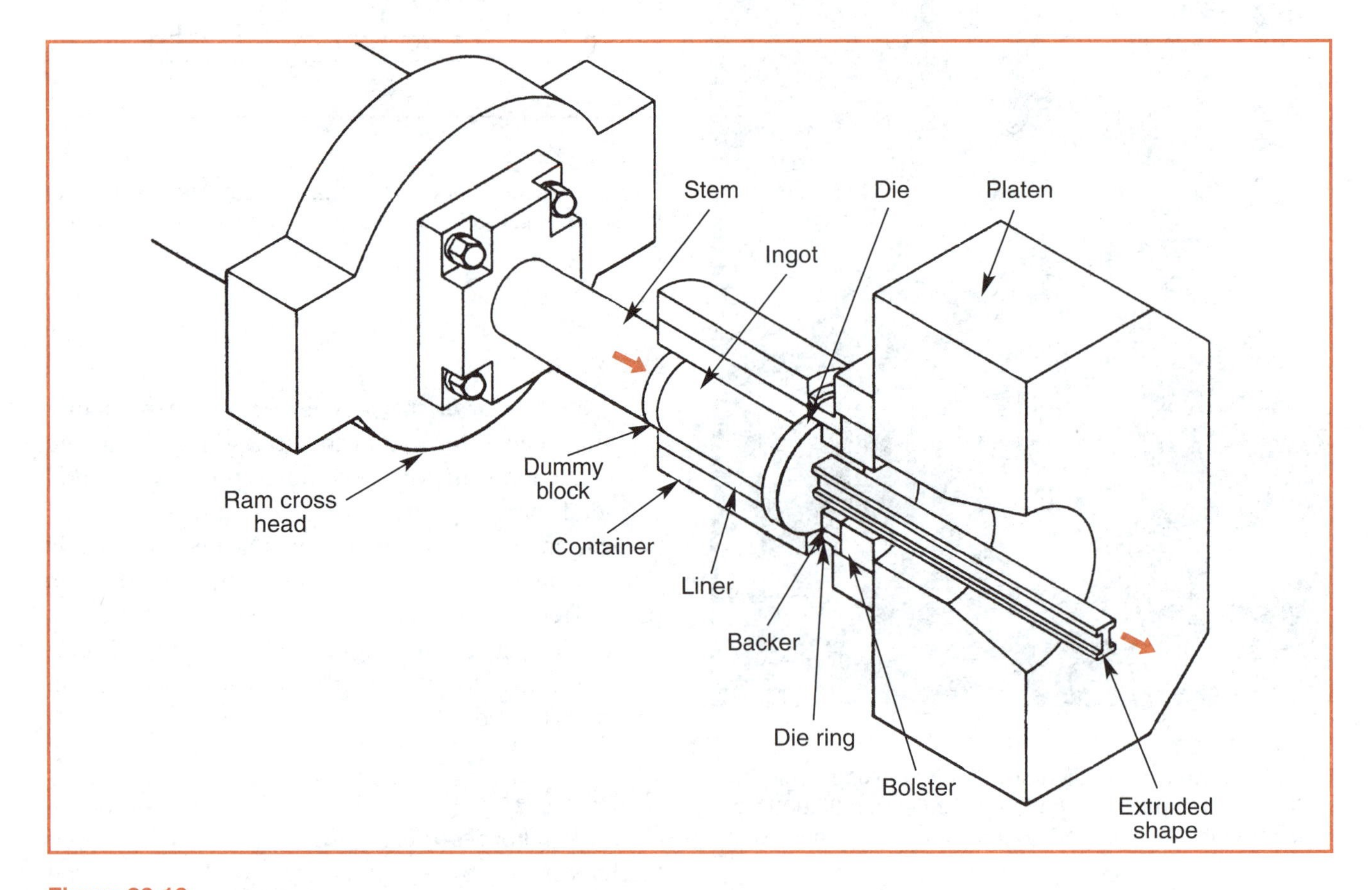

Figure 23-19.
A cross-sectional view of an extrusion press shows the metal (billet) being extruded in the shape of a channel.

Lathe

The lathe, Figure 23-20, operates on the principle of the work being rotated against the edge of a cutting tool, Figure 23-21. The cutting tool can be controlled and moved lengthwise and across the face of the material being turned (machined).

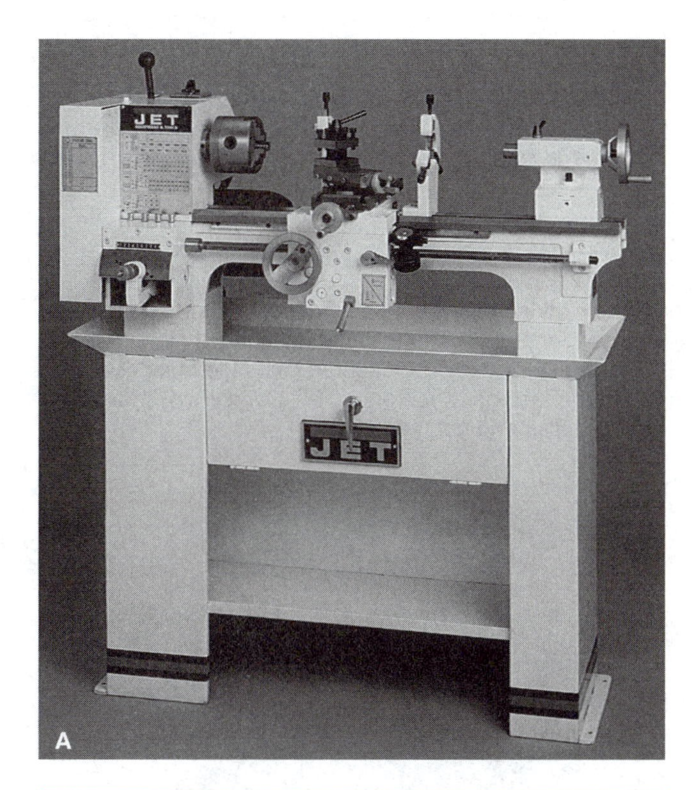

Figure 23-20.
Metal cutting lathes. A—The basic lathe has manually operated controls. It does have power longitudinal and traverse feeds and thread cutting capability. (Jet Equipment & Tools) B—A more advanced computer-controlled lathe is shown. (Harrison/REM Sales Inc.)

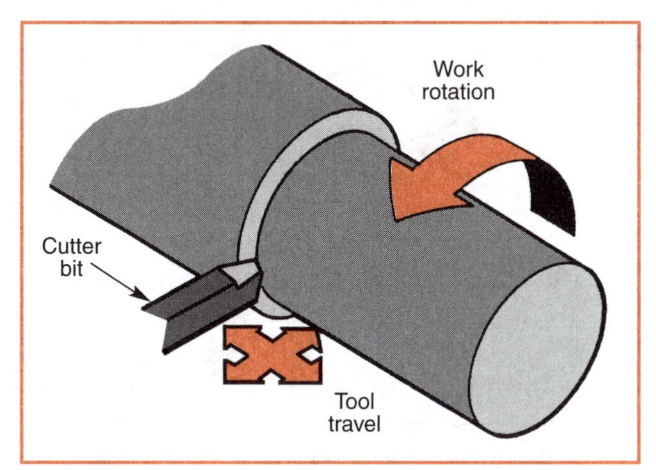

Figure 23-21.
The operating principle of the lathe is illustrated.

Drill press

A drill press, Figure 23-22, uses a cutting tool called a twist drill. The drill press rotates the twist drill against the work with sufficient pressure to cut through the material. Figure 23-23 illustrates drill press action.

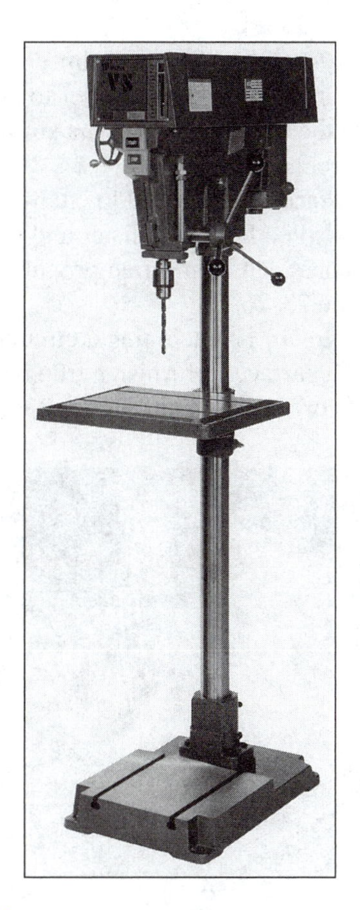

Figure 23-22.
This is a light 15″ variable speed floor model drill press. (Wilton Corp.)

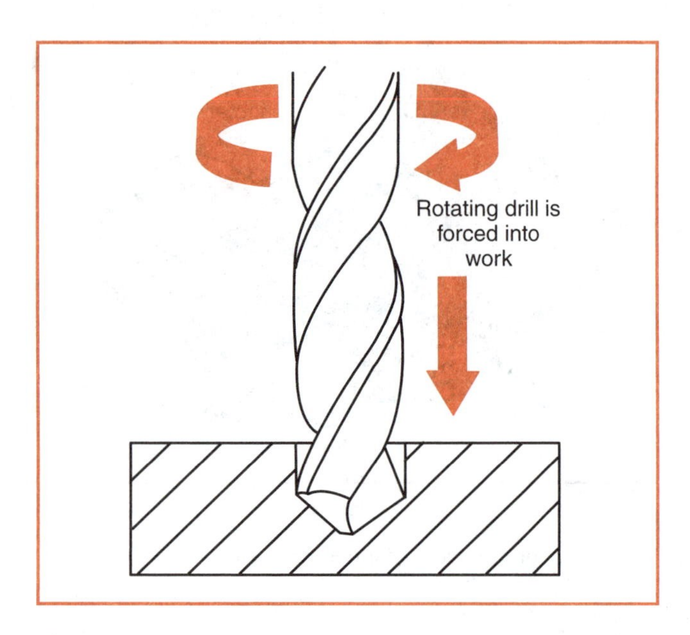

Figure 23-23.
Note how a drill press works.

Other operations performed on the drill press include the following:

- ***Reaming*** improves the accuracy and finish of a drilled hole to close tolerances, Figure 23-24.
- ***Countersinking*** cuts a chamfer on a hole to permit a flat headed fastener to be inserted with the head flush with the surface, Figure 23-25.
- ***Counterboring*** is used to enlarge a portion of a drilled hole so fillister and socket head fasteners can be inserted properly, Figure 23-26.
- ***Spotfacing*** is machining a circular spot on a rough surface to furnish a true bearing surface for a bolt or nut, Figure 23-27.

Figure 23-24.
Shown is reaming on a drill press after drilling a hole.

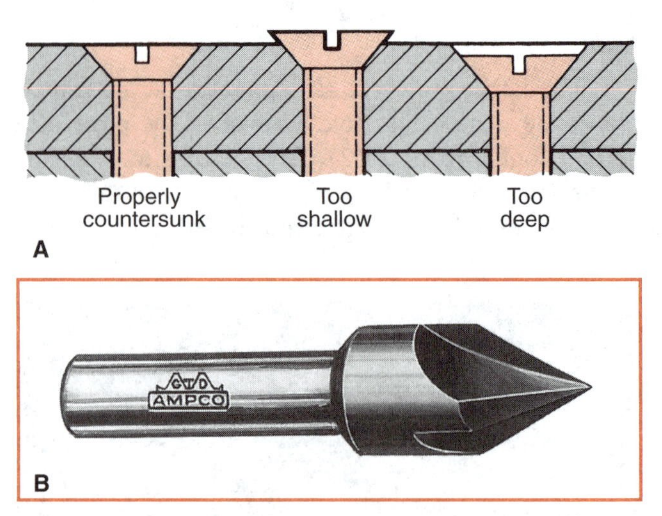

Figure 23-25.
Countersinking allows a flat headed fastener to lie flush with the surface. A—Correctly and incorrectly countersunk holes are illustrated. The countersink angle must also match the fastener head angle. B—Countersinks come in various sizes and point angles.

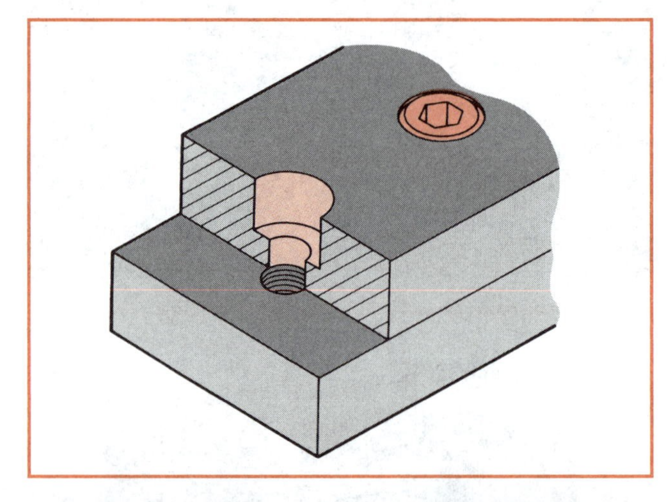

Figure 23-26.
Counterboring is done to prepare a hole to receive a fillister or socket head cap screw. (Clausing Industrial, Inc.)

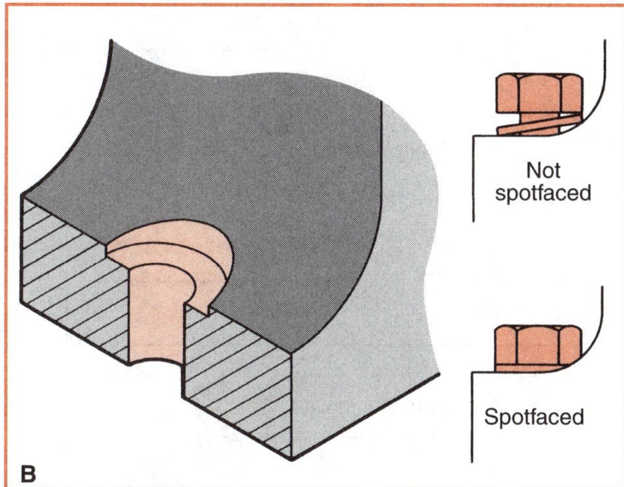

Figure 23-27.
A—Spotfacing machines a true surface to permit a bolt head or nut to bear uniformly over its entire contact area. B—This sectional view of a casting has been spotfaced. Profile drawings show the casting before and after spotfacing. The bolt head cannot be drawn down tightly until the mounting hole has been spotfaced.

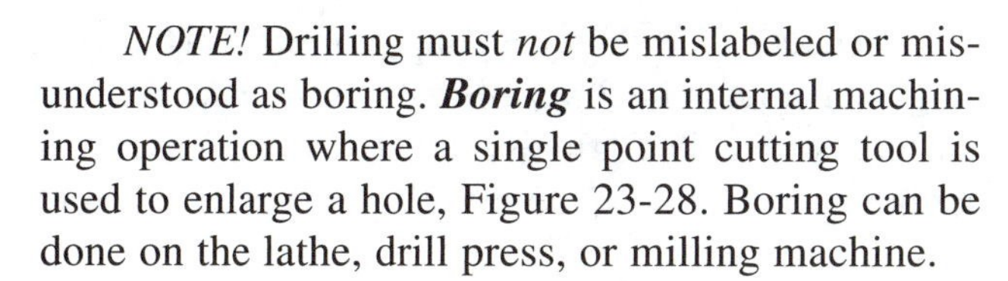
NOTE! Drilling must *not* be mislabeled or misunderstood as boring. ***Boring*** is an internal machining operation where a single point cutting tool is used to enlarge a hole, Figure 23-28. Boring can be done on the lathe, drill press, or milling machine.

Planing Machines

Planing machines are used primarily to machine flat surfaces. The shaper, planer, slotter, and broach are classified this way.

Shaper

While a shaper is primarily used to machine flat surfaces, a skillful machinist can manipulate it to cut curved and irregular shapes, slots, grooves, and keyways, Figure 23-29.

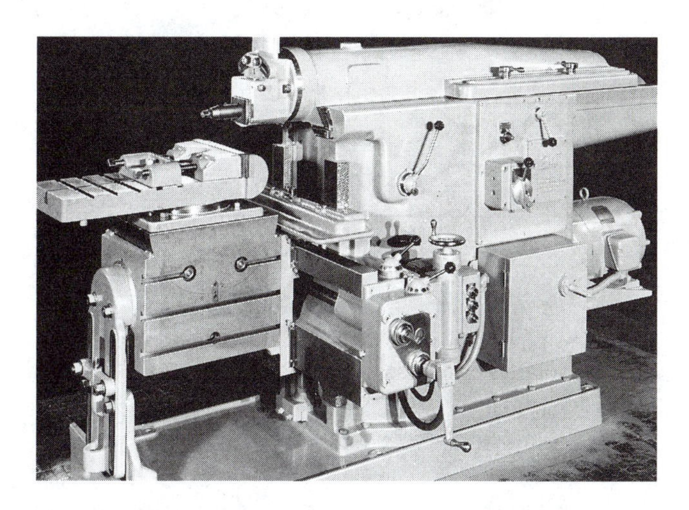

Figure 23-29.
This is a shaper. (Rockford Machine Tool Co.)

Figure 23-28.
A—Drilling is not the same as boring. Here a drill is being used to cut a hole in a casting. B—Boring is a machining operation used to enlarge a drilled hole to exact size and to a much closer tolerance than a drill could accomplish. (Clausing)

Because of the way a shaper operates (the cutting tool travels back and forth over the work), the cutting stroke is normally limited to a maximum length of 36″ (approximately 914 mm). Figure 23-30 shows the action of a shaper.

Planer

A planer can handle work too large to be machined on milling machines, Figure 23-31. Planers are large machine tools, and many are capable of machining surfaces up to 20′ (approximately 6 m) wide and twice as long.

The cutting tool on a planer remains stationary. The work travels back and forth under the cutter, Figure 23-32.

Slotter

The slotter is a vertical shaper, Figure 23-33. The slotter is used to cut slots and keyways (both internal and external). It is also used for jobs such as machining internal and external gears.

The slotter operates similar to the shaper except the cutting tool moves *vertically* rather than horizontally, Figure 23-34. The work is held stationary.

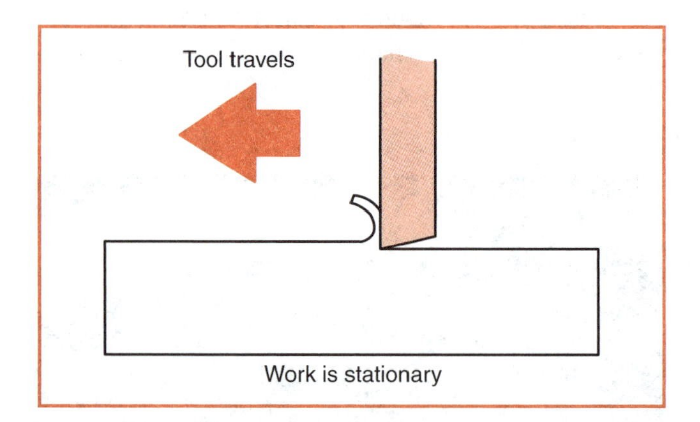

Figure 23-30.
Note how a shaper works. The work is stationary, and a cutting tool moves across it to remove metal.

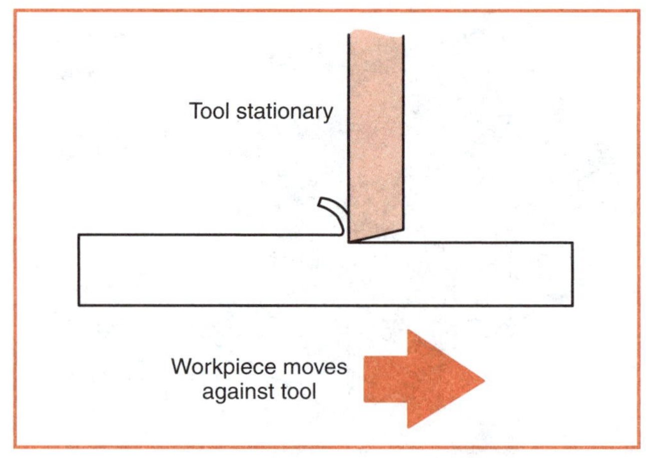

Figure 23-32.
Study how a planer works. The tool remains stationary while the work moves against it.

Figure 23-31.
People look small while standing on a large (144″ × 126″ × 40′) double-housing planer. (G.A. Gray Co.)

Figure 23-33.
This job is being done on a vertical shaper or slotter.

Broach

Broaching is similar to shaping, but instead of a single cutting tool advancing slightly after each stroke, the broach is a long tool with many cutting teeth, Figure 23-35.

Each broach tooth has a cutting edge that is a few thousandths of an inch (hundredths of a millimeter) larger than the one before it. The teeth increase in size to the exact finished dimension required.

The broach is pushed or pulled over the surface being machined. Figure 23-36 shows a modern broaching machine and how the broach travels over the work.

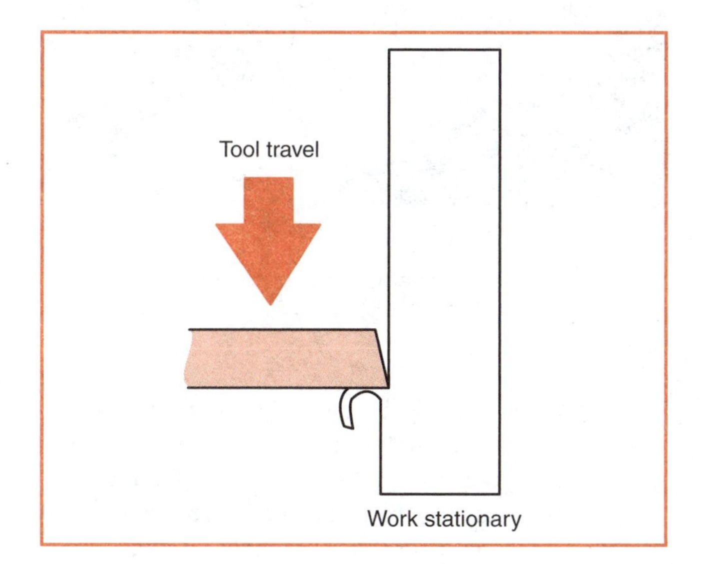

Figure 23-34.
Note how the vertical shaper works. The machine operates in a manner similar to a shaper except the tool moves vertically. The work is held stationary.

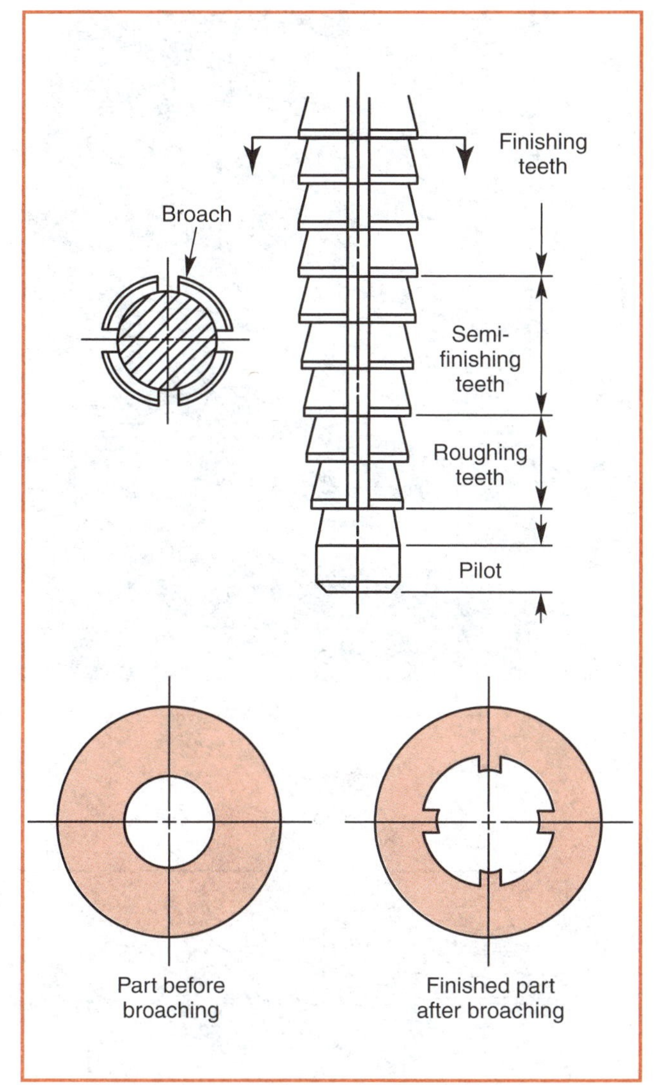

Figure 23-35.
Note the short section of a broaching tool and a cross section of the spline it cuts. A pilot guides the cutter in the work. Each cutting tooth increases slightly in size until the specified size is attained.

Milling Machines

The milling machine removes material by moving a rotating multitoothed cutter into the work, Figure 23-37. It can be used to machine flat and irregularly shaped surfaces and to drill, bore, or cut gears and splines.

With a vertical milling machine, the cutter is mounted *vertical* to the worktable, Figure 23-38. The horizontal milling machine is designed to do peripheral milling, and the cutter is mounted *parallel* to the worktable, Figure 23-39.

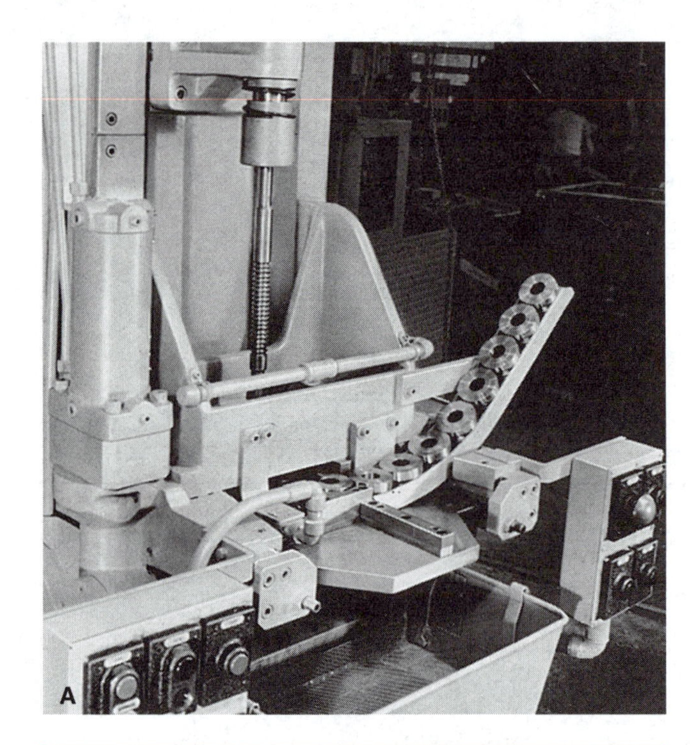

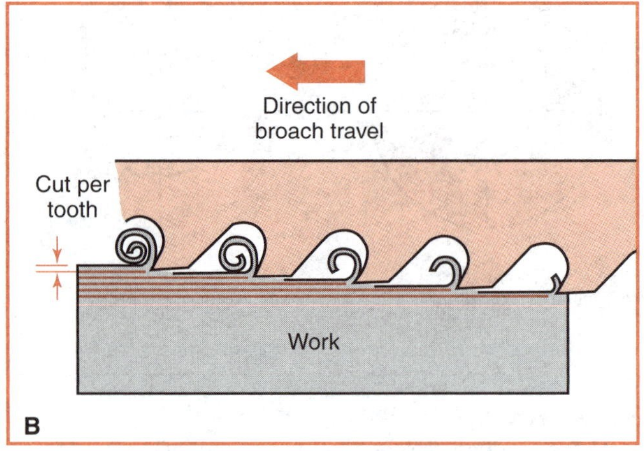

Figure 23-36.
A—The work is moving into position on this broaching machine. (Sundstrand Corp.) B—Each tooth on a broaching tool removes only a small portion of the material being machined.

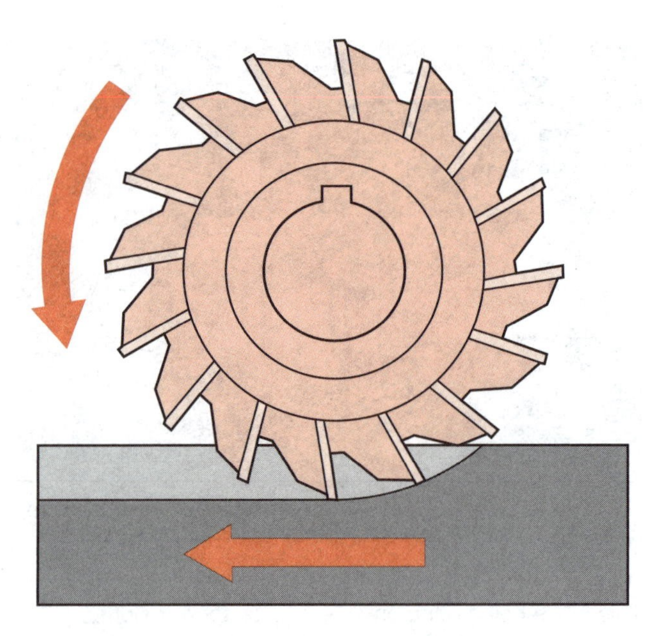

Figure 23-37.
In milling, a rotating multitoothed cutter is fed into the work to remove material. Each tooth removes a small individual chip.

Figure 23-38.
The cutter is mounted vertical to the worktable on a vertical milling machine.

Horizontal spindle machines are also used for precise machining of jigs and fixtures. Figure 23-40 shows a horizontal jig borer. The part is clamped to the table or to a vertical fixture called a knee or tombstone.

Computer Numerical Control (CNC) Machine Tools

The machines described in the previous sections all assume operation by a skilled human. The operator must be familiar with the setup, tool selection, and operation of each machine tool.

However, during the past fifty years, the development of machine tools controlled by computers has continued to improve the capabilities of metal cutting, forming, and shaping machines. The most modern CNC machine tools are capable of unattended machining once the program is loaded into the CNC machine tool controller's memory.

Figure 23-39.
A horizontal milling machine has a cutter mounted parallel to the worktable.

Figure 23-40.
A horizontal jig borer is an example of a milling machine.

Programming a CNC machine tool requires the operator or programmer to enter a series of coded instructions that locate the tool at specific coordinates within the working area of the machine tool.

Figure 23-41 shows a CNC vertical machining center. Notice the working area is enclosed to protect the operator and other workers from flying metal chips. Figure 23-42 shows a mill cutter about to cut a part held in a fixture.

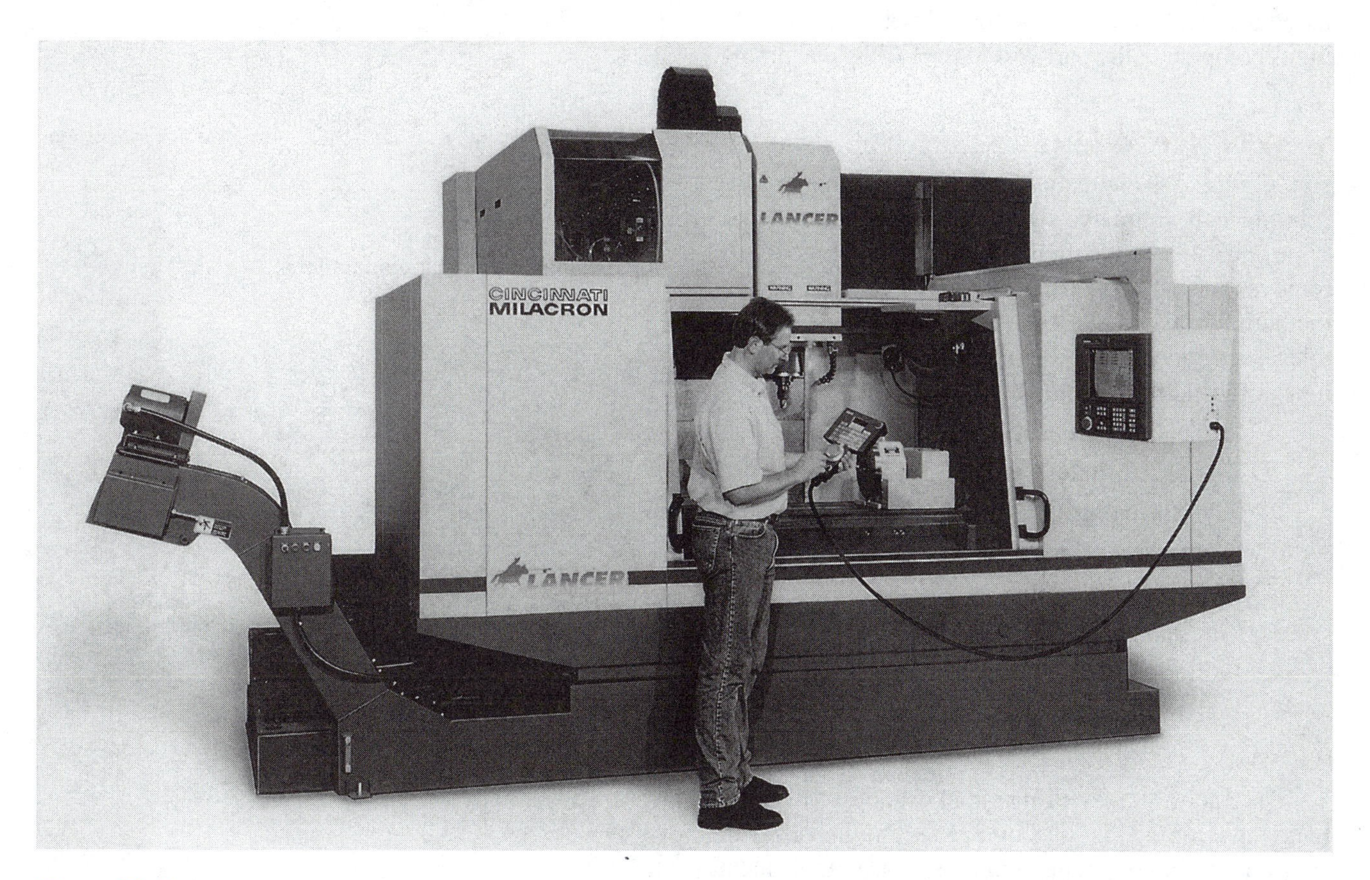

Figure 23-41.
A CNC vertical machining center has an enclosed work area that protects the operator from flying metal chips. (Cincinnati Milacron)

Figure 23-42.
CNC tooling about to cut a part held in a fixture.

Often, many parts that are welded have been prepared using a CNC machine tool. Possibly, after welding is completed, the part will require CNC machining.

Grinding Machines

Grinding is an operation that removes material by rotating an abrasive wheel against the work, Figure 23-43. It is often used to sharpen tools, to remove material too hard to be machined by other methods, or to provide fine surface finishes and close tolerances when they are required. There are many types of grinding machines:

- ❍ Bench and pedestal grinder, Figure 23-44.
- ❍ Portable grinder, Figure 23-45.
- ❍ Cylindrical grinder, Figure 23-46.
- ❍ Centerless grinder, Figure 23-47.
- ❍ Internal grinder, Figure 23-48.
- ❍ Surface grinder, Figure 23-49.
- ❍ Variations of these grinders.

Saws

In addition to cutting stock shapes to length, band machining is a sawing technique used to machine complex shapes from standard stock metal shapes. Refer to Figures 23-50 and 23-51.

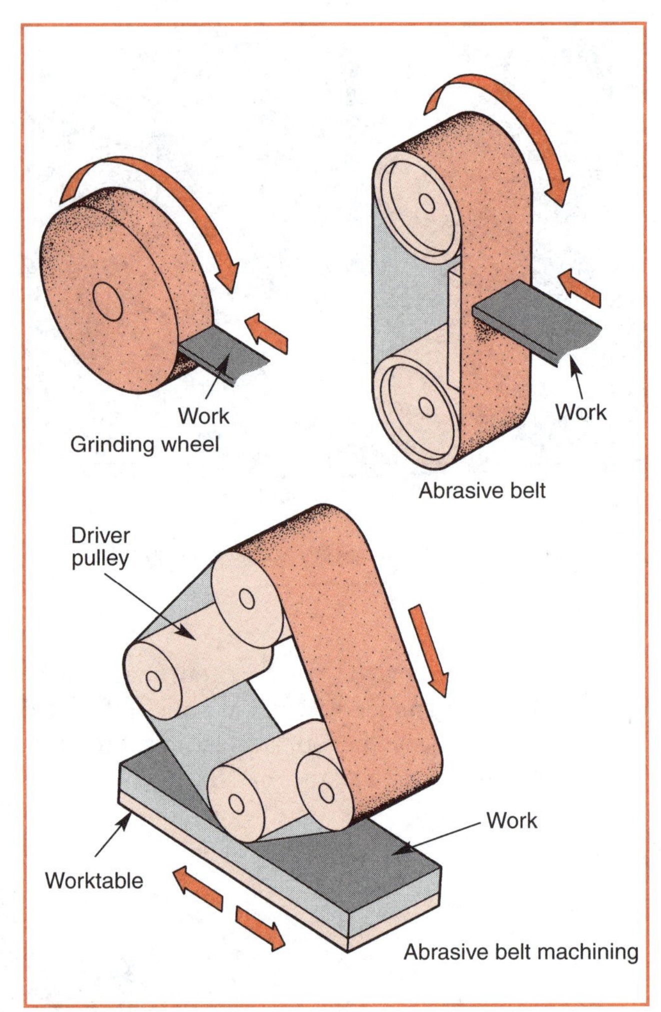

Figure 23-43.
Typical grinding machine operations.

Figure 23-44.
This is a typical bench grinder. (Baldor)

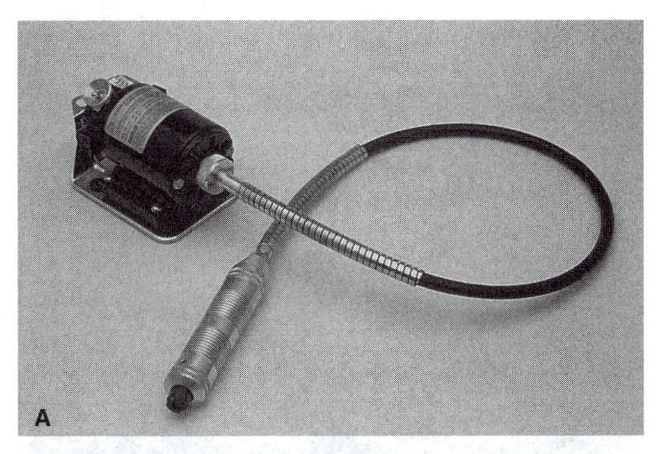

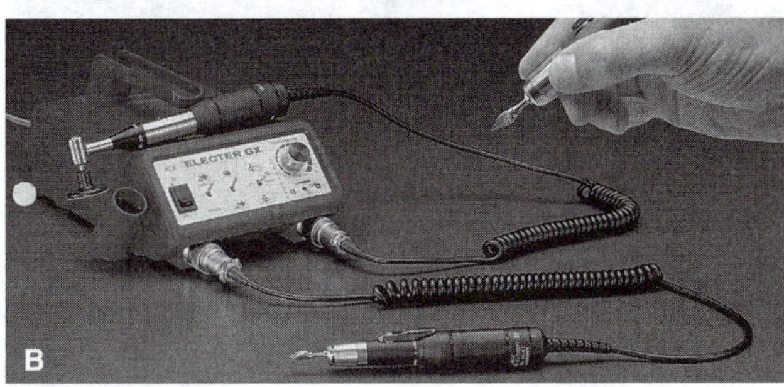

Figure 23-45.
A—A flexible shaft grinding unit. B—Precision electric microgrinder. (The Dumore Co. and NSK America)

Figure 23-46.
Close-up of a cylindrical grinding operation. (Landis Division of Western Atlas)

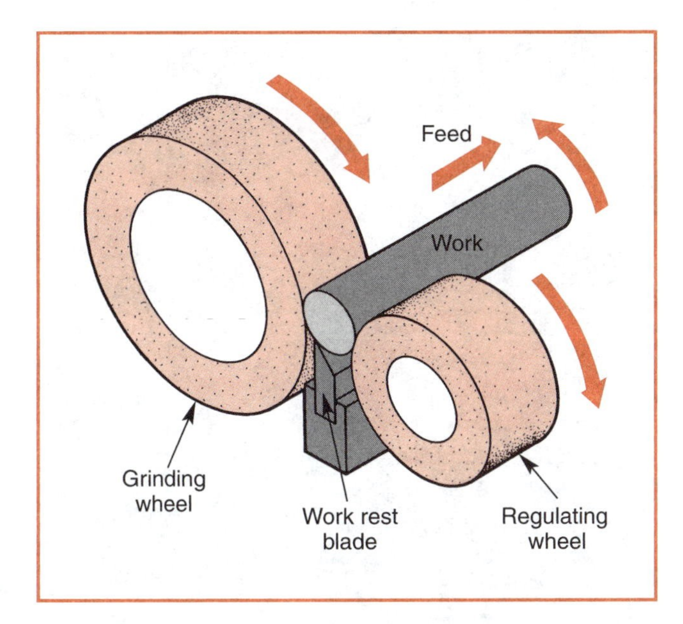

Figure 23-47.
Centerless grinding is done as indicated.

Figure 23-48.
An internal grinding operation is being used to surface the inside diameter of a part. (Norton Co.)

There are three principal types of saws used to cut material to required lengths. One has a reciprocating (back and forth) cutting action, Figure 23-52. It uses a blade similar to the one found in a hand hacksaw. Another saw type uses a continuous or band blade. One is shown in Figure 23-53A and 23-53B. The third type uses a circular blade with either a toothed blade or abrasive wheel, Figure 23-54.

Figure 23-49.
A planer-type surface grinder is used to precision grind flat surfaces. (Jones & Shipman, Inc.)

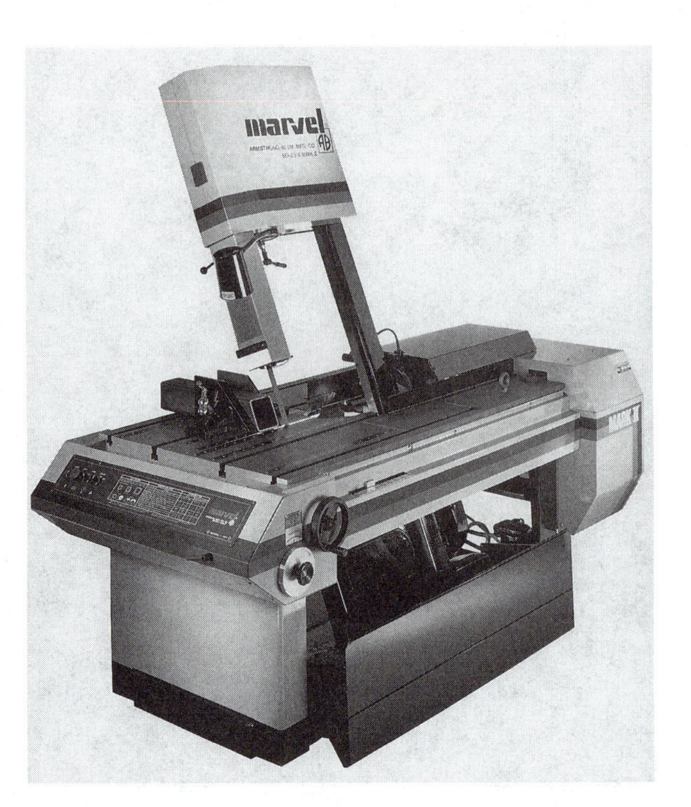

Figure 23-50.
A vertical band saw is commonly used in many weld shops. (Armstrong-Blum Mfg. Co.)

Figure 23-51.
In sawing operations, unwanted material is removed in solid sections instead of being reduced to chips. (DoALL Co.)

Cold Forming Metal

Many products are fabricated from metal forms produced by one or more of the following cold forming techniques. ***Cold forming*** forms cold metal into a desired shape using a series of dies. Welding plays a very important part in the joining together of these components.

Spinning

Spinning is a method of working metal sheet into three-dimensional shapes. An example is in Figure 23-55. Spinning involves, in its simplest form, rotating a disc of metal with a forming block (chuck). The forming block is made to the dimensions specified for the inside of the finished object. As the lathe rotates the metal disc and forming block, pressure is applied. The metal is gradually worked around the form until the disc assumes its shape and size.

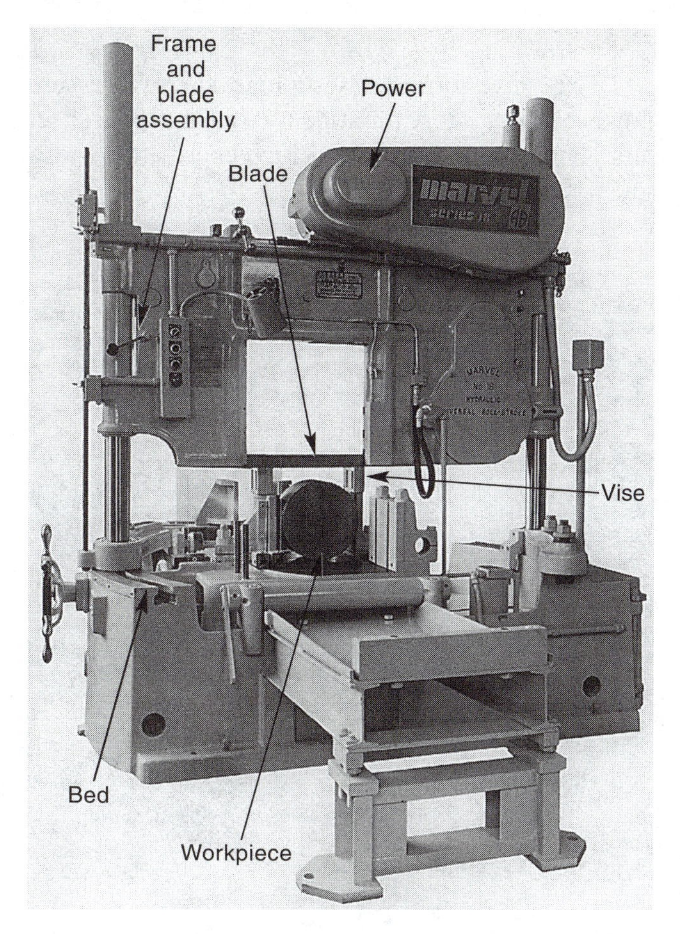

Figure 23-52.
A power hacksaw has a reciprocating cutting action. (Armstrong-Blum Mfg. Co.)

Figure 23-54.
An abrasive cutoff saw is removing sprues from a casting. (American Foundrymen's Society, Inc.)

Shear Spinning

Shear spinning is a metalworking technique similar to conventional spinning in outward appearance only, Figure 23-56. Shear spinning is a cold extrusion process where the parts are shaped by rollers exerting tremendous pressures on a starting blank or preform. This displaces the metal parallel to the center line of the workpiece.

The metal for shear spinning is taken from the thickness of the blank, whereas the metal in spinning is taken from the diameter of the blank.

Figure 23-53.
A—An industrial band saw is shown. B— A continuous or band blade power hacksaw is being used to cut metal stock for weldment. (Armstrong-Blum Mfg. Co., Clausing Industrial, Inc.)

Figure 23-55.
A 42″ × 50″ shear spinning machine is producing the head of a rocket motor case. When it is fully formed, it will be 13½″ deep and 40″ in diameter. (Meta-Dynamics, Cincinnati Milling Machine Co.)

Explosive Forming

Explosive forming uses a high-energy pressure pulse of very short duration to shape metal sheet and plate. It is also known as high energy rate forming (HERF), Figure 23-57.

Figure 23-57.
These are a few of the many parts shaped by explosive forming. (Ryan Aeronautical Co.)

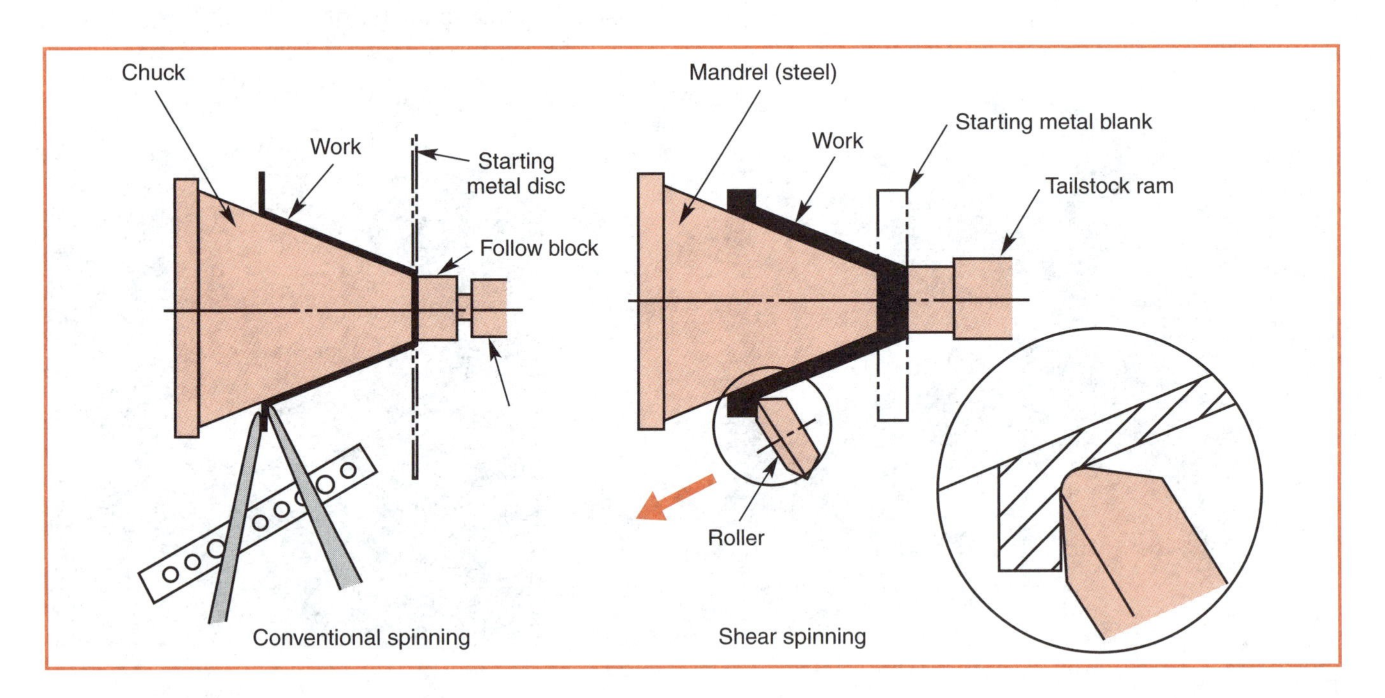

Figure 23-56.
Note how conventional spinning and shear spinning differ.

A chemical explosive or an electrical discharge can generate the pressure pulse for explosive forming. These two methods are shown in Figure 23-58.

Stamping

Stamping is the term used for many press forming operations. Included is a cutting operation called blanking, Figure 23-59. ***Blanking*** involves cutting flat sheets to the shape of the finished part. Cutting is done with a punch and die.

Drawing (or ***forming***), Figure 23-60, takes the flat metal blanks and forms or draws them into three-dimensional shapes. The technique makes use of a draw press and a matched punch and die. There are many variations of the drawing technique.

Shearing

Shearing is common for the preparation of sheet, plate, and some structural shapes. It is normally done while the materials are cold. Shearing generally refers to a process of cutting across the entire length of the material.

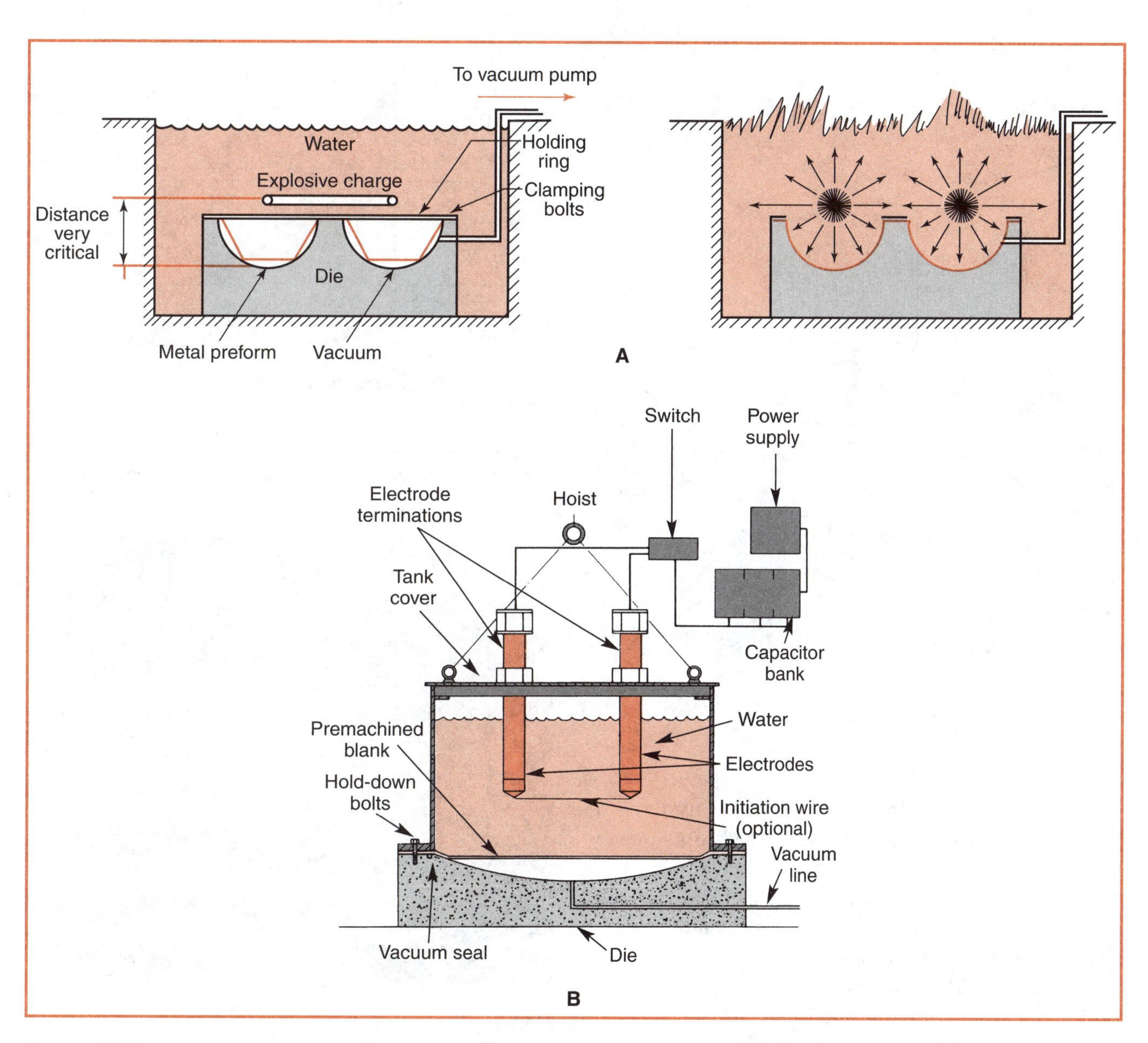

Figure 23-58.
A—This diagram illustrates the principle of the explosive forming process. Fifty cents worth of explosives, in some applications, will do the work of a press costing hundreds of thousands of dollars. B—The diagram shows the setup for electrohydraulic forming, which uses electrical energy as a source of power for HERF operations. (NASA)

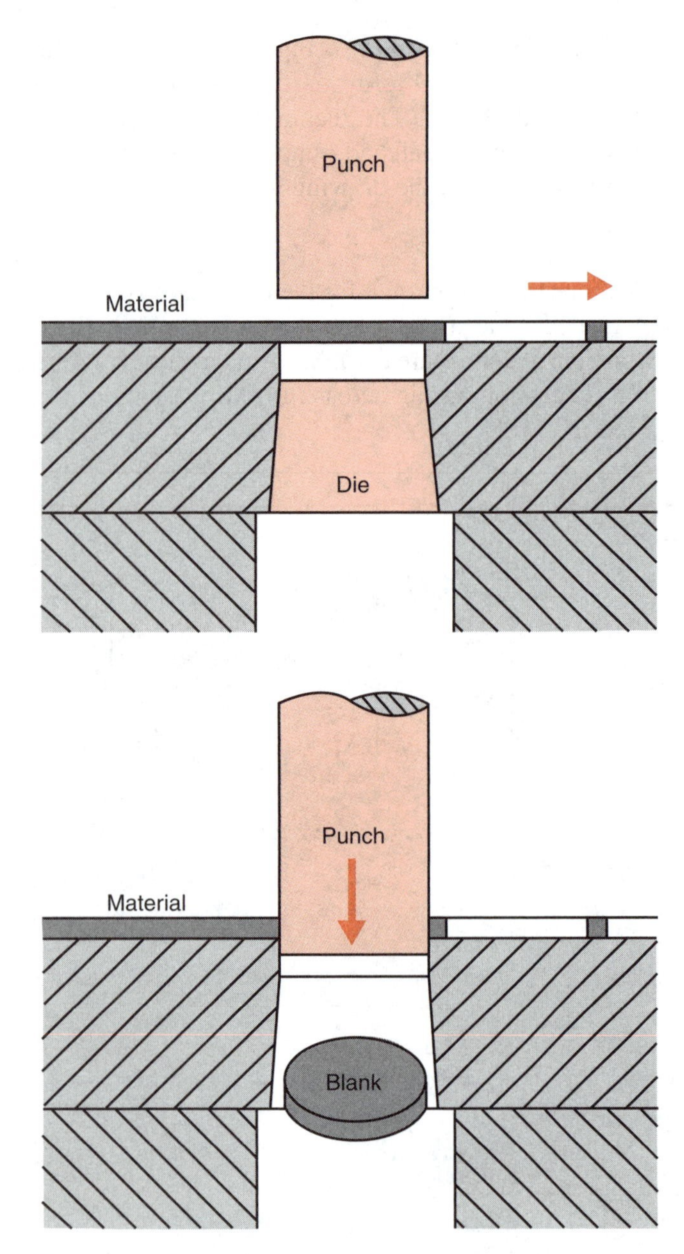

Figure 23-59.
The blanking operation is shown.

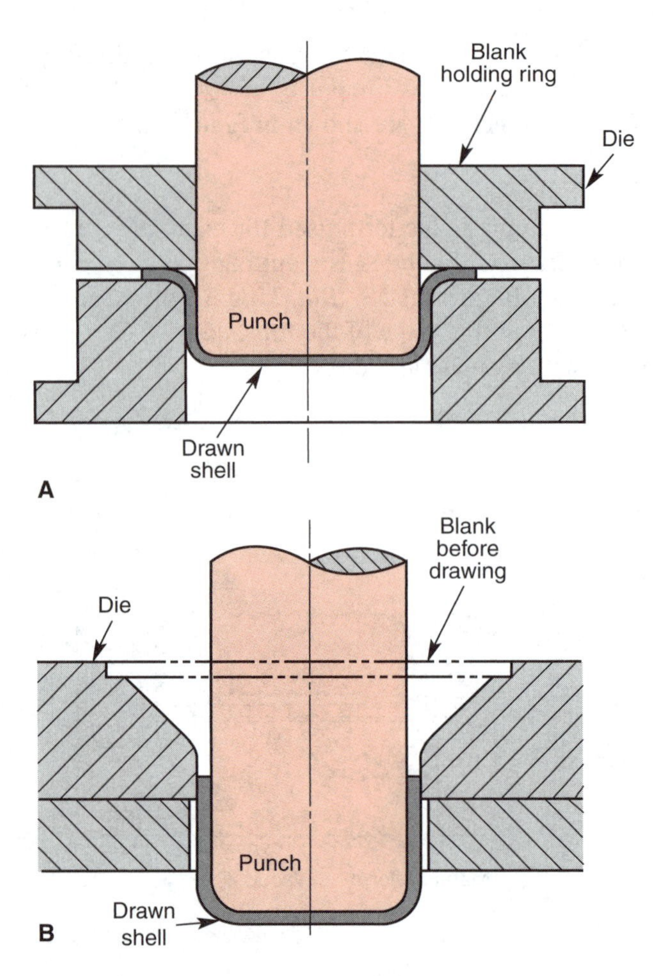

Figure 23-60.
The drawing process uses a matched punch and die set.

The basic shearing action involves moving a top knife or blade (sometimes referred to as punch blade) in opposition to a table or bed blade. The process is a controlled fracture with shearing through a portion of the material, depending upon the material's mechanical properties.

Figure 23-61 shows a basic shear with the material between the upper and lower shear blades. Notice the fracture and deformation shown in Figure 23-62. As a result of the edge quality and deformation of the material, other methods of cutting materials may be used.

Figure 23-61.
Material is between the upper and lower blades of this basic shear.

Stretch Forming

Stretch forming is a process where a metal blank is gripped by inserting opposite edges in clamps and subjecting the blank to a light pull. This causes the blank to hug or wrap around a form block of the required shape, Figure 23-63. The piece is trimmed to final shape after the forming operation.

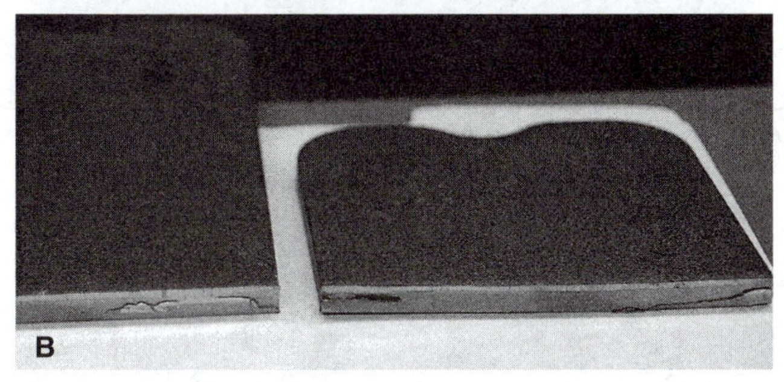

Figure 23-62.
A—Sheared plates. B—Fracture and deformation are evidenced on these sheared plates.

Bending

Bending is an operation in which the surface area of the work is not appreciably changed. A forming brake, Figure 23-64, or forming rolls are used. A few typical bending operations possible on a forming brake are shown in Figure 23-65.

Roll Forming

Roll forming takes a flat metal strip and passes it through a series of rolls that progressively shape it into the required contour, Figure 23-66. The process is extremely rapid, and almost any desired configuration is possible.

Figure 23-64.
This forming brake is used to bend large pieces of a plate. (Clearing-Niagara)

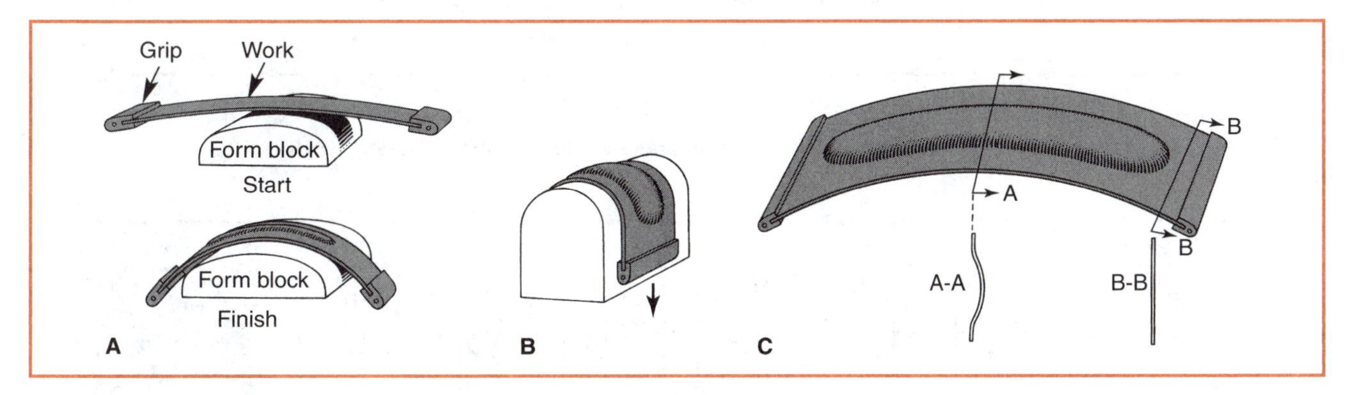

Figure 23-63.
The stretch forming process. A—This is the relative position of the work. The form block and clamps at the start and finish of a typical stretch forming operation produce a raised rib. B— Stretch forming over a form block produces this typical shape. C—A stretch-formed shape is similar to the bottom of a canoe. The section will be riveted or welded to other sections to make the canoe. (Allegheny Ludlum Steel Corp.)

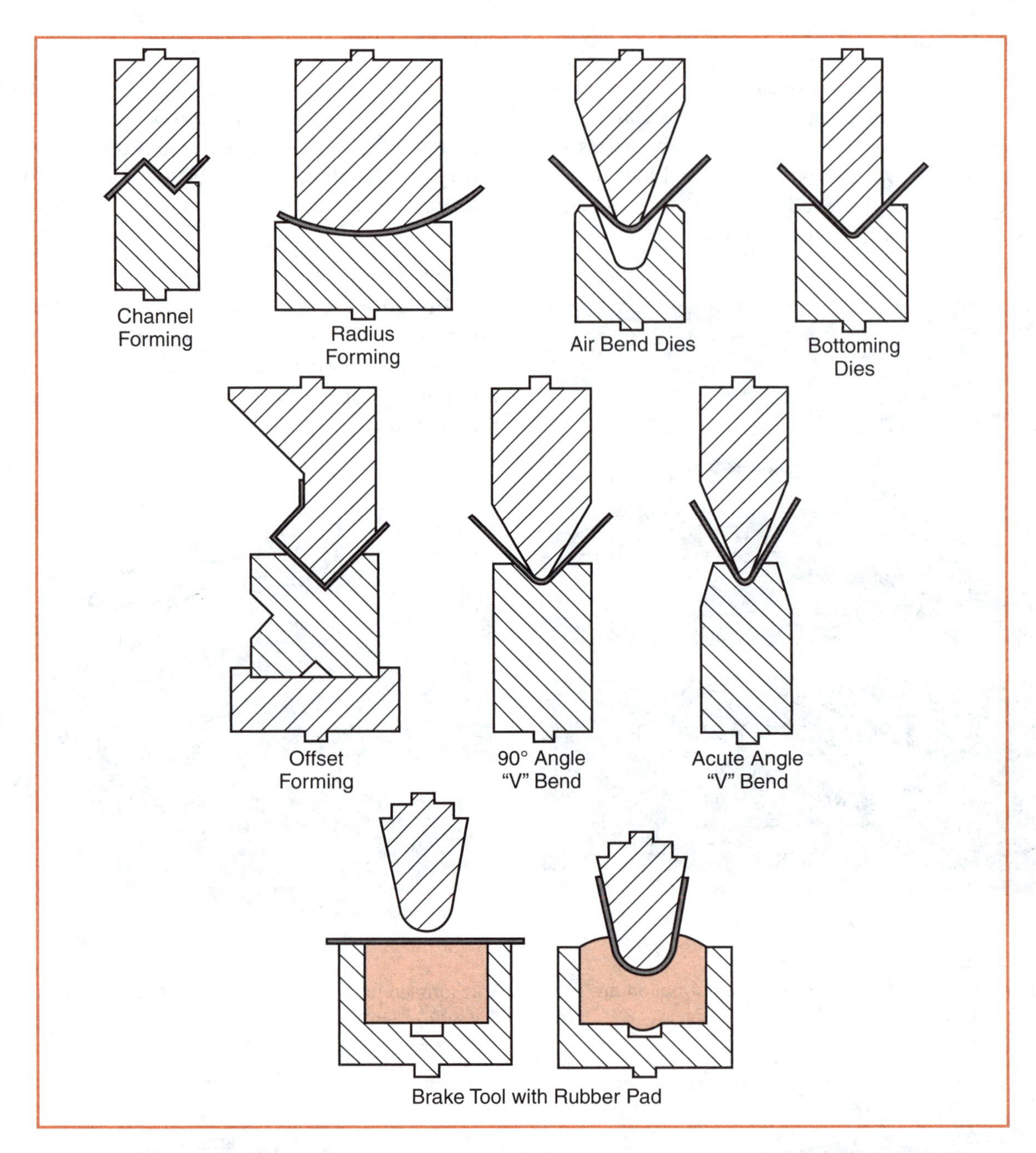

Figure 23-65.
Several typical bending operations possible on press brake equipment are indicated.

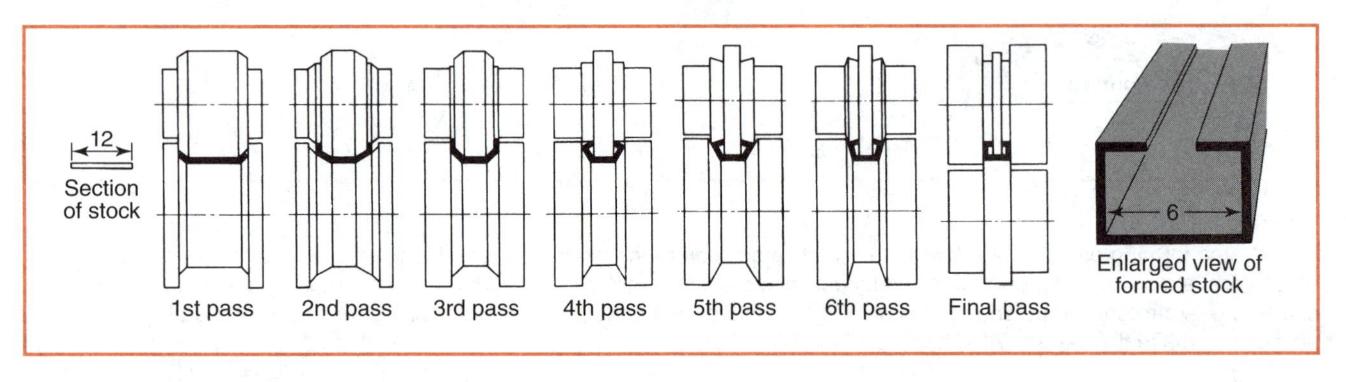

Figure 23-66.
Follow this sequence for roll forming channel.

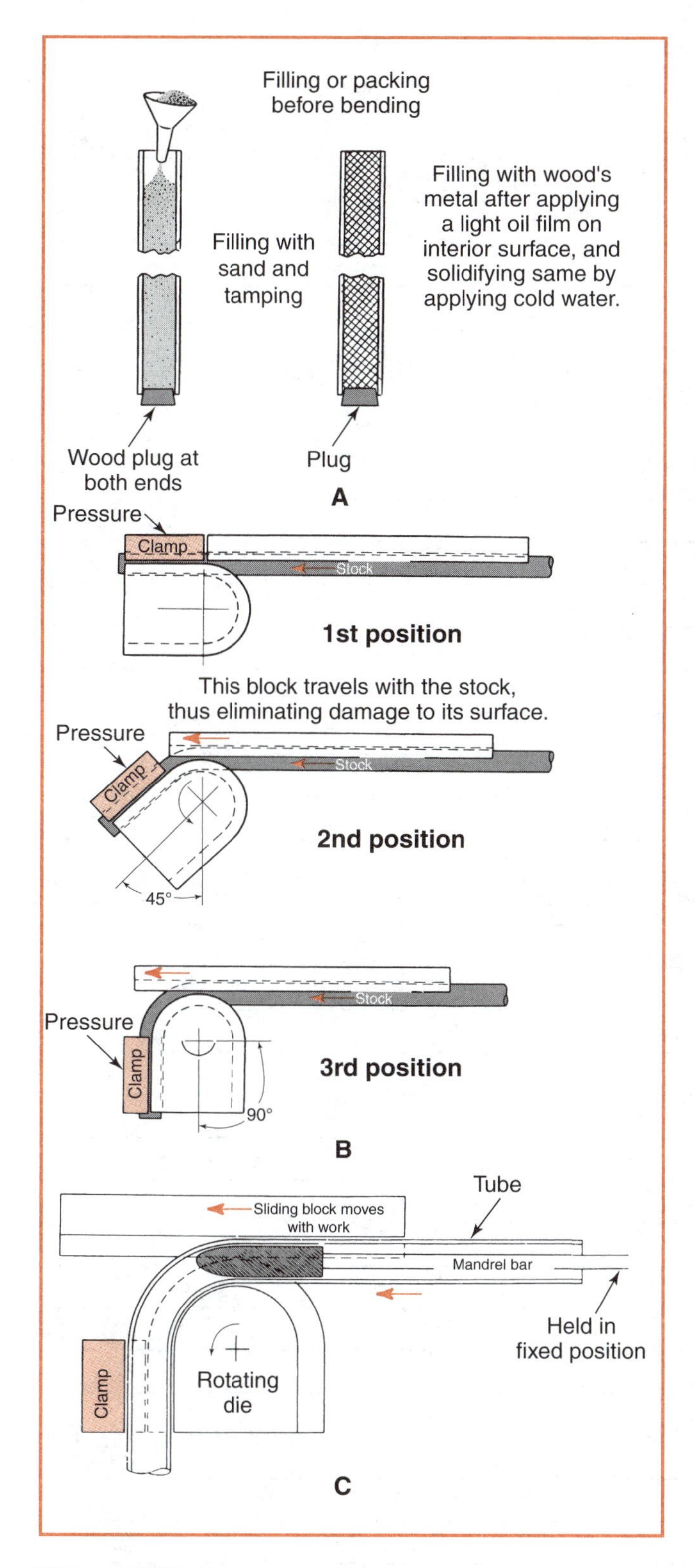

Figure 23-67.
A diagram of the rotating die tube bending technique. A—Preparing the tube for bending. The filler material prevents the tube from collapsing during the bending operation. B—Steps in bending a tube. C—It is not necessary to fill the tube before bending if a reinforcing mandrel is used. (Allegheny Ludlum Steel Corp.)

Tube Bending

Tube bending is used to form radii in tubing without causing the tubing to collapse. Very complex bends on several axes are possible. Figure 23-67 shows this process. A CNC tube bender is illustrated in Figure 23-68.

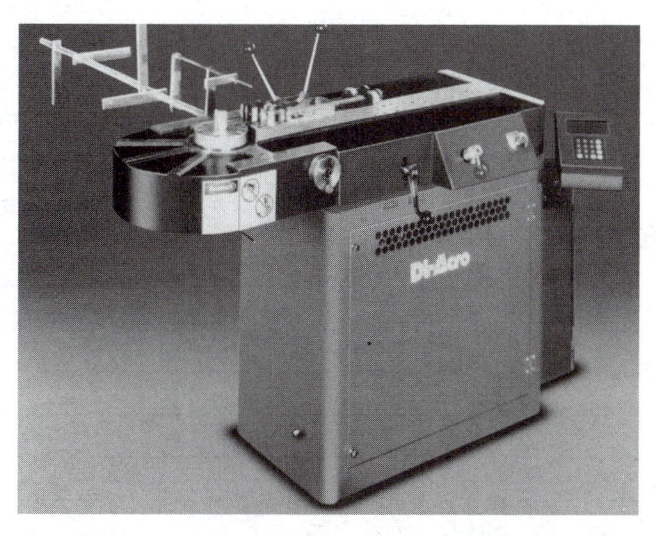

Figure 23-68.
A CNC, hydraulically operated tube bender that can accommodate both tubing and solid stock. (Strippit, Inc.)

Notes

Name ______________________ Date ____________ Class ____________

Review Questions

1. Casting is a metalworking process for making metal objects by _____.

____

2. Briefly describe the following casting techniques:

A. Permanent mold casting ______________________________

B. Centrifugal casting ______________________________

C. Die casting ______________________________

D. Investment casting ______________________________

E. Sand casting ______________________________

Place the letter of the phrase that best describes each metalworking process listed below.

3. _____ Forging
4. _____ Extrusion
5. _____ Spinning
6. _____ Shear spinning
7. _____ Explosive forming
8. _____ Stamping
9. _____ Drawing
10. _____ Stretch forming
11. _____ Roll forming
12. _____ Bending
13. _____ Tube bending

(A) Uses a high-energy pressure pulse of very short duration to shape metal sheet and plate. Pressure can be generated chemically or electrically.

(B) Takes flat sheet metal blanks and forms them into three-dimensional shapes.

(C) Process where a metal blank is gripped by inserting opposite edges into clamps and subjecting the blank to a light pull. This causes blank to hug or wrap around a form block of required shape.

(D) Used to form radii in hollow metal sections without causing section to collapse.

(E) Process that uses pressure to shape metal. Metal may or may not be heated, but *not* to melting point. Pressure is usually applied in a hammering action.

(F) Metal in this process is heated to a plastic state and pressure is applied to squeeze it through die of desired shape.

(G) Method of working sheet metal disc into a three-dimensional shape by forcing it against a rotating forming block (chuck). Pressure is applied and metal is gradually worked around form until it assumes correct shape and size.

(H) Rollers exerting tremendous pressures on a blank or preform shape a part. Rollers displace metal parallel to center line of workpiece. Metal is taken from thickness of blank; whereas, in another similar process, metal is taken from diameter of blank.

(I) An operation in which surface area of metal sheet or plate being worked is *not* appreciably changed.

(J) A term used for many press forming operations.

(K) A process that takes flat metal strip and passes it through a series of rolls that progressively form it into required shape.

Name ________________________________

Place a check next to the correct answer(s).

14. A machine tool is:
    - A._____ a power driven, hand portable cutting tool.
    - B._____ a power driven machine used to shape metal by a cutting process.
    - C._____ manufactured in a limited range of shapes and sizes.
    - D._____ All of the above.
    - E._____ None of the above.

15. Prepare a sketch showing how a lathe operates.

16. The drill press uses a cutting tool called a(n) _____. Explain how a drill press operates.

17. How does boring differ from drilling?

18. Planing machines are used primarily to machine _____ surfaces.

19. How does a milling machine operate?

20. There are two basic types of milling machines. List them. How do they differ? (Use sketches if necessary.)
    A. ________________________________
    B. ________________________________

21. A series of coded instructions used to move the tool to various locations on a part is used by a(n) _____ machine tool.

22. A cutting method used to cut across the entire length of a piece of sheet or plate is _____.

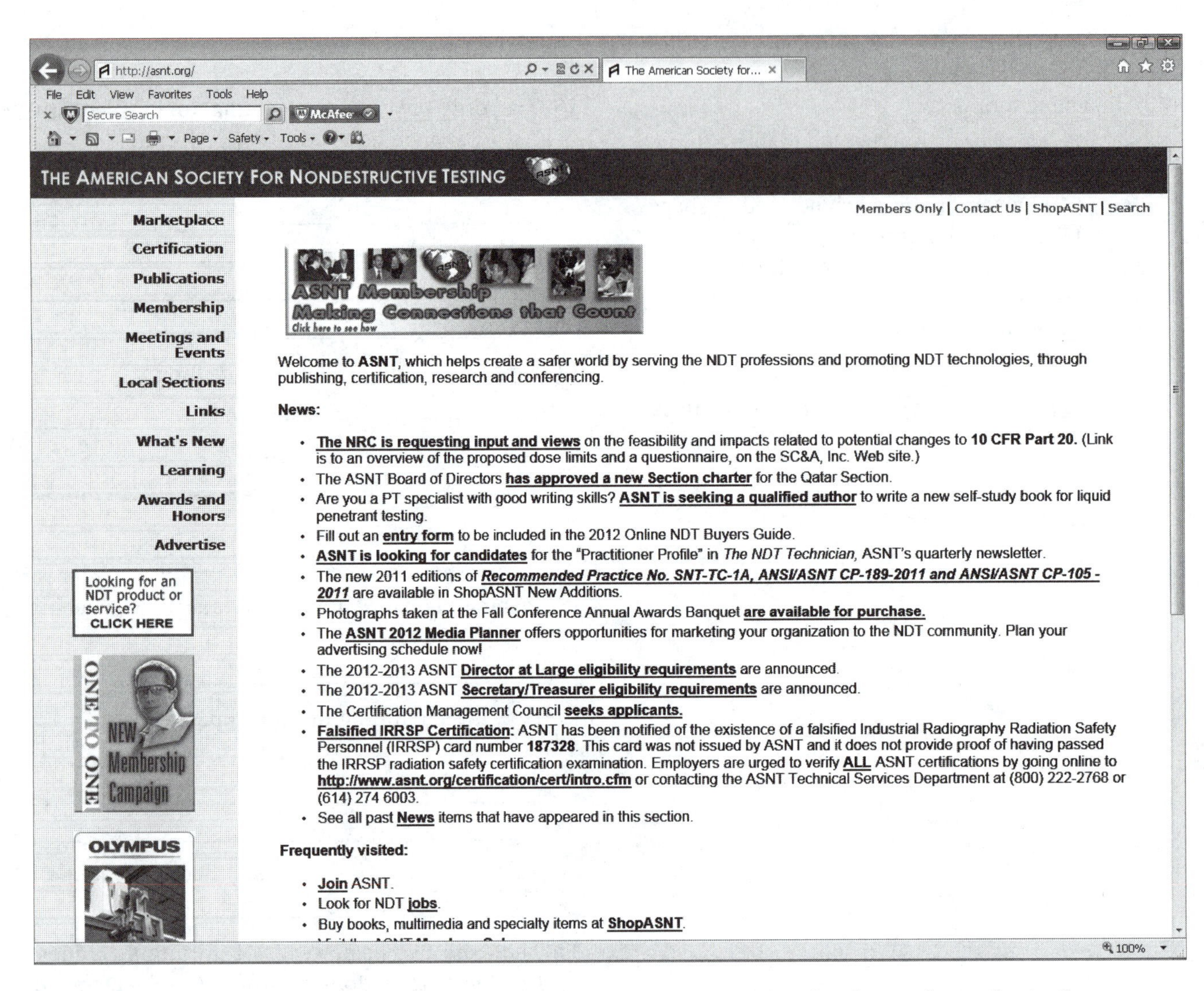

The American Society for Nondestructive Testing (www.asnt.org) is a technical society for nondestructive testing professionals. Their website contains information on how to become certified and find a job as a nondestructive testing professional.

Unit 24
Examining and Testing Welds

After completing Unit 24, you will be able to:

- Explain the purpose of welding codes.
- Describe the destructive test methods used to qualify welders.
- Identify basic nondestructive examination symbols for weldments.
- Interpret elements of an examination symbol.
- Identify the location and amount of testing specified on the print.
- Describe the major nondestructive testing methods.

Key Words

destructive examination
direction of radiation
discontinuities
examination symbol
guided bend test
indication
liquid penetrant testing (PT)
magnetic particle testing (MT)
nondestructive examination (NDE)
pulse echo testing
quality control
radiographic testing (RT)
tensile strength
tensile test
through inspection
transducer
ultrasonic testing (UT)
visual testing (VT)
welding code

The broad area of ***quality control*** includes methods to improve and maintain product quality, Figure 24-1. It is one of the most important segments of industry. Another purpose for quality control is to help reduce costs. It is also an important way to maintain product safety. Quality control plays a vital role in improving the competitive position of the manufacturer, which improves the chances for continued employment for the employees.

Various welding codes used in the United States require the use of nondestructive examination to protect the health and safety of employees and people working and living near potentially dangerous industrial processes and equipment.

A ***welding code*** is a document that specifies how a part or product should be designed, constructed, repaired, and tested. A few of the most common welding codes include the American Society of Mechanical Engineers (ASME) Boiler and Pressure Vessel Code and the American Welding Society (AWS) Structural Steel Code.

Figure 24-1.
An inspector examines a weldment as a quality control measure. (Caterpillar Inc.)

Basic Classifications of Quality Control Techniques

Quality control techniques fall into two basic classifications: destructive and nondestructive.

- With ***destructive examination,*** the part is destroyed during the examination program.
- With ***nondestructive examination,*** the examination is done in such a manner that the usefulness of the part is *not* impaired.

Destructive Examination

Destructive examination is costly and time consuming. The specimen is selected at random from a given number of pieces. Statistically, the results indicate the characteristics of the remaining undestroyed, unexamined pieces. It does not give complete assurance of perfection because a considerable number of defective parts could go undetected.

Guided Bend Test

Various metallurgical and strength tests are destructive. A very common destructive test used for both weld and welder testing is the ***guided bend test.*** This test is used to qualify welding procedures and to qualify welders for code welding. The test uses a standard joint design that is welded using one of the welding processes. Generally, .75″–1.5″ wide specimens are cut from the weldment for testing.

Three types of guided bend tests are used, depending on the thickness of the specimen. The face bend and root bend test are used for material thickness up to .375″. For material thickness over .375″, the specimens are machined down to a thickness of .375″, and a side bend is used.

Once prepared by machining, the specimen is pressed into a standard fixture until bent into a "U" shape or until it breaks. If the sample does not break, it is visually examined after bending for discontinuities (cracks or defects). The sample passes the bend test if no cracks are larger than code allows. Figure 24-2 shows a weld specimen being pressed into a guided bend fixture.

Tensile Test

Another destructive test is the ***tensile test,*** which applies a pulling load until the sample breaks. Tensile strength is used to compare welding filler metals. ***Tensile strength*** is the measure of a material's ability to withstand being pulled apart along a single axis. For an E7018 electrode, the minimum ultimate tensile strength is 70,000 lb. per in^2. Figure 24-3 shows a tensile test for a welded A36 structural steel plate. The machine used in the test pulls on each end of the sample until it stretches and breaks into two pieces. The maximum force applied to break the sample is recorded.

Figure 24-2.
Specimens are examined for cracks or defects after undergoing a guided bend test.

Figure 24-3.
A material's tensile strength is tested when forces are applied to pull and break it into two pieces. The maximum force applied to break it is recorded.

Nondestructive Examination (NDE)

Nondestructive examination is commonly used as a basic tool of the welding industry. It is well adapted to examine products where the performance of *each* part is critical. Each weld can be examined individually or as part of the completed assembly.

Basic Nondestructive Examination Symbols

Basic nondestructive examination symbols approved by the American Welding Society (AWS) are as follows:

Test	Symbol
Acoustic emission test	AET
Electromagnetic testing	ET
Leak testing	LT
Magnetic particle testing	MT
Neutron radiographic testing	NRT
Liquid penetrant testing	PT
Dye penetrant testing	DPT
Fluorescent penetrant testing	FPT
Proof testing	PRT
Radiographic testing	RT
Ultrasonic testing	UT
Visual testing	VT

Elements of the Examination Symbol

An ***examination symbol,*** similar to the welding symbol, is used to indicate which required nondestructive test to perform. Figure 24-4 summarizes the organization of an examination symbol.

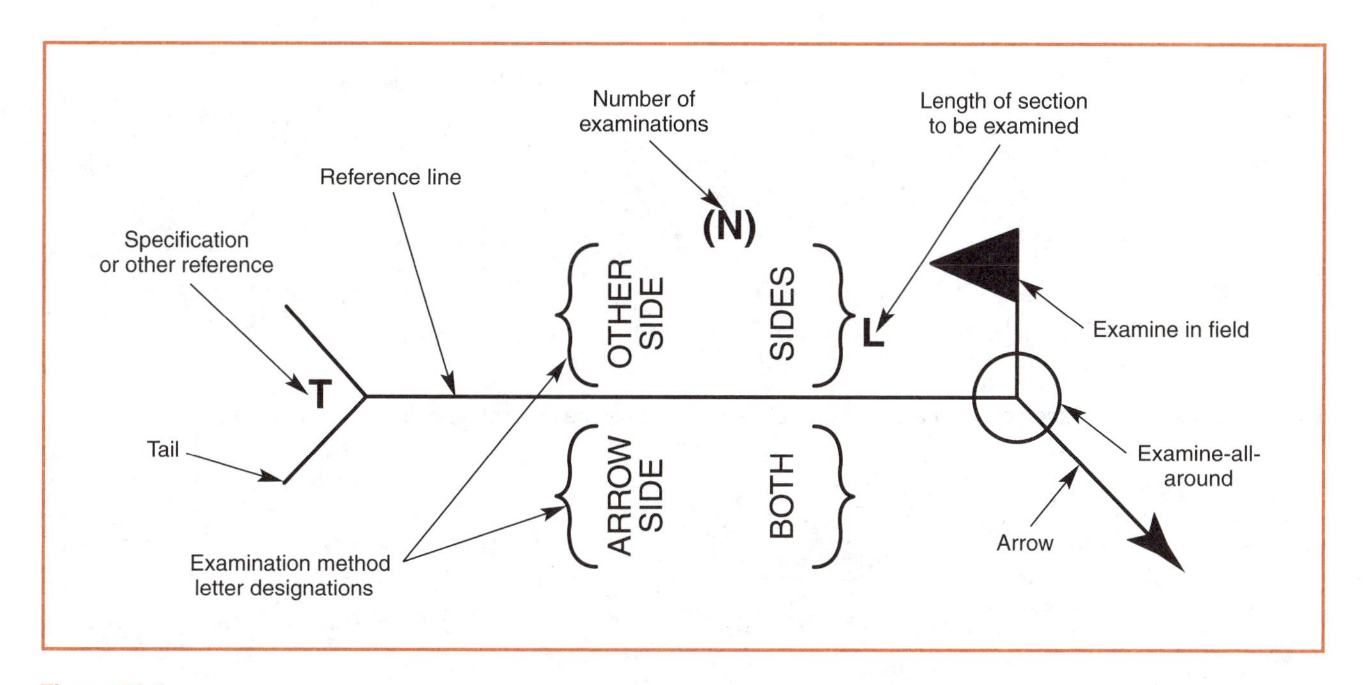

Figure 24-4.
An examination symbol showing nondestructive examinations required on a welded joint resembles a welding symbol. However, additional information must be interpreted correctly.

Location Significance of Arrow

The arrow connects the reference line to the weldment to be examined. The *arrow side* of the joint is the side of the welded joint to which the arrow points. The side opposite the arrow side is considered the *other side.*

Location of Examination Symbol

If the examination symbol is placed on the arrow side of the reference line (toward reader), the examination is to be made on the arrow side of the joint, Figure 24-5.

An examination to be made on the other side is indicated by the symbol being placed on the other side (away from reader) of the reference line, Figure 24-6.

Examinations to be made on both sides of the welded joint are indicated by examination symbols on both sides of the reference line, as in Figure 24-7.

The examination symbol is centered on the reference line when the examination has no arrow or other side significance. Figure 24-8 gives some examples.

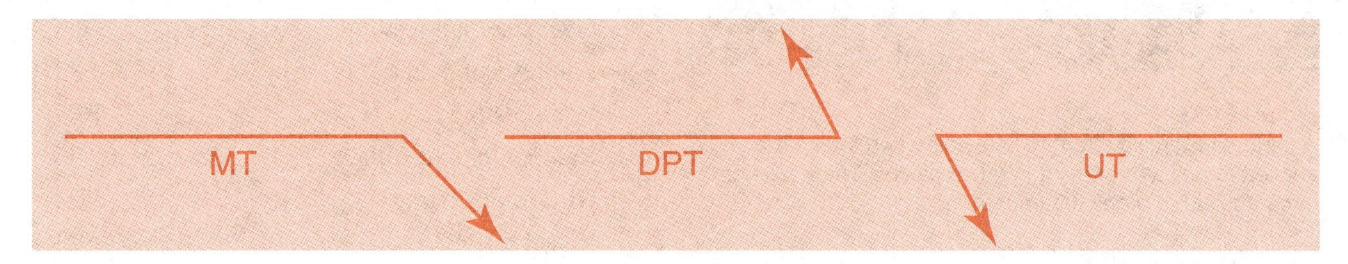

Figure 24-5.
These symbols show the examinations to be performed are on the arrow side of each joint.

Figure 24-6.
These symbols show the examinations to be performed are on the other side of each joint.

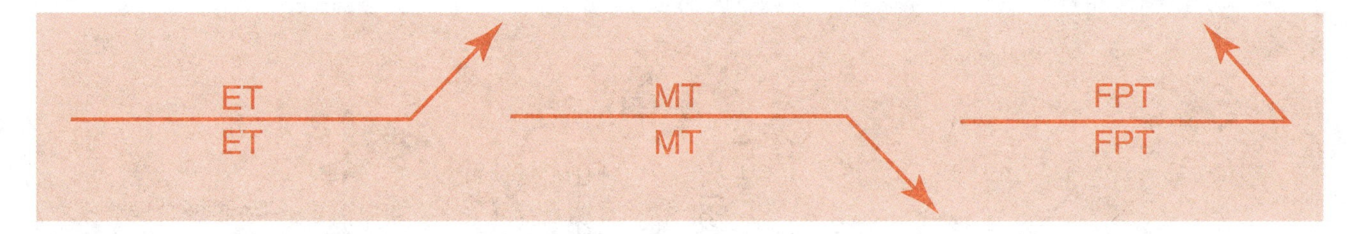

Figure 24-7.
Examination symbols placed on both sides of the reference line indicate inspection on both sides of the joint.

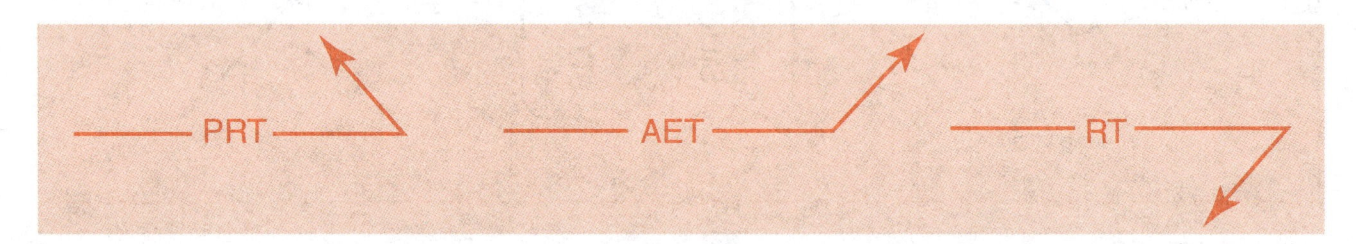

Figure 24-8.
If the examination symbol is centered on the reference line, no arrow or other side significance on the joint is indicated.

Direction of Radiation

The ***direction of radiation*** is specified by a symbol located on the drawing at the desired angle of the examination(s). It is in conjunction with the radiographic and neutron radiographic examination symbols, Figure 24-9.

Combination of Nondestructive Examination Symbols and Welding Symbols

Nondestructive examination symbols and welding symbols may be combined on a print. Two simple examples are illustrated in Figure 24-10.

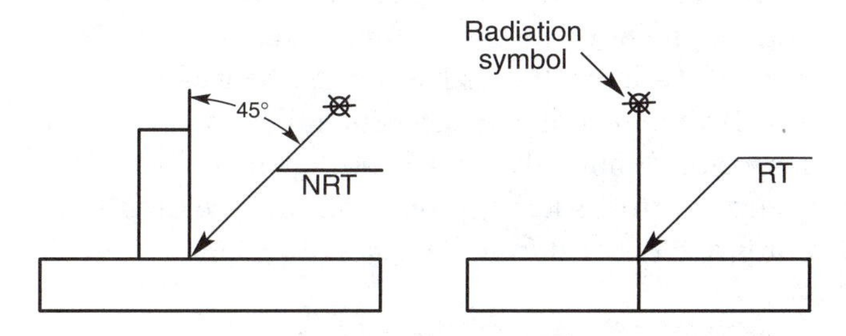

Figure 24-9.
Note the radiation symbol (star) and how direction of radiation is specified.

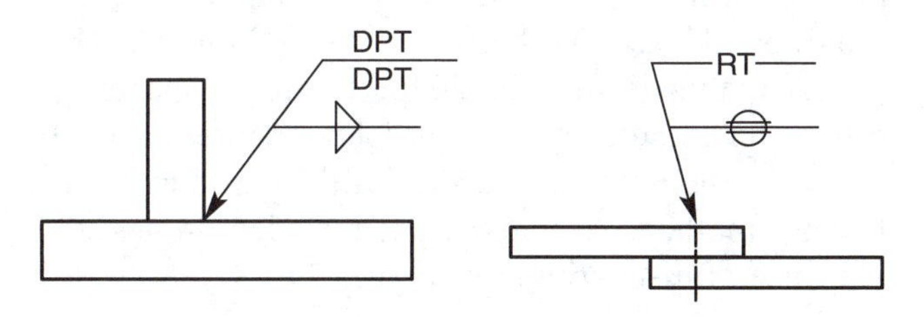

Figure 24-10.
Examination symbols are often used together with weld symbols on multiple reference lines.

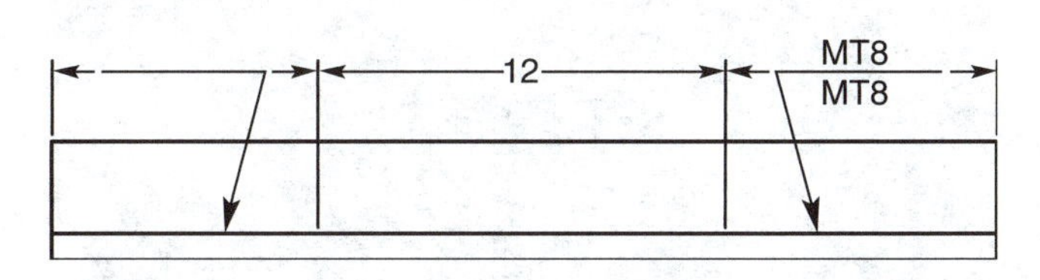

Figure 24-11.
The length of a weld section is specified for examination in a particular way. In this example, only the ends of the weldment are to be examined.

Specifying Length of Section to Be Examined

At times, a specific section of a welded joint must be examined. Dimension lines indicate the exact location and length of the examined section, Figure 24-11.

When the full length of a welded joint requires examination, no dimension is shown on the examination symbol.

If less than 100% of the weld is to be examined, the percentage of the length to be examined is indicated as shown in Figure 24-12. A note elsewhere on the print specifies examination locations.

Should random examinations be required along a weld, the number of examinations is shown in parentheses, Figure 24-13.

The examine-all-around symbol is used when the examination is to be made all around a joint. Appropriate dimensions locate the examination area. Figure 24-14 gives an example.

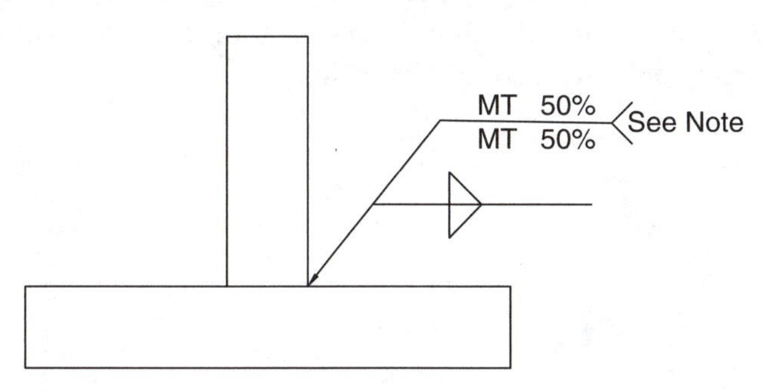

Figure 24-12.
When only a part of the weld length is to be examined, the procedure for selecting examination locations is indicated in the tail of the examination symbol.

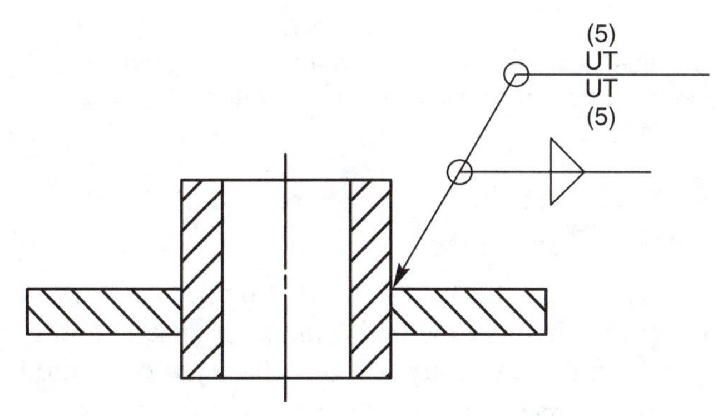

Figure 24-13.
In this example, five random examinations are to be made on each side of the specified weldment.

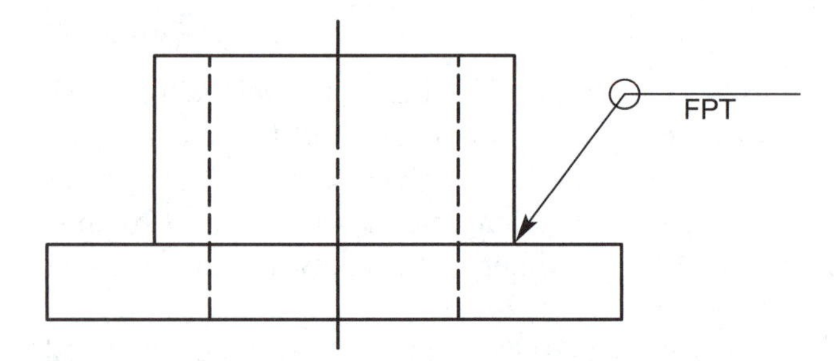

Figure 24-14.
The examine-all-around symbol is used when examination is required completely around the welded joint.

Specifying a Section or Plane Area to Be Examined

Nondestructive examination of an area or part of a unit (before or after being welded) is specified by enclosing the area with a series of straight, broken lines having a circle at each change of direction. The symbol specifying the kind of examination is also shown, Figure 24-15. The examination area may or may not be located by dimensions.

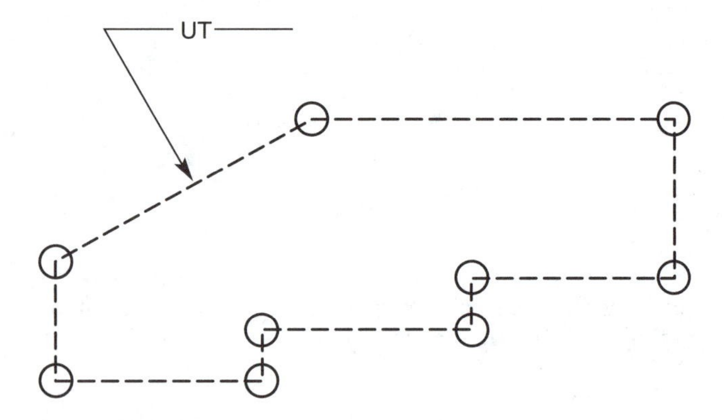

Figure 24-15.
Use this technique for indicating a specific area for nondestructive examination before or after making weld(s).

Major Nondestructive Examination (NDE) Methods

Although the list of nondestructive methods is fairly lengthy, five methods commonly used are discussed in the following sections. The type of method used depends on various factors:

- Type of material tested
- Whether both surfaces of the joint are accessible
- Method required by code
- If the ***discontinuities*** (flaws) are at the surface or subsurface of the weld

Visual Testing (VT)

The most common nondestructive testing method is ***visual testing (VT)***, Figure 24-16. A well-trained and experienced welding inspector can determine whether the weld is made to the specified print size, is in the correct location, and if there are any visual surface discontinuities. Visual testing requires less time than any other inspection technique, and when used during the process of fabricating, visual testing can catch errors at each step of the fabrication.

For other nondestructive inspection techniques, the inspector must interpret various indications present. An ***indication*** tells the inspector there is something present in the weld. The training, experience, and judgment of the inspector are required to determine if the indication is harmful to the weld.

Although other equipment might be used, the inspector commonly uses a pocket magnifier, fillet gauge, mirror, square, protractor, and welding standards and codes during a visual inspection.

Magnetic Particle Testing (MT)

Magnetic particle testing (MT) is a nondestructive testing method used to detect surface and slight subsurface discontinuities in ferromagnetic metals. The process is commonly used to detect cracks, porosity, a lack of fusion, inclusions, and other flaws too fine to be seen with the naked eye, Figure 24-17.

A magnetic field in the part being inspected is established, and magnetic particles (colored iron filings) are applied to the surface of the part during this testing method. Visual inspection determines the presence of discontinuities, Figure 24-18.

Figure 24-16.
This visual test is aided by a fiber optic scope to determine whether certain internal fittings within the aircraft have been made properly.

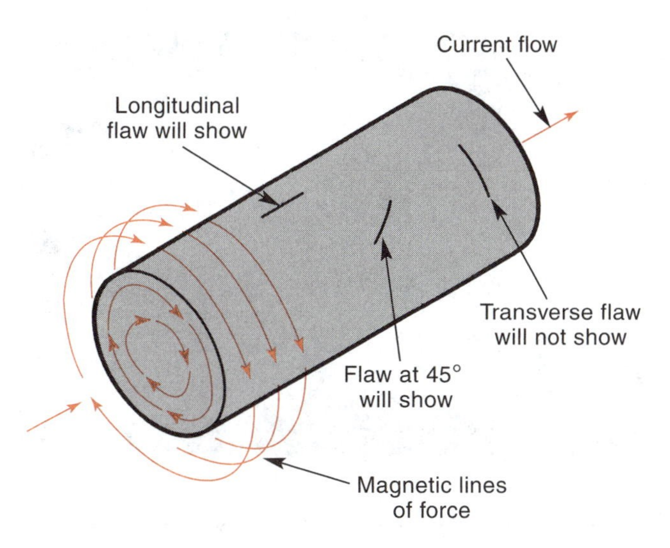

Figure 24-17.
The theory, scope, and limitations of the magnetic particle testing technique are illustrated. (Magnaflux Corp.)

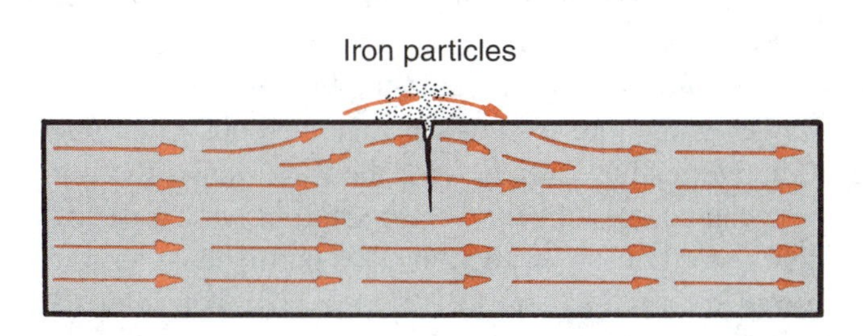

Figure 24-18.
A flaw in a steel bar generates a magnetic field outside the part to hold the iron particles. A buildup of iron particles makes even the tiny flaws visible.

Figure 24-19.
A magnetic yoke used in magnetic particle testing.

Equipment required for the process includes a magnetic yoke, Figure 24-19, or magnetic prods, magnetic particle powder, and a power supply for the magnetic field. The magnetic field may be developed with alternating current (ac) or direct current (dc). The magnetic particles are available in various colors to contrast with the color of the part (known as the dry method), or the particles may be suspended in a liquid (known as the wet method). Another modification is adding fluorescence to the particles. Then ultraviolet light shows formations of the particles around any defect.

The inspection method uses magnetic flux leakage as the principle of operation. Figure 24-20 shows how the magnetic lines of flux are changed by a crack in the part. The particles surround the crack and show as an indication.

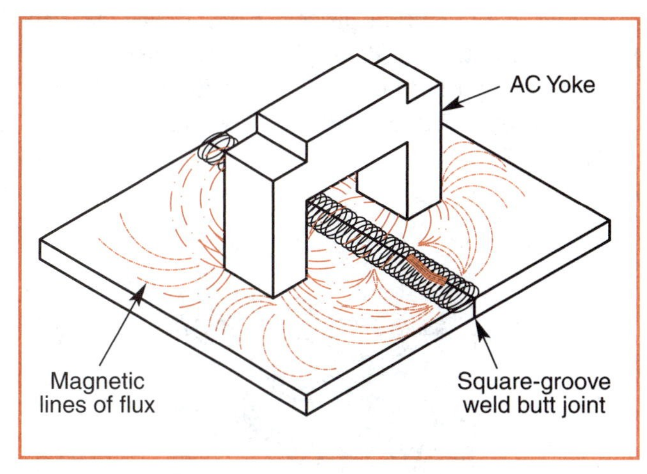

Figure 24-20.
Magnetic lines of flux are changed by a crack in the part.

Liquid Penetrant Testing (PT)

Liquid penetrant testing (PT) is a sensitive testing method used to find discontinuities open to the surface of a part. The testing method is capable of finding cracks, porosity, and other small discontinuities in most clean materials.

To use this inspection method, the surfaces must be clean and free of any loose rust, oil, or other surface material. The inspector coats the part with a dye penetrant (penetrating liquid) and allows it time (called dwell) to seep into any discontinuity, Figure 24-21A. Once the dwell time is complete, the inspector thoroughly removes the excess penetrant and applies a developer. The developer is usually a white powder that draws some of the penetrant out of the discontinuity. The contrast between the developer and the penetrant shows the flaw vividly, Figure 24-21B.

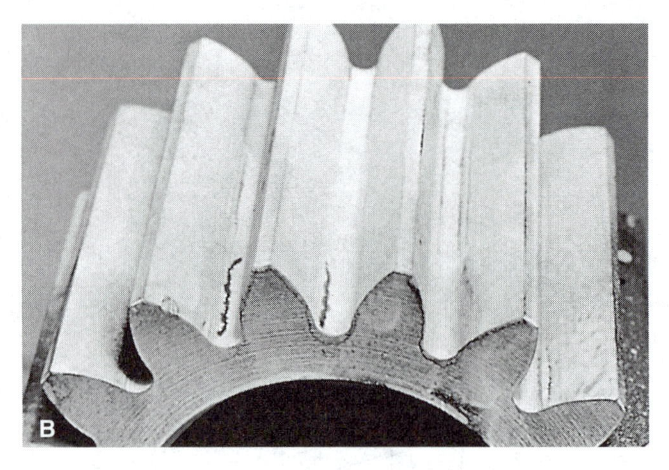

Figure 24-21.
The part must be thoroughly clean for a liquid penetrant test. A—Penetrant may be sprayed on and allowed to seep into any surface crack or flaw. Then it is washed off and the part is allowed to dry. B—Developer is applied, and flaws show up red against the white developer.

The dye penetrant process is relatively inexpensive and easy to use. To provide more sensitivity, the penetrant may be fluorescent and require an ultraviolet light for inspection of any surface flaws, Figure 24-22.

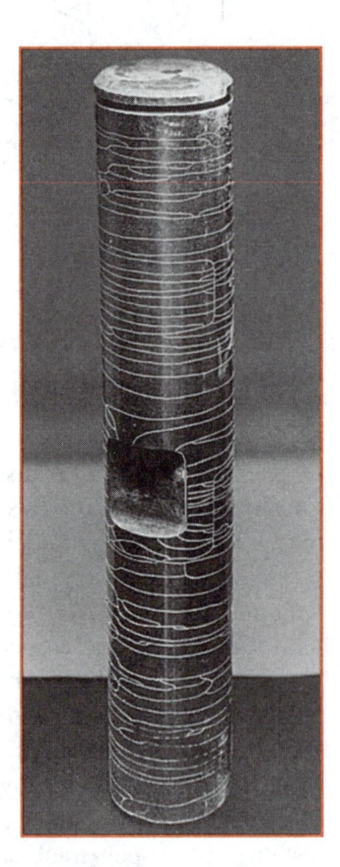

Figure 24-22.
This kingpin from a truck front axle was treated with a fluorescent penetrant testing material and photographed under ultraviolet light. The cracks show very clearly. (Magnaflux Corp.)

Radiographic Testing (RT)

Radiographic testing (RT) is a film exposure method that uses gamma rays or x-rays to detect internal discontinuities. The radiographic testing method is slower and more expensive than methods of nondestructive testing already described. However, the process can detect subsurface flaws and provide a permanent record of the inspection. This permanent record is useful for comparison to future radiographs after the weldment has been in service.

Gamma rays (emitted by a radioactive element) and x-rays (produced by using electricity) pass through materials to expose film placed on the opposite side of the part. The weld discontinuities generally show on the exposed film as darker areas because the discontinuities are less dense than the material being tested, Figure 24-23.

A trained or certified radiographic technician must interpret the indications. Overexposure to the radiation from this testing method is a safety concern. Proper training and the use of exposure badges are required by anyone working with radiographic equipment.

Ultrasonic Testing (UT)

Ultrasonic testing (UT) uses ultrasonic sound wave energy to penetrate into the material and detect flaws. Mechanical vibrations in wave form at frequencies above the range of human hearing produce the ultrasonic energy. An electronic device called a ***transducer*** is placed on the surface of the test part to transmit the waves. In ***pulse echo testing,*** the transducer is used as a transmitter and a receiver of the reflected energy. ***Through inspection*** transmits sound waves through one transducer, and a second transducer picks up the signal at the other end of the piece. See Figure 24-24.

Ultrasonic waves are reflected from the discontinuities in the weld or base metal (as well as the opposite surface of the part) and are displayed on a video screen. Subsurface discontinuities perpendicular to the ultrasonic energy are easily detected. However, depending on the shape, location, and part geometry, the ultrasonic energy may be scattered and a weak signal returned to the transducer.

The ultrasonic instrument must be calibrated before use, and the proper selection of the transducer is critical to the success of the testing process. Therefore, trained (and often certified) technicians must interpret the indications.

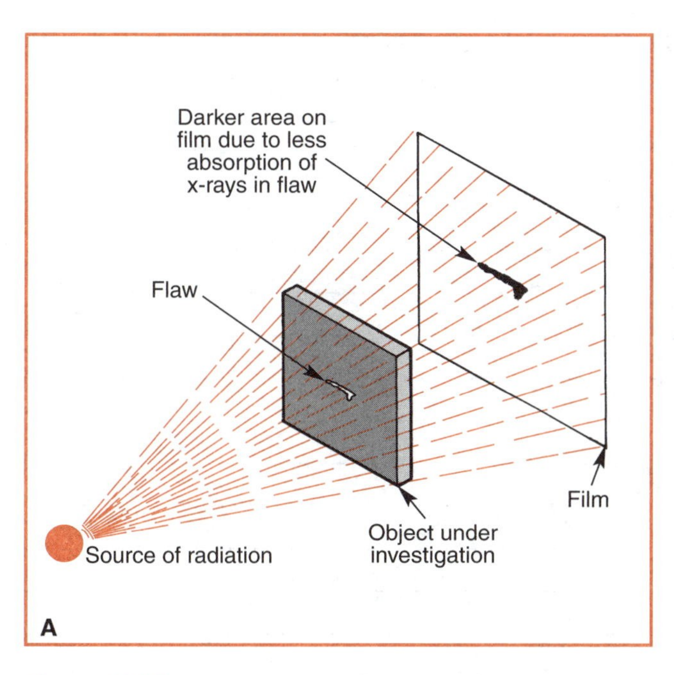

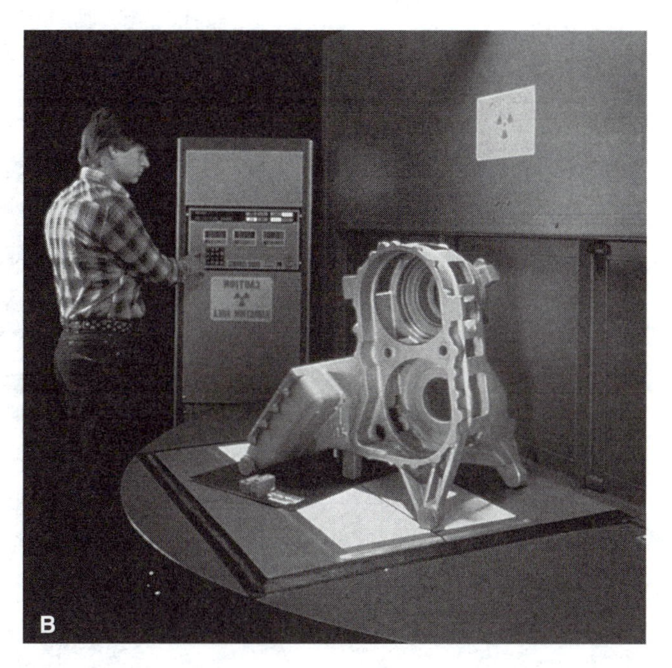

Figure 24-23.
A—The radiographic testing process is illustrated. B—A casting is being prepared for x-ray inspection to detect possible flaws. The quality control technician leaves the area before the part is x-rayed. (American Foundrymen's Society, Inc.)

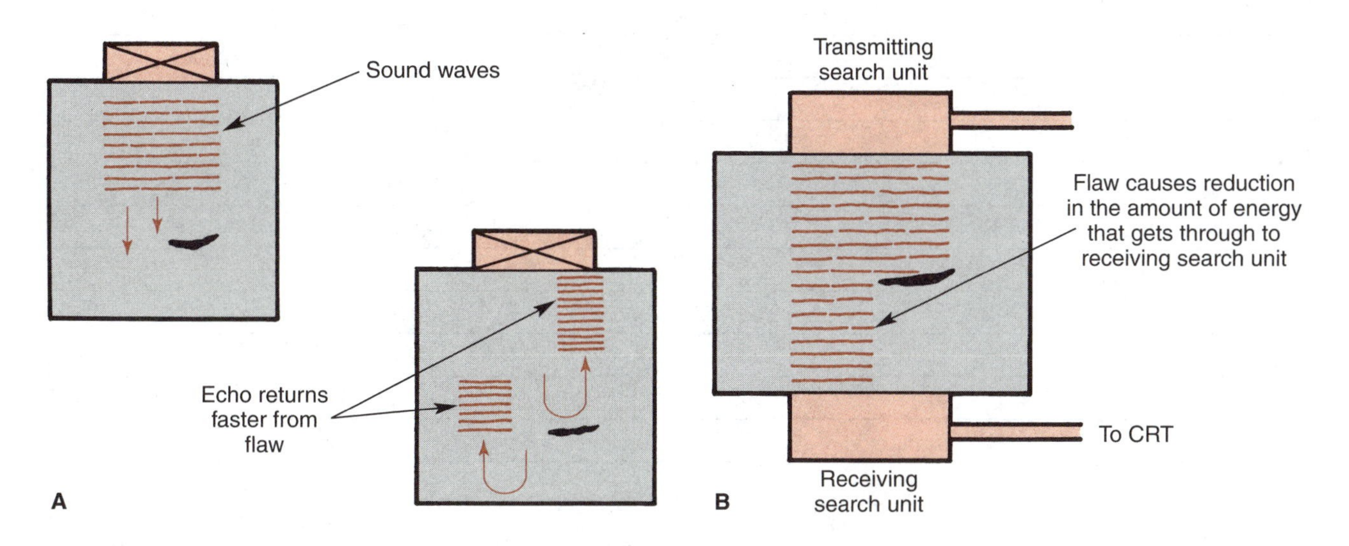

Figure 24-24.
Ultrasonic testing might be pulse echo or through testing. A—With pulse echo ultrasonic testing, reflected waves return sooner when bounced off a flaw. B—With through-type ultrasonic testing, a sensor detects waves on the opposite side of the piece.

Notes

Name ______________________ Date ______________ Class ______________

Review Questions

1. Why is quality control one of the most important segments of industry?

2. There are two basic classifications of quality control techniques. List them and briefly describe each.

3. What do the following mean on an examination symbol? Refer to the drawing.

   a. (N) = ______________________

   b. L = ______________________

   c. T = ______________________

   d. (symbol) = ______________________

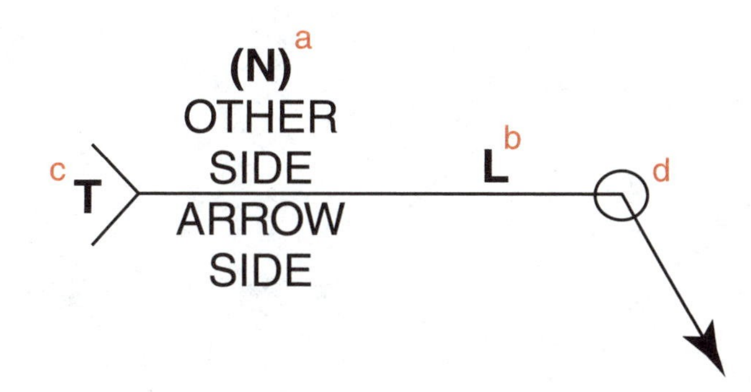

4. If the examination to be performed is placed on the arrow side of the reference line, the test is to be made on the _____.

5. If the examination to be performed is placed on the other side of the reference line, the examination is to be made on the _____.

6. Examinations to be made on both sides of the welded joint are indicated by _____.

7. A welding code is used to _____.

8. The nondestructive testing method that uses colored dye is _____.

9. The guided bend tests used to qualify welders are the face bend, _____, and _____.

10. The type of nondestructive test that uses sound waves is _____.

Many of the prints in Unit 25 are taken from real industry applications. Thrall Car was a family-owned business that began as a railcar repair facility and expanded to a railcar builder before being sold to a larger parent company. (Jorge Moro/ Shutterstock)

Unit 25 Print Reading Activities

This unit consists of additional prints and related questions. These prints are actual welding prints being used in industry today. They are presented as "real-world" activities to help prepare you to work with welding prints in the workplace. These activities are designed to give an opportunity for added practice of your welding print reading skills. It is suggested these activities be performed after the completion of the first 24 units, but they may be used at any time as a review. You should attempt to complete each of the activities before referring back to the first 24 units. Once you have completed all items in an activity, refer to the other units for help with any question you may have. The following is a list of the print reading activities and the page numbers where they can be found:

25A

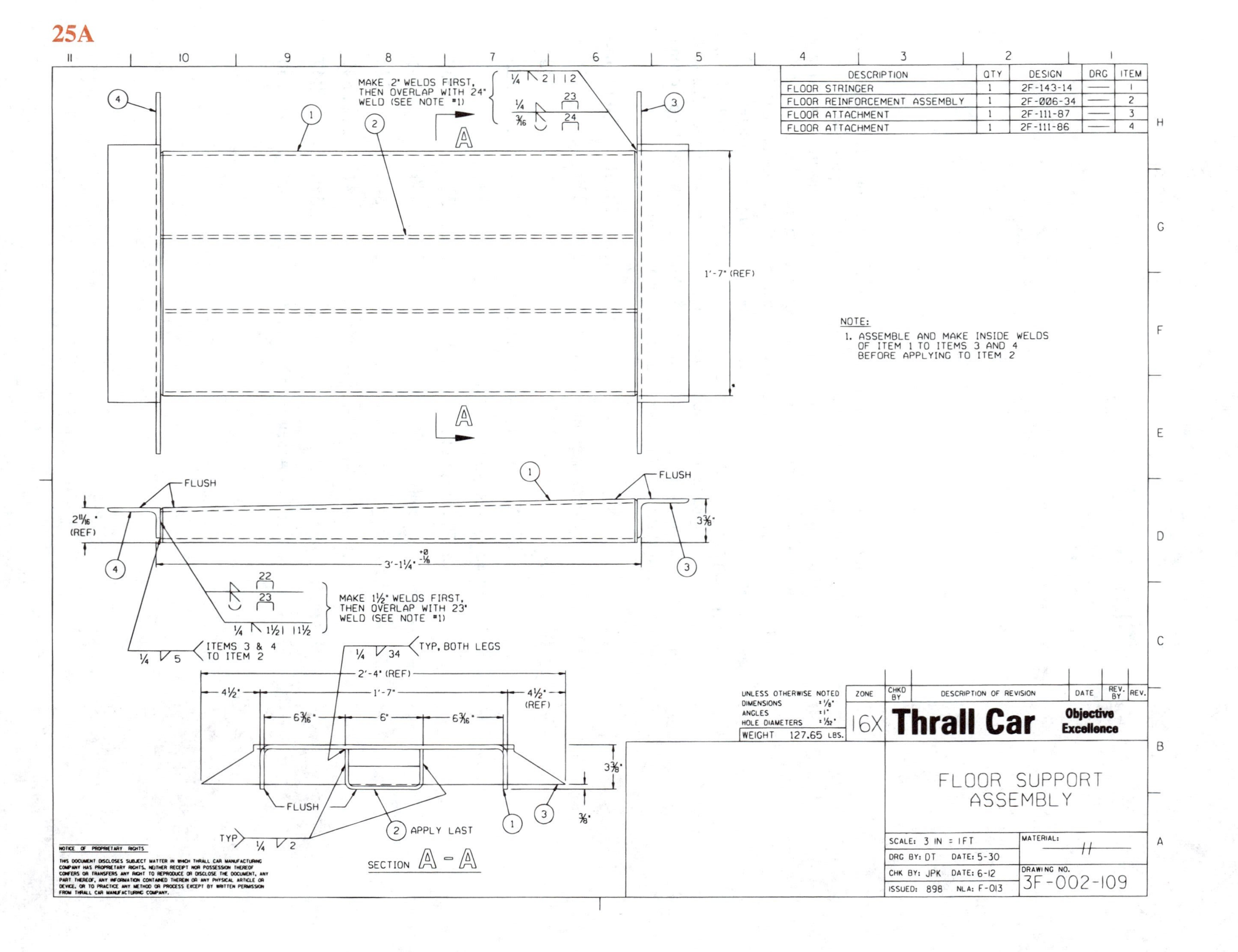

DESCRIPTION | QTY | DESIGN | DRG | ITEM
FLOOR STRINGER | 1 | 2F-143-14 | — | 1
FLOOR REINFORCEMENT ASSEMBLY | 1 | 2F-006-34 | — | 2
FLOOR ATTACHMENT | 1 | 2F-111-87 | — | 3
FLOOR ATTACHMENT | 1 | 2F-111-86 | — | 4
MAKE 2" WELDS FIRST, THEN OVERLAP WITH 24" WELD (SEE NOTE #1)
¼ 2 | 12
¼ 23
3⁄16 24
A
1'-7" (REF)
NOTE:
1. ASSEMBLE AND MAKE INSIDE WELDS OF ITEM 1 TO ITEMS 3 AND 4 BEFORE APPLYING TO ITEM 2
FLUSH
FLUSH
2 11⁄16" (REF)
3⅜"
3'-1¼" +0 -⅛
22
23
MAKE 1½" WELDS FIRST, THEN OVERLAP WITH 23" WELD (SEE NOTE #1)
¼ 1½ | 1½
¼ 5
ITEMS 3 & 4 TO ITEM 2
¼ 34
TYP, BOTH LEGS
2'-4" (REF)
4½"
1'-7"
4½" (REF)
6 3⁄16"
6"
6 3⁄16"
3⅜"
⅜"
FLUSH
TYP
¼ 2
APPLY LAST
SECTION A - A
NOTICE OF PROPRIETARY RIGHTS
THIS DOCUMENT DISCLOSES SUBJECT MATTER IN WHICH THRALL CAR MANUFACTURING COMPANY HAS PROPRIETARY RIGHTS. NEITHER RECEIPT NOR POSSESSION THEREOF CONFERS OR TRANSFERS ANY RIGHT TO REPRODUCE OR DISCLOSE THE DOCUMENT, ANY PART THEREOF, ANY INFORMATION CONTAINED THEREIN OR ANY PHYSICAL ARTICLE OR DEVICE, OR TO PRACTICE ANY METHOD OR PROCESS EXCEPT BY WRITTEN PERMISSION FROM THRALL CAR MANUFACTURING COMPANY.
UNLESS OTHERWISE NOTED
DIMENSIONS ±⅛"
ANGLES ±1°
HOLE DIAMETERS ±1⁄32"
WEIGHT 127.65 LBS.
ZONE
CHKD BY
DESCRIPTION OF REVISION
DATE
REV. BY
REV.
16X
Thrall Car
Objective Excellence
FLOOR SUPPORT ASSEMBLY
SCALE: 3 IN = 1FT
DRG BY: DT DATE: 5-30
CHK BY: JPK DATE: 6-12
ISSUED: 898 NLA: F-013
MATERIAL: //
DRAWING NO. 3F-002-109

Name ______________________________

25A—Floor Support Assembly

Read each question carefully, refer to the corresponding print, and write your answer in the space provided.

1. How many parts make up this assembly? ______________________________
2. Are the Floor Attachment Sections (Items 3 & 4) interchangeable? If they are not interchangeable, how do they differ? ______________________________
3. What is the name of the part welded into place last? ______________________________
4. What condition must be observed on the top surface of the assembly when welding the Floor Attachments (Items 3 & 4) to the Floor Stringer (Item 1)? ______________________________
5. When welding the Floor Attachment (Item 4) to the Floor Stringer (Item 1) a(n) _____ is first made at two places. ______________________________
6. After the above welds are made, they are overlapped with a(n) _____ with a(n) _____ contour. ______________________________
7. When welding the Floor Attachment (Item 3) to the Floor Stringer (Item 1), a(n) _____ is first made in two places. ______________________________
8. After the welds mentioned in Question 7 are made, they are overlapped with a(n) _____ with a(n) _____ contour. ______________________________
9. What must be done before the Floor Reinforcement Assembly (Item 2) can be welded into place? ______________________________
10. The requirement mentioned in Question 9 is specified in _____. ______________________________
11. The Floor Reinforcement Assembly (Item 2) is located _____″ from the interior side of the Floor Stringer (Item 1). ______________________________
12. What are the size, shape, and length of the exterior welds attaching the Floor Reinforcing Assembly (Item 2) to the Floor Attachments (Items 3 & 4)? ______________________________
13. What welds are used to join the legs of the Floor Reinforcing Assembly (Item 2) to the Floor Stringer (Item 1)? ______________________________
14. What are the size, shape, and length of the arrow side welds used to join the Floor Reinforcing Assembly (Item 2) to the Floor Attachments? ______________________________
15. How many places are the welds, mentioned in Question 14, made? ______________________________
16. The length of the assembled Floor Support Assembly is _____′ _____″ long from the exterior surfaces of the vertical legs of the Floor Attachments (Items 3 & 4). ______________________________
17. The tolerances allowed for the dimension mentioned in Question 16 are _____. ______________________________
18. Unless otherwise noted, what tolerances are allowable in the fabrication of the assembly?
    (a) Dimensions _____
    (b) Angles _____
    (c) Hole diameters _____
19. The drawing scale is _____. ______________________________

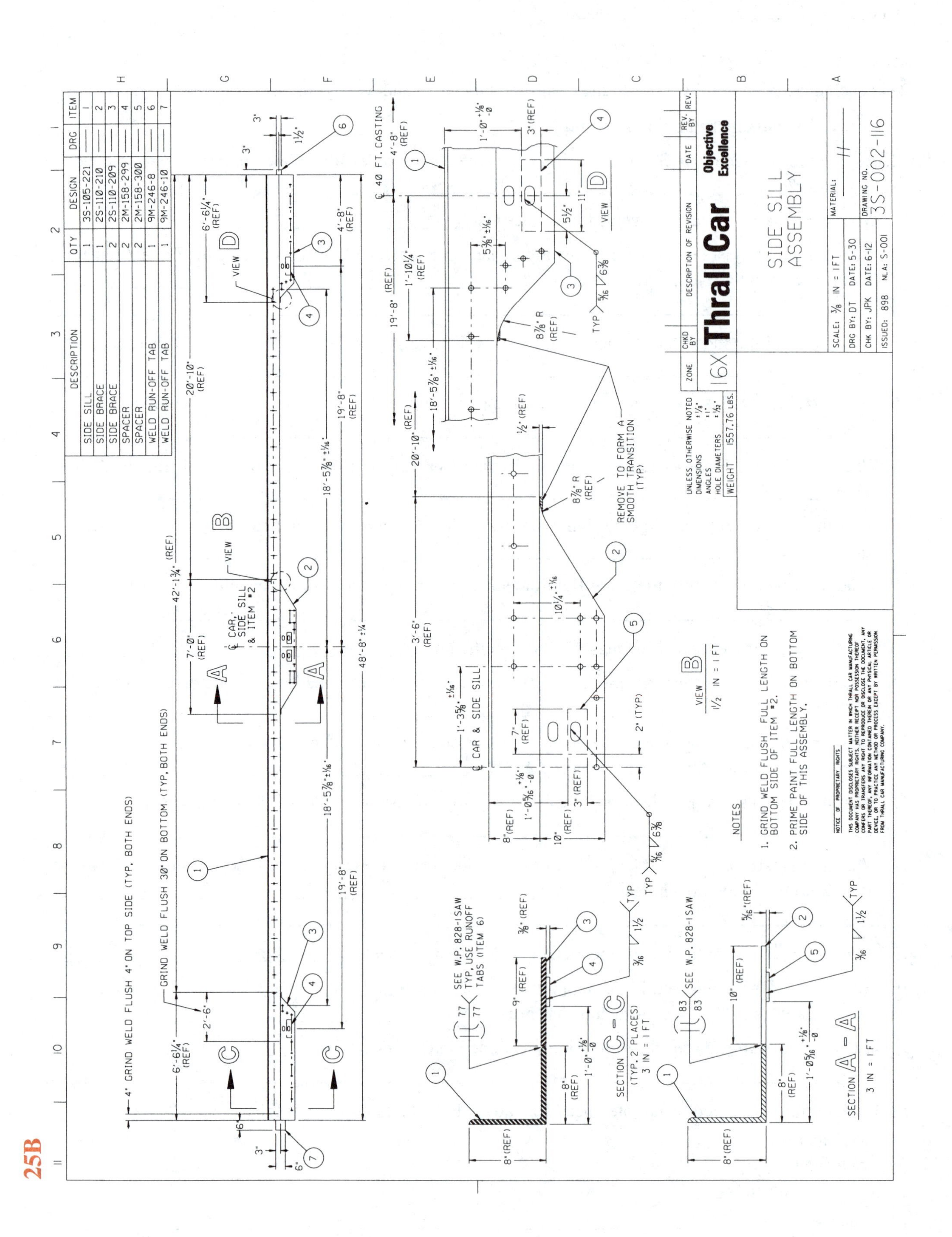
DESCRIPTION | QTY | DESIGN | DRG | ITEM
SIDE SILL | 1 | 3S-105-221 | | 1
SIDE BRACE | 1 | 2S-110-210 | | 2
SIDE BRACE | 2 | 2S-110-209 | | 3
SPACER | 2 | 2M-158-299 | | 4
SPACER | 2 | 2M-158-300 | | 5
WELD RUN-OFF TAB | 1 | 9M-246-8 | | 6
WELD RUN-OFF TAB | 1 | 9M-246-10 | | 7
GRIND WELD FLUSH 4" ON TOP SIDE (TYP, BOTH ENDS)
GRIND WELD FLUSH 30" ON BOTTOM (TYP, BOTH ENDS)
42'-1¾" (REF)
48'-8" ±¼
VIEW B
VIEW D
℄ CAR, SIDE SILL & ITEM #2
℄ 40 FT. CASTING
℄ CAR & SIDE SILL
REMOVE TO FORM A SMOOTH TRANSITION (TYP)
SEE W.P. 828-1SAW TYP, USE RUNOFF TABS (ITEM 6)
SEE W.P. 828-1SAW
SECTION C-C (TYP, 2 PLACES) 3 IN = 1 FT
SECTION A-A 3 IN = 1 FT
VIEW B ½ IN = 1 FT
NOTES
1. GRIND WELD FLUSH FULL LENGTH ON BOTTOM SIDE OF ITEM #2.
2. PRIME PAINT FULL LENGTH ON BOTTOM SIDE OF THIS ASSEMBLY.
UNLESS OTHERWISE NOTED
DIMENSIONS ±⅛"
ANGLES ±1°
HOLE DIAMETERS ±1/32"
WEIGHT 1557.76 LBS.
Thrall Car
Objective Excellence
SIDE SILL ASSEMBLY
SCALE: ⅜ IN = 1 FT
DRG BY: DT DATE: 5-30
CHK BY: JPK DATE: 6-12
ISSUED: 898 NLA: S-001
MATERIAL: //
DRAWING NO. 3S-002-116

25B

Name ____________________

25B—Side Sill Assembly

Read each question carefully, refer to the corresponding print, and write your answer in the space provided.

1. How many parts make up the assembly? ____________________

Refer to Section A-A

2. What are the parts shown in Section A-A? ____________________
3. What welds are specified to join the Side Sill (Item 1) to the Side Brace (Item 2)?
   (a) Arrow side weld ____________ (b) Other side weld ____________
4. What are the contour shapes for the weld mentioned in Question 3?
   (a) Arrow side weld ____________ (b) Other side weld ____________
5. A Spacer (Item 5) is welded to the Side Brace (Item 1). How far in from the edge of the Side Sill (Item 1) is it located? ____________________
6. What tolerances are permitted in the placement of the Spacer (Item 5)? ____________________
7. The first welds to join the Spacer (Item 5) to the Side Brace (Item 2) are made on _____ of the part.
   ____
8. What size, type, and length welds are specified for the weldments mentioned in Question 7?
   ____
9. After all the welds mentioned in Questions 7 and 8 are made, what additional operation is called for in the Notes?
   ____

Refer to Section C-C

10. What parts are shown in Section C-C? ____________________
11. What welds are specified to join the Side Sill (Item 1) to the Side Brace (Item 3)?
    (a) Arrow side weld ____________ (b) Other side weld ____________
12. What contours are specified for the welds mentioned in Question 11?
    (a) Arrow side weld ____________ (b) Other side weld ____________
13. Spacers (Item 4) are welded to the Side Braces (Item 3). They are located _____′ _____″ in from the Side Sill (Item 1) edge. ____________________
14. The tolerances permitted in the placement of the Spacers (Item 4) are _____.
    ____
15. Where are the first welds to be made when joining the Spacers (Item 4) to the Side Braces?
    ____
16. What size, type, and length welds are specified for the above weldments? ____________________
17. Run-off Tabs (Items 6 & 7) are used when joining the Side Braces (Item 3) to the Side Sill (Item 1) because they permit a(n) _____. ____________________

Refer to View B

18. The center of the Side Brace (Item 2) is located _____′ _____″ in from the end of the Side Sill (Item 1).
    ____
19. What size and type welds are indicated in the oblong openings for joining the Spacers (Items 4 & 5) to the Side Braces (Items 2 & 3)? ____________________

25C

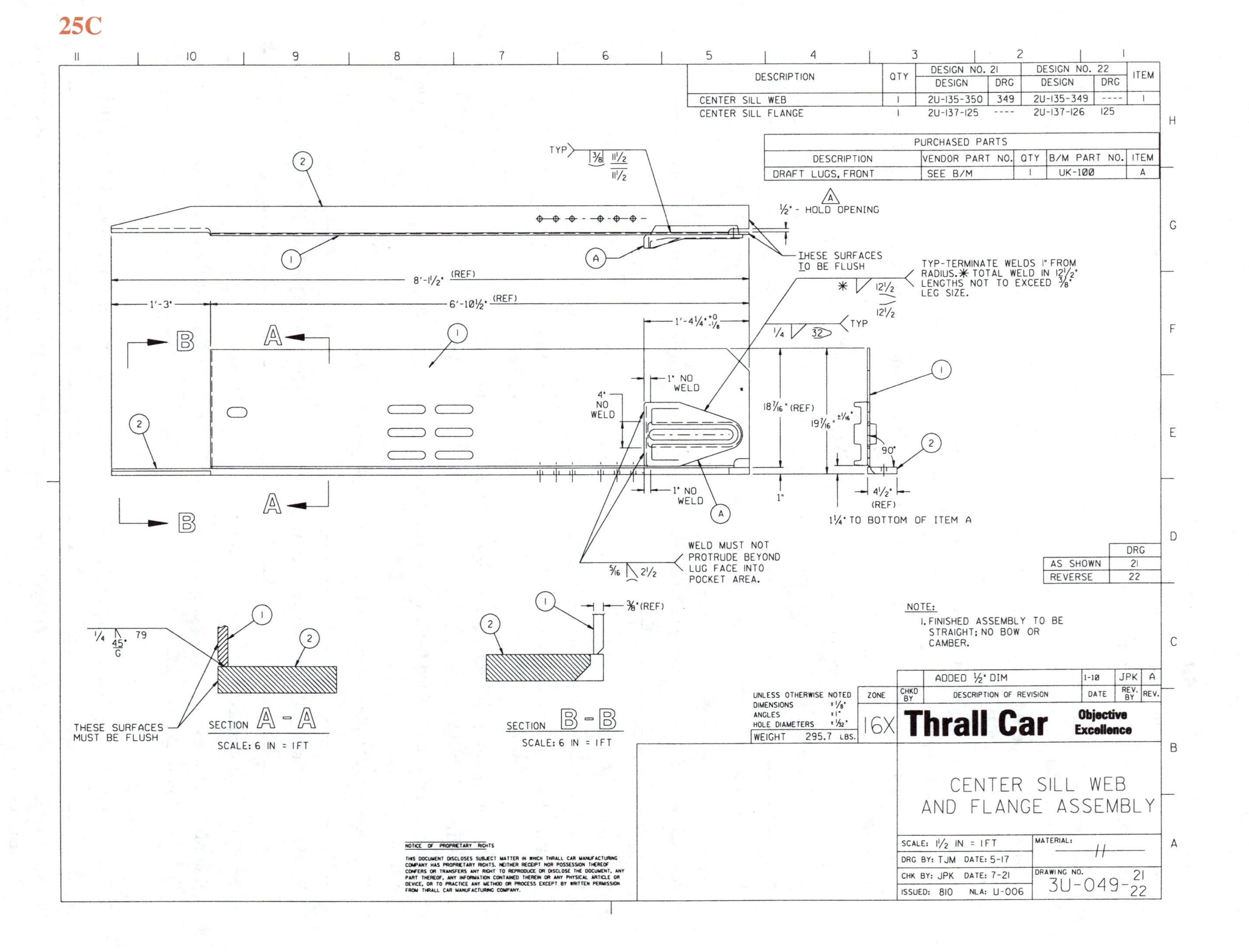

DESCRIPTION
QTY
DESIGN NO. 21
DESIGN NO. 22
ITEM
CENTER SILL WEB
CENTER SILL FLANGE
2U-135-350
349
2U-135-349
2U-137-125
2U-137-126
125
PURCHASED PARTS
VENDOR PART NO.
B/M PART NO.
DRAFT LUGS, FRONT
SEE B/M
UK-100
½" - HOLD OPENING
THESE SURFACES TO BE FLUSH
TYP-TERMINATE WELDS 1" FROM RADIUS. * TOTAL WELD IN 12½" LENGTHS NOT TO EXCEED ⅜" LEG SIZE.
8'-1½" (REF)
6'-10½" (REF)
1'-3"
1" NO WELD
4" NO WELD
18 7/16" (REF)
1¼" TO BOTTOM OF ITEM A
WELD MUST NOT PROTRUDE BEYOND LUG FACE INTO POCKET AREA.
AS SHOWN
REVERSE
NOTE:
1. FINISHED ASSEMBLY TO BE STRAIGHT; NO BOW OR CAMBER.
THESE SURFACES MUST BE FLUSH
SECTION A - A
SCALE: 6 IN = 1 FT
SECTION B - B
SCALE: 6 IN = 1 FT
UNLESS OTHERWISE NOTED
WEIGHT 295.7 LBS.
ADDED ½" DIM
DESCRIPTION OF REVISION
Thrall Car
Objective Excellence
CENTER SILL WEB AND FLANGE ASSEMBLY
SCALE: 1½ IN = 1 FT
DRG BY: TJM DATE: 5-17
CHK BY: JPK DATE: 7-21
ISSUED: 810 NLA: U-006
DRAWING NO.
3U-049-21 22
NOTICE OF PROPRIETARY RIGHTS

Name ___________________________________

25C—Center Sill Web and Flange Assembly

Read each question carefully, refer to the corresponding print, and write your answer in the space provided.

1. How many parts are used to fabricate this assembly? ___________________________
2. What are the names of the parts? ___________________________
3. On the drawing title block there are designations of Design #21 and Design #22. What do these numbers indicate?

   ____
4. What size, type, and length weld is specified to join the Center Sill Web (Item 1) to the Center Sill Flange (Item 2)? ___________________________
5. The Center Sill Web (Item 1) must also be _____° to the Center Sill Flange (Item 2) after welding.

   ____
6. The tolerance allowed for the angle mentioned in Question 5 is _____°. ___________________________
7. What does the Note specify? ___________________________

Refer to the Top View

8. The Draft Lug, Front (Item A) fits through a slot cut in the Center Sill Web (Item 1). Describe the welds specified to be made on the protruding side of the lug. ___________________________

   ____

Refer to the Front View

9. The print specifies two weld segments on the left edge of the Draft Lug, Front (Item A). Describe the arrow side welds. ___________________________

   ____
10. What special instructions are indicated when welding the other edges of the Draft Lug, Front (Item A)?

    ____

    ____

    ____

    ____

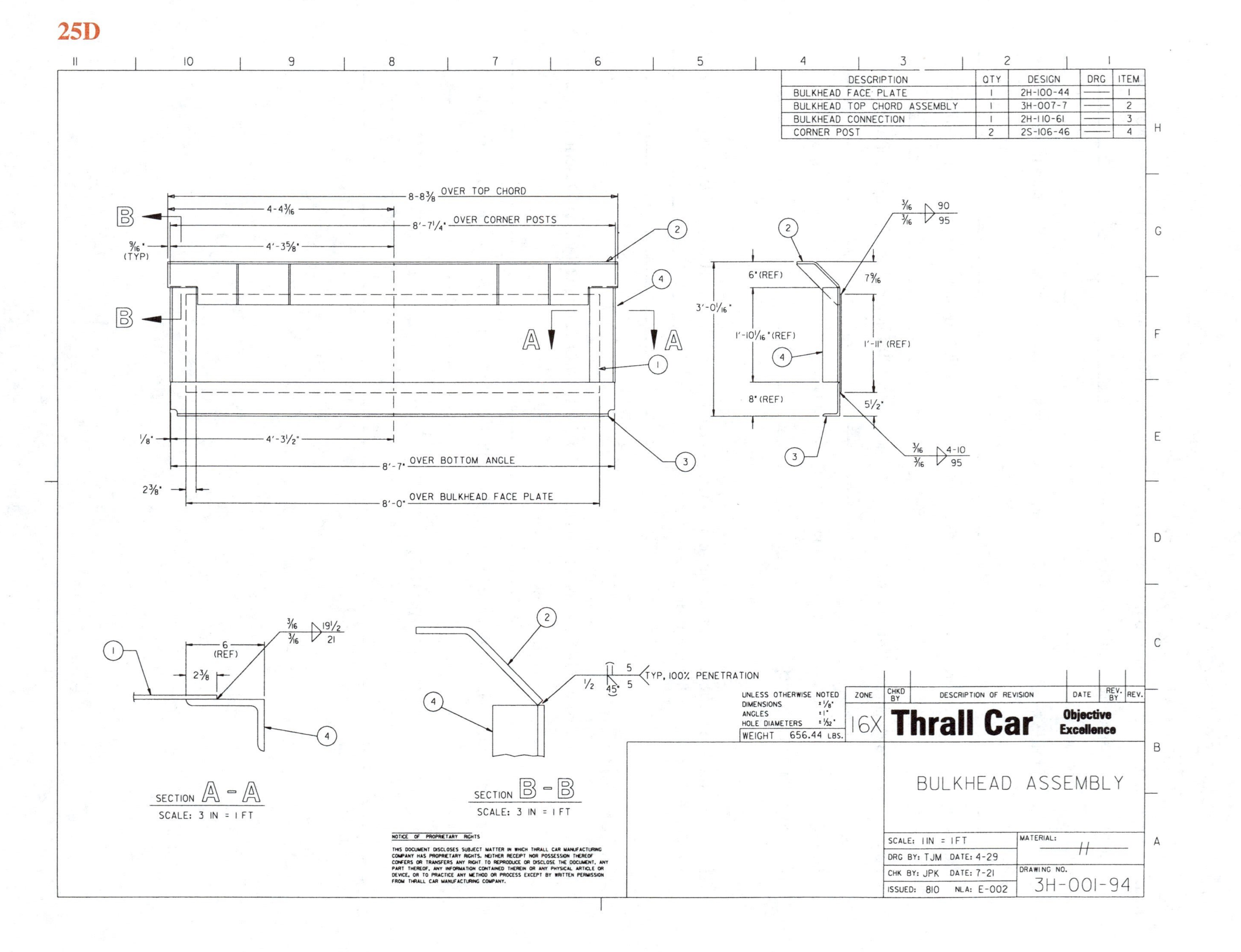

DESCRIPTION	QTY	DESIGN	DRG	ITEM
BULKHEAD FACE PLATE	1	2H-100-44	——	1
BULKHEAD TOP CHORD ASSEMBLY	1	3H-007-7	——	2
BULKHEAD CONNECTION	1	2H-110-61	——	3
CORNER POST	2	2S-106-46	——	4

Name ____

25D—Bulkhead Assembly

Read each question carefully, refer to the corresponding print, and write your answer in the space provided.

1. How many parts are used to fabricate this assembly? ____
2. What are the names of the parts? ____

Refer to Section A-A

3. What parts are shown in Section A-A? ____
4. The edge of the Bulkhead Face Plate (Item 1) is located _____″ from the end of the long leg of the Corner Post (Item 4). The tolerance allowed is _____. ____
5. List the specifications of the welds cited in Section A-A.
   (a) Arrow side weld ____________________ (b) Other side weld ____________________

Refer to Section B-B

6. What parts are shown in Section B-B? ____
7. What are the weld specifications shown in Section B-B?
   (a) Arrow side weld ____________________ (b) Other side weld ____________________
8. What are the contours specified for the welds mentioned in Question 7?
   (a) Arrow side weld ____________________ (b) Other side weld ____________________
9. Weld penetration is specified as _____% for the welds mentioned in Questions 7 and 8.
   ____

Refer to Front and Right Side View

10. In the assembly of this unit, the maximum allowable dimension over the Corner Posts (Item 4) is _____′ _____″. The minimum allowable dimension over the Corner Posts (Item 4) is _____′ _____″.
    ____
11. List the weld specifications for joining the Bulkhead Face Plate (Item 1) across the top of the Bulkhead Top Chord Assembly (Item 2) and Corner Posts (Item 4).
    (a) Arrow side weld ____________________ (b) Other side weld ____________________
12. What are the specifications of the welds joining the Bulkhead Face Plate (Item 1) to the Bulkhead Connection (Item 3)? ____
    (a) Arrow side weld ____________________ (b) Other side weld ____________________
13. What tolerances are allowable in the fabrication of the assembly?
    (a) Dimensions _____
    (b) Angles _____
    (c) Hole diameters _____
14. The sectional views are drawn to a scale of _____. The front and right side views are drawn to a scale of _____.
    ____
15. The Bulkhead Assembly weighs _____ lb. ____

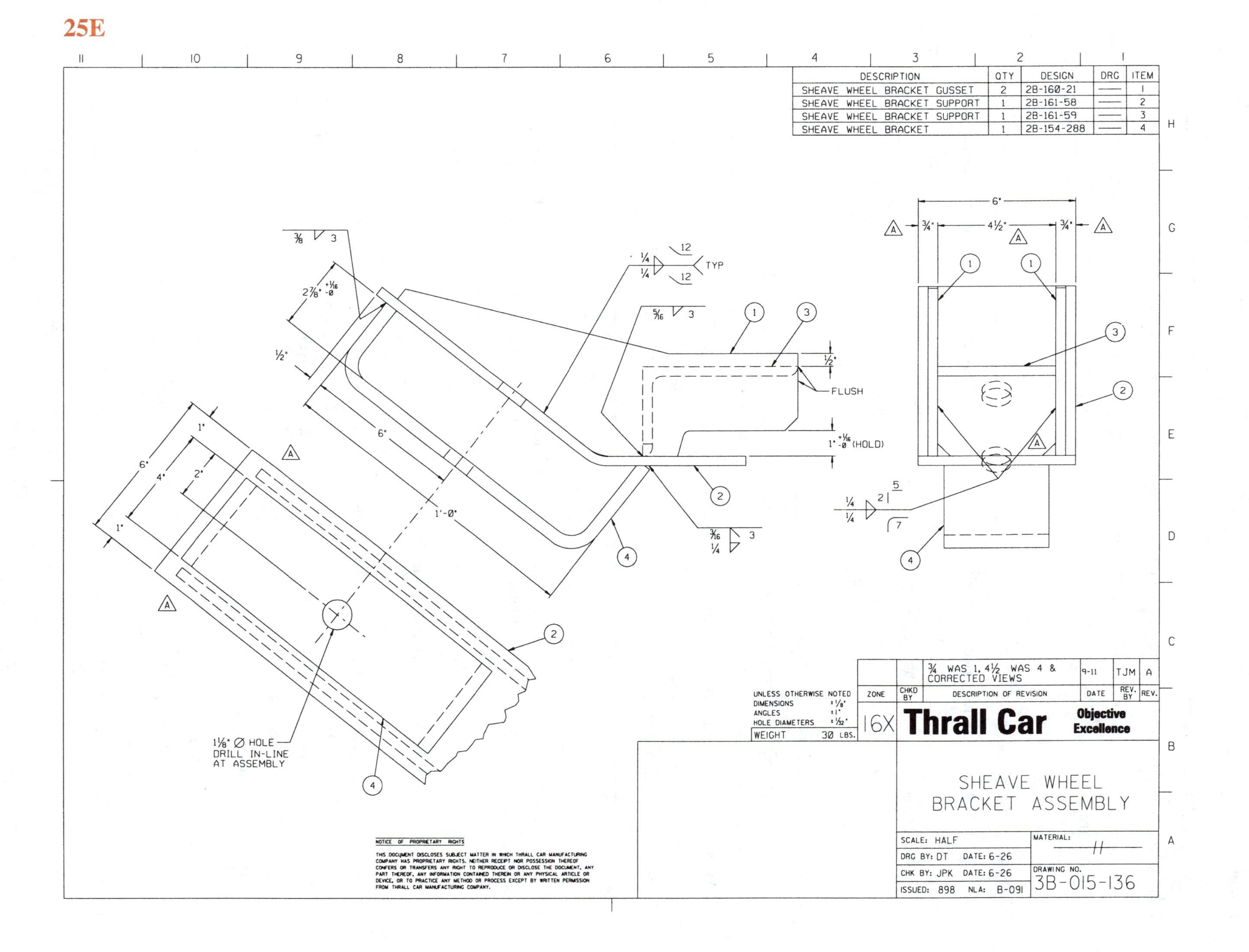

DESCRIPTION	QTY	DESIGN	DRG	ITEM
SHEAVE WHEEL BRACKET GUSSET	2	2B-160-21	—	1
SHEAVE WHEEL BRACKET SUPPORT	1	2B-161-58	—	2
SHEAVE WHEEL BRACKET SUPPORT	1	2B-161-59	—	3
SHEAVE WHEEL BRACKET	1	2B-154-288	—	4

Name ______________________________

25E—Sheave Wheel Bracket Assembly

Read each question carefully, refer to the corresponding print, and write your answer in the space provided.

1. How many parts are used to fabricate this assembly? ______________________________

2. What are the names of the parts? ______________________________

3. The short leg of the Sheave Wheel Bracket (Item 4) is welded to the Sheave Wheel Bracket Support (Item 2) by a(n) _____ weld. ______________________________

4. The long leg of the Sheave Wheel Bracket (Item 4) is welded to the Sheave Wheel Bracket Support (Item 2) by a(n) _____ weld. ______________________________

5. Make a sketch showing a cross section of this weld.

6. The inner floor of the Sheave Wheel Bracket (Item 4) must be located _____″ from the bottom of the Sheave Wheel Bracket Support (Item 2). Tolerances of _____ are permitted. ______________________________

7. The Sheave Wheel Bracket Gussets (Item 1) are joined to the Sheave Bracket Support (Item 3) by _____ welds. The arrow side weld is _____″ long. The other side weld is _____″ long on the vertical legs and _____″ long on the horizontal leg. ______________________________

8. In the welding unit mentioned in Question 7, the horizontal leg of the Sheave Bracket Support (Item 3) must be _____ with the R.H. edge of the Sheave Wheel Bracket Support (Item 1). ______________________________

9. The slot formed when the above assembly is fitted to the Sheave Wheel Bracket Support (Item 2) must be held to _____″. Tolerances of _____ are permitted. ______________________________

10. What welds are used to join the assembly of the Sheave Wheel Bracket Support (Item 3) and the Sheave Wheel Bracket Gussets (Item 1) to the Sheave Wheel Bracket Support (Item 2)?
    (a) Arrow side weld ______________________________ (b) Other side weld ______________________________
    (c) The symbol under the dimension indicates ______________________________.

11. The remainder of the assembly is joined to the Sheave Wheel Bracket Support (Item 2) with a(n) _____. ______________________________

12. The Sheave Wheel Bracket (Item 4) is positioned _____″ in from the L.H. end of the Sheave Wheel Bracket Support (Item 2) and _____ on the width of Item 2. ______________________________

13. Unless they are otherwise noted, what are the tolerances allowed on this assembly?
    (a) Dimensions______________________________ (b) Angles ______________________________
    (c) Hole diameters ______________________________

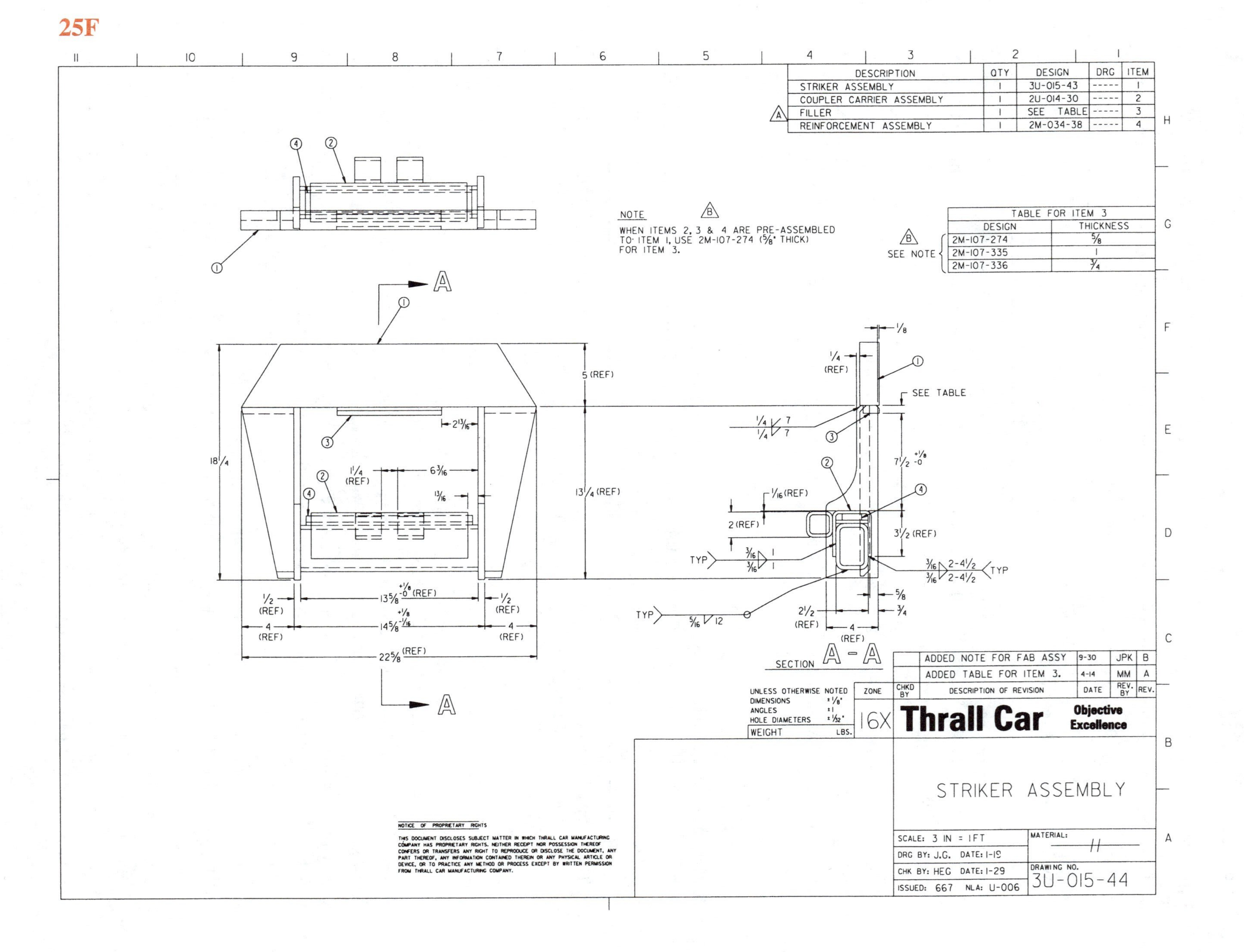

25F
DESCRIPTION | QTY | DESIGN | DRG | ITEM
STRIKER ASSEMBLY | 1 | 3U-015-43 | ----- | 1
COUPLER CARRIER ASSEMBLY | 1 | 2U-014-30 | ----- | 2
FILLER | 1 | SEE TABLE | ----- | 3
REINFORCEMENT ASSEMBLY | 1 | 2M-034-38 | ----- | 4
NOTE
WHEN ITEMS 2, 3 & 4 ARE PRE-ASSEMBLED TO ITEM 1, USE 2M-107-274 (5/8" THICK) FOR ITEM 3.
TABLE FOR ITEM 3
DESIGN | THICKNESS
2M-107-274 | 5/8
2M-107-335 | 1
2M-107-336 | 3/4
SEE NOTE
SEE TABLE
SECTION A-A
TYP
UNLESS OTHERWISE NOTED
DIMENSIONS ±1/8"
ANGLES ±1
HOLE DIAMETERS ±1/32"
WEIGHT LBS.
ADDED NOTE FOR FAB ASSY 9-30 JPK B
ADDED TABLE FOR ITEM 3. 4-14 MM A
ZONE CHKD BY DESCRIPTION OF REVISION DATE REV. BY REV.
16X
Thrall Car
Objective Excellence
STRIKER ASSEMBLY
SCALE: 3 IN = 1 FT
DRG BY: J.G. DATE: 1-19
CHK BY: HEG DATE: 1-29
ISSUED: 667 NLA: U-006
MATERIAL:
DRAWING NO. 3U-015-44
NOTICE OF PROPRIETARY RIGHTS

Name ______________________________

25F—Striker Assembly

Read each question carefully, refer to the corresponding print, and write your answer in the space provided.

1. How many parts are used to fabricate this assembly? ______________________________
2. What are the names of the parts? ______________________________

   ____

   ____
3. What welds are used to join the Filler (Item 3) to the Striker Assembly (Item 1)?

   (a) Arrow side weld ______________ (b) Other side weld ______________
4. The Filler (Item 3) is located _____″ in from the vertical leg of the Striker Assembly (Item 1).
5. When Items 2, 3, and 4 are preassembled to Item 1, the Filler (Item 3) is specified to be _____″ thick.

   ____
6. A _____ weld is used to join the Reinforcement Assembly (Item 4) to the Striker Assembly (Item 1).

   ____
7. The assembly mentioned in Question 6 is located _____″ in from the back edge of the Striker Assembly (Item 1).

   ____
8. The Carrier Assembly (Item 2) is located _____″ down from the bottom surface of the Filler (Item 3). Tolerances allowed are _____. ______________________________
9. What welds are made to join the Carrier Assembly (Item 2) to the Reinforcement Assembly (Item 4)?

   (a) Arrow side weld ______________ (b) Other side weld ______________
10. Describe the horizontal welds specified when joining the assemblies mentioned in Question 9.

    (a) Arrow side weld ______________ (b) Other side weld ______________
11. The total width of the Striker Assembly is _____″. ______________________________
12. The total height of the Striker Assembly is _____″. ______________________________
13. The thickness of the Striker Assembly is _____″. ______________________________
14. What are the allowable tolerances for the Striker Assembly?

    (a) Dimensions ______________ (b) Angles ______________

    (c) Hole diameters ______________
15. The scale of the original drawing is _____. ______________________________

25G

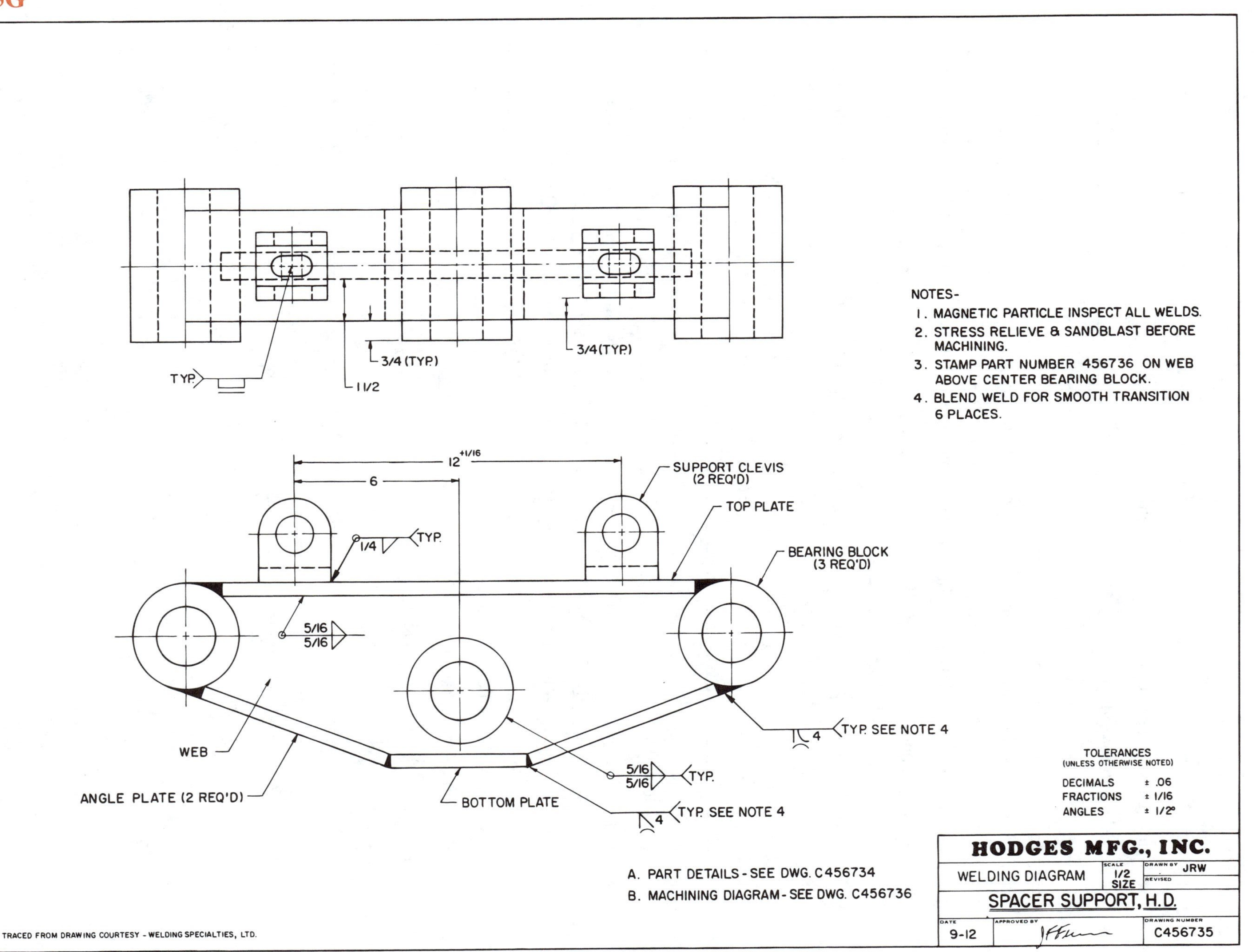
TYP.
3/4 (TYP.)
1 1/2
3/4 (TYP.)
NOTES-
1. MAGNETIC PARTICLE INSPECT ALL WELDS.
2. STRESS RELIEVE & SANDBLAST BEFORE MACHINING.
3. STAMP PART NUMBER 456736 ON WEB ABOVE CENTER BEARING BLOCK.
4. BLEND WELD FOR SMOOTH TRANSITION 6 PLACES.
12 +1/16
6
SUPPORT CLEVIS (2 REQ'D)
TOP PLATE
1/4
TYP.
BEARING BLOCK (3 REQ'D)
5/16
5/16
4
TYP. SEE NOTE 4
WEB
5/16
5/16
TYP.
ANGLE PLATE (2 REQ'D)
BOTTOM PLATE
4
TYP. SEE NOTE 4
TOLERANCES
(UNLESS OTHERWISE NOTED)
DECIMALS ± .06
FRACTIONS ± 1/16
ANGLES ± 1/2°
A. PART DETAILS - SEE DWG. C456734
B. MACHINING DIAGRAM - SEE DWG. C456736
HODGES MFG., INC.
WELDING DIAGRAM
SCALE 1/2 SIZE
DRAWN BY JRW
REVISED
SPACER SUPPORT, H.D.
DATE 9-12
APPROVED BY
DRAWING NUMBER C456735
TRACED FROM DRAWING COURTESY - WELDING SPECIALTIES, LTD.

Name ____

25G—Spacer Support, H.D.

Read each question carefully, refer to the corresponding print, and write your answer in the space provided.

1. How many parts are used to fabricate this assembly? ____________________
2. Describe the welds joining the Support Clevises to the Top Plate.
   (a) ____________________ (b) ____________________
3. The Support Clevises are located _____″ from the center line of the Middle Bearing Block. The full spacing between them is _____″ with a tolerance of _____ allowed. ____________________
4. The Clevises are spaced _____″ in from the long edge of the Top Plate. ____________________
5. What welds and contour are used to join the two Outer Bearing Blocks? ____________________
6. What special instructions are indicated for the completed welds mentioned in Question 5? ____________________
7. What welds are used to join the Center Bearing Block to the Web?
   (a) Arrow side weld ____________________ (b) Other side weld ____________________
8. The ends of the Bearing Blocks project _____″ from the long edge of the Top Plate.
   ____
9. Describe the weld and its contour that join the two Angle Plates to the Bottom Plate.
   ____
10. After completion, what additional operation must be performed on the welds mentioned in Question 9?
    ____
11. What welds are used to join the Top Plate, Angle Plates, Bottom Plate, and the two outer Bearing Blocks to the Web?
    (a) Arrow side weld ____________________ (b) Other side weld ____________________
12. How are the welds to be examined? ____________________
13. What additional operations must be performed on the assembly before the Bearing Blocks can be bored to size?
    ____
14. How is the finished unit identified? ____________________
15. Unless they are otherwise noted, what are the tolerances allowed on this assembly?
    (a) Decimals ____________________ (b) Fractions ____________________
    (c) Angles ____________________

25H

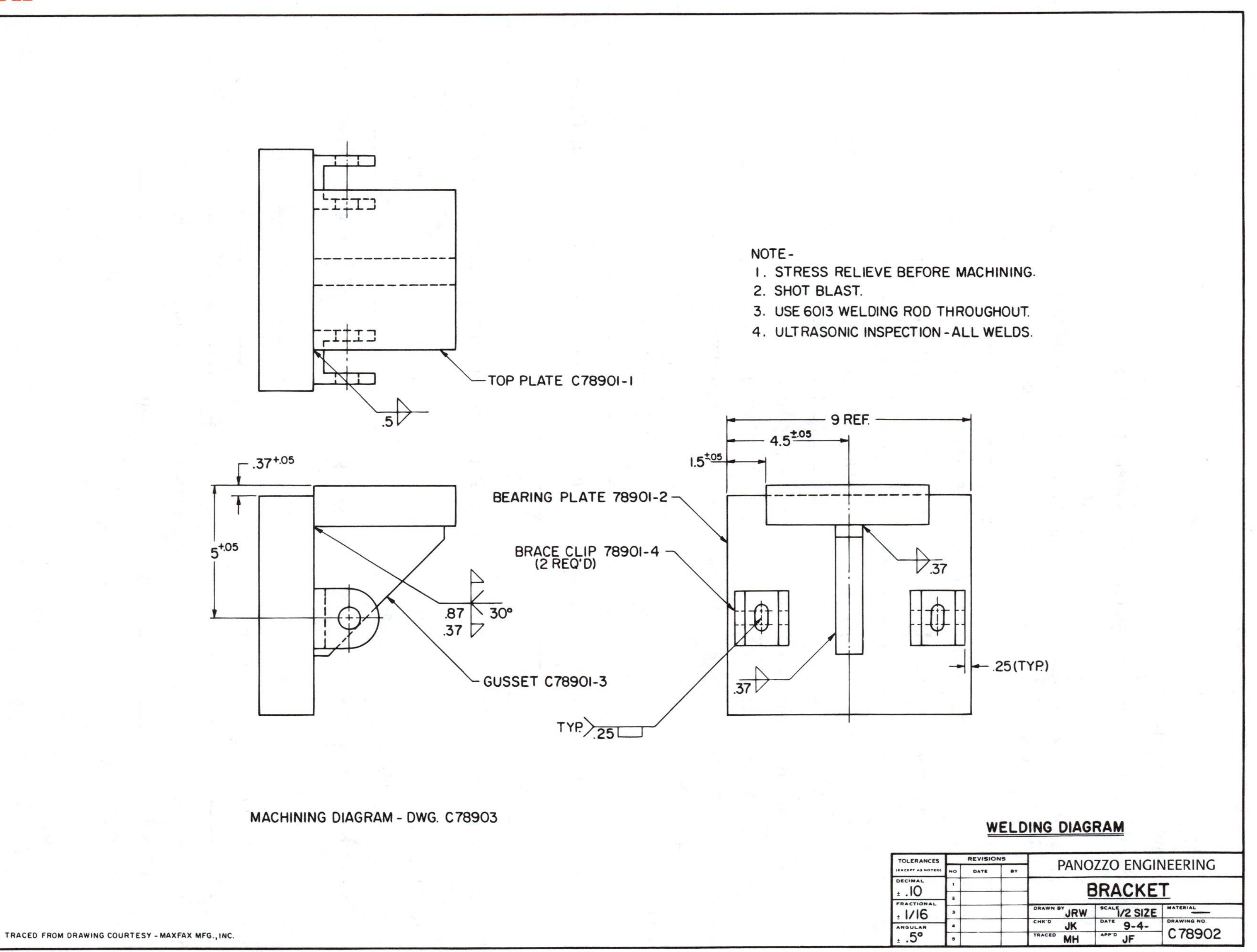
NOTE-
1. STRESS RELIEVE BEFORE MACHINING.
2. SHOT BLAST.
3. USE 6013 WELDING ROD THROUGHOUT.
4. ULTRASONIC INSPECTION-ALL WELDS.
TOP PLATE C78901-1
.5
9 REF.
4.5±.05
1.5±.05
.37+.05
5+.05
BEARING PLATE 78901-2
BRACE CLIP 78901-4
(2 REQ'D)
.37
.87
.37
30°
GUSSET C78901-3
.37
.25 (TYP.)
TYP.
.25
MACHINING DIAGRAM - DWG. C78903
WELDING DIAGRAM
TOLERANCES (EXCEPT AS NOTED)
DECIMAL ± .10
FRACTIONAL ± 1/16
ANGULAR ± .5°
REVISIONS
NO
DATE
BY
PANOZZO ENGINEERING
BRACKET
DRAWN BY JRW
SCALE 1/2 SIZE
MATERIAL —
CHK'D JK
DATE 9-4-
DRAWING NO. C78902
TRACED MH
APP'D JF
TRACED FROM DRAWING COURTESY - MAXFAX MFG., INC.

Name ______________________________

25H—Bracket

Read each question carefully, refer to the corresponding print, and write your answer in the space provided.

1. How many parts are used to fabricate the Bracket? ______________________
2. What type of weld is used to join the two side edges of the Top Plate to the Bearing Plate? ______________________
3. What type welds are used to join the Top Plate to the Bearing Plate?
   (a) Arrow side weld ______________________ (b) Other side weld ______________________
4. Sketch the weld specified in Question 3.
5. The top edge of the Bearing Plate is positioned _____″ below the top surface of the Top Plate. ______________________
6. The Top Plate is also positioned _____″ in from the vertical edge of the Bearing Plate. ______________________
7. The allowable tolerance for the positioning mentioned in Question 6 is _____. ______________________
8. What welds are used to join the Gusset to the Top Plate? ______________________
9. What welds are used to join the Gusset to the Bearing Plate? ______________________
10. The center line of the Brace Clip is located _____″ down from the top surface of the Top Plate. ______________________
11. The allowable tolerance for the positioning mentioned in Question 10 is _____. ______________________
12. The Brace Clips are located _____″ in from each vertical edge of the Bearing Plate. ______________________
13. _____ welds at a depth of _____″ are used to join the Brace Clips to the Bearing Plate. ______________________
14. Omission of a dimension showing the depth of a plug weld would indicate the filling is _____. ______________________
15. What operation must be performed on the Bracket after the welding process? ______________________
16. When the operation mentioned in Question 15 is completed, the Bracket is _____. ______________________
17. How are the welds to be inspected? ______________________
18. What is the welding rod specification? ______________________
19. Unless they are otherwise noted, what are the tolerances allowed on the Bracket?
   (a) Decimals ______________________ (b) Fractions ______________________
   (c) Angles ______________________
20. Scale of the original drawing is _____. ______________________

251

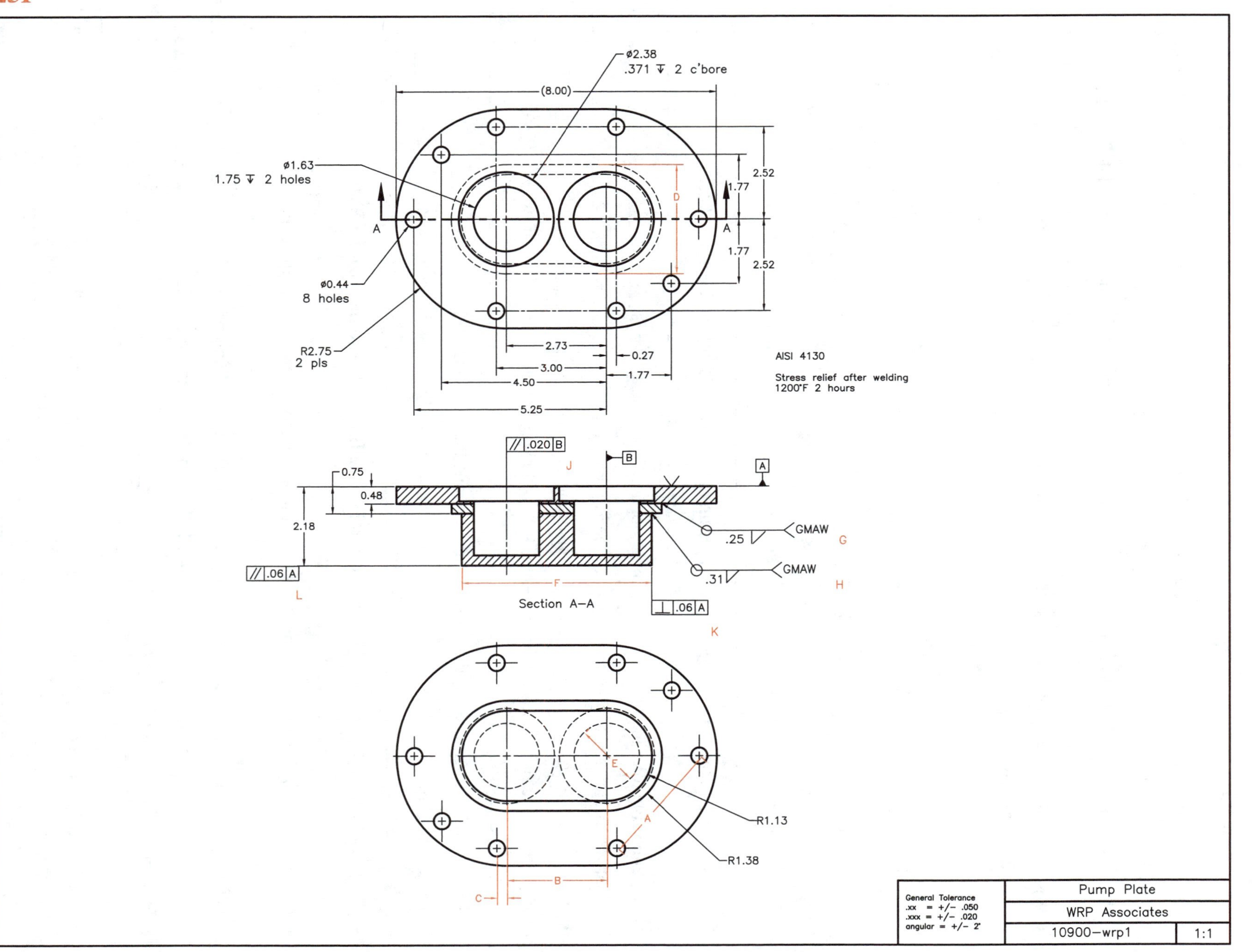
ø2.38
.371 ↧ 2 c'bore
(8.00)
ø1.63
1.75 ↧ 2 holes
A
A
D
2.52
1.77
1.77
2.52
ø0.44
8 holes
R2.75
2 pls
2.73
0.27
3.00
1.77
4.50
5.25
AISI 4130
Stress relief after welding
1200°F 2 hours
.020 B
J
B
A
0.75
0.48
2.18
GMAW
.25
G
GMAW
.31
H
.06 A
L
F
Section A–A
.06 A
K
E
A
R1.13
R1.38
B
C
General Tolerance
.xx = +/- .050
.xxx = +/- .020
angular = +/- 2°
Pump Plate
WRP Associates
10900–wrp1
1:1

Name ________________________________

25I—Pump Plate

Read each question carefully, refer to the corresponding print, and write your answer in the space provided.

1. How many parts are welded together to make the assembly? ________________________________
2. What type of section view is shown? ________________________________
3. List the dimension shown as a reference dimension. ________________________________
4. How many surfaces require finishing machining? ________________________________
5. What process is required after welding? ________________________________
6. What is the depth of the counterbore? ________________________________
7. How many ∅0.44 holes are required? ________________________________
8. List the material specification. ________________________________
9. What is the maximum center-to-center dimension for the ∅1.63 holes? ________________________________
10. List the views shown. ________________________________
11. Determine the dimensions at:

    A. ________________ B. ________________

    C. ________________ D. ________________

    E. ________________ F. ________________
12. Explain the welding symbol at:

    G. ________________________________

    H. ________________________________
13. Explain the feature control frames at:

    J. ________________________________

    K. ________________________________

    L. ________________________________
14. How many datums are listed on the print? ________________________________
15. List the general tolerance for 2-place decimal dimensions. ________________________________

25J

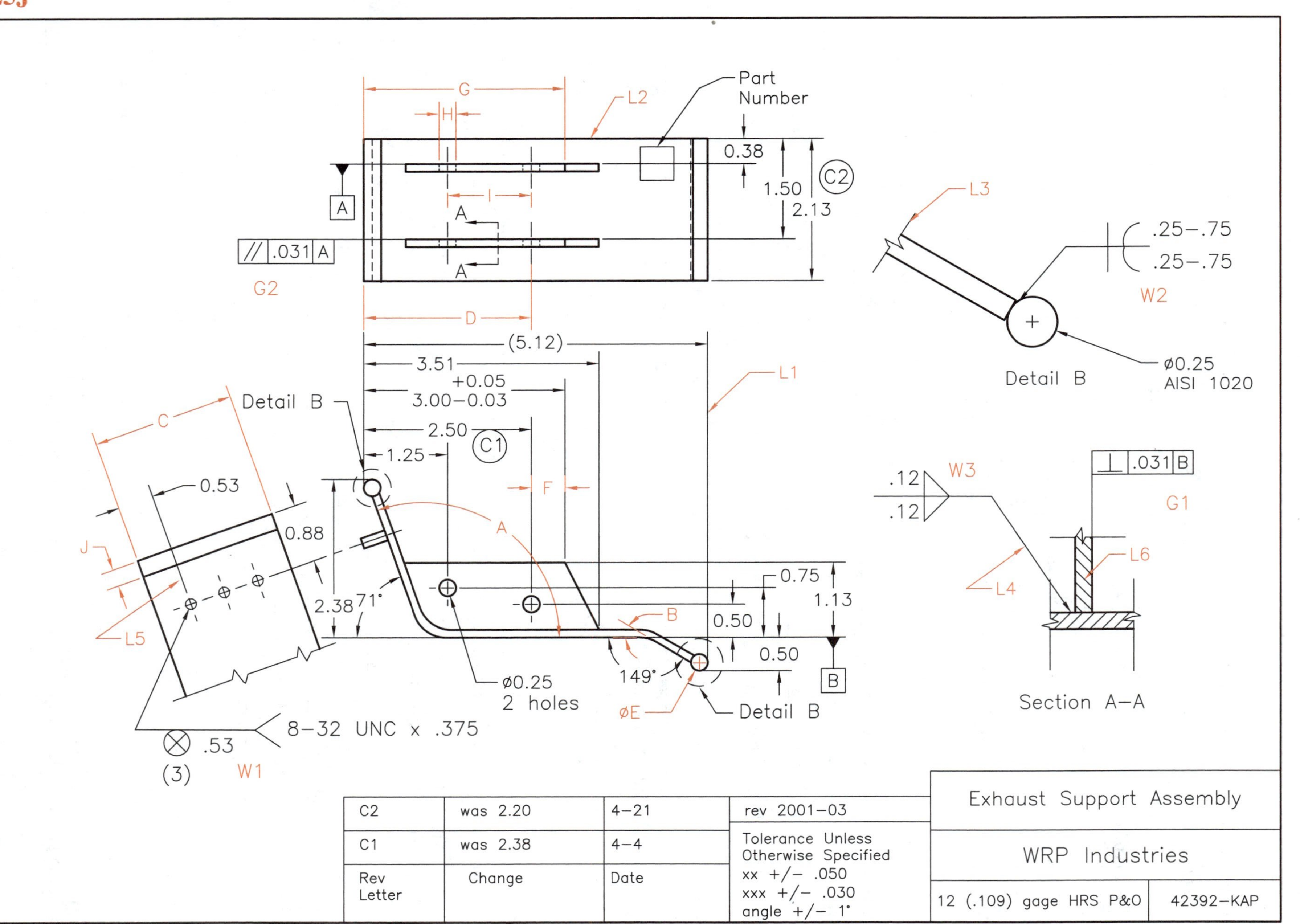
Part Number
G
H
I
A
D
L2
L1
L3
0.38
1.50
2.13
C2
.031 A
G2
(5.12)
3.51
+0.05
3.00–0.03
2.50
C1
1.25
F
C
0.53
0.88
J
2.38
71°
L5
Detail B
B
0.75
1.13
0.50
0.50
ø0.25
2 holes
149°
øE
8–32 UNC x .375
.53
(3)
W1
.25–.75
.25–.75
W2
Detail B
ø0.25
AISI 1020
.12
.12
W3
.031 B
G1
L4
L6
Section A–A
C2 | was 2.20 | 4–21
C1 | was 2.38 | 4–4
Rev Letter | Change | Date
rev 2001–03
Tolerance Unless Otherwise Specified
xx +/– .050
xxx +/– .030
angle +/– 1°
Exhaust Support Assembly
WRP Industries
12 (.109) gage HRS P&O
42392–KAP

Name ______________________________

25J—Exhaust Support Assembly

Read each question carefully, refer to the corresponding print, and write your answer in the space provided.

1. List the two (2) main views shown on the drawing. ______________________
2. List the tolerance for a 2-place decimal dimension. ______________________
3. What is the maximum size for the 3.00 dimension? ______________________
4. What type of section view is shown in A-A? ______________________
5. List the material specification for the part. ______________________
6. List the dimensioning system used on the print. ______________________
7. How many datums are indicated? ______________________
8. What is the length of the stud? ______________________
9. List the reference dimension. ______________________
10. What type of line is found at:

    L1. ______________ L2. ______________
    L3. ______________ L4. ______________
    L5. ______________ L6. ______________

11. Determine the following dimensions or angles:

    A. ______________ B. ______________
    C. ______________ D. ______________
    E. ______________ F. ______________
    G. ______________ H. ______________
    I. ______________ J. ______________

12. Measure and list the angles and determine their supplement.

	Angle	Supplement
A.	______________	______________
B.	______________	______________

13. List the hole diameter and number of holes specified for the part.

    Diameter ______________ Number of Holes ______________

14. Determine limits for the 3.00 dimension.

    Lower limit ______________ Upper limit ______________

15. Explain the feature control frame found at:

    G1 ______________ G2 ______________

(Continued)

25J—Exhaust Support Assembly (continued)

16. Explain the welding symbols found at:

    W1 ______________________ W2 ______________________

    W3 ______________________

17. List the thickness of the sheet used.

    Thickness ______________________ Gage ______________________

18. List the thread note for the welded stud. ______________________

19. List the view that indicates where the part number is stamped. ______________________

20. List the material specification for the ∅.25 round. ______________________

Reference Section

Welding Safety Checklist		
Hazard	**Factors to Consider**	**Precautionary Summary**
Electric shock can kill	• Wetness • Welder in or on workpiece • Confined space • Electrode holder and cable insulation	• Insulate welder from workpiece and ground using dry insulation, rubber mat, or dry wood. • Wear dry, hole-free gloves. (Change as necessary to keep dry.) • Do not touch electrically "hot" part or electrode with bare skin or wet clothing. • If wet area and welder cannot be insulated from workpiece with dry insulation, use a semiautomatic, constant-voltage welding machine or stick welding machine with voltage-reducing device. • Keep electrode holder and cable insulation in good condition. Do not use if insulation is damaged or missing.
Fumes and gases can be dangerous	• Confined area • Positioning of welder's head • Lack of general ventilation • Electrode types, i.e., manganese, chromium, etc. • Base metal coatings, galvanizing, paint	• Use ventilation or exhaust to keep air-breathing zone clear and comfortable. • Use helmet and position of head to minimize fume breathing zone. • Read warnings on electrode container and material safety data sheet for electrode. • Provide additional ventilation/exhaust where special ventilation requirements exist. • Use special care when welding in a confined area. • Do not weld unless ventilation is adequate.
Welding sparks can cause fire or explosion	• Containers that have held combustibles • Flammable materials	• Do not weld on containers that have held combustible materials (unless strict AWS F4.1 explosion procedures are followed). Check before welding. • Remove flammable materials from welding area or shield from sparks, heat. • Keep a fire watch in area during and after welding. • Keep a fire extinguisher in the welding area. • Wear fire-retardant clothing and hat. Use earplugs when welding overhead.
Arc rays can burn eyes and skin	• Process: gas-shielded arc most severe	• Select a filter lens that is comfortable for you while welding. • Always use helmet when welding. • Provide nonflammable shielding to protect others. • Wear clothing that protects skin while welding.
Confined space	• Metal enclosure • Wetness • Restricted entry • Heavier-than-air gas • Welder inside or on workpiece	• Carefully evaluate adequacy of ventilation, especially where electrode requires special ventilation or where gas may displace breathing air. • If basic electric shock precautions cannot be followed to insulate welder from work and electrode, use semiautomatic, constant-voltage equipment with cold electrode or stick welding machine with voltage-reducing device. • Provide welder helper and method of welder retrieval from outside enclosure.
General work area hazards	• Cluttered area	• Keep cables, materials, tools neatly organized.
	• Indirect work (welding ground) connection	• Connect work cable as close as possible to area where welding is being performed. Do not allow alternate circuits through scaffold cables, hoist chains, ground leads.
	• Electrical equipment	• Use only double insulated or properly grounded equipment. • Always disconnect power to equipment before servicing.
	• Engine-driven equipment	• Use only in open, well-ventilated areas. • Keep enclosure complete and guards in place. • Refuel with engine off. • If using auxiliary power, OSHA may require GFCI protection or assured grounding program (or isolated windings if less than 5 KW).
	• Gas cylinders	• Never touch cylinder with the electrode. • Never lift a machine with cylinder attached. • Keep cylinder upright and chained to support.

Safe Limits for Specific Welding Fumes			
Material	**Gases**		**Million Parts**
	ppm	**mg/m³**	**per ft³**
Acetylene	1000		
Beryllium		.002	
Cadmium oxide fumes		.1	
Carbon dioxide	5000		
Copper fumes		.1	
Iron oxide fumes		10.0	
Lead		.2	
Manganese		5.0	
Nitrogen dioxide	5.0		
Oil mist		5.0	
Ozone	.1		
Titanium oxide		15.0	
Zinc oxide fumes		5.0	
Silica, crystalline			2.5
Silica, amorphous			20.0
Silicates:			
Asbestos			5.0
Portland cement			50.0
Graphite			15.0
Nuisance dust			50.0

A table of safe limits for some welding fumes is shown. All gases tend to reduce oxygen by replacement. Such gases as argon, helium, carbon dioxide, etc., present this danger.

Metric Size Drafting Sheets

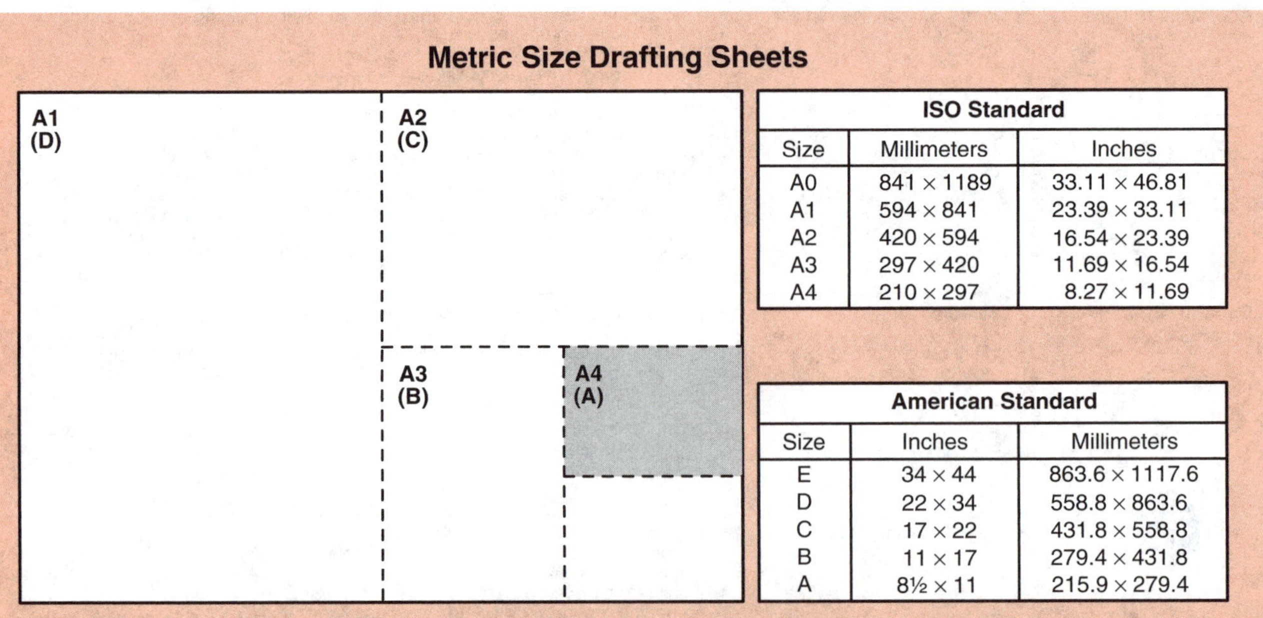

ISO Standard		
Size	Millimeters	Inches
A0	841 × 1189	33.11 × 46.81
A1	594 × 841	23.39 × 33.11
A2	420 × 594	16.54 × 23.39
A3	297 × 420	11.69 × 16.54
A4	210 × 297	8.27 × 11.69

American Standard		
Size	Inches	Millimeters
E	34 × 44	863.6 × 1117.6
D	22 × 34	558.8 × 863.6
C	17 × 22	431.8 × 558.8
B	11 × 17	279.4 × 431.8
A	8½ × 11	215.9 × 279.4

Left. Metric size drafting paper sheets are exactly proportional in size. Right. Charts show millimeter sizes and customary inch equivalent sizes of ISO papers.

Standard Abbreviations for Use on Drawings

A

Abrasive	ABRSV
Accessory	ACCESS
Accumulator	ACCUMR
Acetylene	ACET
Across Flats	ACR FLT
Actual	ACT
Actuator	ACTR
Addendum	ADD
Adhesive	ADH
Adjust	ADJ
Advance	ADV
Aeronautic	AERO
Alclad	CLAD
Alignment	ALIGN
Allowance	ALLOW
Alloy	ALY
Alteration	ALT
Alternate	ALT
Alternating Current	AC
Aluminum	AL
American National Standards Institute	ANSI
American Wire Gauge	AWG
Ammeter	AMM
Amplifier	AMPL
Anneal	ANL
Anodize	ANOD
Antenna	ANT
Approved	APPD
Approximate	APPROX
Arrangement	ARR
As Required	AR
Assemble	ASSEM
Assembly	ASSY
Automatic	AUTO
Auxiliary	AUX
Average	AVG

B

Babbitt	BAB
Base Line	BL
Battery	BAT
Bearing	BRG
Bend Radius	BR
Bevel	BEV
Bill of Material	B/M
Blueprint	BP or B/P
Bolt Circle	BC
Bracket	BRKT
Brass	BRS
Brazing	BRZG
Brinell Hardness Number	BHN
Bronze	BRZ
Brown & Sharpe (Gage)	B&S
Burnish	BNH
Bushing	BUSH

C

Cabinet	CAB
Calculated	CACL
Cancelled	CANC
Capacitor	CAP
Capacity	CAP
Carburize	CARB
Case Harden	CH
Casting	CSTG
Cast Iron	CI
Cathode-Ray Tube	CRT
Celsius	C
Center	CTR
Center to Center	C to C
Centimeter	CM
Centrifugal	CENT
Chamfer	CHAM
Circuit	CKT
Circular	CIR
Circumference	CIRC
Clearance	CL
Clockwise	CW
Closure	CLOS
Coated	CTD
Cold-Drawn Steel	CDS
Cold-Rolled Steel	CRS
Color Code	CC
Commercial	COMM
Concentric	CONC
Condition	COND
Conductor	CNDCT
Contour	CTR
Control	CONT
Copper	COP
Counterbore	CBORE
Counterclockwise	CCW
Counterdrill	CDRILL
Countersink	CSK
Countersunk head	CSK H
Cubic	CU
Cylinder	CYL

D

Datum	DAT
Decimal	DEC
Decrease	DECR
Degree	DEG
Detail	DET
Detector	DET
Developed Length	DL
Developed Width	DW
Deviation	DEV
Diagonal	DIAG
Diagram	DIAG
Diameter	DIA
Diameter Bolt Circle	DBC
Diametral Pitch	DP
Dimension	DIM
Direct Current	DC
Disconnect	DISC
Double-Pole Double-Throw	DPDT
Double-Pole Single-Throw	DPST
Dowel	DWL
Draft	DFT
Drafting Room Manual	DRM
Drawing	DWG
Drawing Change Notice	DCN
Drill	DR
Drop Forge	DF
Duplicate	DUP

E

Each	EA
Eccentric	ECC
Effective	EFF
Electric	ELEC
Enclosure	ENCL
Engine	ENG
Engineer	ENGR
Engineering	ENGRG
Engineering Change Order	ECO
Engineering Order	EO
Equal	EQ
Equivalent	EQUIV
Estimate	EST

F

Fabricate	FAB
Figure	FIG
Fillet	FIL
Finish	FIN
Finish All Over	FAO
Fitting	FTG
Fixed	FXD
Fixture	FIX
Flange	FLG
Flat Head	FHD
Flat Pattern	F/P
Flexible	FLEX
Fluid	FL
Forged Steel	FST
Forging	FORG
Furnish	FURN

G

Gage	GA
Gallon	GAL
Galvanized	GALV
Gasket	GSKT
Generator	GEN
Grind	GRD
Ground	GRD

H

Half-Hard	1/2H
Handle	HDL
Harden	HDN

Standard Abbreviations for Use on Drawings *(continued)*

Head	HD
Heat Treat	HT TR
Hexagon	HEX
Hexagonal head	HEX HD
High Carbon Steel	HCS
High Frequency	HF
High Speed	HS
Horizontal	HOR
Hot-Rolled Steel	HRS
Hour	HR
Housing	HSG
Hydraulic	HYD
I	
Identification	IDENT
Inch	IN
Inclined	INCL
Include, Including, Inclusive	INCL
Increase	INCR
Independent	INDEP
Indicator	IND
Information	INFO
Inside Diameter	ID
Installation	INSTL
International Standards Organization	ISO
Interrupt	INTER
J	
Joggle	JOG
Junction	JCT
K	
Keyway	KWY
L	
Laboratory	LAB
Lacquer	LAQ
Laminate	LAM
Left Hand	LH
Length	LG
Letter	LTR
Limited	LTD
Limit Switch	LS
Linear	LIN
Liquid	LIQ
List of Material	L/M
Long	LG
Low Carbon	LC
Low Voltage	LV
Lubricate	LUB
M	
Machine(ing)	MACH
Magnaflux	M
Magnesium	MAG
Maintenance	MAINT
Major	MAJ
Malleable	MALL
Malleable Iron	MI
Manual	MAN
Manufacturing(ed, er)	MFG
Mark	MK
Master Switch	MS
Material	MATL
Maximum	MAX
Measure	MEAS
Mechanical	MECH
Medium	MED
Meter	MTR
Middle	MID
Military	MIL
Millimeter	MM
Minimum	MIN
Miscellaneous	MISC
Modification	MOD
Mold Line	ML
Motor	MOT
Mounting	MTG
Multiple	MULT
N	
Nickel Steel	NS
Nomenclature	NOM
Nominal	NOM
Normalize	NORM
Not to Scale	NTS
Number	NO.
O	
Obsolete	OBS
Opposite	OPP
Oscilloscope	SCOPE
Ounce	OZ
Outside Diameter	OD
Over-All	OA
P	
Package	PKG
Parting Line (Castings)	PL
Parts List	P/L
Pattern	PATT
Piece	PC
Pilot	PLT
Pitch	P
Pitch Circle	PC
Pitch Diameter	PD
Plan View	PV
Plastic	PLSTC
Plate	PL
Pneumatic	PNEU
Port	P
Positive	POS
Potentiometer	POT
Pounds Per Square Inch	PSI
Pounds Per Square Inch Gage	PSIG
Power Amplifier	PA
Power Supply	PWR SPLY
Pressure	PRESS
Primary	PRI
Process, Procedure	PROC
Product, Production	PROD
Q	
Quality	QUAL
Quantity	QTY
Quarter-Hard	1/4H
R	
Radar	RDR
Radio	RAD
Radio Frequency	RF
Radius	RAD or R
Ream	RM
Receptacle	RECP
Reference	REF
Regular	REG
Regulator	REG
Release	REL
Required	REQD
Resistor	RES
Revision	REV
Revolutions Per Minute	RPM
Right Hand	RH
Rivet	RIV
Rockwell Hardness	RH
Round	RD
S	
Schedule	SCH
Schematic	SCHEM
Screw	SCR
Screw Threads	
American National Coarse	NC
American National Fine	NF
American National Extra Fine	NEF
American National 8 Pitch	8N
American Standard Taper Pipe	NTP
American Standard Straight Pipe	NPSC
American Standard Taper (Dryseal)	NPTF
American Standard Straight (Dryseal)	NPSF
Unified Screw Thread Coarse	UNC
Unified Screw Thread Fine	UNF

Standard Abbreviations for Use on Drawings *(continued)*

Unified Screw Thread Extra Fine	UNEF
Unified Screw Thread 8 Thread	8UN
Section	SECT
Sequence	SEQ
Serial	SER
Serrate	SERR
Sheathing	SHTHG
Sheet	SH
Silver Solder	SILS
Single-Pole Double-Throw	SPDT
Single-Pole Single-Throw	SPST
Society of Automotive Engineers	SAE
Solder	SLD
Solenoid	SOL
Speaker	SPKR
Special	SPL
Specification	SPEC
Spotface	SF
Spring	SPG
Square	SQ
Stainless Steel	SST
Standard	STD
Steel	STL
Stock	STK
Support	SUP
Switch	SW
Symbol	SYM
Symmetrical	SYM
System	SYS
T	
Tabulate	TAB
Tangent	TAN
Tapping	TAP
Technical Manual	TM
Teeth	T
Television	TV
Temper	TEM
Temperature	TEM
Tensile Strength	TS
Thick	THK
Thread	THD
Through	THRU
Tolerance	TOL
Tool Steel	TS
Torque	TOR
Total Indicator Reading	TIR
Transformer	XFMR
Transistor	XSTR
Transmitter	XMTR
Tungsten	TU
Typical	TYP
U	
Ultra-High Frequency	UHF
Unit	U
Universal	UNIV
Unless Otherwise Specified	UOS
V	
Vacuum	VAC
Vacuum Tube	VT
Variable	VAR
Vernier	VER
Vertical	VERT
Very High Frequency	VHF
Vibrate	VIB
Video	VD
Void	VD
Volt	V
Volume	VOL
W	
Washer	WASH
Watt	W
Weatherproof	WP
Weight	WT
Wide, Width	W
Wire Wound	WW
Wood	WD
Wrought Iron	WI
Y	
Yield Point (PSI)	YP
Yield Strength (PSI)	YS

Location of Elements of a Welding Symbol

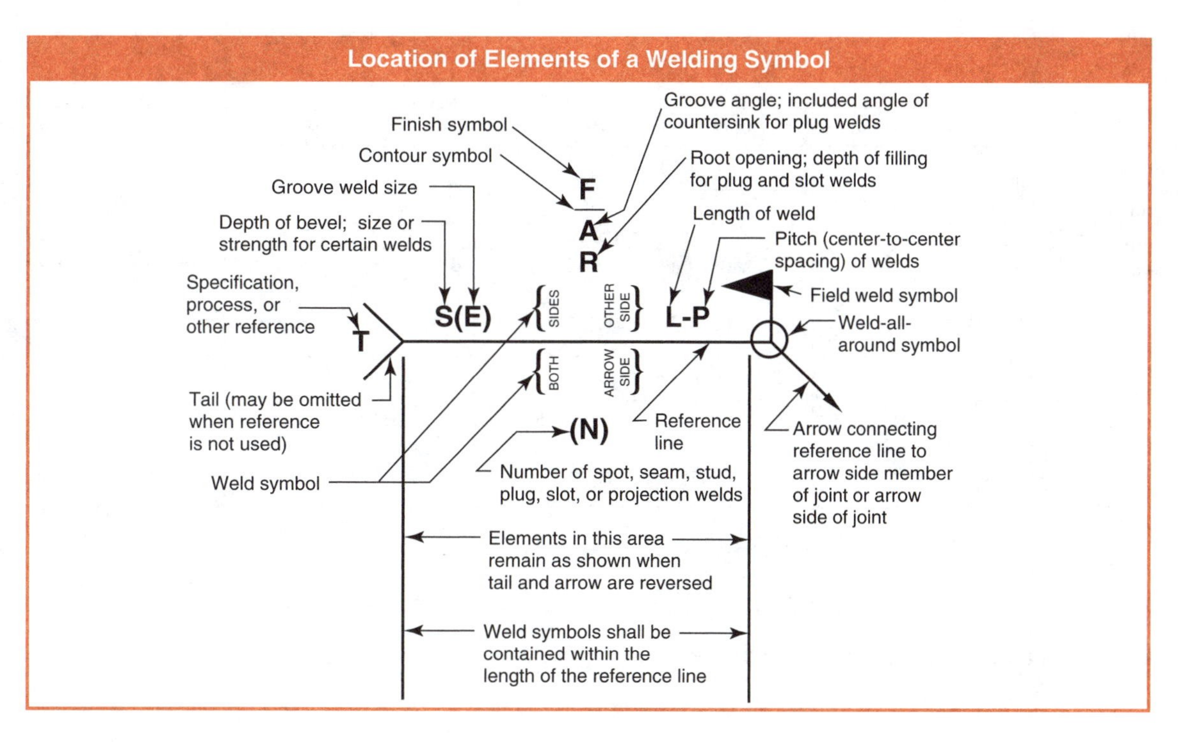

A welding symbol is a graphic explanation needed to fully specify weld requirements.

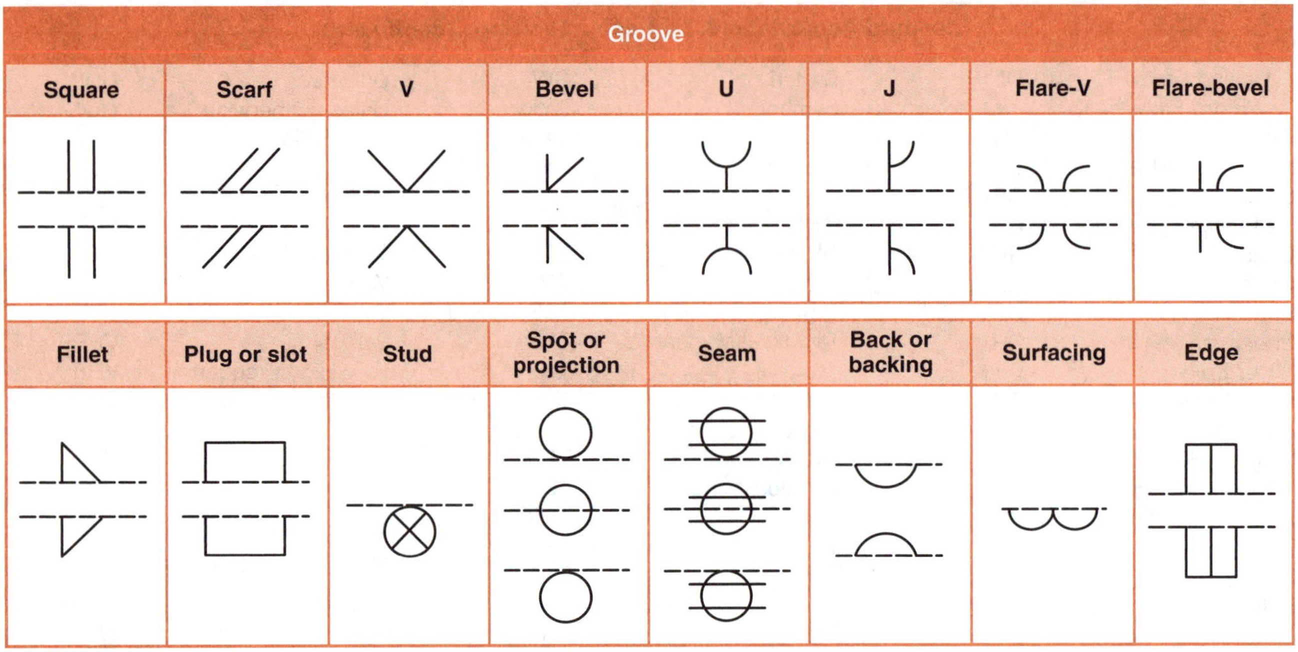

Groove							
Square	Scarf	V	Bevel	U	J	Flare-V	Flare-bevel
Fillet	Plug or slot	Stud	Spot or projection	Seam	Back or backing	Surfacing	Edge

Names and shapes of basic weld symbols are shown.

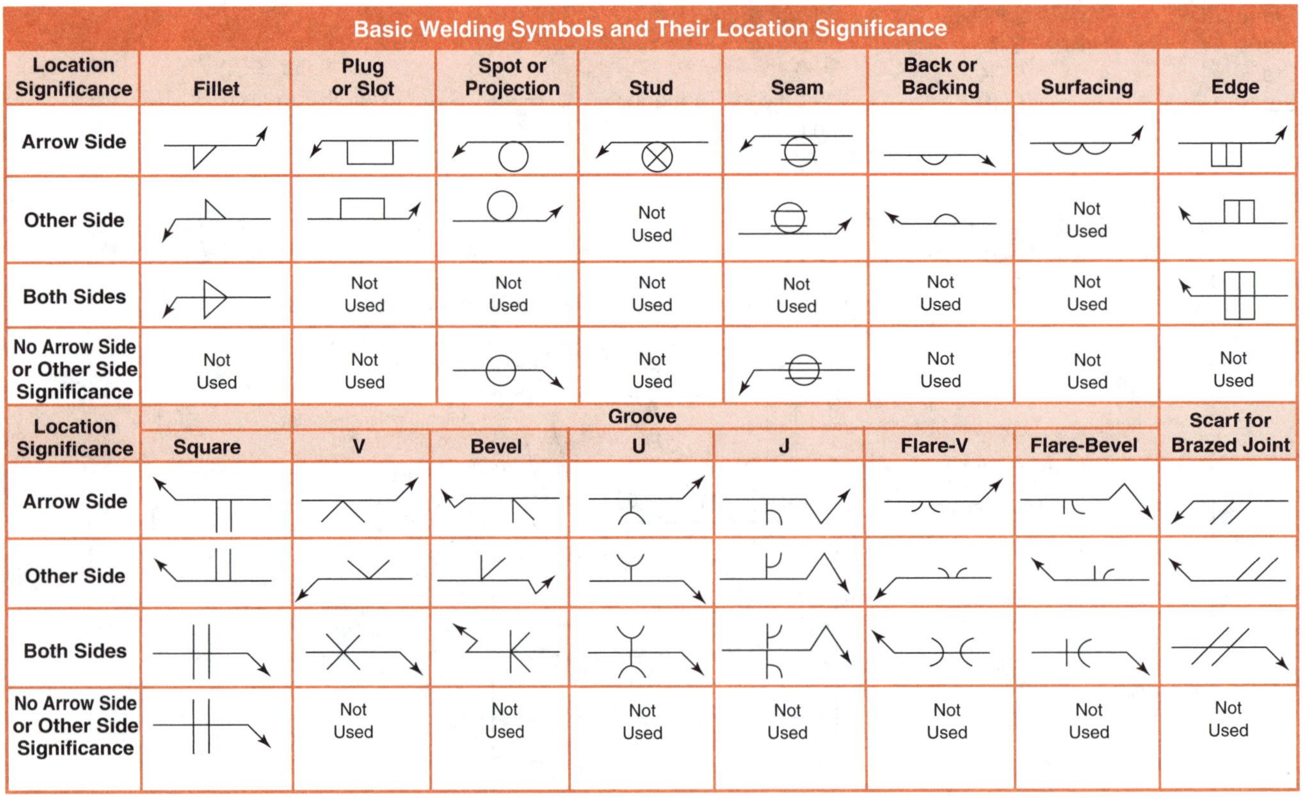

Basic Welding Symbols and Their Location Significance								
Location Significance	Fillet	Plug or Slot	Spot or Projection	Stud	Seam	Back or Backing	Surfacing	Edge
Arrow Side								
Other Side				Not Used			Not Used	
Both Sides		Not Used	Not Used	Not Used	Not Used	Not Used	Not Used	
No Arrow Side or Other Side Significance	Not Used	Not Used		Not Used		Not Used	Not Used	Not Used
Location Significance	Groove							Scarf for Brazed Joint
	Square	V	Bevel	U	J	Flare-V	Flare-Bevel	
Arrow Side								
Other Side								
Both Sides								
No Arrow Side or Other Side Significance		Not Used	Not Used	Not Used	Not Used	Not Used	Not Used	Not Used

This chart shows basic welding symbols and their location significance.

National Coarse and National Fine Threads and Tap Drills							
Threads per Inch	Major Dia.	Minor Dia.	Pitch Dia.	Tap Drill 75 Percent Thread	Decimal Equivalent	Clearance Drill	Decimal Equivalent
56	.0860	.0628	.0744	50	.0700	42	.0935
64	.0860	.0657	.0759	50	.0700	42	.0935
48	.099	.0719	.0855	47	.0785	36	.1065
56	.099	.0758	.0874	45	.0820	36	.1065
40	.112	.0795	.0958	43	.0890	31	.1200
48	.112	.0849	.0985	42	.0935	31	.1200
32	.138	.0974	.1177	36	.1065	26	.1470
40	.138	.1055	.1218	33	.1130	26	.1470
32	.164	.1234	.1437	29	.1360	17	.1730
36	.164	.1279	.1460	29	.1360	17	.1730
24	.190	.1359	.1629	25	.1495	8	.1990
32	.190	.1494	.1697	21	.1590	8	.1990
24	.216	.1619	.1889	16	.1770	1	.2280
28	.216	.1696	.1928	14	.1820	2	.2210
20	.250	.1850	.2175	7	.2010	G	.2610
28	.250	.2036	.2268	3	.2130	G	.2610
18	.3125	.2403	.2764	F	.2570	21/64	.3281
24	.3125	.2584	.2854	I	.2720	21/64	.3281
16	.3750	.2938	.3344	5/16	.3125	25/64	.3906
24	.3750	.3209	.3479	Q	.3320	25/64	.3906
14	.4375	.3447	.3911	U	.3680	15/32	.4687
20	.4375	.3725	.4050	25/64	.3906	29/64	.4531
13	.5000	.4001	.4500	27/64	.4219	17/32	.5312
20	.5000	.4350	.4675	29/64	.4531	33/64	.5156
12	.5625	.4542	.5084	31/64	.4844	19/32	.5937
18	.5625	.4903	.5264	33/64	.5156	37/64	.5781
11	.6250	.5069	.5660	17/32	.5312	21/32	.6562
18	.6250	.5528	.5889	37/64	.5781	41/64	.6406
10	.7500	.6201	.6850	21/32	.6562	25/32	.7812
16	.7500	.6688	.7094	11/16	.6875	49/64	.7656
9	.8750	.7307	.8028	49/64	.7656	29/32	.9062
14	.8750	.7822	.8286	13/16	.8125	57/64	.8906
8	1.0000	.8376	.9188	7/8	.8750	1–1/32	1.0312
14	1.0000	.9072	.9536	15/16	.9375	1–1/64	1.0156
7	1.1250	.9394	1.0322	63/64	.9844	1–5/32	1.1562
12	1.1250	1.0167	1.0709	1–3/64	1.0469	1–5/32	1.1562
7	1.2500	1.0644	1.1572	1–7/64	1.1094	1–9/32	1.2812
12	1.2500	1.1417	1.1959	1–11/64	1.1719	1–9/32	1.2812
6	1.5000	1.2835	1.3917	1–11/32	1.3437	1–17/32	1.5312
12	1.5000	1.3917	1.4459	1–27/64	1.4219	1–17/32	1.5312

Tap Drill Sizes for ISO Metric Threads				
Series				
	Coarse		Fine	
Nominal Size mm	**Pitch mm**	**Tap Drill mm**	**Pitch mm**	**Tap Drill mm**
1.4	0.3	1.1	—	—
1.6	0.35	1.25	—	—
2	0.4	1.6	—	—
2.5	0.45	2.05	—	—
3	0.5	2.5	—	—
4	0.7	3.3	—	—
5	0.8	4.2	—	—
6	1.0	5.0	—	—
8	1.25	6.75	1	7.0
10	1.5	8.5	1.25	8.75
12	1.75	10.25	1.25	10.50
14	2	12.00	1.5	12.50
16	2	14.00	1.5	14.50
18	2.5	15.50	1.5	16.50
20	2.5	17.50	1.5	18.50
22	2.5	19.50	1.5	20.50
24	3	21.00	2	22.00
27	3	24.00	2	25.00

Metric Prefixes, Exponents, and Symbols					
Decimal Form	**Exponent or Power**	**Prefix**	**Pronunciation**	**Symbol**	**Meaning**
1 000 000 000 000 000 000	$= 10^{18}$	exa	ex'a	E	quintillion
1 000 000 000 000 000	$= 10^{15}$	peta	pet'a	P	quadrillion
1 000 000 000 000	$= 10^{12}$	tera	tĕr'á	T	trillion
1 000 000 000	$= 10^{9}$	giga	jĭ'gá	G	billion
1 000 000	$= 10^{6}$	mega	mĕg'á	M	million
1 000	$= 10^{3}$	kilo	kĭl'ō	k	thousand
100	$= 10^{2}$	hecto	hĕk'to	h	hundred
10	$= 10^{1}$	deka	dĕk'a	da	ten
1					base unit
0.1	$= 10^{-1}$	deci	dĕs'ĭ	d	tenth
0.01	$= 10^{-2}$	centi	sĕn'tĭ	c	hundredth
0.001	$= 10^{-3}$	milli	mĭl'ĭ	m	thousandth
0.000 001	$= 10^{-6}$	micro	mi'krō	µ	millionth
0.000 000 001	$= 10^{-9}$	nano	năn'ō	n	billionth
0.000 000 000 001	$= 10^{-12}$	pico	péc'ō	p	trillionth
0.000 000 000 000 001	$= 10^{-15}$	femto	fĕm'tō	f	quadrillionth
0.000 000 000 000 000 001	$= 10^{-18}$	atto	ăt'tō	a	quintillionth

Kilo 1000 ×
Hecto 100 ×
Deka 10 ×
Base Unit 1
Deci 1/10 0.1
Centi 1/100 0.01
Milli 1/1000 0.001
Smaller
Larger

Conversion Table: US Conventional to SI Metric

When You Know	Multiply By: Very Accurate	Multiply By: Approximate	To Find
Length			
inches	* 25.4		millimeters
inches	* 2.54		centimeters
feet	* 0.3048		meters
feet	* 30.48		centimeters
yards	* 0.9144	0.9	meters
miles	* 1.609344	1.6	kilometers
Weight			
grains	15.43236	15.4	grams
ounces	* 28.349523125	28.0	grams
ounces	* 0.028349523125	.028	kilograms
pounds	* 0.45359237	0.45	kilograms
short ton	* 0.90718474	0.9	tonnes
Volume			
teaspoons		5.0	milliliters
tablespoons		15.0	milliliters
fluid ounces	29.57353	30.0	milliliters
cups		0.24	liters
pints	* 0.473176473	0.47	liters
quarts	* 0.946352946	0.95	liters
gallons	* 3.785411784	3.8	liters
cubic inches	* 0.016387064	0.02	liters
cubic feet	* 0.028316846592	0.03	cubic meters
cubic yards	* 0.764554857984	0.76	cubic meters
Area			
square inches	* 6.4516	6.5	square centimeters
square feet	* 0.09290304	0.09	square meters
square yards	* 0.83612736	0.8	square meters
square miles		2.6	square kilometers
acres	* 0.40468564224	0.4	hectares
Temperature			
Fahrenheit	* 5/9 (after subtracting 32)		Celsius

* = Exact

Conversion Table: SI Metric to US Conventional

When You Know	Multiply By: Very Accurate	Multiply By: Approximate	To Find
Length			
millimeters	0.0393701	0.04	inches
centimeters	0.3937008	0.4	inches
meters	3.280840	3.3	feet
meters	1.093613	1.1	yards
kilometers	0.621371	0.6	miles
Weight			
grains	0.00228571	0.0023	ounces
grams	0.03527396	0.035	ounces
kilograms	2.204623	2.2	pounds
tonnes	1.1023113	1.1	short tons
Volume			
milliliters		0.2	teaspoons
milliliters	0.06667	0.067	tablespoons
milliliters	0.03381402	0.03	fluid ounces
liters	61.02374	61.024	cubic inches
liters	2.113376	2.1	pints
liters	1.056688	1.06	quarts
liters	0.26417205	0.26	gallons
liters	0.03531467	0.035	cubic feet
cubic meters	61023.74	61023.7	cubic inches
cubic meters	35.31467	35.0	cubic feet
cubic meters	1.3079506	1.3	cubic yards
cubic meters	264.17205	264.0	gallons
Area			
square centimeters	0.1550003	0.16	square inches
square centimeters	0.00107639	0.001	square feet
square meters	10.76391	10.8	square feet
square meters	1.195990	1.2	square yards
square kilometers		0.4	square miles
hectares	2.471054	2.5	acres
Temperature			
Celsius	* 9/5 (then add 32)		Fahrenheit

* = Exact

Formulas

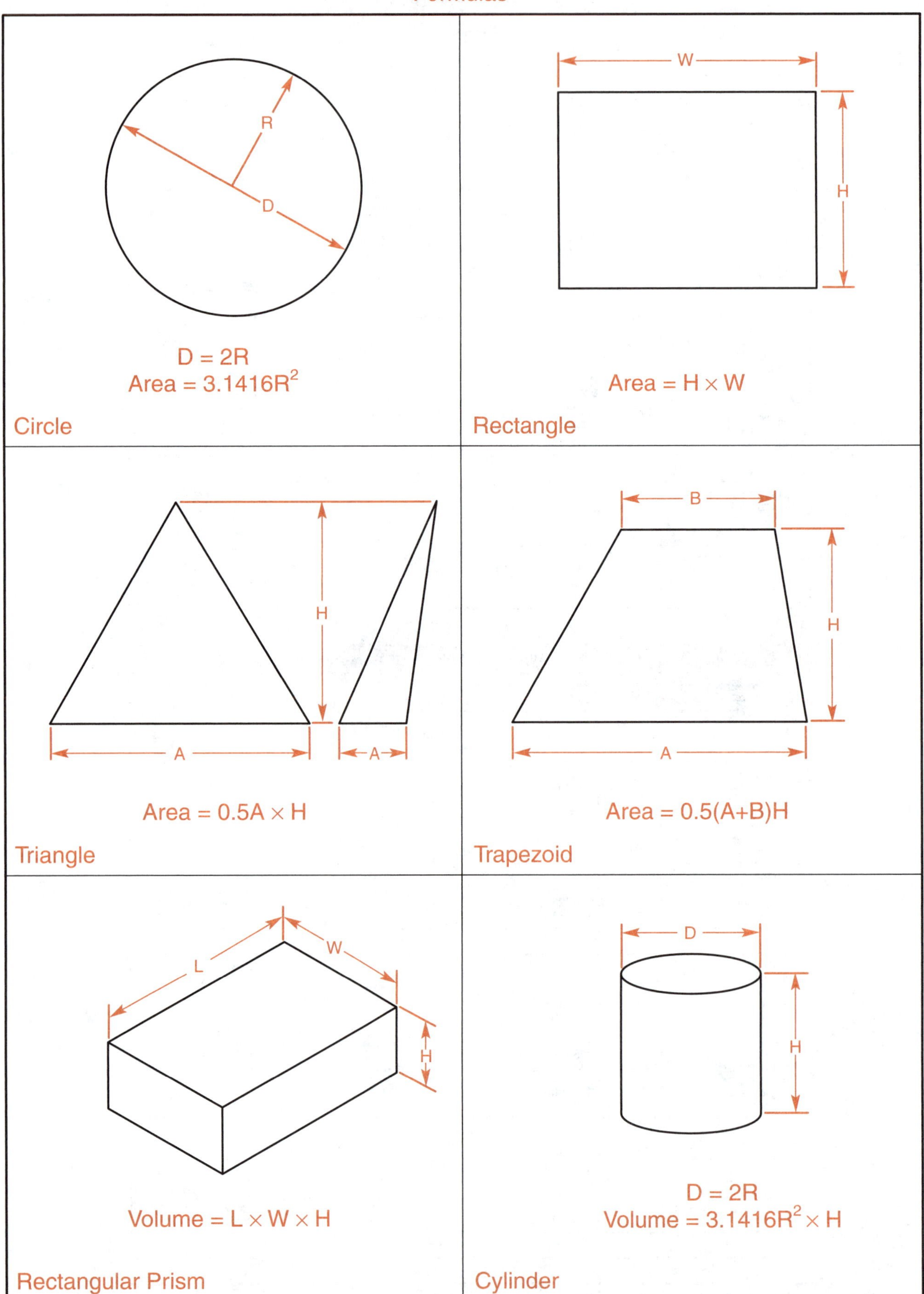
R
D
D = 2R
Area = 3.1416R²
Circle
W
H
Area = H × W
Rectangle
H
A
A
Area = 0.5A × H
Triangle
B
H
A
Area = 0.5(A+B)H
Trapezoid
L
W
H
Volume = L × W × H
Rectangular Prism
D
H
D = 2R
Volume = 3.1416R² × H
Cylinder

Formulas

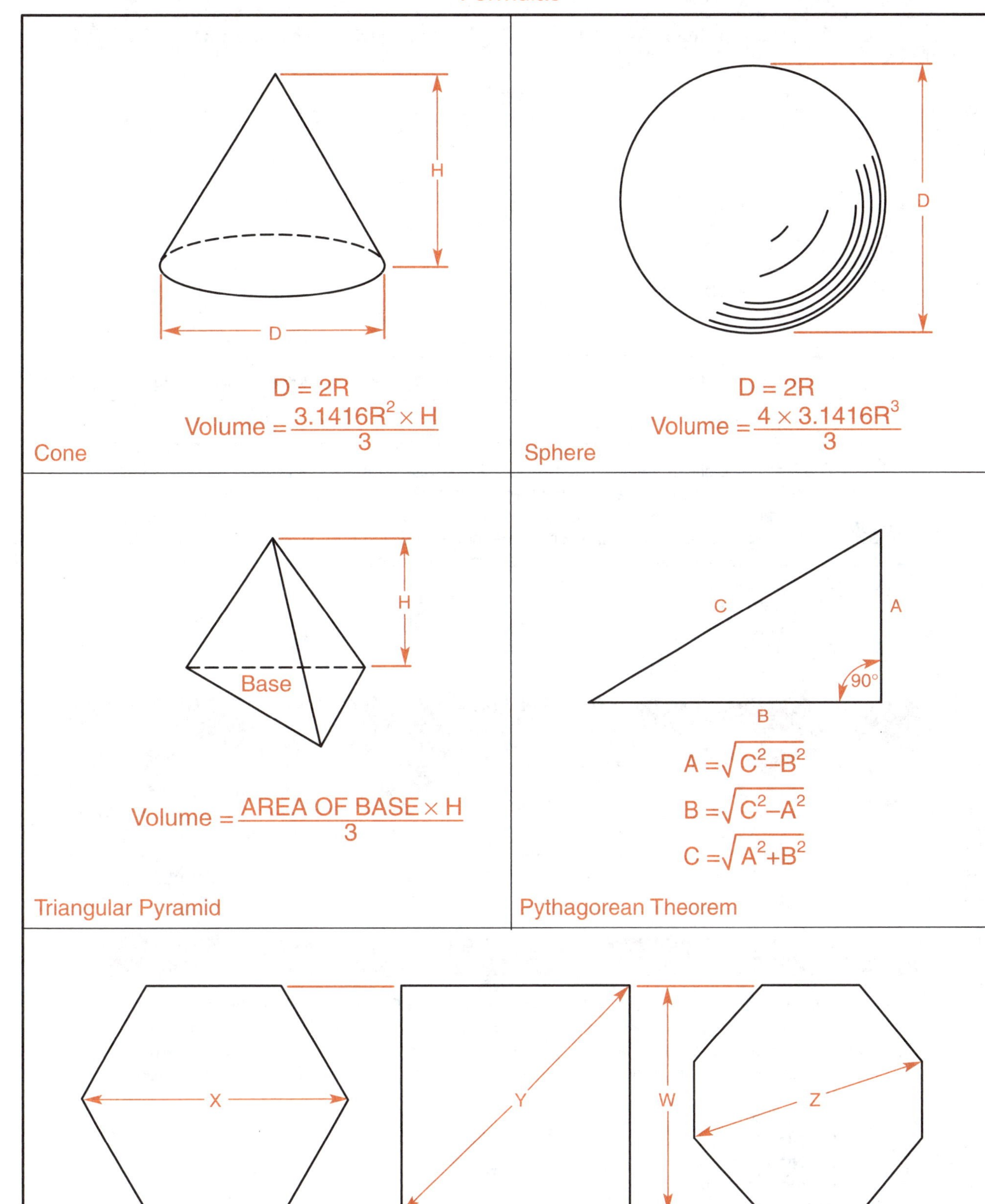

H
D
D = 2R
Volume = 3.1416R2 × H / 3
Cone
D
D = 2R
Volume = 4 × 3.1416R3 / 3
Sphere
H
Base
Volume = AREA OF BASE × H / 3
Triangular Pyramid
C
A
90°
B
A = √(C2–B2)
B = √(C2–A2)
C = √(A2+B2)
Pythagorean Theorem
X
Y
W
Z
W = WIDTH
X = 1.1574W
Y = 1.4142W
Z = 1.0824W

Aluminum Association Designation System	
Number Group	**Principal Alloying Element**
1XXX	Aluminum—99% purity or greater
2XXX	Copper
3XXX	Manganese
4XXX	Silicon
5XXX	Magnesium
6XXX	Magnesium and silicon
7XXX	Zinc
8XXX	An element other than mentioned above
9XXX	Unassigned

The last two digits in this system indicate similar alloys before the present identification was adopted. For example, the alloy 5052 was formerly 52S and 7075 was known as 75S.

The letter *H,* when used to designate temper, is followed by two numbers. For example, 3003-H14. The first digit following the *H* denotes the process used to produce the temper. The second number indicates the actual temper (degree of hardness):

2 ¼ hard (2/8)
4 ½ hard (4/8)
6 ¾ hard (6/8)
8 Full hard (8/8)

Table I Carbon Steels	
Type	**Carbon Content**
Low	0.05 to approximately 0.30% carbon
Medium	0.30 to approximately 0.60% carbon
High	0.60 to approximately 1.00% carbon

Table II SAE-AISI Code Classification (First Digit)

The first number of the SAE-AISI Code Classification System frequently, but not always, indicates the basic type of steel. When carbon or alloy steel contains the letter *L* in the code, it contains from 0.15–0.35% lead to improve machinability. These steels are also known as free-machining steel. The prefix *E* before the alloy steel designation indicates it is made only by electric furnace.

1 – Carbon
2 – Nickel
3 – Nickel-chrome
4 – Molybdenum
5 – Chromium
6 – Chromium-vanadium
7 – Tungsten
8 – Nickel-chromium-molybdenum
9 – Silicomanganese

Color Codes for Marking Steels

S.A.E. Number	Code Color	S.A.E. Number	Code Color	S.A.E. Number	Code Color	S.A.E. Number	Code Color
	Carbon steels	2115	Red and bronze	T1340	Orange and green	3450	Black and bronze
1010	White	2315	Red and blue	T1345	Orange and red	4820	Green and purple
1015	White	2320	Red and blue	T1350	Orange and red		Chromium steels
X1015	White	2330	Red and white		Nickel-chromium steels	5120	Black
1020	Brown	2335	Red and white	3115	Blue and black	5140	Black and white
X1020	Brown	2340	Red and green	3120	Blue and black	5150	Black and white
1025	Red	2345	Red and green	3125	Pink	52100	Black and brown
X1025	Red	2350	Red and aluminum	3130	Blue and green		Chromium-vanadium steels
1030	Blue	2515	Red and black	3135	Blue and green	6115	White and brown
1035	Blue		Molybdenum steels	3140	Blue and white	6120	White and brown
1040	Green	4130	Green and white	X3140	Blue and white	6125	White and aluminum
X1040	Green	X4130	Green and bronze	3145	Blue and white	6130	White and yellow
1045	Orange	4135	Green and yellow	3150	Blue and brown	6135	White and yellow
X1045	Orange	4140	Green and brown	3215	Blue and purple	6140	White and bronze
1050	Bronze	4150	Green and brown	3220	Blue and purple	6145	White and orange
1095	Aluminum	4340	Green and aluminum	3230	Blue and purple	6150	White and orange
	Free cutting steels	4345	Green and aluminum	3240	Blue and aluminum	6195	White and purple
1112	Yellow	4615	Green and black	3245	Blue and aluminum		Tungsten steels
X1112	Yellow	4620	Green and black	3250	Blue and bronze	71360	Brown and orange
1120	Yellow and brown	4640	Green and pink	3312	Orange and black	71660	Brown and bronze
X1314	Yellow and blue	4815	Green and purple	3325	Orange and black	7260	Brown and aluminum
X1315	Yellow and red	X1340	Yellow and black	3335	Blue and orange		Silicon-manganese steels
X1335	Yellow and black		Manganese steels	3340	Blue and orange	9255	Bronze and aluminum
	Nickel steels	T1330	Orange and green	3415	Blue and pink	9260	Bronze and aluminum
2015	Red and brown	T1335	Orange and green	3435	Orange and aluminum		

Structural Metals

Shapes	Length	How measured	*How Purchased
Sheet less than ¼″ thick	to 144″	Thickness × width, widths to 72″	Weight, foot, or piece
Plate more than ¼″ thick	to 20′	Thickness × width	Weight, foot, or piece
Band	to 20′	Thickness × width	Weight or piece
Rod	12-20′	Diameter	Weight, foot, or piece
Square	12-20′	Width	Weight, foot, or piece
Flats	Hot rolled 20-22′ Cold finished	Thickness × width	Weight, foot, or piece
Hexagon	12-20′	Distance across flats	Weight, foot, or piece
Octagon	12-20′	Distance across flats	Weight, foot, or piece
Angle	Lengths to 40′	Leg length × leg length × thickness of legs	Weight, foot, or piece
Channel	Lengths to 60′	Depth × web thickness × flange width	Weight, foot, or piece
American standard beam	Lengths to 60′	Height × web thickness × flange width	Weight, foot, or piece

*Charge made for cutting to other than standard lengths.

Physical Properties of Metals						
Metal	**Symbol**	**Specific Gravity**	**Specific Heat**	**Melting Point***		**Lbs. per Cubic Inch**
				C	**F**	
Aluminum (cast)	Al	2.56	.2185	658	1217	.0924
Aluminum (rolled)	Al	2.71	–	658	1217	.0978
Antimony	Sb	6.71	.051	630	1166	.2424
Bismuth	Bi	9.80	.031	271	520	.3540
Boron	B	2.30	.3091	2300	4172	.0831
Brass	–	8.51	.094	–	–	.3075
Cadmium	Cd	8.60	.057	321	610	.3107
Calcium	Ca	1.57	.170	810	1490	.0567
Chromium	Cr	6.80	.120	1510	2750	.2457
Cobalt	Co	8.50	.110	1490	2714	.3071
Copper	Cu	8.89	.094	1083	1982	.3212
Columbium	Cb	8.57	–	1950	3542	.3096
Gold	Au	19.32	.032	1063	1945	.6979
Iridium	Ir	22.42	.033	2300	4170	.8099
Iron	Fe	7.86	.110	1520	2768	.2634
Iron (cast)	Fe	7.218	.1298	1375	2507	.2605
Iron (wrought)	Fe	7.70	.1138	1500-1600	2732-2912	.2779
Lead	Pb	11.37	.031	327	621	.4108
Lithium	Li	.057	.941	186	367	.0213
Magnesium	Mg	1.74	.250	651	1204	.0629
Manganese	Mn	8.00	.120	1225	2237	.2890
Mercury	Hg	13.59	.032	–39	–38	.4909
Molybdenum	Mo	10.2	.0647	2620	47.48	.368
Monel metal	–	8.87	.127	1360	2480	.320
Nickel	Ni	8.80	.130	1452	2646	.319
Phosphorus	P	1.82	.177	43	111.4	.0657
Platinum	Pt	21.50	.033	1755	3191	.7767
Potassium	K	0.87	.170	62	144	.0314
Selenium	Se	4.81	.084	220	428	.174
Silicon	Si	2.40	.1762	1427	2600	.087
Silver	Ag	10.53	.056	961	1761	.3805
Sodium	Na	0.97	.290	97	207	.0350
Steel	–	7.858	.1175	1330-1378	2372-2532	.2839
Strontium	Sr	2.54	.074	769	1416	.0918
Tantalum	Ta	10.80	–	2850	5160	.3902
Tin	Sn	7.29	.056	232	450	.2634
Titanium	Ti	5.3	.130	1900	3450	.1915
Tungsten	W	19.10	.033	3000	5432	.6900
Uranium	U	18.70	–	1132	2070	.6755
Vanadium	V	5.50	–	1730	3146	.1987
Zinc	Zn	7.19	.094	419	786	.2598

*Circular of the Bureau of Standards No.35, Department of Commerce and Labor

Stainless, Carbon, and Alloy Aluminum Nickel

PIPE SIZE	O.D. IN INCHES	PIPE SCHEDULES																
		5s	5	10s	10	20	+	30	40s & STD	40	(+)	60	80s & E.H.	80	100	120	140	160	DBL E.H.
1/8	.405		.035 .1383	.049 .1863	.049 .1863				.068 .2447	.068 .2447 .0850			.095 .3145	.095 .3145 .1090			WALL THICKNESS IN INCHES		
1/4	.540		.049 .2570	.065 .3297	.065 .3297				.088 .4248	.068 .4248 .1470			.119 .5351	.119 .5351 .1850			STEEL WEIGHT PER FOOT IN POUNDS		
3/8	.675		.049 .3275	.065 .4235	.065 .4235				.091 .5676	.091 .5676 .1960			.126 .7388	.126 .7388 .2560			ALUMINUM WEIGHT PER FOOT IN POUNDS		
1/2	.840	.065 .5383	.065 .5383 .1860	.083 .6710	.083 .6710 .2320				.109 .8510	.109 .8510 .2940			.147 1.088	.147 1.088 .3760				.187 1.304 .4510	.294 1.714
3/4	1.050	.065 .6838	.065 .6838 .2370	.083 .8572	.083 .8572 .2970				.113 1.131	.113 1.131 .3910			.154 1.474	.154 1.474 .5100				.218 1.937 .6700	.308 2.441
1	1.315	.065 .8678	.065 .8678 .3000	.109 1.404	.109 1.404 .4860				.133 1.679	.133 1.679 .5810			.179 2.172	.179 2.172 .7510				.250 2.844 .9840	.358 3.659
1 1/4	1.660	.065 1.107	.065 1.107 .3830	.109 1.806	.109 1.806 .6250				.140 2.273	.140 2.273 .7860			.191 2.997	.191 2.997 1.037				.250 3.765 1.302	.382 5.214
1 1/2	1.900	.065 1.274	.065 1.274 .4410	.109 2.085	.109 2.085 .7210				.145 2.718	.145 2.718 .9400			.200 3.631	.200 3.631 1.256				.281 4.859 1.681	.400 6.408
2	2.375	.065 1.604	.065 1.604 .5550	.109 2.638	.109 2.638 .9130				.154 3.653	.154 3.653 1.264			.218 5.022	.218 5.022 1.737				.344 7.462 2.575	.436 9.029
2 1/2	2.875	.083 2.475	.083 2.475 .8560	.120 3.531	.120 3.531 1.221				.203 5.793	.203 5.793 2.004			.276 7.661	.276 7.661 2.650				.375 10.01 3.464	.552 13.70
3	3.500	.083 3.029		.120 4.332	.120 4.332 1.498				.216 7.576	.216 7.576 2.621			.300 10.25	.300 10.25 3.547				.438 14.32 4.945	.600 18.58
3 1/2	4.000	.083 3.472	.083 3.472 1.201	.120 4.937	.120 4.937 1.720				.226 9.109	.226 9.109 3.151			.318 12.51	.318 12.51 4.326					.636 22.85
4	4.500	.083 3.915	.083 3.915 1.354	.120 5.613	.120 5.613 1.942				.237 10.79	.237 10.79 3.733			.337 14.98	.337 14.98 5.183		.438 19.00 6.560		.531 22.51 7.786	.674 27.54
4 1/2	5.000								2.47 12.54				.355 17.61						
5	5.563	.109 6.349	.109 6.349 2.196	.134 7.770	.134 7.770 2.688				.258 14.62	.258 14.62 5.057			.375 20.78	.375 20.78 7.188		.500 27.04 9.353		.625 32.96 11.40	.750 38.55
6	6.625	.109 7.585	.109 7.585 2.624	.134 9.289	.134 9.290 3.213				.280 18.97	.280 18.97 6.564			.432 28.57	.432 28.57 9.884		.562 36.39 12.59		.719 43.35 15.67	.864 53.16
7	7.625								.301 23.54			.500 38.04						.875 63.08	

PIPE SIZE	O.D. IN INCHES	PIPE SCHEDULES (continued)																	
		5s	5	10s	10	20	+	30	40s & STD	40	⊕	60	80s & E.H.	80	100	120	140	160	DBL E.H.
8	8.625		1.09 9.914 3.429	.148 13.40	.148 13.40 4.635	.250 22.36 7.735	.175 5.463	.277 24.70 8.543	.322 28.55	.322 28.55 9.878		.406 35.64 12.33	.500 43.39	.500 43.39 15.01	.594 50.95 17.60	.719 60.71 20.97	.812 67.76 23.44	.906 74.79 25.84	.875 72.42
9	9.625								.342 33.91				.500 48.73						
10	10.75		.134 15.19	.165 18.65	.165 18.70 6.453	.250 28.04 9.698	.279 10.79	.307 34.24 11.34	.365 40.48	.365 40.48 14.00		.500 54.74 18.93	.500 54.74	.500 64.43 22.25	.719 77.03 26.61	.844 82.29	1.000 104.1	1.125 115.6	1.000 104.1
11	11.75								.375 45.56				.500 60.08						
12	12.75	.156 21.07	.165 22.18	.180 24.16	.180 24.16 8.359	.250 33.38 11.55	.375 17.14	.330 43.77 15.14	.375 49.56	.406 53.52 18.52	.500 22.63	.562 73.15 25.31	.500 65.42	.688 88.63 30.62	.844 107.3	1.000 125.5	1.125 136.7	1.312 160.3	1.000 125.5
14	14.00	.156 23.07		.188 27.73	.250 36.71	.312 45.61		.375 54.57	.375 54.57	.438 63.44		.594 85.05	.500 72.09	.750 106.1	.938 130.9	1.094 150.8	1.250 170.2	1.406 189.1	
16	16.00	.156 27.90		.188 31.75	.250 42.05	.312 52.27		.375 62.58	.375 62.58	.500 82.77		.656 107.5	.500 82.77	.844 136.6	1.031 164.8	1.219 192.4	1.438 223.6	1.594 245.3	
18	18.00	.156 31.43		.188 35.76	.250 47.39	.312 58.94		.438 82.15	.375 70.59	.562 104.7		.750 138.2	.500 93.45	.938 170.9	1.156 208.0	1.375 244.1	1.562 274.2	1.781 308.5	
20	20.00	.188 39.78		.218 46.05	.250 52.73	.375 78.60		.500 104.1	.375 78.60	.594 123.1		.812 166.4	.500 104.1	1.031 208.9	1.281 256.1	1.500 296.4	1.750 341.1	1.969 379.2	
24	24.00	.218 55.37		.250 63.41	.250 63.41	.375 96.42		.562 140.7	.375 94.62	.688 171.3		.969 238.4	.500 125.5	1.219 296.6	1.531 367.4	1.812 429.4	2.062 438.1	2.344 542.1	
26	26.00				.312 85.60	.500 136.17			.375 102.63				.500 136.17						
28	28.00				.312 92.26	.500 146.85		.625 182.73	.375 110.64										
30	30.00	.250 79.43		.312 98.93	.312 98.93	.500 157.53		.625 196.08	.375 118.65				.500 157.53						
32	32.00				.312 105.59	.500 168.21		.625 209.43	.375 126.66	.688 230.08			.500 168.21						
34	34.00				.312 112.25	.500 178.89		.625 222.78	.375 134.67	.688 244.77									
36	36.00				.312 118.92			.625 236.13	.375 142.68	.750 282.35			.500 189.57						

Factors applicable to other products.
For nickel and alloy produced to these pipe sizes, apply these factors to the red numbers.

Nickel 200	1.1343	Monel 400	1.1272
Bucjek 201	1.1378	Inconel 600	1.0742
Incoloy* 800	1.0247	Incoloy 825	1.0389

*Registered Trade Mark of INCO

For aluminum TUBING produced to the listed pipe sizes–apply these factors for each grade

1100 Wt. as shown	2024 Wt. times 1.02
6061 Wt. as shown	3003 Wt. times 1.01
6063 Wt. as shown	5086 Wt. times .98
2014 Wt. times 1.03	7075 Wt. times 1.03

COLUMNS + AND ⊕ ARE WALL THICKNESS PRODUCED TO PIPE TOLERANCES

Number and Letter Size Drills Conversion Chart

Drill No. or Letter		Inch	mm
		.001	0.0254
		.002	0.0508
		.003	0.0762
		.004	0.1016
		.005	0.1270
		.006	0.1524
		.007	0.1778
		.008	0.2032
		.009	0.2286
		.010	0.2540
		.011	2.2794
		.012	0.3048
		.013	0.3302
80	.0135	.014	0.3556
79	.0145	.015	0.3810
	1/64	.0156	0.3969
78		.016	0.4064
		.017	0.4318
77		.018	0.4572
		.019	0.4826
76		.020	0.5080
75		.021	0.5334
		.022	0.5588
74	.0225	.023	0.5842
73		.024	0.6096
72		.025	0.6350
71		.026	0.6604
		.027	0.6858
70		.028	0.7112
		.029	0.7366
69	.0292	.030	0.7620
68		.031	0.7874
	1/32	.0312	0.7937
67		.032	0.8128
66		.033	0.8382
		.034	0.8636
65		.035	0.8890
64		.036	0.9144
63		.037	0.9398
62		.038	0.9652
61		.039	0.9906
		.0394	1.0000
60		.040	1.0160
59		.041	1.0414
58		.042	1.0668
57		.043	1.0922
		.044	1.1176
		.045	1.1430
56	.0465	.046	1.1684
	3/64	.0469	1.1906
		.047	1.1938
		.048	1.2192
		.049	1.2446
		.050	1.2700
		.051	1.2954
55		.052	1.3208
		.053	1.3462
		.054	1.3716
54		.055	1.3970
		.056	1.4224
		.057	1.4478
		.058	1.4732
		.059	1.4986
53	.0595	.060	1.5240
		.061	1.5494
		.062	1.5748
	1/16	.0625	1.5875
52	.0635	.063	1.6002
		.064	1.6256
		.065	1.6510
		.066	1.6764
51		.067	1.7018
		.068	1.7272
		.069	1.7526
50		.070	1.7780
		.071	1.8034
		.072	1.8288
49		.073	1.8542
		.074	1.8796
		.075	1.9050
48		.076	1.9304
		.077	1.9558
47	.0785	.078	1.9812
	5/64	.0781	1.9844
		.0787	2.0000
		.079	2.0066

Drill No. or Letter		Inch	mm
		.080	2.0320
46		.081	2.0574
45		.082	2.0828
		.083	2.1082
		.084	2.1336
		.085	2.1590
44		.086	2.1844
		.087	2.2098
		.088	2.2352
43		.089	2.2606
		.090	2.2860
		.091	2.3114
		.092	2.3368
42	.0935	.093	2.3622
	3/32	.0937	2.3812
		.094	2.3876
		.095	2.4130
41		.096	2.4384
		.097	2.4638
40		.098	2.4892
		.099	2.5146
39	.0995	.100	2.5400
		.101	2.5654
38	.1015	.102	2.5908
		.103	2.6162
37		.104	2.6416
		.105	2.6670
36	.1065	.106	2.6924
		.107	2.7178
		.108	2.7432
		.109	2.7686
	7/64	.1094	2.7781
35		.110	2.7490
34		.111	2.8194
		.112	2.8448
33		.113	2.8702
		.114	2.8956
		.115	2.9210
32		.116	2.9464
		.117	2.9718
		.118	2.9972
		.1181	3.0000
		.119	3.0226
31		.120	3.0480
		.121	3.0734
		.122	3.0988
		.123	3.1242
		.124	3.1496
	1/8	.125	3.1750
		.126	3.2004
		.127	3.2258
		.128	3.2512
30	.1285	.129	3.2766
		.130	3.3020
		.131	3.3274
		.132	3.3528
		.133	3.3782
		.134	3.4036
		.135	3.4290
29		.136	3.4544
		.137	3.4798
		.138	3.5052
		.139	3.5306
28	.1405	.140	3.5560
	9/64	.1406	3.5719
		.141	3.5814
		.142	3.6068
		.143	3.6322
27		.144	3.6576
		.145	3.6830
		.146	3.7484
26		.147	3.7338
		.148	3.7592
25	.1495	.149	3.7846
		.150	3.8100
		.151	3.8354
24		.152	3.8608
		.153	3.8862
23		.154	3.9116
		.155	3.9370
		.156	3.9624
	5/32	.1562	3.9687
22		.157	3.9878
		.1575	4.0000
		.158	4.0132
21		.159	4.0386

Drill No. or Letter		Inch	mm
		.160	4.0640
20		.161	4.0894
		.162	4.1148
		.163	4.1402
		.164	4.1656
		.165	4.1910
19		.166	4.2164
		.167	4.2418
		.168	4.2672
		.169	4.2926
18	.1635	.170	4.3180
		.171	4.3434
	11/64	.1719	4.3656
		.172	4.3688
17		.173	4.3942
		.174	4.4196
		.175	4.4450
		.176	4.4704
16		.177	4.4958
		.178	4.5212
		.179	4.5466
15		.180	4.5720
		.181	4.5974
14		.182	4.6228
		.183	4.6482
		.184	4.6736
13		.185	4.6990
		.186	4.7244
		.187	4.7498
	3/16	.1875	4.7625
		.188	4.7752
12		.189	4.8006
		.190	4.8260
11		.191	4.8514
		.192	4.8768
		.193	4.9022
10	.1935	.194	4.9276
		.195	4.9530
9		.196	4.9784
		.1969	5.0000
		.197	5.0038
		.198	5.0292
		.199	5.0546
8		.200	5.0800
7		.201	5.1054
		.202	5.1308
		.203	5.1562
	13/64	.2031	5.1594
6		.204	5.1816
5	.2055	.205	5.2070
		.206	5.2324
		.207	5.2578
		.208	5.2832
4		.209	5.3086
		.210	5.3340
		.211	5.3594
		.212	5.3848
3		.213	5.4102
		.214	5.4356
		.215	5.4610
		.216	5.4864
		.217	5.5118
		.218	5.5372
	7/32	.2187	5.5562
		.219	5.5626
		.220	5.5880
2		.221	5.6134
		.222	5.6388
		.223	5.6642
		.224	5.6896
		.225	5.7150
		.226	5.7404
		.227	5.7658
1		.228	5.7912
		.229	5.8166
		.230	5.8410
		.231	5.8674
		.232	5.8928
		.233	5.9182
A		.234	5.9436
	15/64	.2344	5.9531
		.235	5.9690
		.236	5.9944
		.2362	6.0000
		.237	6.0198
B		.238	6.0452

Drill No. or Letter		Inch	mm
		.239	6.0706
		.240	6.0960
		.241	6.1214
C		.242	6.1468
		.243	6.1722
		.244	6.1976
		.245	6.2230
D		.246	6.2484
		.247	6.2738
		.248	6.2992
		.249	6.3246
E	1/4	.250	6.3500
		.251	6.3754
		.252	6.4008
		.253	6.4262
		.254	6.4516
		.255	6.4770
		.256	6.5024
F		.257	6.5278
		.258	6.5532
		.259	6.5786
		.260	6.6040
G		.261	6.6294
		.262	6.6548
		.263	6.6802
		.264	6.7056
		.265	6.7310
	17/64	.2656	6.7469
H		.266	6.7564
		.267	6.7818
		.268	6.8072
		.269	6.8326
		.270	6.8580
		.271	6.8834
I		.272	6.9088
		.273	6.934
		.274	6.9596
		.275	6.9850
		.2756	7.0000
		.276	7.0104
		.277	7.0358
		.278	7.0612
		.279	7.0866
		.280	7.1120
K		.281	7.1374
	9/32	.2812	7.1437
		.282	7.1628
		.283	7.1882
		.284	7.2136
		.285	7.2390
		.286	7.2644
		.287	7.2898
		.288	7.3152
		.289	7.3406
L		.290	7.3660
		.291	7.3914
		.292	7.4168
		.293	7.4422
		.294	7.4676
M		.295	7.4930
	19/64	.2969	7.5406
		.297	7.5438
		.298	7.5692
		.299	7.5946
		.300	7.6200
		.301	7.6454
N		.302	7.6708
		.303	7.6962
		.304	7.7216
		.305	7.7470
		.306	7.7724
		.307	7.7978
		.308	7.8232
		.309	7.8486
		.310	7.8740
		.311	7.8994
		.312	7.9248
	5/16	.3125	7.9375
		.313	7.9502
		.314	7.9756
		.3150	8.0000
		.315	8.0010
O		.316	8.0264
		.317	8.0518
		.318	8.0772
		.319	8.1026

Number and Letter Size Drills Conversion Chart (continued)

Drill No. or Letter	Inch	mm
	.320	8.1280
	.321	8.1534
	.322	8.1788
P	.323	8.2042
	.324	8.2296
	.325	8.2550
	.326	8.2804
	.327	8.3058
	.328	8.3312
21/64	.3281	8.3344
	.329	8.3566
	.330	8.3820
	.331	8.4074
Q	.332	8.4328
	.333	8.4582
	.334	8.4836
	.335	8.5090
	.336	8.5344
	.337	8.5598
	.338	8.5852
R	.339	8.6106
	.340	8.6360
	.341	8.6614
	.342	8.6868
	.343	8.7122
11/32	.3437	8.7312
	.344	8.7376
	.345	8.7630
	.346	8.7884
	.347	8.8138
S	.348	8.8392
	.349	8.8646
	.350	8.8900
	.351	8.9154
	.352	8.9408
	.353	8.9662
	.354	8.9916
	.3543	9.0000
	.355	9.0170
	.356	9.0424
	.357	9.0678
T	.358	9.0932
	.359	9.1186
23/64	.3594	9.1281
	.360	9.1440
	.361	9.1694
	.362	9.1948
	.363	9.2202
	.364	9.2456
	.365	9.2456
	.366	9.2964
	.367	9.3218
U	.368	9.3472
	.369	9.3726
	.370	9.3980
	.371	9.4234
	.372	9.4488

Drill No. or Letter	Inch	mm
	.373	9.4488
	.374	9.4996
3/8	.375	9.5250
	.376	9.5504
V	.377	9.5758
	.378	9.6012
	.379	9.6266
	.380	9.6520
	.381	9.6774
	.382	9.7028
	.383	9.7282
	.384	9.7536
	.385	9.7790
W	.386	9.8044
	.387	9.8298
	.388	9.8552
	.389	9.8806
	.390	9.9060
25/64	.3906	9.9219
	.391	9.9314
	.392	9.9568
	.393	9.9822
	.3937	10.0000
	.394	10.0076
	.395	10.0330
	.396	10.0584
X	.397	10.0838
	.398	10.1092
	.399	10.1346
	.400	10.1600
	.401	10.1854
	.402	10.2108
	.403	10.2362
Y	.404	10.2616
	.405	10.2870
	.406	10.3124
13/32	.407	10.3378
	.408	10.3632
	.409	10.3886
	.410	10.4140
	.411	10.4394
	.412	10.4648
Z	.413	10.4902
	.414	10.5156
	.415	10.5410
	.416	10.5664
	.417	10.5918
	.418	10.6182
	.419	10.6426
	.420	10.6680
	.421	10.6934
27/64	.4219	10.7156
	.422	10.7188
	.423	10.7442
	.424	10.7696
	.425	10.7950
	.426	10.8204

Drill No. or Letter	Inch	mm
	.427	10.8458
	.428	10.8712
	.429	10.8966
	.430	10.9220
	.431	10.9474
	.432	10.9728
	.433	10.9982
	.4331	11.0000
	.434	11.0236
	.435	11.0490
	.436	11.0744
	.437	11.0998
7/16	.4375	11.1125
	.438	11.1252
	.439	11.1506
	.440	44.1760
	.441	11.2014
	.442	11.2268
	.443	11.2522
	.444	44.2776
	.445	11.3030
	.446	11.3284
	.447	11.3538
	.448	11.3792
	.449	11.4046
	.450	11.4300
	.451	11.4554
	.452	11.4808
	.453	11.5062
29/64	.4531	11.5094
	.454	11.5316
	.455	11.5570
	.456	11.5824
	.457	11.6078
	.458	11.6332
	.459	11.6586
	.460	11.6840
	.461	11.7094
	.462	11.7348
	.463	11.7602
	.464	11.7856
	.465	11.8110
	.466	11.8364
	.467	11.8618
	.468	11.8872
15/32	.4687	11.9062
	.469	11.9126
	.470	11.9380
	.471	11.9634
	.472	11.9888
	.4724	12.0000
	.473	12.0142
	.474	12.0396
	.475	12.0650
	.476	12.0904
	.477	12.1158
	.478	12.1412

Drill No. or Letter	Inch	mm
	.479	12.1666
	.480	12.1920
	.481	12.2174
	.482	12.2428
	.483	12.2682
	.484	12.2936
31/64	.4844	12.3031
	.485	12.3190
	.486	12.3444
	.487	12.3698
	.488	12.3952
	.489	12.4206
	.490	12.4460
	.491	12.4714
	.492	12.4936
	.493	12.5222
	.494	12.5476
	.495	12.5730
	.496	12.5984
	.497	12.6238
	.498	12.6492
	.499	12.6746
1/2	.500	12.7000
33/64	.516	13.0969
17/32	.531	13.4938
35/64	.547	13.8907
9/16	.562	14.2875
37/64	.578	14.6844
19/32	.594	15.0813
39/64	.609	15.4782
5/8	.625	15.875
41/64	.641	16.2719
21/32	.656	16.6688
43/64	.672	17.0657
11/16	.688	17.4625
45/64	.703	17.8594
23/32	.719	18.2563
47/64	.734	18.6532
3/4	.750	19.05
49/64	.766	19.4469
25/32	.781	19.8438
51/64	.797	20.2407
13/16	.813	20.6375
53/64	.828	21.0344
27/32	.844	21.4313
55/64	.859	21.8282
7/8	.875	22.225
57/64	.891	22.6219
29/32	.906	23.0188
59/64	.922	23.4157
15/16	.938	23.8125
61/64	.953	24.2094
31/32	.969	24.6063
63/64	.984	25.0032
1	1.000	25.4000

Glossary

A

Actual throat: The shortest distance between the weld root and the face of the weld.

Acute angle: An angle with a measurement of less than 90°.

Less than 90°

acute angle

Alloy: A mixture of two or more metals.

Alphabet of lines: A term used to describe the characteristics of various lines universally accepted throughout industry to interpret prints.

Angle: The intersection of two lines or sides measured in degrees (∠).

Angle dimension: An arc drawn through the center of the angle at its uppermost point with the resulting angle dimensioned using degrees.

Anneal: A term used to describe the softening of metals during heat treatment.

Application: See *Next assembly.*

Arc: A portion of a circumference.

Arc stud welding (SW): A welding process where fusion is produced by an electric arc between a metal stud, or similar part and the other work part. No shielding is used when surfaces are joined, heated, and brought together under pressure.

Arc welding: See *Shielded metal arc welding (SMAW).*

Area: The number of unit squares equal to the entire surface measure of an object.

Arrow: A symbol used to connect the reference line of a welding symbol to one side of the joint to be welded.

Arrow side: A term referring to the lower side of the reference line that indicates the same side or near side of the joint.

Assembly drawing: One type of working drawing that shows where and how the parts described in the detail drawings fit into the complete assembly of the product.

Automatic welding: A welding system that uses computer-controlled machines to ensure consistent welds and fits between parts.

Auxiliary line: An additional line used to help visualization of the relationship between lines and angles.

Auxiliary view: A type of view chosen to show the true shape and size of an angular surface.

B

Back weld: A weld made to ensure full weld material through the joint. It is made after the required weld indicated by the symbol and is sometimes called complete joint penetration (CJP).

Backing: Material placed against the back side of a joint to withstand molten weld metal.

Backing weld symbol: A symbol used to indicate a bead-type backing weld on the opposite side of the regular weld.

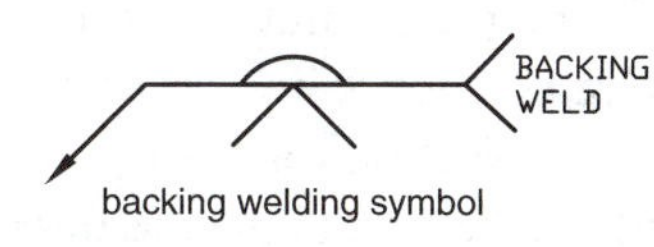

backing welding symbol

Balloons: Sometimes added to a print to identify each component of a part. Consists of a divided circle with an attached line leading to the component. The item number from the parts list and the quantity needed are written in the circle.

Base line dimensioning: See *Datum dimensioning.*

Base material: Refers to the material being welded. Sometimes called the parent metal.

Base metal: The metal or alloy that is welded, brazed, soldered, or cut.

Bead weld: See *Surfacing weld.*

Bending: A cold forming technique that uses a forming brake or forming rolls in which the surface area of the work is not greatly changed.

Bent leader: A term used to describe the arrow for a bevel or J-groove weld. The arrow head points to the particular section to be prepared.

Bevel angle: The angle formed between the bevel of one piece and a plane perpendicular to the surface of the piece. Usually, refers to an angle made on the base metal before welding on a single plate.

Bilateral tolerance: Describes a variance from the basic dimension in both directions, plus (+) and minus (-).
Bill of materials: See *Parts list.*
Blanking: A stamping operation involving cutting flat sheets to the shape of the finished part.
Blow-backs: A microfilm image retrieved from the files and printed on photographic paper.
Bluelines: See *White prints.*
Blueprint: A common term used to refer to a copy of an original drawing.
Blueprint process: A technique used to duplicate drawings with white lines on a blue background.
Boring: An internal machining operation where a single point cutting tool is used to enlarge a hole. It is not to be confused with drilling.
Bracket method: A method of displaying dual dimensions on a drawing where metric dimensions are placed within square brackets if the drawing is to be used primarily in the United States.
Braze welding: A manual process of adding the filler metal to the joint.
Brazing (B): A group of joining processes using nonferrous alloys with melting temperatures above 800°F (427°C). The filler metal's melting point is lower than the melting point of the metals being joined.
Break line: A line used on a print to remove or "break out" a section for clarity, often providing clearer detail in viewing parts that lie directly below a removed part.
Broach: A long planing machine tool with many cutting teeth that are pushed or pulled over the surface being machined.
Broken leader: See *Bent leader.*
Broken-chain dimensioning: Dimensioning system where the undimensioned feature accumulates the tolerance.
Broken-out section: A view used when a small portion of the sectional view will provide all necessary information.
Buildup: A surfacing variation in which surfacing material is deposited to obtain the desired dimensions.
Buttering: A process where a weld is used to improve the fit of a joint or to increase the dimensions of a part.

C

CAD system: An acronym for computer-aided design, this system is a combination of hardware and software that allows designs to be drawn and viewed from any angle.
Calculated dimension: A method used by the print reader when various dimensions on the print are incorporated to determine a dimension not directly stated.
Capillary action: Refers to the force by which the liquid filler metal is drawn between fitted, mating surfaces during the brazing process.
Caret mark: A symbol (^) used in math problems to aid in the proper placement of the decimal point.
Casting: A metalworking process of making objects by pouring molten metal into a mold.
Center line: A fine, dark line composed of alternate long and short dashes with spaces between the dashes. Used to indicate the center of symmetrical objects.

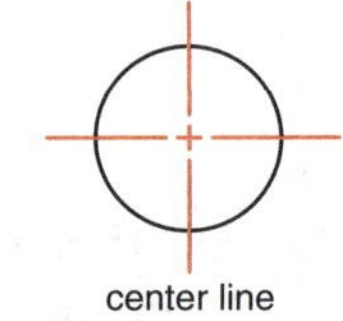
center line

Centrifugal casting: A form of permanent mold casting where centrifugal force holds the molten metal against the wall of the mold until it solidifies.
Chain dimension method: A method that does not include an overall dimension. The accumulation of tolerances for each feature often exceeds the tolerance of the overall dimension.
Chord: A line having endpoints on a circle, but the line does not pass through the center of the circle.
Circle: A geometric figure where all points on the shape are the same distance from the center.
Circumference: The distance around a circle.
Class of fit: A standard working tolerance for thread accuracy, indicated by the last number on a thread description.
Cold forming: A technique forming cold metal into a desired shape using a series of dies.
Cold welding (CW): A process where pressure is used to join the two metal surfaces and special tools direct the flow of metal into a true weld.
Common factor: Refers to a number that will divide evenly into both the numerator and denominator of a fraction.
Common fraction: A fraction with one number placed above the other.
Complementary angles: Two angles formed by three lines with the sum of the angles equal to 90°.

58°
32°
complementary angles

Complete joint penetration: A joint root condition in a groove weld in which weld metal extends through the joint thickness. Sometimes abbreviated as CJP.

Computer numerical control machine tools (CNC): Machine tools controlled by computers capable of unattended machining once the program is loaded into the CNC machine tool controller's memory.

Contour symbol: A symbol used with the weld symbol when the finished shape of the weld is important.

Conventional break: Permits elongated objects to be shortened so a large enough drawing scale can be used to present details with clarity.

Cope: Refers to the top half of the mold used in a sand casting.

Cores: Inserts of sand positioned in the mold cavity that create openings required in a sand casting.

Counterboring: A drill press operation used to enlarge a portion of a drilled hole so fillister and socket head fasteners can be inserted properly.

Countersinking: A process that enlarges the end of a hole conically so it is able to fit the head of a flat head screw.

Crosshatching: See *Section lining.*

Cutting-plane line: A heavy line used with section views to indicate an imaginary cut made through the object to reveal its interior characteristics.

D

Datum: Defined as the exact point, axis, or plane from which the location or geometric features of a part are located.

Datum dimensioning: A dimensioning method where a single side of a part is used to start all of the dimensions.

Datum feature: A feature used as a datum.

Datum identification symbol: A boxed-in letter referencing a datum.

Datum targets: Used to establish consistent locations on parts with rough or irregular surfaces. The target is usually circular and indicates where the part should contact the fixture.

Decimal rule: Refers to 10 divisions per inch (one division equals 0.1 inch) on one edge of a rule. The opposite edge has either 50 divisions or 100 divisions.

Denominator: The bottom number in a common fraction. The denominator indicates how many parts a whole unit has been divided.

Destructive examination: A quality control technique that destroys the part under inspection during the examination process.

Detail assembly drawing: A working drawing where the details and assembly of a mechanism appear on the same print.

Detail drawing: One type of working drawing that includes a print of the part (one or more views) with dimensions and other information needed to make the part.

Diameter: A line segment through the center of a circle having its endpoints on the circle.

d
diameter

Diazo process: A copying technique for making direct positive prints (dark lines on a white background).

Die casting: A variation of permanent mold casting. A die cast object has a smooth surface finish, fine details, and great accuracy.

Dimension line: A fine, solid line used on a print to indicate the extent and direction of dimensions.

Dimensions: Numerical values describing the size, location, geometric characteristics, or surface texture of a part or its features.

Direct dimensions: Dimensions directly specified for features used to build a part.

Direction of radiation: An element of an examination symbol specified by a symbol located on the drawing at the desired angle of the examination.

Discontinuities: Internal or external flaws on a weld.

Divisor: The number by which the dividend is divided.

Drag: Refers to the bottom half of the mold used in a sand casting.

Drawing: A stamping operation that takes the flat metal blanks and forms or draws them into three-dimensional shapes. Also called forming.

Drawing title: Information given in a print title block that gives the correct name of the part.

Drill press: A machining tool that rotates a twist drill against the work to cut through material. Drill press operations include reaming, countersinking, counterboring, and spotfacing.

Dual dimensions: System used where dimensions on a drawing are given in both inches (usually decimal inches) and in the metric system (usually millimeters).

E

Edge weld: Welds used to join the edges of two or more members. Primarily used with light gage (sheet metal) metal.

Effective throat: The minimum distance (minus any convexity) between the weld root and the face of the weld.

Electron beam welding (EBW): A process using a beam of fast-moving electrons to supply the energy needed to melt and fuse the base metals.

Electroslag welding (ESW): A welding process that uses molten slag from flux that is held in place by water-cooled copper shoes to complete single pass welds.

Electrostatic process. See *Xerography.*

Engineering copiers: Can be used to create reproductions from original drawings.

Equal fractions: Fractions with the same value but may have different forms, such as ½ and 2/4.

Erection drawing: A specialized drawing including all of the detail information needed to complete a specific process or group of processes on a part.

Examination symbol: Similar to a welding symbol, it is used to indicate which required nondestructive test to perform.

Explicit tolerance: A tolerance with the maximum and minimum limits stated.

Explosive forming: Process using a high-energy pressure pulse of very short duration to shape metal sheet and plat. *Also* known as high energy rate forming (HERF).

Extension and dimension line method: A traditional method using an overall dimension located farthest out from the view and broken-chain dimensions, or chain dimensioning.

Extension line: A fine, solid line used on a print to indicate the termination of a dimension.

Extrusion: A metalworking process used to manufacture irregular shapes. Metal heated to a plastic state is placed in the extrusion press, and pressure exerted on the metal squeezes it through a die.

F

Fastener: A device used to hold together parts that may be assembled and disassembled easily.

Faying: Term used to describe surfaces on one member in contact or close proximity to another surface to which it is to be joined.

Feature control frame: Used to link a feature or group of features to the datum once the datum reference is established.

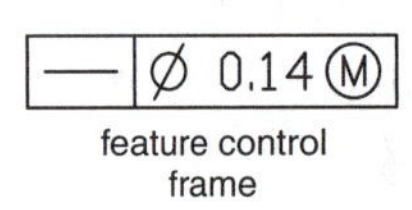

feature control frame

Feature dimension: Used to show individual features, such as the center of a hole or the location of individual parts welded into an assembly.

Ferrous metal: A metal containing iron.

Field weld symbol: A symbol used to indicate the weld is to be done at the field site and not where the unit is first made.

Fillet gauge: A tool used to measure the distance of the weld leg.

Fillet weld: A type of weld performed when joining two surfaces at an angle. It is approximately triangular in shape.

Finish: Information given in a print title block that specifies general finish requirements, such as sand blasting, plating, etc.

Finish symbol: This symbol is included with a contour symbol if a weld is to be finished.

First angle projection: A type of projection used in European drawings where the object is drawn as if it were placed on each side of the glass box.

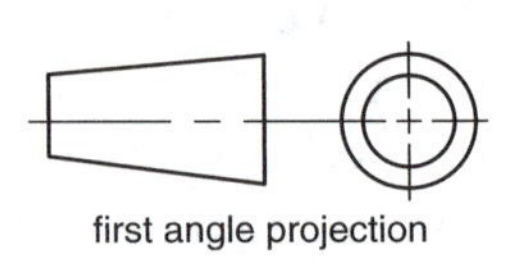
first angle projection

Flame spraying (FLSP): A term used to describe the process where a metal is brought to its melting temperature and sprayed onto a surface to produce a coating. Processes include metallizing, thermo spraying, and plasma flame spraying.

Flanged butt joint: A joint used when both members to be joined are flanged.

Flanged corner joint: A joint used when only one of the members to be joined is flanged.

Flash welding (FW): A process where fusion is produced simultaneously over the entire area of abutting surfaces by welding heat and pressure application.

Flask: A wood or metal box used in sand casting. Sand is packed around a pattern positioned in the flask.

Flux cored arc welding (FCAW): A high-quality welding process with high deposit rates, it uses an arc produced between a continuous wire electrode and the weld pool. The coating that produces the gas to cover the weld pool and the flux used to clean the metal is found inside the wire.

Forging: The metalworking process of using pressure (usually a hammering action) to shape heated metal in an effort to improve physical characteristics of the metal.

Form geometric tolerances: Used to control flatness, straightness, circularity (roundness), and cylindricity.

Forming: See *Drawing.*

Formula: A mathematical equation containing a fact, rule, or principle.

Fraction: A number that tells what part of a whole unit is taken.

Friction stir welding (FSW): A solid-state welding process that uses a rotating tool to create heat through friction to mix the base metals.

Friction welding (FRW): A welding process using frictional heat and pressure to produce full strength welds in a matter of seconds.

Full section: A sectional view shown when the cutting-plane line passes entirely through the object and the interior features are revealed.

G

Gas metal arc welding (GMAW): A welding process that uses an arc formed between continuous filler metal and the weld pool. Filler metal is supplied from a spool and fed into the weld pool.

Gas tungsten arc welding (GTAW): A process using an arc produced between a tungsten electrode that is *not* consumed in the welding process and the weld pool.

General tolerance: Tolerances stated on the print in the title block or in a note with the number of *x's* representing the number of decimal places to which the tolerance applies.

Geometric dimensioning and tolerancing: A method used by the print designer to specify the requirements of a part by the actual function and relationship of the features.

Geometry: The branch of mathematics dealing with points, lines, angles, planes, and shapes.

Graduation: Refers to each mark or division on the edge of a ruler.

Grinding: A machine tool operation that removes material by rotating an abrasive wheel against the work.

Groove: The opening provided between the two metal pieces being joined by a groove weld.

Groove angle: The total angle formed between the groove face on one workpiece and the groove face on the other workpiece.

Groove face: The joint member surface included in the groove.

Groove radii: Used to form the shape of J- or U-groove welds and shown by cross section, detail, or other means with a reference on the welding symbol.

Groove weld: A weld made in a groove on one or both surfaces to be joined.

Guided bend test: A common destructive test used to check welds and to qualify welders for code welding. The sample is bent into a "U" shape or until it breaks, and then it is checked for cracks or defects if it does not break.

H

Half section: A view used primarily on symmetrical objects. The shape of one-half of the interior features and one-half of the exterior features of the object are shown.

Hardfacing: Refers to materials generally added to improve the properties of a part.

Hardness: A property allowing a part to resist abrasion and wear.

Hatching: A graphic tool used on a print to show the extent and location of a fillet weld.

Heat treatment: Information given in a print title block that includes a number of processes involving the controlled heating and cooling of a metal or alloy in an effort to obtain desirable changes in its physical characteristics.

Helix: The basic shape of a screw thread, formed by a point curve that wraps around a cylinder in a spiral.

Hidden line: A medium weight line composed of short dashes used to indicate a hidden edge or internal feature of an object. *Also* known as hidden object line.

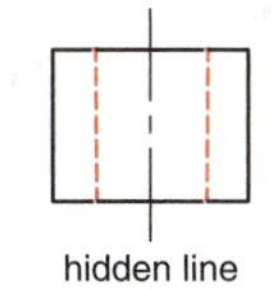

Hidden object line: See *Hidden line.*

High energy rate forming (HERF): See *Explosive forming.*

Higher terms: A method of changing a fraction to an equal fraction by multiplying the numerator and denominator by the same number. The value of the fraction is not changed.

Hole diameter: Indicated on a SI unit drawing with the number and symbol ∅.

Hypotenuse: The side opposite the right angle.

I

Identifying number: See *Print number.*

Improper fraction: Describes a fraction with a numerator equal to, or greater than, its denominator.

Inch rule: Divisions on a ruler edge as small as 1/64″ but more likely to be 1/16″ for general welding work.

India ink: Dense, black, waterproof ink used for inked tracings.

Indication: Something present in the weld, which the welding inspector must interpret as being harmful or not to the weld.

Ink jet printer: An inexpensive printer that produces good quality output.

Inside diameter (ID): Nominal measurement used to determine pipe size.

Interface zone: The point where spot weld members are joined. The diameter of the weld at this point determines spot weld size.

Intermittent weld: A weld broken by recurring unwelded spaces. The nonstandard term is skip weld.

Investment casting: A foundry process that produces an accurate and intricately designed casting. *Also* called lost wax casting.

Isometric view: A pictorial with lines showing width, length, and depth drawn full size (or to scale). The object is shown as it is.

J

Job shop welder: A welder who commonly makes single, special order weldments for individual jobs, seldom doing the same job twice.
Joint: A basic way of arranging metal pieces in relation to each other so they can be welded together.
Joint penetration: A term referring to the distance the weld metal extends from the weld face into a joint, not including weld reinforcement.
Joint root: Refers to the part of a joint to be welded where the members align closest to each other.
Joint spacers: Metal parts inserted in the joint root as backing and to keep the root opening during welding.

L

Laser beam welding (LBW): A process in which a light beam is used to vaporize the work at its point of focus. Molten metal surrounds the point of vaporization when the beam is moved along the path to be welded.
Laser printer: A printer that works on the electrostatic process and produces high quality images.
Lathe: A machining tool that operates on the principle of the work being rotated against the edge of a cutting tool.
Leader: A dimension line with an arrowhead on only one end, used to indicate a dimension or for adding a note.

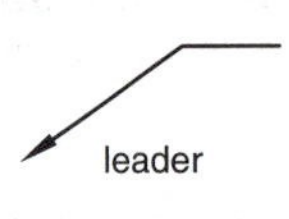

Legs: The name for the two sides of a right triangle that form the right angle.
Limits: Describes the maximum distance allowed in the tolerance.
Linear measure: Refers to the straight-line distance between two points.
Liquid penetrant testing (PT): A nondestructive test used to find discontinuities open to the surface of a part. Penetrant is allowed to seep into any discontinuities and then removed. A white developer is applied, and any discontinuity is shown vividly in contrast between the developer and the penetrant.
Location tolerances: See *Positional tolerances.*
Lost wax casting: See *Investment casting.*
Lowest terms: A) A fraction in its most reduced form where no common factor for the numerator and denominator is possible. B) A method of changing a fraction by dividing the numerator and denominator by the largest number common to both. The value of the fraction is not changed.

M

Machine tool: A power driven machine manufactured in a wide range of styles and sizes and used to shape metal by a cutting process.
Magnetic particle testing (MT): A nondestructive testing method used to detect surface and slight subsurface discontinuities in ferromagnetic metals. A magnetic field is established and colored magnetic particles are applied to the surface to determine the presence of any flaws.
Maintenance welder: A welder who repairs machine parts, fabricates pipelines, builds structural components for manufacturing facilities, and modifies existing equipment.
Material list: See *Parts list.*
Material specification: Information given in a print title block that gives the exact grade or type of substance to be used in the weldment.
Mechanized welding: A welding system using a machine (usually supervised by a welding operator) to move the welding gun along the weld joint.
Melt-through symbol: A symbol used when complete joint penetration is required in a weld made from only one side.

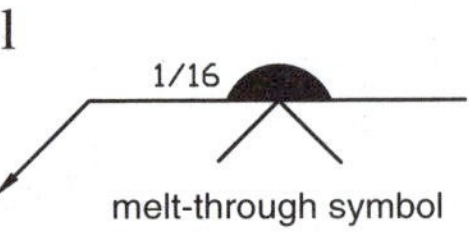

Metal specifications: Instructions for the type of metal to use, usually located in a section of the title block and provided in one or more accepted standards.
Metallizing: Describes the flame spraying process involving the use of metal in wire form. Wire drawn through a spray gun is melted in the gas flame and atomized by compressed air.
Metalworking processes: Refers to processes used to fabricate most welded products, such as casting, forging, extruding, machining, and forming.
Meter: A basic unit of the metric system (m). Equal to 1000 mm or 100 centimeters.
Metric rule: Refers to graduations on a rule edge in millimeter (mm) divisions.
Microfilm process: A technique originally designed to reduce storage facilities and to protect prints from loss.
Microwelding: A technology developed to attach leads to microcircuits or chips with very little heat produced in making the welds.
MIG: Informal term for gas metal arc welding (GMAW), it stands for metal inert gas.
Milling machine: A planing machine that removes metal by moving a rotating multitooth cutter into the work. It is used to machine flat and irregularly shaped surfaces and can also drill, bore, or cut gears and splines.
Mixed number: Consists of a whole number and a common fraction, such as $2\frac{2}{3}$.
Mold: A cavity made in a material suitable for holding the molten metal until it cools and solidifies.

Multiview drawings: Drawings used when more than one view is required to give an accurate shape description.

N

Next assembly: Information given in a print title block necessary to provide the next step in the manufacturing and assembly operations. *Also* called application.

Nondestructive examination (NDE): A basic quality control technique that does not destroy the part under examination.

Nonferrous metal: A metal containing no iron.

Nonpreferred weld symbols: A term referring to symbols that have been replaced by new ones. Although not used on newer prints, they might still be found on older prints.

Notation: Refers to information called out in the tail of a welding symbol.

Note: Additional information placed on the drawing.

Numerator: The top number in a common fraction. The numerator indicates how many parts of the whole unit are taken.

O

Object line: See *Visible line.*

Oblique view: A pictorial showing the front view in true shape and size and the other views similar to isometric drawings.

Obtuse angle: An angle measuring more than 90°.

Greater than 90°

obtuse angle

Offset section: Used to show the shape of features that lie in more than one plane, indicated by a stepped cutting-plane line.

Ordinate dimensioning: See *Zero plane dimensioning.*

Orientation geometric tolerances: Tolerances that control the degree of parallelism, perpendicularity, or angularity of a feature with respect to one or more datums.

Orthographic projection: A technique of envisioning objects enclosed in a hinged box with the views projected on the sides of the box. It permits a three-dimensional object to be described on a flat sheet of paper having only two dimensions.

Other side: A term referring to the upper side of the reference line that indicates the far side of the joint.

Outside diameter (OD): Used to measure tubes, shafts, and round parts.

Overall dimension: A dimension stating the largest linear distance of a part.

Oxyfuel gas welding (OFW): A group of welding processes that use burning gases or hydrogen mixed with oxygen. The heat produced causes the base metal to melt and fuse.

P

Parent metal: See *Base material.*

Partial auxiliary view: A term used to describe when one of the principal views is removed from an auxiliary view to avoid confusion and unnecessary sections of the part. Only the angular surface is shown and break lines indicate where part of the view was removed.

Parts list: Information on a print that includes all of the parts required in the manufacture of a product. Also included is a description of each part, the quantity of each part needed per assembly, part number, and the number of the drawing used to manufacture each part.

Pen plotter: A device that duplicates the motion of a human hand while drawing, it is a traditional way of making paper drawings from CAD systems.

Permanent mold casting: A casting made in a metal mold that is not destroyed when the casting is removed.

Phantom line: A thin, dark line made of long dashes alternated with pairs of short dashes. Used to show adjacent positions of related parts, show alternate positions of moving parts, show repeated details, or show filleted and rounded corners.

phantom line

Pipe schedule: A way to specify pipe, it indicates the wall thickness of pipe as listed in the schedule by nominal diameter.

Pipe size: Measured by its nominal inside diameter (ID).

Piping drawings: Drawings showing the assembly of a pipe system.

Piping system: An assembly of pipe or tube (thin wall pipe) sections with the fittings necessary to direct and control fluids in either liquid or gaseous form.

Pitch: Described as center-to-center spacing of an intermittent weld expressed as the distance between centers of increments on one side of the joint.

Plane figure: A flat figure with no depth.

Planer: A large machine tool capable of machining surfaces up to 20′ (6 m) wide and twice as long.

Planing machines: A machine tool used to machine flat surfaces, including the shaper, planer, slotter, and broach.

Plasma flame spray: A flame spraying process where the spray gun utilizes an electric arc contained within a water-cooled jacket. An inert gas is passed through the arc until a temperature up to 30,000°F is reached.

Plug weld: A type of weld made through one circular or elongated hole in a piece of metal to join it to another piece of metal.

Position method: Method used primarily in the United States, where dual dimensions on a drawing are displayed with the inch dimension above or to the left of the millimeter dimension.

Positional tolerances: Tolerances used to establish the location of features.

Primary datum: The first datum reference specified in the datum identification symbol.

Principal views: Commonly shown views on a print, including the front, top, and right side views.

Print: Refers to a reproduction, or duplicate, of an original drawing.

Print number: A number provided on a drawing for convenience in filing and location. Also known as an identifying number.

Process drawing: A specialized drawing that includes all of the detail information needed to complete a specific process or group of processes on a part.

Product: The term used to describe the answer to a multiplication problem.

Production welder: A welder who usually does the same welding procedure over and over, often during the mass production of a weldment.

Projection welding (PW): A resistance welding technique where heat is produced by the flow of an electric current through the work parts held under pressure by the electrodes.

Projection welds: A weld produced from the heat created by the resistance to the welding current flow.

Proper fraction: A fraction with a numerator smaller than the denominator, such as $7/8$.

Pulse echo testing: An ultrasonic testing method where the transducer is used as both the transmitter and the receiver of the reflected energy.

Pythagorean Theorem: The theorem stating the square of the hypotenuse is equal to the sum of the squares of the two legs.

Q

Quality control: An important segment of the welding industry used to improve and maintain product quality. Also used to reduce costs, maintain product safety, and improve the competitive position of the manufacturer.

Quotient: The term used to describe the answer to a division problem.

R

Radiographic testing (RT): A film exposure method of nondestructive examination that uses gamma rays or x-rays to detect internal discontinuities. It can detect subsurface flaws and provide a permanent record of inspection.

Radius: The distance from the center of a circle to any point on the circle (one-half the diameter).

Reaming: A drill press operation that improves the accuracy and finish of a drilled hole to close tolerances.

Reference dimension: A reference stated on a print for information of convenience only. No tolerances apply to a reference dimension.

Reference line: The required central element of a welding symbol. Other elements describing weld requirements are located on or near the reference line.

Refractory mold: A mold used in investment or lost wax casting that can withstand high temperatures.

Reinforcement: A name for the weld buildup above the surface of the base material on a butt joint.

Removed section: Used when it is not possible to show the sectional views on one of the principal views.

Repeating decimal fractions: Describes a division problem that cannot be completed when a common fraction is converted to a decimal fraction. Decimal places must be used and the problem rounded off.

Reproduction: Refers to a print of the original drawing.

Resistance seam weld: A series of overlapping spot welds made progressively along the joint by rotating electrodes.

Resistance seam welding (RSEW): A resistance welding process where fusion is produced by the heat obtained from resistance to the flow of an electric current through the work parts.

Resistance spot weld: A type of individually formed weld where the shape and size of the weld nugget is limited by the size and contour of the welding electrodes.

Resistance spot welding (RSW): The most common resistance welding technique where welds can be made directly between the metal parts being joined. Adding filler metal is not required.

Resistance weld: Used as a general welding process classification to make spot, seam, and projection welds.

Resistance welding (RW): A group of welding processes where fusion heat is obtained from the electrical resistance of the work.

Revision block: Portion of a print title block indicating what changes have been made to the original drawing.

Revolved section: Section that rotates or turns the cut section 90°. Primarily used to show the shape of such things as spokes, ribs, and stock metal shapes.

Right angle: An angle formed when two lines intersect perpendicular to each other.

Right triangle: The name for a triangle with one 90° angle.

right triangle

Robotic welding: See *Automatic welding.*

Roll forming: A rapid process that takes a flat metal strip and passes it through a series of rolls, progressively forming it into the required shape.

Root faces: Parts of the groove face within the joint root shown by cross section, detail, or other means with a reference on the welding symbol.

Root opening: The gap at the joint root workpieces.

Rounded off: Describes when the value of a number is closely approximated.

Routing sheet: Often accompanies a work order and lists specific processes required to make the part, subassembly, or assembly. *Also* supplies the order of operation. *Also* known as a shop traveler.

Runoff weld tabs: Used to provide an extension of the groove beyond the pieces being joined when a full length groove weld is specified.

Runout tolerance: A tolerance controlling the variation of straightness, circularity, cylindricity, or angularity of surfaces with respect to the center line or axis of cylindrical parts.

S

Sand casting: A casting method used with a mold composed of a mixture of sand, clay, and a plastic and/or oil binder. Patterns are wood or metal.

Scale drawings: Describes views on a drawing made other than actual size.

Schedule of parts: See *Parts list.*

Seam welds: Continuous welds made between or upon overlapping members with coalescence starting at the mating surfaces or proceeding from the outer surface of one member.

Section line: A fine, dark line used on a print to show the cut surfaces of the object in section views.

Section lining: Represents the type of exposed cut surface of a section. General-purpose section lining (cast iron) is usually used on drawings when exact material specifications are located elsewhere on a print. *Also* called crosshatching.

Sectional view: View permitting the true internal shape of a complex object to be shown without the confusion caused by too many hidden lines.

Security classification: Information noted on the top of the sheet and below the title block if the drawing is top secret, confidential, or restricted.

Shaft diameter: Indicated on SI unit drawings with the number and symbol ∅.

Shape: The shape of an object is described through the combination of views and features listed on the print.

Shaper: A planing machine used to shape and cut curved and irregular shapes, slots, grooves, and keyways.

Shear spinning: A cold extrusion process where parts are shaped by rollers exerting great pressures on a starting blank or preform, displacing the metal parallel to the center line of the workpiece.

Shearing: A forming technique commonly used during the preparation of (usually cold) sheet, plate, and some structural shapes. Shearing generally refers to a process of cutting across the entire length of the material.

Shell molding: A casting method that uses a thin sand shell mold.

Shielded metal arc welding (SMAW): An arc welding process using an arc between a covered electrode and the weld pool. The process is used with shielding and filler metal from the electrode.

Shop traveler: See *Routing sheet.*

Size: A size description of a part includes all needed dimensions (English or metric measure) listed on a drawing.

Skip weld: See *Intermittent weld.*

Slot weld: A type of weld made through one circular or elongated hole in a piece of metal to join it to another piece of metal.

Slotter: A vertical shaper used to cut slots and keyways. Also used to machine internal and external gears.

Soft soldering: See *Soldering.*

Soldering (S): A low-temperature process of joining metals together. Only the filler metal is melted.

Specialized drawing: A drawing made when a product is produced in quantity. It shows separate detail drawings usually prepared for each specific manufacturing process.

Spinning: A cold forming technique of working metal sheet into three-dimensional shapes, often involving the rotation of a disc of metal with a forming block.

Spot welds: Welds made between or upon overlapping members with coalescence beginning at the mating surfaces or proceeding from the outer surface of one member.

Spotfacing: A drill press operation that machines a circular spot on a rough surface to furnish a true bearing surface for a bolt or nut.

Staggered intermittent weld: A weld that occurs on both sides of the joint, with weld increments on one side alternated with respect to those on the other side.

Stamping: A cold forming technique term used for many press forming operations, including blanking and drawing (or forming).

Stick welding: See *Shielded metal arc welding (SMAW).*

Straight angle: An angle that forms a straight line and measures 180°.

Stretch forming: A cold forming process where a metal blank is gripped at opposite edges in clamps and given a light pull, causing the blank to hug or wrap around a form block of the required shape.

Structural shapes: Customary shapes identified by group symbols and designated on drawings.

Stud arc welding (SW): Nonstandard term for arc stud welding.

Stud welding: A general term used to describe joining a metal stud to a workpiece.

Subassembly drawing: A drawing frequently used on large or complicated products and showing the assembly of a small portion or section of the complete product.

Submerged arc welding (SAW): A process where fusion is produced by heating with an electric arc between a bare metal electrode and the work. Flux shields the welding arc, no pressure is used, and filler metal is obtained from the electrode or a supplementary welding rod into the weld pool from an additional coil of wire.

Supplementary angles: The name for two angles formed by three lines and creating a straight line.

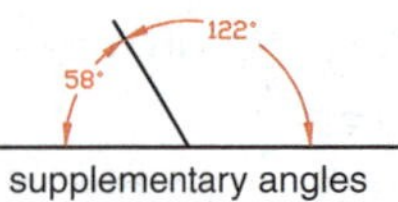

supplementary angles

Supplementary symbols: Additional symbols often included with basic weld symbols to provide more specific weld data not provided by other elements in the welding symbol.

Surfacing symbol: A supplementary symbol used to indicate the surface is to be built up by single or multiple pass welding.

Surfacing weld: A type of weld consisting of a narrow layer (or layers) of metal deposited in an unbroken puddle on the surface of the metal.

Symbols: Lines and figures having specific and standardized meanings to accurately describe the shape, size, material, finish, and fabrication of an object.

T

Tail: The part of a welding symbol containing any notes relevant to the process, filler metal, or any standards needed to establish specific weld requirements.

Tensile strength: The measure of a material's ability to withstand being pulled apart along a single axis.

Tensile test: A quality control destructive test that applies a pulling load until the sample breaks. The tensile strength is used to compare welding filler metals.

Theorem: A mathematical truth that can be proven.

Thermo spray: A term used to describe the flame spraying equipment that involves the application of metals and other materials that cannot be drawn into wire.

Thin section: A section not thick enough for conventional section lining (such as sheet metal) and shown as a solid black line.

Third angle projection: A type of projection commonly used in the United States with the object drawn as viewed in a glass box and the views projected to the six sides of the box.

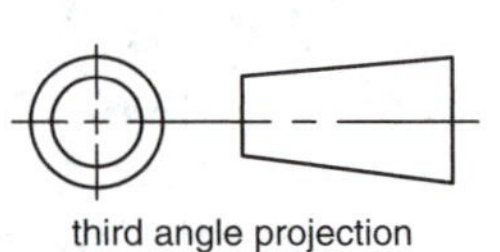
third angle projection

Through inspection: An ultrasonic testing method where sound waves are transmitted through one transducer, and a second transducer picks up the signal at the other end of the piece.

TIG: Informal term for gas tungsten arc welding (GTAW), it stands for tungsten inert gas.

Title block: The part of the print that gives general information, such as the title part, material, scale, and tolerance requirements.

Tolerance: An acceptable variance in a dimension in the size of a part or object. It indicates how much larger or smaller a part may be made and still be within specifications.

Tracing: Drawing on translucent material commonly made for manual drafting when prints are to be produced.

Transducer: An electronic device used to transmit sound wave energy during ultrasonic testing.

Triangles: Three-sided geometric figures with three interior angles equal to 180°.

Tube bending: A cold forming process used to form radii in tubing without causing the tubing to collapse.

U

Ultrasonic testing (UT): A nondestructive test that uses ultrasonic sound wave energy to penetrate into material and detect flaws. The ultrasonic waves are reflected from any discontinuities in the weld and displayed on a video screen.

Ultrasonic welding (USW): A process for joining metals without the use of solders, fluxes, or filler metals, and usually without applying external heat. Metals are clamped between welding tips, and ultrasonic energy is applied briefly, which forms a strong metallurgical bond.

Unidirectional dimensioning: A method of placing dimensions where dimensions are read from the bottom of the print.

Unified National Coarse (UNC): Thread series used to specify standards for coarse screw threads.

Unified National Fine (UNF): Thread series used to specify standards for fine screw threads.

Unilateral tolerance: Describes a variance in a dimension in only one direction, either plus (+) or minus (-).

Upset welding (UW): A resistance welding process where pieces to be joined are butted together under pressure. Current flows through the pieces until the joint is heated to the fusion point.

V

Visible line: A thick, continuous line used on a print to outline the visible edges or contours of the object. Also known as visible object line or object line.

Visible object line: See *Visible line.*

Visual testing (VT): A common nondestructive testing method used by an experienced welding inspector to determine if a weld is made to the specified print size, is in the correct location, and if there are any visual surface discontinuities.

W

Weld face: Refers to the exposed surface of the weld.

Weld leg: The vertical or horizontal distance from the base material to the toe of the weld.

Weld length: A dimension shown in inches/fractions of an inch, millimeters, or degrees (an angle). The weld length dimension is shown on the right of a weld symbol.

Weld root: Describes the deepest penetration of the weld into the base material.

Weld size: A dimension shown in inches/fractions of an inch, millimeters, or degrees (an angle). The weld size dimension is placed on the left side of the weld symbol.

Weld symbol: One part of the welding symbol, a basic weld symbol shows the cross-sectional shape of the weld or joint.

Weld toe: A term referring to where the weld face and base metal meet.

Weld-all-around symbol: A supplementary symbol that signifies the weld is to be made completely around the joint without interruption.

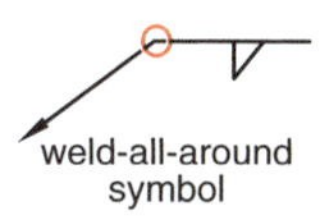

Welding: Heating metals to a suitable temperature, causing them to melt and fuse together.

Welding code: A document specifying how a part or product should be designed, constructed, repaired, and tested.

Welding symbol: A graphic assembly of the elements needed to fully specify weld requirements. Used on drawings, the welding symbol contains the data required to communicate the type of weld wanted.

White print: A copy of an original drawing with dark lines on a light background.

Whole numbers: These numbers include zero and any positive number containing no fractional parts.

Work order: Included with each job received by the shop, it indicates the total number of units to be manufactured.

Working drawings: Drawings that supply information needed to make and assemble the many pieces and parts that make up a product.

X

Xerography: A printmaking process that uses an electrostatic charge to duplicate an original. Commonly called electrostatic process.

Z

Zero plane dimensioning: A dimensioning method using a single plane, usually at the edge of the part or intersecting planes, as datums from which dimensions are derived. Features are dimensioned from a zero plane indicated by an extension line ending with a box.

Zoning: A technique used to aid in locating details on larger size drawings.

Index

A

B

C

Q

R

S